Astronomy and Space Science

Astronomy and Space Science

Editor: Audria Baldwin

R CALLISTO
REFERENCE
www.callistoreference.com

Callisto Reference,
118-35 Queens Blvd., Suite 400,
Forest Hills, NY 11375, USA

Visit us on the World Wide Web at:
www.callistoreference.com

ISBN: 978-1-63239-988-5 (Hardback)

Cataloging-in-Publication Data

Astronomy and space science / edited by Audria Baldwin.
    p. cm.
Includes bibliographical references and index.
ISBN 978-1-63239-988-5
1. Astronomy. 2. Space sciences. I. Baldwin, Audria.
QB51 .A88 2018
520--dc23

Table of Contents

Permissions

List of Contributors

Index

Preface

This book has been an outcome of determined endeavour from a group of educationists in the field. The primary objective was to involve a broad spectrum of professionals from diverse cultural background involved in the field for developing new researches. The book not only targets students but also scholars pursuing higher research for further enhancement of the theoretical and practical applications of the subject.

Space science refers to the study of space exploration. It includes the fields like astronomy, astrophysics, planetary science, physical cosmology, etc. Astronomy, being a part of space science, is concerned with the study of the evolution and origin of celestial bodies like comets, gamma ray bursts, stars, galaxies, planets, cosmic microwave background radiation, moons, etc. Different approaches, evaluations, methodologies and advanced studies on astronomy and space science have been included in this book. It presents researches and studies performed by experts across the globe. The book will help the readers in keeping pace with the rapid changes in this field.

It was an honour to edit such a profound book and also a challenging task to compile and examine all the relevant data for accuracy and originality. I wish to acknowledge the efforts of the contributors for submitting such brilliant and diverse chapters in the field and for endlessly working for the completion of the book. Last, but not the least; I thank my family for being a constant source of support in all my research endeavours.

Editor

1

Cluster Physics with Merging Galaxy Clusters

Sandor M. Molnar *

Institute of Astronomy and Astrophysics, Academia Sinica, Taipei, Taiwan

Collisions between galaxy clusters provide a unique opportunity to study matter in a parameter space which cannot be explored in our laboratories on Earth. In the standard ΛCDM model, where the total density is dominated by the cosmological constant (Λ) and the matter density by cold dark matter (CDM), structure formation is hierarchical, and clusters grow mostly by merging. Mergers of two massive clusters are the most energetic events in the universe after the Big Bang, hence they provide a unique laboratory to study cluster physics. The two main mass components in clusters behave differently during collisions: the dark matter is nearly collisionless, responding only to gravity, while the gas is subject to pressure forces and dissipation, and shocks and turbulence are developed during collisions. In the present contribution we review the different methods used to derive the physical properties of merging clusters. Different physical processes leave their signatures on different wavelengths, thus our review is based on a multifrequency analysis. In principle, the best way to analyze multifrequency observations of merging clusters is to model them using N-body/hydrodynamical numerical simulations. We discuss the results of such detailed analyses. New high spatial and spectral resolution ground and space based telescopes will come online in the near future. Motivated by these new opportunities, we briefly discuss methods which will be feasible in the near future in studying merging clusters.

Keywords: clusters of galaxies, observations of clusters, numerical simulations

Edited by:
Mauro D'Onofrio,
University of Padova, Italy

Reviewed by:
Dominique Eckert,
University of Geneva, Switzerland
Alberto Cappi,
INAF - Osservatorio Astronomico di Bologna, Italy

*Correspondence:
Sandor M. Molnar
sandor@asiaa.sinica.edu.tw

Specialty section:
This article was submitted to Milky Way and Galaxies, a section of the journal Frontiers in Astronomy and Space Sciences

1. INTRODUCTION

According to our most successful structure formation models, the cosmological constant (Λ) dominated cold dark matter (ΛCDM) models, structures form in a hierarchical process due to gravitational instabilities. Small positive density fluctuations collapse first, then massive structures form by merging and smooth accretion. This model predicts that numerous galaxy clusters are in some stage of merging at any epoch. During this merging process the gravitational potential energy of the infalling groups and subclusters is dissipated, and heats up the gas. Large clusters gain most of their mass by merging with less massive halos, as opposed to secondary infall of unprocessed gas (e.g., Muldrew et al., 2015).

Galaxy clusters are usually classified into two broad categories: relaxed and non-relaxed clusters. Many different quantitative methods have been proposed to classify galaxy clusters and identify dynamically active clusters. Based on X-ray observations, the most common parameters used to classify clusters have been power ratios, the ratios of the squares of the multiple moments of the 2D gravitational potential of the cluster to that of the monopole term (Buote and Tsai, 1995), and centroid shifts in the X-ray flux (Mohr et al., 1993).

Recently, Parekh et al. (2015) carried out an extensive study of galaxy clusters classification based on X-ray observations. They found that the most useful parameters are: the Gini, M20, and Concentration. The Gini parameter describes the non-uniformity of X-ray flux distribution in the image pixels, for uniform distribution, $G = 0$ (Lotz et al., 2004). M20 is based on the second moment of the flux distribution in pixels; it is a measure of the relative contribution of the brightest 20% pixels (Lotz et al., 2004). The Concentration depends on the ratio between radii enclosing 80 and 20% of the total X-ray flux from a cluster (Bershady et al., 2000). Parekh et al. found that non-relaxed clusters exhibit low Gini and Concentration (< 0.4 and < 1), numbers and high M20 (> -1.4) values.

Radio observations have also been used to identify dynamically active clusters. Radio halos and relics have been associated with merging clusters, while mini-halos were found in relaxed clusters (for a review see Feretti et al., 2012).

Optical spectroscopic surveys have been used to study the line of sight (LOS) velocity distribution of cluster galaxies, and infer the dynamical state of clusters. One of the most popular indicators of dynamical activity is the offset between the brightest cluster galaxy (BCG) and the X-ray surface brightness peak. Large offset signals a merging cluster or a recent merger (Poole et al., 2007). Recently, Wen and Han (2013) used different statistical parameters to describe the dynamical states of 2092 clusters using the Sloan Digital Sky Survey (SDSS). They classified only 28% of clusters as relaxed. Wen and Han found that the dominance (the difference between the brightest and second brightest cluster galaxies), the absolute magnitude of the brightest cluster galaxies, and the amplitude of the flux from radio halos correlate well with the dynamical state of clusters.

Merging of galaxy clusters and groups generate turbulence and bulk flows in the intracluster gas (ICG). Infalling massive halos loose gas due to ram pressure stripping. Assuming a spherical geometry, a simple approximation suggests that gas is removed at large where the ram pressure is greater than the pressure in the ICG of the infalling cluster/group, $P_{ram} \propto \rho_{gas} v_{rel}^2 \geq P_{gas}(r)$, where ρ_{gas} and v_{rel} are the gas density of the infalling halo and its relative velocity to the ICG of the main cluster (e.g., Gunn and Gott, 1972).

If the mass ratio is large (~ 10), cold fronts may be generated in cool core clusters due to sloshing induced by the passage of the infalling halo (Ascasibar and Markevitch, 2006; for a review see Markevitch and Vikhlinin, 2007). Massive cluster mergers with small mass ratios may produce a bow shock in front of the infalling cluster followed by a contact discontinuity ("cold front"), which separates the gas of the two colliding components. In the opposite direction a shock is propagating outward in the ICG of the subcluster. These two shocks are thought to be responsible for the double radio relics found in some merging galaxy clusters (e.g., van Weeren et al., 2011a,b; Bonafede et al., 2012).

In the cases of large velocity, the dark matter surface density peaks, the X-ray, and the centroids of the Sunyaev-Zel'dovich (SZ) signal (Section 2.3) do not coincide. Offsets between them are generated by the different behavior of the collisionless component, the dark matter, and the collisional gas during the merging process: the dark matter responds only to gravity, and therefore the core of the infalling cluster passes through that of the main cluster relatively unchanged, but the gas is subject to hydrodynamical effects, which cause large disturbances, shocks and contact discontinuities.

Major mergers are the most energetic events since the Big Bang, ($\sim 10^{63}$ ergs). The ΛCDM models predict that the infall velocities of massive merging systems are typically 500–700 $\mathrm{km\,s^{-1}}$, with a tail extending to higher velocities (Thompson and Nagamine, 2012). Recent observations of several massive merging clusters suggest a high-velocity tail in the velocity distribution extending out to $\gtrsim 3000$ $\mathrm{km\,s^{-1}}$. The most famous example of these high-velocity mergers is the so called "Bullet cluster" (1E0657-56). We discuss this merging cluster in more detail in Section 6. The Bullet cluster was the first merging system which provided direct evidence for the existence of dark matter (Markevitch et al., 2002; Markevitch, 2005). In the left panel in **Figure 1** we show a multi-frequency image of the Bullet cluster with the mass surface density reconstructed from weak lensing superimposed (Clowe et al., 2006). The mass surface densify is dominated by the most massive component, the dark matter, which is about 82% of the total mass. The X-ray emission is due to the ICG the most massive component of the ordinary matter ($\sim 16\%$), conventionally called baryonic matter in astrophysics, because the mass of ordinary matter is dominated by baryons (protons and neutrons), and not electrons, which are actually leptons. The optical emission is dominated by the least massive component, the cluster galaxies ($\sim 2\%$). The displacement of the gas from the dark matter is clearly visible.

Since the discovery of the Bullet cluster, several other merging clusters have been found with a bullet-cluster-like morphology. As an example, we show a multifrequency (X-ray and optical) image of "El Gordo" (ACT-CT J0102-4915) in the right panel of **Figure 1**. El Gordo is a recently discovered very massive high redshift merging cluster (Marriage et al., 2011; Menanteau et al., 2012) showing a unique, bullet-like X-ray morphology, followed by a "twin tailed" structure. Also, several other merging clusters have been found with significant offsets between the mass surface density and peaks of the X-ray emission and SZ signal (e.g., MACS J0025.4-1222: Bradač et al., 2008; CL J0152-1347: Massardi et al., 2010; A2163: Bourdin et al., 2011; MACS J0744.8+3927 and CL J1226.9+3332: Korngut et al., 2011; MACS J0717.5+3745: Mroczkowski et al., 2012; DLSCL J0916.2+2951: Dawson et al., 2012; ZwCl 1234.0+02916: Dahle et al., 2013; SL2S J08544-0121: Gastaldello et al., 2014).

2. MULTIFREQUENCY OBSERVATIONS OF GALAXY CLUSTERS

2.1. Optical/Near-Infrared Observations

Galaxy clusters were discovered as galaxy overdensities in the sky. The first statistically complete sample of galaxy clusters based on image plates obtained by the Palomar Observatory Sky Survey was published by Abell (1958). For many years catalogs of clusters were assembled using only optical observations.

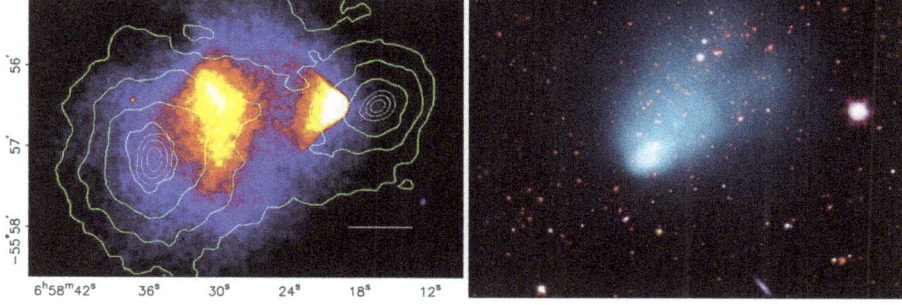

FIGURE 1 | Multi-frequency/component images of merging clusters of galaxies. Left panel: Chandra image of the "Bullet cluster" with green contours showing the projected total mass distribution from gravitational lensing observations (from Clowe et al., 2006). The white contours show the errors in the mass peaks (68.3%, 95.5%, and 99.7% CL). **Right panel:** A multi-frequency image of "El Gordo". X-ray emission from NASA's Chandra X-ray Observatory shown in blue, optical image from the European Southern Observatory's Very Large Telescope (VLT) and infrared emission from the NASA's Spitzer Space Telescope in red and orange (from Menanteau et al., 2012).

The most straightforward quantities we can measure based on O/NIR observations of a cluster are the luminosities, angular positions, shapes, and redshifts of galaxies located in the cluster region. From these quantities it is straightforward to derive the number and luminosity distribution of the cluster galaxies, and their radial velocities relative to the cluster center assuming that they are all at the distance of the cluster. Note that the distance to the cluster is much larger than the distance between cluster galaxies, thus the errors in radial velocities are negligible due to this assumption.

From the distortion of the shapes of background galaxies due to gravitational lensing, the surface mass distribution can be derived, which is a LOS integral of the total density:

$$\Sigma(x, y) = \int_{z_1}^{z_2} \rho_t(\mathbf{r}) dz', \qquad (1)$$

where $\rho_t(\mathbf{r})$ is the total mass density, the sum of the dark matter, the ICG and stars, $\rho_t(\mathbf{r}) = \rho_d(\mathbf{r}) + \rho_g(\mathbf{r}) + \rho_s(\mathbf{r})$, and the physical position within the cluster, $\mathbf{r} = (x, y, z')$ is expressed as angular coordinates here. A derivation of the mass surface density based on gravitational lensing needs more assumptions due to the low signal-to-noise (S/N) ratio of lensing observations, thus we will discuss it in more details in Section 3.1.

2.2. X-Ray Observations

The most massive baryonic component of galaxy clusters is the ICG. The low density, high temperature gas trapped in the gravitational potential of the cluster is heated up to $\sim 10^8$ K during cluster formation as a result of conversion of gravitational potential to heat. At these temperatures the most efficient radiation mechanism is thermal bremsstrahlung in the X-ray band. X-ray emission from clusters also contain emission lines from partially ionized atoms, which can be used to study the chemical composition, and certain physical properties of the

ICG. Galaxy clusters can be identified as large scale extended sources in the X-ray band.

X-ray observations measure the flux density, $F_X(x, y)$ (in, e.g., $\mathrm{ergs\,s^{-1}\,cm^{-2}\,Hz^{-1}}$), in the direction, x, y, in the sky, which is related to the integrated X-ray emissivity, ϵ, (in, e.g., $\mathrm{ergs\,s^{-1}\,cm^{-3}\,Hz^{-1}}$), over the LOS:

$$F_X(x, y) = \frac{1}{4\pi D_L^2} \int_{\nu_1/(1+z)}^{\nu_2/(1+z)} \int_{z_1}^{z_2} \epsilon(\rho, T_g, Z_{ab}; \nu)\, dz'\, d\nu, \qquad (2)$$

where the inner integral is along the LOS (z' axis), the outer integral is over the observed frequency band, (ν_1, ν_2) the $(1 + z)$ factors convert observed frequencies to emitted ones at the source at redshift z, the inner integral is over the LOS through the cluster, from z_1 to z_2, D_L is the luminosity distance to the cluster, $\rho(\mathbf{r})$ and $T_g(\mathbf{r})$ are the gas density and temperature, $Z_{ab}(\mathbf{r})$ is the abundance of heavy elements, and the emissivity at frequency ν can be expressed as

$$\epsilon(\rho, T_g, Z_{ab}; \nu) = n_e \sum_{i, Z} n_{i,Z}\, \Lambda(T_e, Z_i; \nu), \qquad (3)$$

where electron and ion number densities, $n_e(\mathbf{r})$ and $n_{i,Z}(\mathbf{r})$ are known functions of ρ, T_e (usually assuming $T_e = T_g$), and the cooling function, $\Lambda(T_e, Z_i; \nu)$, depends only the frequency, the electron temperature and physical constants (we ignore the telescope response function here). The angular coordinates can be expressed as a function of the distance to the cluster. In a narrow spectral band, from Equation (2), we obtain an approximation for the X-ray intensity,

$$I_X(x, y; \nu) = \frac{F_X(x, y)}{\Delta\nu}, \qquad (4)$$

where $\Delta\nu$ is the size of the frequency bin. The frequency bin has to be fine enough to resolve the narrowest spectral lines expected. If the geometry (the shape) of the cluster is known, $I_X(x, y; \nu)$ can be inverted and we can derive $\rho(\mathbf{r})$, $T_g(\mathbf{r})$, and $Z_{ab}(\mathbf{r})$. In practice, we usually assume a simple geometry for the cluster, spherical or ellipsoidal, however, this approximation breaks down for dynamically active clusters.

2.3. Radio Observations

The distribution of the ICG can also be studied using radio observations. The change in the cosmic microwave background (CMB) due to inverse-Compton scatterings (IC) of cold photons of the CMB off electrons in the hot ICG is called the SZ effect (Sunyaev and Zel'dovich, 1980; for recent reviews see Rephaeli, 1995; Birkinshaw, 1999; Carlstrom et al., 2002). If the cluster is not moving relative to the Hubble flow and the ICG is in thermal equilibrium (or in local thermodynamic equilibrium, LTE), this is called the thermal SZ (TSZ) effect. Additional changes in the CMB may be caused by radial motion of the cluster or bulk velocity in the ICG, which is called the kinematic SZ (KSZ) effect. We measure the sum of these two effects,

$$T_{SZ}(x, y; \nu) = T_{TSZ}(x, y; \nu) + T_{KSZ}(x, y; \nu), \quad (5)$$

where, using the single scattering approximation, the TSZ effect can be expressed as

$$\Delta T_{TSZ}(x, y, \nu) = T_{CMB}\frac{\sigma_T k_B}{m_e c^2} \int_{z_1}^{z_2} g_{TSZ}(x_\nu, T_e)\, n_e(\mathbf{r}) T_e(\mathbf{r}) dz', \quad (6)$$

and the KSZ effect as

$$\Delta T_{KSZ}(x, y, \nu) = T_{CMB}\frac{\sigma_T}{c} \int_{z_1}^{z_2} g_{KSZ}(x_\nu, T_e) n_e(\mathbf{r}) v_r(\mathbf{r}) dz', \quad (7)$$

where $T_{CMB} = 2.72548 \pm 0.00057$ K is the temperature of the CMB (Fixsen, 2009), $v_r(\mathbf{r})$ is the radial (LOS) velocity of the ICG, σ_T is the Thomson cross section, k_B is the Boltzmann constant, m_e and c is the mass of the electron and the speed of light, and the frequency dependences are:

$$g_{TSZ}(x_\nu, T_e) = g_T(x_\nu) + \delta_{RT}(T_e, x_\nu), \quad (8)$$

and

$$g_{KSZ}(x_\nu, T_e) = g_K(x_\nu) + \delta_{RK}(T_e, x_\nu), \quad (9)$$

where

$$g(x_\nu) = x_\nu \frac{e^{x_\nu} + 1}{e^{x_\nu} - 1} - 4, \quad (10)$$

and

$$g_K(x_\nu) = -\frac{x_\nu^2 e^x}{(e^{x_\nu} - 1)^2}, \quad (11)$$

where the dimensionless frequency is $x_\nu = h_P\nu/(k_B T_{CMB})$, where h_P is the Planck constant.

The relativistic corrections, δ_{RT} and δ_{RK}, can be found in, e.g., Nozawa et al. (1998a) and Nozawa et al. (1998b). The effect of multiple scatterings due to the relativistic SZ effect was studied using analytic approximations by Itoh et al. (2001), and performing Monte Carlo simulations by Molnar and Birkinshaw (1999). These corrections were found to be negligible.

In most applications the relativistic corrections can be ignored, and, assuming an isothermal static cluster, in the Rayleigh-Jeans (RJ) limit we obtain:

$$\frac{\Delta T_{RJ}}{T_{CMB}}(x, y) = -2y_C(x, y) - \tau(x, y)\frac{v_r}{c}, \quad (12)$$

where v_r is the radial peculiar velocity of the cluster, and y_C is the Compton y parameter, $y_C(x, y) = \tau(x, y)k_B T_e(x, y)/(m_e c^2)$, where $\tau(x, y)$ is the LOS optical depth.

Similarly to X-ray observations, we need to know the geometry of the cluster to derive the three-dimensional (3D) distribution of $n_e(\mathbf{r})$, $T_e(\mathbf{r})$, and $v_r(\mathbf{r})$. From Equation (12) it is clear that in the RJ limit we measure the sum of both TSZ and KSZ effects, and, in some cases of merging clusters some care is needed in analyzing the SZ data (e.g., Mroczkowski et al., 2012). For example, for a 10 keV subcluster moving with $v_r = 3000$ km s^{-1}, $[\Delta T_{KSZ}/\Delta T_{TSZ}]_{RJ} \propto (v_r/c)/(2k_B T_e(x, y)/(m_e c^2) \sim 0.25$, i.e., the KSZ effect is 25% of the TSZ effect.

3. PHYSICS OF DARK MATTER FROM MERGING GALAXY CLUSTERS

3.1. Total Mass Distribution from Gravitational Lensing

In principle, the best way to derive the total mass distribution of a galaxy cluster is to use gravitational lensing, the bending of light rays due to mass concentrations (Schneider et al., 1992, for a more recent review see e.g., Bartelmann, 2010). In the case of gravitational lensing by a galaxy cluster, the distances between the source, the lens, and the observer are large, thus we can assume that all the mass of the deflector is in a plane perpendicular to the LOS, the lens plane. This is called the thin lens approximation. In this case, choosing our z' axis parallel to the LOS, the bending angle becomes:

$$\boldsymbol{\alpha}(\mathbf{r}_{2D}) = -\frac{4G}{c^2} \int \frac{\Sigma(\mathbf{r}'_{2D})}{b^2}\, \mathbf{b}\, dx' dy', \quad (13)$$

where $\mathbf{b} = \mathbf{r}_{2D} - \mathbf{r}'_{2D}$, where $\mathbf{r}_{2D}$ and $\mathbf{r}'_{2D}$ are 2D vectors, $\mathbf{r}'_{2D} = (x', y')$, Σ is the surface mass density (Equation 1). The bending angle can be expressed using the 2-dimensional gravitational deflection potential, $\phi_{2D}(x, y)$, as

$$\boldsymbol{\alpha}(x, y) = \frac{2}{c^2} \nabla\phi_{2D}(x, y), \quad (14)$$

where ∇ is the 2D gradient operator in the lens plane, and the 2D gravitational deflection potential is defined as the LOS integral of the 3D Newtonian potential,

$$\phi_{2D}(x, y) = \int \phi(x, y, z')dz'. \tag{15}$$

Using the thin lens approximation, we can project the distances between the source and its image onto the source plane, and obtain the lens equation, which, in angular coordinates, becomes:

$$\theta^I = \theta^S + \frac{D_{LS}}{D_{OS}}\alpha = \theta^S + \hat{\alpha}, \tag{16}$$

where θ^S and θ^I are the source and image coordinates, and we defined the reduced bend-angle, $\hat{\alpha}$. Defining the 2D scaled deflection potential, ψ_{2D}, as

$$\psi_{2D}(\theta) = \frac{2D_{LS}}{D_{OL}D_{OS}} \int \phi(x, y, z')dz', \tag{17}$$

where $x = D_{OL}\theta_1$ and $y = D_{OL}\theta_2$, the lens equation becomes

$$\theta^S = \theta^I - \hat{\alpha} = \theta^I - \nabla_\theta \psi_{2D}(\theta^I), \tag{18}$$

where ∇_θ is to be taken with respect to the angular coordinates, θ_1 and θ_2, thus $\nabla_\theta = D_{OL}\nabla$, where ∇ acts on the Cartesian coordinates x and y in the lens plane.

It is convenient to define the convergence field, $\kappa(\theta)$, the dimensionless measure of the surface mass density, as

$$\kappa(\theta) = \frac{1}{\Sigma_{crit}}\Sigma(\theta), \tag{19}$$

where we defined the critical surface mass density, Σ_{crit},

$$\Sigma_{crit} = \frac{c^2}{4\pi G}\frac{D_{OS}}{D_{LS}D_{OL}}, \tag{20}$$

where D_{OS} and D_{OL} are the angular diameter distances from the observer to the source and the lens (defecting cluster), and D_{LS} is the distance from the lens to the source. For a lens at a redshift of 0.3, a source at $z = 1$, and $H_0 = 70$ km s^{-1} Mpc^{-1}, the critical surface mass density is $\Sigma_{crit} = 1.18$ g cm^{-2}. The scaled deflection angle can be expressed with these new variables as

$$\hat{\alpha}(\theta) = \frac{1}{\pi}\int \frac{\kappa(\theta')}{|\theta'|^2}\theta'\, d\theta', \tag{21}$$

and the scaled 2D deflection potential becomes

$$\psi_{2D}(\theta) = \frac{1}{\pi}\int \kappa(\theta')\ln|\theta'|d\theta'. \tag{22}$$

The gravitational bending of the light depends only on the mass distribution of the lensing object, therefore these measurements are not sensitive to the dynamical state of galaxy clusters. As a consequence, for a merging clusters of galaxy, the total mass

can be derived from lensing independently from the viewing angle of the system. The mass is dominated by dark matter, thus the gravitational lensing derived dark matter can be used as an approximation for the dark matter distribution. In some cases of merging clusters the gas is removed from their dark matter potential wells, and thus lensing measurements give directly the dark matter distribution (e.g., Bullet cluster; see left panel in **Figure 1**). In general, lensing observations of merging clusters allow us to derive the total masses and locations of the projected centers of subclusters, which can be used to constrain the phase of the collision (e.g., the time after the first core passage), subject to projection effects.

Gravitational lensing has two main regimes: (1) strong lensing, which is based on strong, non-linear image distortions of background galaxies, is sensitive to the mass distribution only in the core of clusters; (2) weak lensing, which is the linear regime of image distortion, it is a result of small distortions of background galaxies caused by slowly varying weak gravitational field outside of the core out to the virial radius of clusters.

Using strong lensing to reconstruct the mass distribution in a galaxy cluster, the deflection field is calculated using Equation (13) assuming models of the mass distributions for the background galaxies and the cluster, and the positions of the observed and calculated images of background galaxies are compared. The process is iterated until convergence (e.g., Broadhurst et al., 2005; for non-parametric methods see, e.g., Diego et al., 2005).

In the case of weak lensing, we can rewrite the lens equation using physical coordinates in the 2D lens plane, where it is straightforward to represent the observations. In component notation, the lens equation becomes:

$$y_i = x_i - \hat{\alpha}_i(x_1, x_2), \tag{23}$$

where y_i and x_i ($i = 1, 2$) are the 2D coordinates in the source and lens planes. This equation clearly describes a coordinate transformation between the lens plane to the source plane. This is a non-linear transformation due to the deflection angle field, $\hat{\alpha}_i(x_1, x_2)$. However, in the case of weak lensing, the distortions are small and we can do a Taylor expansion to first order of this transformation, which can be expressed as:

$$y_i(x_1, x_2) \approx \left.\frac{\partial y_i}{\partial x_j}\right|_0 x_j = A_{ij}x_j, \tag{24}$$

where the Jacobian of the transformation, A_{ij}, using Equation (23), is

$$A_{ij} = \delta_{ij} - \frac{\partial \hat{\alpha}_i}{\partial x_j} = \delta_{ij} - \frac{\partial^2\psi}{\partial x_i \partial x_j}. \tag{25}$$

Using the convergence field, $\kappa(x_1, x_2)$, and introducing the shear tensor field, $\Gamma(x_1, x_2)$, with components γ_1 and γ_2, also referred to as a two component pseudo-vector field, γ, the isotropic and traceless part of the lensing Jacobian can be separated, and we obtain,

$$\kappa = \frac{1}{2}(\psi_{11} + \psi_{22}), \tag{26}$$

$$\gamma_1 = \frac{1}{2}(-\psi_{11} + \psi_{22}), \tag{27}$$

$$\gamma_2 = \frac{1}{2}(\psi_{12} + \psi_{21}), \tag{28}$$

and

$$\Gamma = \begin{pmatrix} \gamma_1 & \gamma_2 \\ \gamma_2 & -\gamma_1 \end{pmatrix}. \tag{29}$$

With these variables, the Jacobian of the transformation becomes

$$A = (1 - \kappa)\begin{pmatrix} 1 & 0 \\ 0 & 1 \end{pmatrix} - \begin{pmatrix} \gamma_1 & \gamma_2 \\ \gamma_2 & -\gamma_1 \end{pmatrix} = (1-\kappa)I_{2D} - \Gamma, \tag{30}$$

where I_{2D} is the 2D identity matrix. Since A is a real and symmetric matrix, it can be diagonalized and written as

$$A = (1 - \kappa')\begin{pmatrix} 1 & 0 \\ 0 & 1 \end{pmatrix} - \gamma'\begin{pmatrix} 1 & 0 \\ 0 & -1 \end{pmatrix}, \tag{31}$$

where $(\gamma')^2 = (\gamma_1')^2 + (\gamma_2')^2$. This form clearly shows that the convergence is responsible for isotropic distortion of a background galaxy, (magnifying or de-magnifying it, i.e., changes the surface area of the galaxy) and γ' describes a shear distortion.

The shear coefficients, $\gamma_{1,2}$, can be derived from observations defining complex ellipticities using

$$\epsilon = \epsilon_1 + i\epsilon_2 = \frac{1 - b/a}{1 + b/a}e^{2i\phi}, \tag{32}$$

where a and b are the major and minor axis of the ellipse and ϕ is the position angle. The average ellipticities measured from observations are related to the shear and convergence in that region as

$$\langle \epsilon \rangle = \frac{\gamma}{1 - \kappa}, \tag{33}$$

where the angle brackets refer to averages over an area of the sky. In the limits of small distortions, i.e., the weak lensing approximation, $\gamma \ll 1$ and $\kappa \ll 1$, thus the averaged ellipticities directly give the shear fields $\gamma_1(\theta) \approx \langle \epsilon_1(\theta) \rangle$ and $\gamma_2(\theta) \approx \langle \epsilon_2(\theta) \rangle$. The convergence can be derived from

$$\kappa(\theta) = \frac{1}{\pi}\int d^2\theta' D^*(\theta - \theta')\gamma(\theta'), \tag{34}$$

where D^* is the complex conjugate of the convolution kernel, $\gamma = \gamma_1 + i\gamma_2$,

$$D(\theta) = \frac{(\theta_2^2 - \theta_1^2) - 2i\theta_1\theta_2}{\theta^4}, \tag{35}$$

and the mass surface density as $\Sigma(\theta) = \Sigma_{cr}\kappa(\theta)$ (this is called the Kaiser-Squires algorithm; Kaiser and Squires, 1993).

The angular resolution of strong lensing is much higher than that of the weak lensing, because in order to measure the weak lensing signal, it is necessary to average out the intrinsic ellipticities of background galaxies. The best way to derive the mass distribution in clusters using gravitational lensing is to combine these two methods.

Gravitational lensing measurements are very difficult. Massive components can be mapped accurately, but the S/N ratio is much smaller for less massive components, and, as a consequence, the errors in their amplitudes and locations are higher. The 3D distribution of the total mass can be derived either by assuming a geometry for the mass distribution applicable to relaxed clusters, or using N-body/hydrodynamical simulations to model the system, which is the best method for merging clusters (see Section 6).

Large observing programs using the Hubble Space Telescope (*HST*) provide deep galaxy cluster observations for gravitational lensing analysis: the Cluster Lensing And Supernova survey with Hubble (CLASH, Postman et al., 2012) and the Frontiers Fields (http://www.stsci.edu/hst/discretionary-campaigns/discretionary-frontier-fields/). These new data make a more accurate derivation of the mass distribution in the core of clusters possible by increasing the number of multiple lensed background galaxies (CLASH: Umetsu et al., 2014, 2015b; Merten et al., 2015; Frontier Fields: Lam et al., 2014; Diego et al., 2015a,b; Jauzac et al., 2015).

3.2. Physical Properties of Dark Matter: Cross Section

Overwhelming evidence shows that the mass of galaxy clusters is dominated by dark matter (DM). However, as of today, no DM particles have been detected. The most popular candidates for DM are weakly interacting massive particles (WIMPs) and axions. WIMPs are predicted by R-parity-conserving supersymmetry (SUSY) theory, a very popular extension to the standard model of particle physics predicting a number of hypothetical particles (e.g., Kolb and Turner, 1990). WIMPs are assumed to be interacting via gravity and the weak interaction only, and have a large mass (> 1 GeV). Axions are hypothetical particles introduced to explain the strong charge–parity problem associated with the non-trivial nature of the vacuum in quantum chromodynamics (QCD), which is part of our standard model of particle physics. Sterile (right handed) neutrinos, hypothetical particles which do not take part in the weak interaction, suggested by some extensions of the standard model, were also proposed as candidates for dark matter (e.g., Dodelson and Widrow, 1994). One of the main arguments against these heavy particles is that they should decay and produce line emission in the X-ray band at a frequency depending on their mass, which has not been found yet.

Recently, analyzing XMM observations of a sample of galaxy clusters and M31, some evidence has been found for the existence of an unidentified X-ray line at 3.5 keV, which is suggested to be a decay line of a sterile neutrino with a mass of 7 keV in energy units (Boyarsky et al., 2014; Bulbul et al., 2014). This line was identified in the *Chandra* data as well, but it has not been found in the ACIS-I spectrum of the Virgo cluster. At this point it cannot be ruled out that this line is due to instrumental effects or to an unidentified atomic line. The suggested decay line is very weak (with an equivalent width of only about 1 eV) in the vicinity of known atomic lines.

Merging galaxy clusters provided the first direct evidence for the existence of dark matter. Large offsets between the X-ray emission due to the collisional baryonic matter in the ICG, and the mass concentrations derived from gravitational lensing marking the centers of the two colliding clusters in the Bullet cluster clearly showed that most of the mass in clusters consists of collisionless particles, as we would expect based on our ΛCDM models. It was soon realized that, in principle, the offsets between the dark matter centers and those of the baryonic component (ICG and galaxies) provide a possibility to measure the self-interacting cross section of dark matter. The offsets in the Bullet cluster were the first to derive upper limits on the self-interacting cross section (Markevitch et al., 2004; Randall et al., 2008).

Since the first constraint, several merging clusters have been used to put upper limits on the cross section of the dark matter (e.g., MACSJ0025.4-1222: Bradač et al., 2008; A520: Clowe et al., 2012; Jee et al., 2012, A2744: Merten et al., 2011; DLSCL J0916.2+2951: Dawson et al., 2012). The best constraints on the interaction cross section of DM particles using these merging clusters were $\sigma_D/m_D \lesssim 1 \, \text{cm}^2 \, \text{g}^{-1}$. It is also possible to use a sample of minor mergers and derive constraints on the dark matter cross section. Recently Harvey et al. (2015), applying statistical methods to analyze the offset between the gas, the dark matter, and the galaxy component of merging clusters observed with the *HST* and *Chandra* X-ray observatory, derived the tightest limit for the dark matter cross section: $\sigma_D/m_D \leq 0.47 \, \text{cm}^2 \, \text{g}^{-1}$.

Using offsets between bright central galaxies and the centroids of the DM mass surface density in dense cores of galaxy clusters can also be used to study dark matter cross section. Using *HST* imaging of four BCGs in the core region of A3827 and VLT/MUSE integral field spectroscopy data, Massey et al. (2015) claimed to find the most accurate offset between stellar and dark matter centroids of $1.62^{+0.47}_{-0.49}$ kpc. Assuming that this is due to DM scatterings exclusively, they derived $\sigma_D/m_D \sim 1.07 \pm 0.7 \times 10^{-4}(t_{in}/10^9\text{yrs})^{-2} \, \text{cm}^2 \, \text{g}^{-1}$, where t_{in} is the time interval of the infall. However, they acknowledged that at these length scales the gravitational forces do not dominate and other interactions may also cause these offsets. Using the same data, Kahlhoefer et al. (2015) derived a much larger cross section, $\sigma_D/m_D \sim 1.5$ or $3 \, \text{cm}^2 \, \text{gr}^{-1}$ (depending on whether assuming contact interaction or an effective force for the interaction). As opposed to Massey et al. (2015), Kahlhoefer et al. (2015) assumed that the effective drag force on the DM subhalo is not constant during the evolution of the system. Note, however, that this DM cross section is larger than some of the upper limits derived earlier based on cluster mergers.

The explanation of offsets between peaks of mass surface density and the gas distribution based on the merging scenario assuming ΛCDM cosmological models is questionable in some merging clusters. Since the DM and the galaxies are nearly collisionless, we expect that they move together during merging. For example. it was found that, in spite of the violent collision taking place in the Bullet cluster, the galaxies associated with the infalling subcluster still track its DM center well. However, in A520, a DM peak was found without an overdensity of galaxies in the middle of the cluster near the X-ray peak associated with the ICG, surrounded by three DM peaks which coincide with galaxy density peaks (Mahdavi et al., 2007). The physical reason for the existence of this DM peak without a galaxy density enhancement in A520 is still a subject of debate. One possible explanation would be that it is due to collisions between DM particles (Jee et al., 2012, 2014; however see Clowe et al., 2012). In A2744, an extended X-ray emission was found at large physical distances from the DM mass peaks (Merten et al., 2011). As of today, no merging scenario can explain all observed features of this object (Merten et al., 2011; Medezinski et al., 2016). Clearly, detailed numerical simulations are necessary to find out whether an explanation based on ΛCDM is possible.

4. PHYSICS OF THE INTRACLUSTER GAS

The ICG is the most massive baryonic component of galaxy clusters. As we mentioned, we observe integrated physical quantities of the ICG, thus, in general we need to apply deprojection techniques to derive their 3D distribution. However, some physical quantities can be derived from observations of merging clusters independently of the viewing angle. We discuss these first.

4.1. Total Thermal Energy of Electrons

The total thermal energy of electrons can be derived from observations without any further assumptions. The *TSZ* effect (Equation 6) may be expressed, ignoring relativistic corrections, as a line of sight integral of the electron pressure:

$$\frac{\Delta T_{TSZ}}{T_{CMB}}(\nu) = \frac{\sigma_T g_T(\nu)}{m_e c^2} \int_{-R_{vir}}^{R_{vir}} n_e(\mathbf{r}) k_B \, T_e(\mathbf{r}) \, dz' = \frac{2\sigma_T g_T(\nu)}{3m_e c^2} U_e,$$

(36)

where $U_e = \int (3/2)n_e(\mathbf{r})k_B T_e(\mathbf{r}) \, dV$, is the total thermal energy of the electrons, and we assumed that the cluster gas is bounded by the virial radius, R_{vir}. This non-relativistic approximation is valid for most part of the cluster at low radio frequencies. The total pressure of the ICG can be derived from the electron pressure if the gas composition and temperature are known.

4.2. Thermodynamical Properties of the ICG from X-Ray and SZ Observations

The derivation of the 3D intrinsic physical properties of the ICG, the density, and temperature, $\rho_g(\mathbf{r})$, $T_g(\mathbf{r})$, is not trivial for merging clusters. The reason for this is that we observe integrated quantities along the LOS, and in order to derive the 3D physical quantities we need to deproject the observed quantities, which can be done exactly only if we know the geometry of the cluster. This deprojection is usually done assuming a distribution with a simple symmetry, such as spherical or ellipsoidal, which works well only for relaxed clusters (e.g., Sarazin, 1988; Reiprich et al., 2013; Molnar, 2015).

In the case of merging clusters, the 3D geometry is usually guessed from the observed 2D image. Menanteau et al. (2012) developed a method to deproject X-ray images assuming cylindrical symmetry with the axis lying in the plane of the sky, and applied their method to analyze *Chandra* observations of El Gordo. They used the geometry derived from X-ray observations to model the SZ signal in this system. This is a better approximation than a spherical model for the geometry of the gas in merging clusters, but it is only realistic if the impact parameter of the collision is negligible. Korngut et al. (2011) analyzed Mustang (SZ) and *Chandra* (X-ray) observations of MACS J0744.8+3927, and modeled the shock assuming elliptical geometry. SZ observations of merging clusters have also been analyzed using "pseudo-pressure" maps derived from X-ray observations (e.g., Mroczkowski et al., 2012).

Relativistic corrections to the SZ effect measured at different frequencies depend on the temperature of the ICG. In principle, this can also be used to derive the temperature distribution in the ICG (e.g., Prokhorov et al., 2011; Chluba et al., 2013). Prokhorov et al. (2011) discussed how merging clusters can be studied using the relativistic SZ effect. Chluba et al. (2013) showed that variations in the temperature along the LOS can introduce bias in the derived temperature and peculiar velocity.

If it is possible, the best way to carry out the deprojection is to perform full N-body/hydrodynamical simulations tailored to the observations of the merging cluster. We discuss these methods in detail in Section 6.

4.3. Constraining Transport Processes of the ICG using Merging Clusters

High angular resolution X-ray observations provided evidence that the transport processes (heat conduction and viscosity) in the ICG have been suppressed relative to those expected based on the mean fee path due to the Coulomb interaction for particles crossing the cold front from the inner to the outer part of the cold front. The mean fee path of these particles can be expressed as

$$\lambda_C = 15 \, \text{kpc} \left(\frac{T_{e,1}}{T_{e,2}} \right) \left(\frac{G(1)}{G(\sqrt{T_{e,1}/T_{e,2}}\,]} \right) \left[\frac{T_{e,2}}{7 \, \text{keV}} \right]^2 \left[\frac{n_{e,2}}{10^{-3} \, \text{cm}^{-3}} \right],$$

(37)

where n_e and T_e are the electron densities and temperatures in the regions 1 or 2, and $G(x) = (\mathscr{E}(x) - x \mathscr{E}(x))/(2x^2)$, where $\mathscr{E}(x)$ is the error function (Markevitch and Vikhlinin, 2007), sometimes also been referred to as the "Spitzer value" (Spitzer, 1962).

High-resolution X-ray observations carried out by the *Chandra* mission found shocks in the ICG, as well as unexpected sharp density and temperature discontinuities, called "cold fronts" (for reviews see Markevitch and Vikhlinin, 2007 and Ghizzardi et al., 2010). Cold fronts were found to be contact discontinuities: two regions of gas with different temperatures and densities but with no pressure jump between them separated by a boundary surface. Two types of cold fronts have been identified. One is interpreted as a result of sloshing of the ICG in the main cluster with a cool core due to its displacement from equilibrium in its DM potential well due to minor mergers. In this case the gas on the two sides of the cold font belong to the same cluster, and their motion is tangential (Keshet et al., 2010). The other type of cold front separates the two regions of ICG of the main and infalling subcluster following the bow shock in the main cluster due to major mergers. In the rest of this review we refer to cold fronts following bow shocks in major mergers as "contact discontinuities" for easy reference.

X-ray observations suggest that the width of cold fronts may be significantly smaller than the Coulomb free mean path. The cold front found in A3667 has a width of $\lesssim 3$ kpc, while the local Coulomb mean free path is estimated to be $\lambda_C \approx 12$ kpc (Vikhlinin et al., 2001). Similar results were obtained analyzing contact discontinuities in other clusters (e.g., Ettori and Fabian, 2000). In the merging cluster A2146, the width of the bow shock and the reverse shock were found to be $\lesssim 12$ kpc and $\lesssim 6$ kpc, but the they found that $\lambda_C \approx 21$ kpc and $\lambda_C \approx 23$ kpc (Russell et al., 2012). In galaxy group merger RXJ0751.3+5012, the width of the CD was found to be 2.4±0.7 kpc (leading edge in the NE core component; Russell et al., 2014). The width of contact discontinuities suggest a large gradient of the temperature across them within less than the Coulomb mean free path.

The heat conductivity has also ben found to be suppressed relative to an estimate based on the Coulomb interaction, usually refereed to as the "Spitzer conduction,"

$$\kappa_S = 4.6 \times 10^{13} \, \text{ergs s}^{-1} \, \text{cm}^{-1} \, \text{K}^{-1} \left[\frac{T_e}{10^8 \, \text{K}} \right]^{5/2} \left[\frac{\ln \Lambda_C}{40} \right]^{-1},$$

(38)

where T_e is the electron temperature (conductivity is mainly due to electrons), and $\ln \Lambda_C$ is the Coulomb logarithm (Spitzer, 1962; Sarazin, 1988). The effective conductivity is often expressed as a fraction of the Spitzer conductivity, $\kappa = f_S \kappa_S$. Eckert et al. (2014), using *XMM-Newton* observations of a galaxy group falling into the massive cluster A2142, found that $f_C \lesssim 0.0025$. Eckert et al. used the time scale of diffusion due to conductivity, which smoothes sharp temperature edges, to the time scale the ram pressure stripped tail of the infalling group survived ($\gtrsim 600$ Myr).

The first explanation for the narrow width of cold fronts and shocks was that the flow of the ambient ICM around the dense subcluster core will stretch the initially tangled magnetic field lines to form a draping layer with a magnetic field parallel to the front. Transport processes are significantly suppressed perpendicular to magnetic field lines in a plasma because charged particles get trapped and circle around the field liens with a very small gyroradius, much smaller than the Coulomb mean free path. This assumption was used to explain the narrow width of contact discontinuities and shocks in merging clusters and groups: in A3667 (Vikhlinin et al., 2001), A2146 (Russell et al., 2012), and RXJ0751.3+5012 (Russell et al., 2014). This explanation was supported by MHD numerical simulations including anisotropic heat conduction and assuming a uniform magnetic field perpendicular to the path of the infalling cluster carried out by Asai et al. (2005). They suggested that the magnetic field lines wrapping around the infalling subcluster suppress the heat conduction across them. Asai et al. argue that without magnetic field in the ICG a cold front cannot be maintained because isotropic heat conduction from the hot ambient plasma would rapidly heats the cold subcluster plasma.

The latest detailed study using long exposure (500 k) *Chandra* observations of the cold fronts in the core of the Virgo cluster found that the northern part of the front is <2.5 kpc

(99% confidence level, CL). Comparing the observations with MHD numerical simulations, the presence of Kelvin-Helmholtz instability would imply that the effective viscosity of the ICG is suppressed by more than an order of magnitude with respect to the isotropic Spitzer viscosity (Werner et al., 2015).

Detailed numerical simulations question the magnetic draping scenario for suppression of transport processes at cold fronts and shocks in the ICG. MHD simulations suggest that magnetic draping suppresses the conduction only by a factor of a few (ZuHone et al., 2013). Simulations also suggest, that magnetic draping is broken near the cold front due to tangled magnetic fields and the increased magnetic pressure (Ruszkowski et al., 2014).

A more straightforward explanation for the significantly suppressed transport processes is that a thin shock surface is developed where the kinetic energy of the colliding plasma is dissipated due to collective plasma instabilities. Fluctuations in the electron and ion distributions lead to electric currents which generate magnetic fields even in an unmagnetized plasma. As a consequence of these processes we expect the width of a shock in the ICG to be in the order of the ion inertial length,

$$\lambda_{pi} = c/\omega_{pi} = 2.33 \times 10^{-10} \text{ pc} \left[\frac{n_i}{10^{-3} \text{ cm}^{-3}} \right]^{-1/2}, \quad (39)$$

where ω_{pi} and n_i are the plasma frequency and number density of ion i (e.g., Treumann, 2009; Bykov et al., 2015). We expect the width of cold fronts and shocks to be a few times this inertial length, which is many orders of magnitude less than the Coulomb mean free path (Equation 37). In addition, if external magnetic fields exist, the width of the shocks will be limited perpendicular to the magnetic field, by the gyroradius, for an ion is

$$r_{gyro} \approx 2 \times 10^{-9} \text{ pc} \left[\frac{T_e}{10^8 \text{ K}} \right]^{1/2} \left[\frac{B}{1 \, \mu\text{G}} \right]^{-1}, \quad (40)$$

which is of similar order as the inertial length (Equation 39).

4.4. Constraining Turbulence in the ICG from X-Ray Observations

As a consequence of the hierarchical structure formation, merging of massive objects generate large scale turbulence and bulk motion. Largest eddies in the order of 100s of kpc with velocities ~ 400 km s^{-1} were found in cosmological simulations (e.g., Norman and Bryan, 1999). Similar results were found in numerical simulations of binary cluster mergers (e.g., Ricker and Sarazin, 2001). Turbulence plays an important role in determining the structure of clusters by transporting heat, and providing non-thermal pressure to the ICG.

Numerical simulations suggest that turbulent motions of the ICG caused by mergers and shocks provide a significant non-thermal pressure support which varies as a function of radius (e.g., Norman and Bryan, 1999; Dolag et al., 2005; Rasia et al., 2006; Vazza et al., 2006, 2009; Iapichino and Niemeyer, 2008; Lau et al., 2009; Molnar et al., 2010). Observations confirm that non-thermal pressure support is important in clusters at the core

and the outer regions (e.g., Chiu and Molnar, 2012; Morandi and Limousin, 2012; Morandi et al., 2012; Umetsu et al., 2015a).

The contributions to non-thermal pressure support from turbulence and residual bulk motion generated by hierarchical structure formation, if not accounted for, introduce bias in cluster mass determinations, and therefore in cosmological parameters derived using clusters. Turbulence may also accelerate relativistic electrons, which are responsible for the emission in diffuse radio sources in clusters and amplify magnetic fields (Dolag et al., 2008; Ferrari et al., 2008). Turbulence generated by cluster mergers and AGN feedback is thought to be able to heat the cluster cores of cool core clusters and prevent them from overcooling and forming too many stars. Numerical simulations demonstrated that a large amount of energy is deposited in the central region of clusters due to a major merger providing a significant turbulent pressure support (about 10% of the total pressure; Paul et al., 2011).

Hydrodynamical flows become turbulent, when their Reynolds number is high. The Reynolds number is defined as the ratio of inertial forces to viscous forces. It can be approximated as $Re \approx ML/\lambda$, where M is the Mach number, L is the characteristic size of the cluster and λ is the relevant free mean path. We expect the ICG, where M is a few, $L \approx 1$ Mpc and $\lambda \approx 1$ kpc, Re is the order of 1000 assuming the Coulomb free mean path, or even larger if we take the reduction of the mean free path by magnetic field, is highly turbulent (e.g., Sarazin, 1988; Molnar, 2015).

With the new generation of high angular resolution X-ray telescopes on board Chandra and XMM-Newton it has become possible to study turbulence in the ICG for the first time. In the case of isotropic and homogeneous turbulence, a statistical model is possible, and it can be shown that the velocity energy spectrum, E_v, the kinetic energy per unit mass at wave number k, is $E_v(k) \sim \epsilon^{2/3} k^{-5/3}$, where ϵ is the kinetic energy per unit mass (Kolmogorov, 1941). Schuecker et al. (2004) used the result for the gas pressure fluctuation spectrum, $E_p(k) \sim \epsilon^{4/3} k^{-7/3}$, to derive constraints on turbulence based on X-ray observations of the Coma cluster of galaxies, and found that it is compatible with a Kolmogorov type turbulence spectrum ($E_v(k) \sim k^{-5/3}$).

Recently Gaspari and Churazov (2013) used the power spectrum of the gas density fluctuations ($\delta = \delta\rho/\rho$) derived from X-ray observations of the Coma cluster to constrain turbulence. They used N-body/hydrodynamical simulations (FLASH) including conduction with no viscosity to derive an empirical relation between the amplitude of the gas density fluctuation ($\delta = \delta\rho/\rho$) spectrum, $A_\delta(k)$, defined as $A_\delta(k) = \sqrt{P_\delta(k) 4\pi k^3}$, where $P_\delta(k)$ is the power spectrum, as a function of wavelength, k, and the characteristic Mach number of the turbulence, M_{ch}. They found that the normalization of the characteristic amplitude of the density fluctuation spectrum, is linearly proportional to the Mach number: $A_\delta(k)|_{max} \propto M_{ch}$. The characteristic Mach number was found to be $M_{ch} \sim 0.45$.

4.5. Constraining Turbulence and Resonant Line Scatterings from X-Ray Lines

It is very difficult to characterize turbulence in the ICG based on observations of X-ray emission lines. Since thermal broadening is proportional to $T_i^{-1/2}$, and the amplitude of turbulence is independent of temperature, the best way to measure turbulence is to use emission lines of heavy elements with high abundance (to get as much photons in the line as possible). The different ionization stages (depending on the temperature of the ICG) of Iron seems to be the best choice to study turbulence.

Turbulence causes line broadening and a unique line profile, as well as shifts in the line centroid (Inogamov and Sunyaev, 2003). Turbulence suppresses the optical depth of resonant lines due to a shift in frequency caused by the Doppler effect, thus observations of resonant lines can also be used to study turbulence (Churazov et al., 2004; Gastaldello and Molendi, 2004). We illustrate the effect of turbulence on emission lines in the left panel in **Figure 2** (from Zhuravleva et al., 2013). In this panel we show simulated spectra of the emerging line from the Perseus cluster with and without resonant scattering turbulence.

The maximum optical depth of the continuum radiation and most emission lines are very low ($\tau_0 \lesssim 0.01$), thus the emergent radiation is thin. However, as it has been pointed out by Gil'fanov et al. (1987), resonant lines of ions of heavy elements may have optical depths larger than unity in clusters. Strong resonance lines have large transition probabilities, and thus high probabilities of scattering in the high density core of clusters, and subsequently being reemitted in a different direction, depending on their scattering distribution function. Photons in these lines suffer multiple scatterings, and as a result, the line shapes and intensities change (Gil'fanov et al., 1987; Shigeyama, 1998); moreover, the line amplitudes are reduced along the line of sight through the core of the cluster and enhanced through the outer regions.

Resonant line scatterings change the observed X-ray surface brightness distribution in a cluster as a function of the LOS optical depth. The X-ray surface density along a LOS close to the center of the cluster, for example, is reduced. Resonant scatterings

also change the shape of the emergent line from clusters. We illustrate the effect of resonant line scattering on the line shape in the right panel in **Figure 2** (for a detailed description see Molnar et al., 2006). As it can be seen from this figure, the line flattens at the line center, and at high central optical depths, τ_0, a unique double peaked line profile develops toward the cluster center.

Ignoring scatterings can cause an underestimate in the abundances based on resonance lines (see for example, Shigeyama, 1998; Akimoto et al., 2000). Scatterings can also polarize the emission lines (Sazonov et al., 2002). The line shape is also affected by gas motion in clusters. This can reduce the optical depth of all lines by Doppler broadening (Churazov et al., 2004). Therefore, resonance scattering on emission lines can be used to probe velocity fields and turbulence in clusters (for a review of the effects of resonant line scatterings on X-ray spectra see Churazov et al., 2010).

Using XMM-*Newton* observations, Sakelliou et al. (2002) found no sign for resonant scatterings in M87. On the other hand, Xu et al. (2002) found evidence for resonant line scatterings in NGC 4636 in the outskirts of the Virgo cluster. In the Perseus cluster resonant scatterings were ruled out based on an analysis of the Fe XXVI K lines (Churazov et al., 2004; Gastaldello and Molendi, 2004). However, Sanders et al. (2005) argue that the existence of Hα filaments in the core of Perseus suggests no turbulence. Based on resonance lines of Fe XXIV and Fe XXIII, Sanders et al. (2005) conclude that, in Perseus, resonance scattering may be the cause of the apparent central drop in metal abundances.

Constraints on turbulent velocities were derived by measuring the suppression of resonant line amplitudes due to scatterings (e.g., Werner et al., 2009). It is difficult to measure turbulence with the spectral resolution of today's X-ray detectors. Turbulence broadens the line profile due to the Doppler effect.

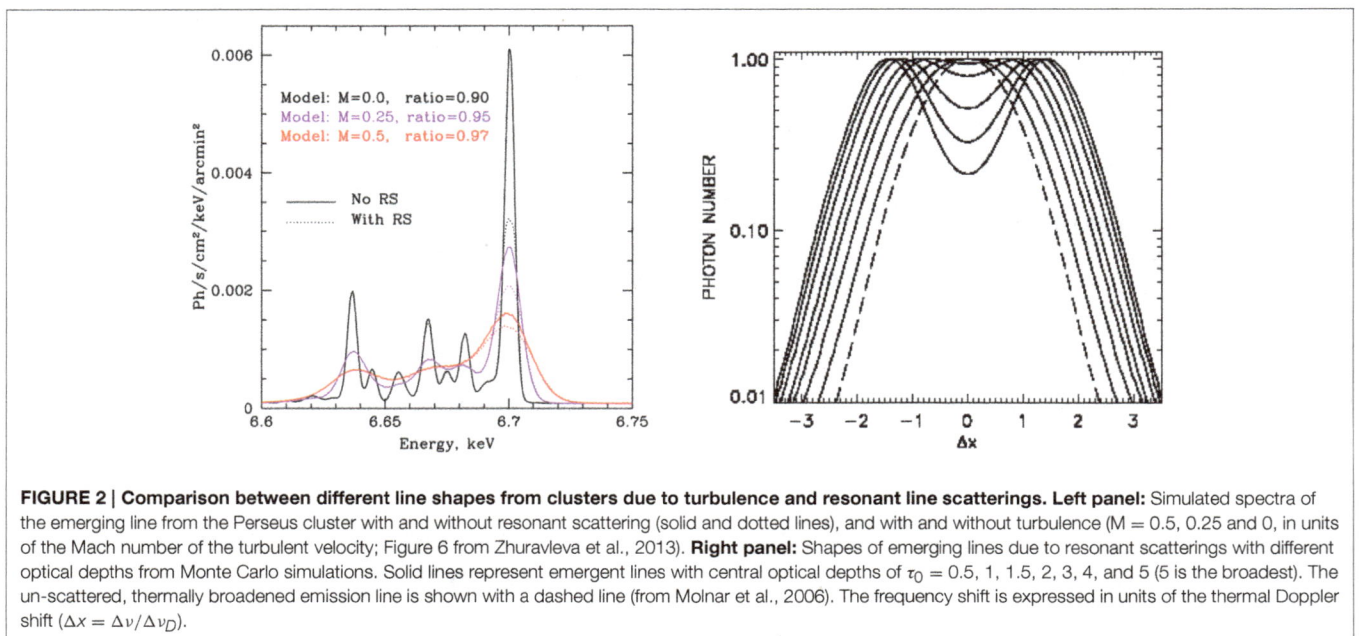

FIGURE 2 | Comparison between different line shapes from clusters due to turbulence and resonant line scatterings. Left panel: Simulated spectra of the emerging line from the Perseus cluster with and without resonant scattering (solid and dotted lines), and with and without turbulence (M = 0.5, 0.25 and 0, in units of the Mach number of the turbulent velocity; Figure 6 from Zhuravleva et al., 2013). **Right panel:** Shapes of emerging lines due to resonant scatterings with different optical depths from Monte Carlo simulations. Solid lines represent emergent lines with central optical depths of $\tau_0 = 0.5$, 1, 1.5, 2, 3, 4, and 5 (5 is the broadest). The un-scattered, thermally broadened emission line is shown with a dashed line (from Molnar et al., 2006). The frequency shift is expressed in units of the thermal Doppler shift ($\Delta x = \Delta \nu / \Delta \nu_D$).

Currently only the Reflection Grating Spectrometers (RGS1 and RGS2) on board of *XMM-Newton* provide the necessary effective area and spectral resolution to possibly measure line broadening due to turbulence. Spectroscopic observations of clusters with bright central emission (cool core clusters) were used to put upper limits on turbulence using RGS1 and RGS2 (Sanders et al., 2010, 2011). Recently Pinto et al. (2015) derived upper limits on turbulence in 44 groups and clusters of galaxies using *XMM-Newton* data. They found an upper limit of 500 km s^{-1} for half of their sources, while a few objects were consistent with large turbulence with upper limits of about 500 km s^{-1}. Pinto et al. found a velocity broadening of $\gtrsim 1000$ km s^{-1} (± 1000 km s^{-1}) in one group of galaxies, NGC507, which is most likely due to previous merging activity. Turbulence <500 km s^{-1} may be caused by bubbles inflated by previous AGN activity (e.g., Brüggen et al., 2005). Larger turbulence is probably generated by merging due to hierarchical structure formation. Note that these results are subject to systematic errors due to the fact that the *XMM-Newton* RGS instruments are slitless spectrographs, and thus line broadening can also be caused by finite extension of the source. Observations carried out by the future mission *Astro-H* will not have this systematic effect.

Separation of the effect of line broadening due to turbulence and resonant line scattering can be performed based on line ratios of resonant and thin lines (e.g., Shang and Oh, 2013; Zhuravleva et al., 2013), as well as using their different line shapes (as demonstrated by Molnar et al., 2006). Mild turbulence may leave some lines with optical depth larger than unity, so that it remains possible to determine the optical depth from the line shapes. In such cases we could use the shapes of optically-thin lines as a template and calculate the broadening due to resonance line scattering. Resonant line scatterings can also be identified using line ratios between resonant lines and other lines, for example using intensity ratios between triplet lines of helium-like ions. The He like FeXXV K line triplet at 6.7 keV is especially useful due to the high relative abundance of Fe. In the left panel in **Figure 3** we show the complex line structure of the FeXXV K line complex with and without resonant scattering (from Zhuravleva et al., 2013). The strongest line is the resonance line at 6.7 keV. Since the amplitude of the resonant line is reduced as a function of its optical depth, the ratio between these lines is proportional to the optical depth to resonant line scatterings. In the right panel in **Figure 3** we illustrate the effects of resonant line scatterings and turbulence on the FeXXV K line complex. Here we show theoretical model spectra with and without resonant line scattering, as well as with and without turbulence (Zhuravleva et al., 2013). Points with error bars represent simulated data based on a 100 ks *Astro-H* observation of the core region ($0.5' \leq r \leq 1'$) of the Perseus cluster.

Gaspari et al. (2014) using high-resolution hydrodynamical simulations concluded that *Astro-H* should be able to detect turbulence in the ICG, and suggested that, in massive clusters a 200 km s^{-1} broadening of the Fe XXV emission line would be detectable. Ota et al. (2015) studied how well the *Astro-H* mission is going to be able to constrain bulk motion and turbulence in clusters. The did not take into account resonant line scatterings, which are not important outside of the core of clusters. They

demonstrated that the spherically averaged velocity profiles can be derived within 20% error from *Astro-H* observations of the 6.7 keV Fe XXV line complex in clusters, however, they pointed out that even relaxed clusters, the significant azimuthal variations in the ICM velocities should be taken into account. Ota et al. also showed that *Astro-H* should be able to directly measure the hydrostatic mass bias with an accuracy of $\lesssim 5\%$.

4.6. Large Scale Diffuse Radio Emission from Galaxy Clusters

The large scale diffuse radio emission due to synchrotron radiation has been observed in many galaxy clusters. This diffuse synchrotron emission indicates the existence of non-thermal electrons accelerated to relativistic energies by magnetic fields in the ICG. Diffuse radio emission in clusters can be divided into three main categories: halos, mini halos, and relics (for a recent review see Feretti et al., 2012). Radio halos and relics seem to be associated with merging clusters, mini-halos are concentrated at the center of some cool-core clusters usually containing a powerful radio galaxy. Non-thermal relativistic particles can be generated in cluster mergers by direct particle acceleration at the merger shocks or by indirect processes as a result of turbulence.

Radio halos are found at the center of merging clusters extending out to large radii, $\gtrsim 1$ Mpc. Their smooth radio emission is unpolarized with a steep power law spectrum, $\propto \nu^{-\alpha}$ ($\alpha = 1.2 - 2$). Radio halo emission is probably due to turbulent acceleration of electrons, which is a random process similar to the second-order Fermi acceleration. This enhanced turbulence is thought to be generated by ongoing or a recent merger event (e.g. Fujita et al., 2003; Cassano and Brunetti, 2005). Brunetti and Lazarian (2011) carried out a detailed study of particle acceleration due to compressible hydrodynamical turbulence in magnetized ICG.

Radio relics extend to Mpc scales showing an elongated shape. Relics are located in the outer regions of merging clusters (e.g., the "Sausage": CIZAJ2242.8+ 5301; van Weeren et al., 2010, and the "Toothbrush": 1RXS 0603.3+4214; van Weeren et al., 2012). Radio emission from relics is highly polarized and have a steep spectrum. Relics indicate the existence of non-thermal electrons in the outskirts of clusters with field strength in the order of $\sim 1\mu$G. The radio emission from relics is likely due to a diffusive shock acceleration, a first order Fermi acceleration (e.g., Drury, 1983), generated by merger shocks. This mechanism assumes that the particles scatter back and forth crossing the shock surface and gain energy at each crossing, which is comparable to the energy of the particle ($\Delta E \sim E$). The acceleration efficiency is mainly determined by the Mach number of the shock. The spectral index of the synchrotron emission, α, which can be derived from radio observations, is related to the index of the energy distribution as $\alpha = (p-1)/2$. From the index of the energy distribution, p, the compression ratio, $r = \rho_2/\rho_1$, can be derived, because it is related to the index of the electron energy distribution, p, $N = E^{-p}$, as $p = (r+2)/(r-1)$. Finally, the Mach number of the merging shock can be derived using the Rankine-Hugoniot jump conditions (for details see Section 5.1.1). Cosmological simulations suggest that emission from relics may be powered by reaccelerated fossil electrons (Pinzke et al., 2013).

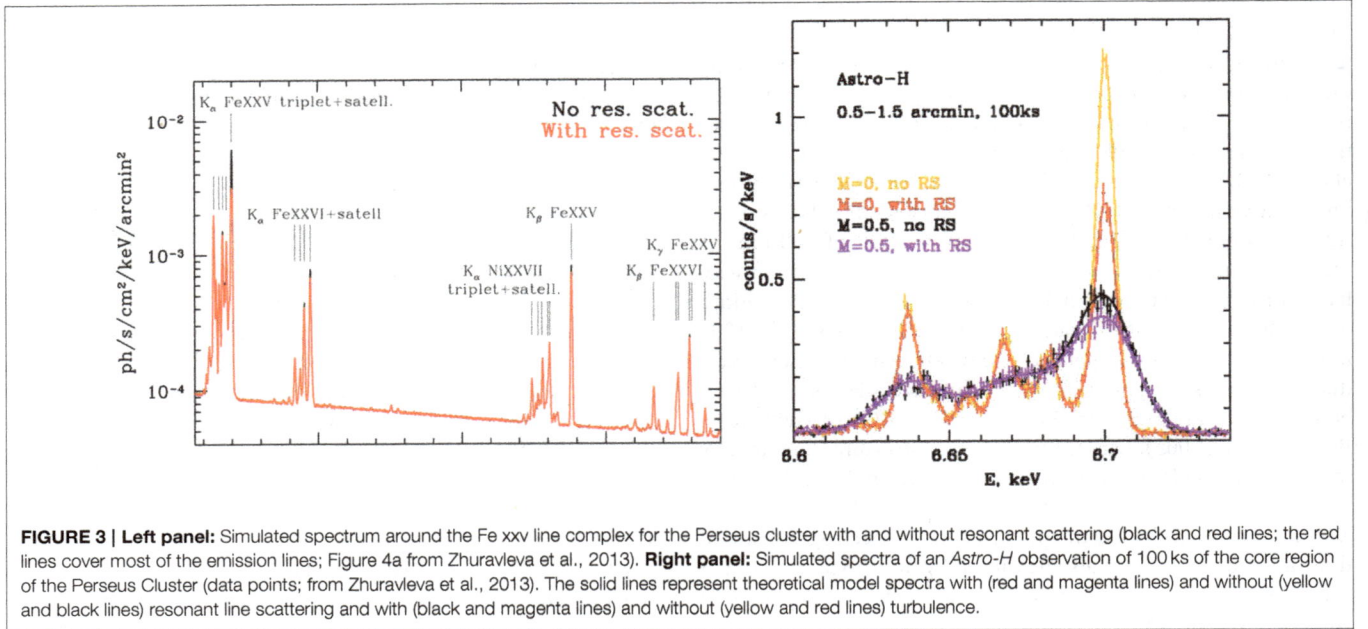

FIGURE 3 | Left panel: Simulated spectrum around the Fe xxv line complex for the Perseus cluster with and without resonant scattering (black and red lines; the red lines cover most of the emission lines; Figure 4a from Zhuravleva et al., 2013). **Right panel:** Simulated spectra of an *Astro-H* observation of 100 ks of the core region of the Perseus Cluster (data points; from Zhuravleva et al., 2013). The solid lines represent theoretical model spectra with (red and magenta lines) and without (yellow and black lines) resonant line scattering and with (black and magenta lines) and without (yellow and red lines) turbulence.

Recently, the diffusive shock acceleration explanation for the origin of emission from radio relics has been challenged. The Mach numbers derived from spectral indexes of the radio emission from the relics in two clusters, 1RXS J0603.3+4214 ("Toothbrush cluster") and CIZA J2242.8+5301, were found to be about 4, twice as large as those derived from *XMM-Newton* and *Chandra* observations (1RXS J0603.3+4214, the "Toothbrush cluster": Ogrean et al., 2013 and CIZA J2242.8+5301: Ogrean et al., 2014). Also, Vazza et al. (2015) found that diffusive shock acceleration cannot explain the synchrotron radio emission from relics and the observed upper limits for the expected γ-ray emission from them simultaneously, unless unrealistically large magnetic fields are assumed (much larger than $10\,\mu$G). However, note that, the Mach number derived from the gas temperature jumps at the shock based on X-ray observations may be significantly biased (Molnar et al., 2013b). The results of Vazza et al. (2015) depend on a lot of assumptions, and they used a simplified spherically symmetric semi-analytical model of expanding merger shocks in clusters to reconstruct the propagation history of shocks. Clearly, more work is necessary to clarify the particle acceleration mechanisms and their role in the generation of radio relics.

In principle, combining the synchrotron radio emission form relics and their γ,-ray emission, the ratio of kinetic energies between electrons and protons due to the acceleration, K_e/K_p, can be derived. We expect that relativistic protons accelerated by shocks will collide with thermal protons in the ICG, and generate pions and other mesons, $p + p \rightarrow p + p + n\pi$, and the charged pions will decay to produce electrons and positrons, $\pi^{\pm} \rightarrow e^{\pm} + \nu_{[e,\mu]} + \bar{\nu}_{[\mu,e]}$, where e and μ is square brackets refer to electron and muon neutrinos, and the line over ν refer to their antiparticles. As of today, γ-ray emission has not been observed from galaxy clusters (e.g., Ackermann et al., 2014; Griffin et al., 2014; Zandanel and Ando, 2014).

4.7. Cluster Magnetic Fields from Radio Observations

The existence of magnetic fields in galaxy clusters has been established based on synchrotron emission from diffuse radio sources and Faraday rotation observations of polarized radio galaxies within or in background of clusters (e.g., Carilli and Taylor, 2002). A magnetized plasma is birefringent, i.e., the photon propagation speed depends on the orientation of its circular polarization (e.g., Spitzer, 1978). Synchrotron emission in a background radio galaxy produces linearly polarized photons, which can be represented as a sum of circularly polarized photons. Since circularly polarized photons propagate with different velocities in magnetized plasma, a rotation of the polarization plane will result. This is called Faraday rotation. The angle of the Faraday rotation is proportional to the rotation measure, R_B, and the wavelength of the photon,

$$\Delta\varphi = R_B\lambda^2, \tag{41}$$

where the rotation measure is defined, and expressed in radians m^{-2}, as

$$R_B = \frac{e^3}{2\pi\, m_e^2 c^4}\int dz\, n_e\, B_{\parallel} = 8.12\times10^{-6}$$

$$\int \left[\frac{dz}{1\text{kpc}}\right]\left[\frac{n_e}{10^{-2}\text{cm}^{-3}}\right]\left[\frac{B_{\parallel}}{1\mu\text{G}}\right], \tag{42}$$

where n_e is the electron number density, $B_{\parallel}$ is the magnitude of the magnetic field parallel to the LOS integral over z. The original polarization angle of the background radio emission is unknown, but since the amount of Faraday rotation depends on the photon wavelength, the difference in the polarization can be measured, and thus the cluster magnetic field estimated.

The first Faraday rotation measurements established that clusters of galaxies have a rotation measure of $R_B \approx 100$ m^{-2}, which, assuming a characteristic electron density of $n_e = 10^{-3}$ cm^{-3} and length of 500 kpc, gives $B_\parallel \approx 0.1\mu$ G (Jaffe, 1980). In general, observations of the Faraday rotation indicate that the intra-cluster magnetic fields are of the order of 1–10 μ G decreasing as a function of the distance from the center with high filling factors up to Mpc scale (e.g., Andernach et al., 1988; Giovannini et al., 1993; Bonafede et al., 2010, 2013; for a review see Feretti et al., 2012).

The luminosity of the synchrotron radiation can also be used to derive a lower bound on the magnetic field assuming the minimum energy of the emitting electrons required to produce the measured luminosity (Burbidge, 1959; for a detailed discussion see Miley, 1980). This minimum energy requirement is close to equipartition between the kinetic energy of electrons and the energy of the magnetic field, $U_{kin} = (3/4)U_B$, thus equipartition is often used as an approximation to the magnetic field. It is possible to estimate the magnetic field without the minimum energy requirement. Using radio and X-ray spectra of the Coma cluster and assuming particle acceleration models, the magnetic field was found to be smoothly declining from $2\pm1\mu$ G in the cluster center, to $0.3 \pm 0.1\mu$ G at 1 Mpc, confirming the results based on the minimum energy requirement (Brunetti et al., 2001).

In principle, the magnetic field in clusters can be derived directly from observations of the radio synchrotron emission and the inverse Compton emission from the same non-thermal electron population, because the inverse Compton flux depends on the electron density, while synchrotron flux depends on the electron density and the magnetic field (Rephaeli, 1977; for a recent review see Rephaeli et al., 2008). The inverse Compton emission in the hard X-ray/γ-ray band has to be measured as an excess emission relative to the thermal emission from the cluster. This is a very difficult measurement, even with the new generation of satellites. Observations carried out using *Swift* and *NuSTAR* have not confirmed the detection of non-thermal emission from clusters (Wik et al., 2009, 2012, 2014; Gastaldello et al., 2015).

Magnetic fields play an important role in the ICG. They suppress transport processes (heat conduction, viscosity) and instabilities (Kelvin-Helmholtz) by reducing the mean free path from that derived from Coulomb collisions. Magnetic pressure, $P_M = B^2/4\pi$, where B is the magnetic field strength, may also contribute to the pressure equilibrium at the core and the outer regions of clusters (e.g., Laganá et al., 2010).

4.8. Non-Thermal Pressure from Multi-Wavelength Observations

As a consequence of ongoing hierarchical structure formation and various feedback mechanisms (AGN, star formation), we expect that the ICG is not in perfect hydrostatic equilibrium. Cosmological simulations demonstrated that the ICG may not be in strict hydrostatic equilibrium even in relaxed clusters (Evrard, 1990; Norman and Bryan, 1999). Simulations suggest that the non-thermal pressure support depends on the radial distance from the cluster center (Norman and Bryan, 1999; Rasia et al.,

2004; Dolag et al., 2005; Rasia et al., 2006; Vazza et al., 2006, 2009; Iapichino and Niemeyer, 2008; Lau et al., 2009; Burns et al., 2010; Molnar et al., 2010; Suto et al., 2013; Nelson et al., 2014). Based on simulations, Kravtsov, et al. (2002) found that in relaxed clusters the non-thermal pressure support increases with distance from the cluster center reaching about 20–30% at the outer regions. Analyzing high-resolution cosmological simulations, Molnar et al. (2012) found a significant non-thermal pressure contribution to the equilibrium pressure, $\sim$30%, at the center of relaxed clusters, decreasing to a minimum in the core region ($0.1R_{\rm vir}$), and an increasing up to $\sim$30–45% at the virial radius.

High resolution cosmological simulations were used to parameterize the contribution to the equilibrium pressure from gas turbulence as a function of radius (Shaw et al., 2010; Battaglia et al., 2012). It has been shown recently, that if the radial coordinate is scaled by the mass over density relative to the mean mass density and not the critical density, the contribution to the non-thermal pressure from turbulence can be derived with no direct dependence of the redshift, however, a small correction is necessary due to the mass accretion history of clusters (Nelson et al., 2014).

Random turbulent flows remaining from previous merger events were suggested to be the main contributions to the non-thermal pressure (Rasia et al., 2004; Lau et al., 2009). Acceleration of the gas was also suggested as the origin of the lack of hydrostatic equilibrium in clusters based on the Euler equation instead of the Jeans equation (Suto et al., 2013).

Multiwavelength observations confirm the lack of strict hydrostatic equilibrium in relaxed galaxy clusters. The most straightforward way to test the assumption of hydrostatic equilibrium and quantify the non-thermal pressure contribution is to derive the pressure necessary for hydrostatic equilibrium from the total mass determined from gravitational lensing observations using the equation of hydrostatic equilibrium,

$$\nabla P_{HE}(\mathbf{r}) = -\rho_g \nabla \Phi(\mathbf{r}), \qquad (43)$$

where $\Phi(\mathbf{r})$ is the potential generated by all matter. The gas thermal pressure, P_{th} is derived from gas density and temperature distribution using X-ray and/or SZ observations, in practice (in principle high resolution multi-wavelength SZ observations could also be applied), assuming the ideal gas law $P \propto \rho_g T_g$. The difference defines the non-thermal pressure contribution: $P_{nth} = P_{HE} - P_{th}$. This method was used to test the strict hydrostatic equilibrium in galaxy clusters A1689 (Molnar et al., 2010; Morandi et al., 2011; Limousin et al., 2013; Umetsu et al., 2015a), A383 (Morandi and Limousin, 2012), A1835 (Morandi et al., 2012), and MS 2137 (Chiu and Molnar, 2012).

The exact physical origin of the lack of hydrostatic equilibrium in clusters is currently a subject of active research. The most important mechanisms proposed to provide non-thermal pressure support in galaxy clusters are turbulence, magnetic and cosmic ray pressure. These contributions may be separated, for example, assuming a functional form for them and fitting the profiles to observations. Assuming spherical geometry, Laganá et al. (2010) adopted this method to separate the different contributions to non-thermal pressure using X-ray and lensing observations of clusters. Their results suggest that all three components contribute significantly to the non-thermal pressure, contributions from magnetic and cosmic ray pressure dominating in the center of cool core clusters. However, as they note, their results may be subject to systematic effects due to the parameterization used for these contributions. Non-thermal pressure support from different physical mechanisms may be related to each other. Magneto-hydrodynamical simulations assuming anisotropic conduction suggest that magneto-thermal instability can drive turbulence which can provide 5–30% of the hydrostatic equilibrium pressure to balance gravity in the outer parts of clusters (Parrish et al., 2012).

The physical understanding of non-thermal pressure support is essential in using clusters of galaxies as cosmological probes. Ignoring non-thermal pressure support when constructing scaling relations which link the cluster observables (X-ray luminosity, temperature, etc). to the mass distribution predicted by cosmological models leads to biases in the derived cosmological parameters (e.g., Rasia et al., 2006; Battaglia et al., 2012; Nelson et al., 2014). Understanding non-thermal pressure support is also important when using the angular power spectrum of the thermal SZ effect for cosmology (e.g., Bhattacharya et al., 2012; Hill and Sherwin, 2013; Crawford et al., 2014; George et al., 2015). The contribution originating from non-thermal pressure to the amplitude of the thermal SZ power spectrum may reach 60% (Battaglia et al., 2010; Shaw et al., 2010; Trac et al., 2011).

We discuss how the non-thermal pressure support can be constrained using detailed numerical simulations and multi-frequency observations of merging galaxy clusters in Section 6.1.

5. KINEMATICS OF MERGING GALAXY CLUSTERS

In this section we discuss how the velocity of the different mass components in merging clusters can be derived. We start with an approximation which can be used to derive the 3D velocity of the infalling cluster, then we describe different methods to derive the tangential and radial velocity components, and shock velocities of the gas.

5.1. Relative 3D Velocity Based on O/NIR Spectroscopy: an Approximation

The 3D relative velocity at the phase of observation of a merging cluster can be approximated using a simplified model of gravitational collapse introduced by Beers et al. (1982). This approximation is very useful when no additional information is available for the merging cluster, and it can also be used as a check for other methods (e.g., Molnar et al., 2013b).

This dynamical model assumes that a spherical perturbation with the total mass of the merging cluster is expanding with the universe and then collapses. This model has a physical meaning until the first core passage, however, assuming time symmetry, it may be used after that as a rough approximation. If no bound model can be found, there is still an unbound solution,

also considered sometimes when applying this model. However, this out-going solution describes an overdensity perturbation expanding with the universe and never collapsing, which, in a case of a merging cluster, has no physical meaning. The lack of a bound solution means that the observed merging system may be in contradiction with the ΛCDM models of structure formation.

This dynamical model treats the infalling cluster as a test particle moving in the gravitational field of a main cluster with a total mass of the system in an expanding background. It assumes that the test particle start out at zero spatial separation on a radial orbit ignoring a likely finite angular momentum of the system and any tidal forces due to the large scale structure. In this case the system can be described by the well-known parametric solution of the spherical collapse model based on the general relativistic Einstein's field equations,

$$R = \frac{R_{max}}{2}(1 - \cos\chi) \tag{44}$$

$$V = \left[\frac{2GM}{R_{max}}\right]^{1/2} \frac{\sin\chi}{1 - \cos\chi} \tag{45}$$

$$t = \left[\frac{R_{max}^3}{8GM}\right]^{1/2}(\chi - \sin\chi), \tag{46}$$

where M is the total mass of the system, R and V are the 3D distance and the relative velocity between the two components, t is the time elapsed since zero separation, $R(t = 0) = 0$, which is assumed to be equal to the age of the universe at the redshift of the merging system, R_{max} is the distance at maximum separation, G is the gravitational constant, and χ is the development angle (e.g., Peebles, 1993).

The rotation angle between the line of collision and the plane of the sky, θ, connects the observables: R_p, the distance between the two components projected to the plane of the sky and relative radial velocity, V_r, to the equations of the spherical collapse model: $R_p = R\cos\theta$ and $V_r = V\cos\theta$. Substituting the observables into the equations above, we can derive a relation between the rotation angle, θ and the development angle (Gregory and Thompson, 1984), we find

$$\tan\theta = \frac{tV_r}{R_p}\frac{(\cos\chi - 1)^2}{\sin\chi(\chi - \sin\chi)}. \tag{47}$$

Using Equations (44–47), the development angle, χ and the 3D velocity, V, can be derived.

This model has been used to estimate the relative velocities of merging clusters recently (e.g., Bourdin et al., 2011; Girardi et al., 2011; Maurogordato et al., 2011; Molnar et al., 2013b). Molnar et al. (2013b) compared 3D relative velocities derived from this simplified dynamical model and from full N-body/hydrodynamical simulations for A1750, a merging cluster before the first core passage (see Discussion in Section 6.2), and found that the simplified model provided a good approximation.

5.1.1. Shock Velocities of the Gas Component from Observations of Radio Relics

The 3D shock velocity of the gas in merging clusters can be derived using spectroscopic observations of synchrotron radiation from radio relics (for a recent review see, e.g., Feretti et al., 2012). Radio relics, located in the outer regions of merging clusters extend to Mpc scales showing an elongated shape (e.g., the "Sausage": CIZAJ2242.8+ 5301 van Weeren et al., 2010, and the "Toothbrush": 1RXS 0603.3+4214 van Weeren et al., 2012). Radio emission from relics is highly polarized and has a steep spectrum. Relics indicate the existence of non-thermal electrons accelerated to relativistic energies by magnetic fields located in the outskirts of clusters with field strength in the order of $\sim$ $1\mu G$. The radio emission from relics is likely due to first order Fermi acceleration generated by merging shocks. Cosmological simulators suggest that emission from relics may be powered by reaccelerated fossil electrons (Pinzke et al., 2013).

Assuming that the electrons are accelerated via the first–order Fermi acceleration mechanism (e.g., Drury, 1983), the compression ratio, $r = \rho_2/\rho_1$, is related to the index of the electron energy distribution, p, $N = E^{-p}$, as $p = (r + 2)/(r - 1)$. This is a diffusive mechanism, the particles scatter back and forth crossing the shock surface and gain energy at each crossing, which is comparable to the energy of the particle ($\Delta E \sim E$). The acceleration efficiency is mostly determined by the Mach number of the shock. The spectral index of the synchrotron emission, which can be derived from radio observations, is related to the index of the energy distribution as $\alpha = (p - 1)/2$. Thus, the compression ratio can be derived from radio observations.

Using the Rankine-Hugoniot jump conditions for the shock, the compression ratio can be expressed as a function of the Mach number of the shock as

$$\frac{1}{r} = \frac{2}{\gamma + 1}\frac{1}{M^2} + \frac{\gamma - 1}{\gamma + 1}, \tag{48}$$

where the Mach number, $M = M_1 = v_1/c_{s1}$, where v_1 and c_{s1} are the gas velocity relative to the shock and the sound speed in the pre-shocked gas. The ICG is assumed to be monatomic in most applications, thus $\gamma = 5/3$. The speed of the pre-shocked gas v_1 (and thus the shock speed in the frame of the pre-shocked gas, $v_{sh} = |v_1|$) and the speed of the shocked gas, v_2, can be derived from the compression ratio, $r = v_1/v_2$. This method was used by Lindner et al. (2014) to estimate the relative velocity of the infalling subcluster in El Gordo assuming that the NW relic is associated with the merger shock. Note, however, that our simulations based on the observed X-ray morphology (Molnar and Broadhurst, 2015) show that the NW relic is on the opposite side of the merging system, thus it is not related to the merging shock directly. Also, the shock velocity may not be a good approximation for the relative velocity of the dark matter components of a merging system. It has been demonstrated, that the velocity of the shock front relative to the gas of the main cluster may be very different from the velocity of the infalling cluster relative to the main cluster (e.g., Springel and Farrar, 2007).

5.2. Tangential Velocities

5.2.1. Tangential Velocities of the Dark Matter from Radio Observations

Tangential velocities of the components of merging clusters, in principle, can be derived based on the Birkinshaw-Gull (BG) effect (first derived by Birkinshaw and Gull, 1983). The BG effect, the changes in the CMB caused by frequency shifts due to moving gravitational lenses, can be expressed as

$$\frac{\Delta \nu}{\nu_0} = \gamma_v \beta_T^L \cdot \delta, \tag{49}$$

where $\gamma_v^2 = 1/(1 - (\beta_T^L)^2)$, $\beta_T^L = v_T^L/c$, and δ is the bend angle due to gravitational lensing. In the case of a spherical cluster, this reduces to

$$\frac{\Delta \nu}{\nu_0} = \gamma_v \beta_T^L \delta \cos \varphi, \tag{50}$$

where φ is the angle in the sky measured from the direction of the tangential velocity of the moving lens.

The frequency change due to the BG effect, taking the motion of the source and the observer into account, in first order in velocity (FOV), was derived by Wucknitz and Sperhake (2004). They obtained:

$$\frac{\Delta \nu}{\nu_0} = \left[\beta_T^L - \frac{D_{LS}}{D_{OS}}\beta_T^O - \frac{D_{OL}}{D_{OS}}\frac{1 + z_L}{1 + z_S}\beta_T^S \right] \cdot \delta, \tag{51}$$

where $\beta_T = v_T/c$ and D denote tangential velocity and angular diameter distance, and the indices, O, L, and S refer to the observer, the lens, and the source. Equation (51) clearly shows that the frequency shift depends on the relative velocities between the source, the lens, and the observer, and it is zero when the quantity in the bracket vanishes. We expect that the peculiar velocities of the field galaxies, especially at large redshifts, are quite small. The peculiar velocity of the observer (our Heliocentric velocity relative to the CMB) is only $c\beta^O = v^O = 369.0 \pm 0.99$ km s^{-1} (Hinshaw et al., 2009), and we can measure only tangential velocities on the order of a few 1000 km s^{-1}, thus we can ignore the last two terms in most applications, and use Equation (49) (e.g., Molnar et al., 2013a). In principle, the BG effect can be used to derive our peculiar velocity relative to CMB using a sample of massive galaxy clusters.

However, it is very difficult to measure the BG effect using individual clusters because the bending angle generated by a massive galaxy cluster is $\lesssim 1'$, thus the frequency shift is very small: $\Delta \nu/\nu_0 \lesssim 10^{-6} v_T/(1000$ km s^{-1}), or $\Delta T_{BG} \lesssim \pm 3\mu K$ in temperature units (e.g., Molnar and Birkinshaw, 2000). The sensitivity of the SZ instruments on *Planck* is in the order of $1\mu K$, but have low angular resolution. A statistical measurement of clusters in a bulk flow may still be possible by stacking clusters. The BG effect in individual clusters can be measured only with high-angular resolution radio telescopes sensitive to μK signals on large angular scales, which may be possible based on imaging observations by the next generation ground based radio interferometers, such as, for example, the Atacama Large Millimeter/Submillimeter Array (ALMA).

Although the frequency change due to the BG effect is small, it still would be important to measure it because no other methods exist to constrain tangential motion of the main mass component, the dark matter, of clusters. Velocity fields of massive objects derived from observations can be compared to those predicted by different cosmological models, and thus they provide an important test for cosmological models.

5.2.2. Tangential Velocities of the Dark Matter from O/NIR Observations

Optical/Near infrared (O/NIR) observations can be used to derive tangential velocities of the main mass components (essentially the dark matter components), in principle, based on frequency changes due to the BG effect. The frequency shifts of emission lines from individual background galaxies caused by the BG effect cannot be measured, because frequency shifts due to other effects are not known (e.g., cosmological redshift, peculiar velocity). However, Molar and Birkinshaw (2003) suggested that the BG effect may be used to derive tangential velocities of clusters by measuring frequency shifts between multiple images of the same background galaxy using high-resolution spectroscopic observations.

Following up this idea, Molnar et al. (2013a) demonstrated that the tangential velocity of the "bullet" in the Bullet cluster may be measurable using spectra observed by the new generation high-spectral resolution ground based interferometer, ALMA. In merging galaxy clusters, the frequency shifts due to the different moving components can be added together, as a good approximation, because the shifts are small. In a binary merging cluster, such as the Bullet cluster, Equation (49), becomes, in velocity units,

$$V(x, y) = v_{T1} \cdot \delta_1 + v_{T2} \cdot \delta_2, \qquad (52)$$

where the 2D vector fields $v_{T1,2}$ and $\delta_{1,2}$ represent the tangential velocities and the deflection angles of the main cluster (1) and the bullet (2). The difference between the maximum and minimum frequency shifts due to the tangential motion of the bullet is 1.2 km s^{-1}.

5.3. Radial Velocities

5.3.1. Radial Velocities of the Galaxy Components from Optical Spectroscopy

Spectroscopic observations with the new generation of ground based observatories make it possible to study substructure and galaxy kinematics in galaxy clusters based their phase-space distribution. Due to observational constraints, the phase space coordinates for each galaxy is reduced to two coordinates in the configuration space (the angular coordinates in the sky) and the LOS velocity derived from the redshift of the galaxy (e.g., Owers et al., 2011; Boschin et al., 2013; Girardi et al., 2015).

The most straightforward way to derive relative radial velocities of the galaxy components in merging clusters is to measure the Doppler shift using O/NIR spectroscopy. The Doppler shift between two emission lines is proportional to the

LOS relative velocity, v_r:

$$\frac{v_2 - v_1}{v_1} = \frac{\Delta v_{1,2}}{v_1} = \frac{v_r}{c}, \qquad (53)$$

where v_1 and v_2 are the frequencies of the line centers. If we assume that the subclusters are at the same distance from us, the observed redshift of a galaxy can be decomposed to a cosmological term due to the expansion of the universe and to a Doppler shift due to the LOS velocity of the galaxy. This is a good approximation, because, if two clusters seem to interact, the cluster centers should be only a few Mpc away form each other, and ignoring this distance introduces only ~ 100 km s^{-1} error in the measured radial velocities. Note, that in principle, it is possible to correct for this effect if we model the merging system using full N-body/hydrodynamical simulations. Combining these spectroscopic measurements with a dynamical analysis, we can determine which component is closer to us and which is farther.

As an example, we show the radial velocity distribution in two merging clusters: A2744 and A1750 in **Figure 4** (left and right panels; from Owers et al., 2011 and Molnar et al., 2013b). Gaussian fits were used to determine the relative radial velocities of the subclusters in both clusters. A Gaussian distribution is often assumed for galaxy velocities in clusters because they are expected to relax to an equilibrium configuration. This may be a good assumption around the core region even in merging clusters. Owers et al. (2011) used the Kayes Mixture Model algorithm to fit Gaussians to the galaxy velocity distributions near the center of the two components of A2744. We show their results on the left panel in **Figure 4** for the main cluster and the northern component. The northern component is closer to us moving away from us with a radial velocity of ~ 2900 km s^{-1} relative to the main component. In the right panel, we show the velocity distribution of all galaxies in the field of A1750, as well as Gaussian fits to the galaxy velocities extracted near the core of the two subclusters. A dynamical analysis of A1750 based on N-body/hydrodynamical simulations shows that the southern subcluster is closer to us moving away form us toward the northern component with a radial velocity of ~ 960 km s^{-1} (Molnar et al., 2013b).

5.3.2. Radial Velocities of the ICG from X-Ray Spectroscopy

The relative radial velocity of the gas components in merging clusters can be derived from high resolution X-ray spectroscopy based on the Doppler shift between emission lines using Equation (53), where the frequencies of the line centers, v_1 and v_2, are usually expressed in keV in X-ray astronomy (e.g., as in the left panel in **Figure 5**). However, with the present spectral resolution of X-ray instruments, this is a difficult measurement with substantial systematic errors.

The frequency change due to Doppler shift was used to derive the relative velocity between the main and infalling cluster in A2256 by Tamura et al. (2011) using the XIS instrument on *Suzaku*, which is appropriate for extended sources. They found a relative velocity of $v_r = 1500 \pm 300(stat) \pm 300(sys)$ km s^{-1} (with

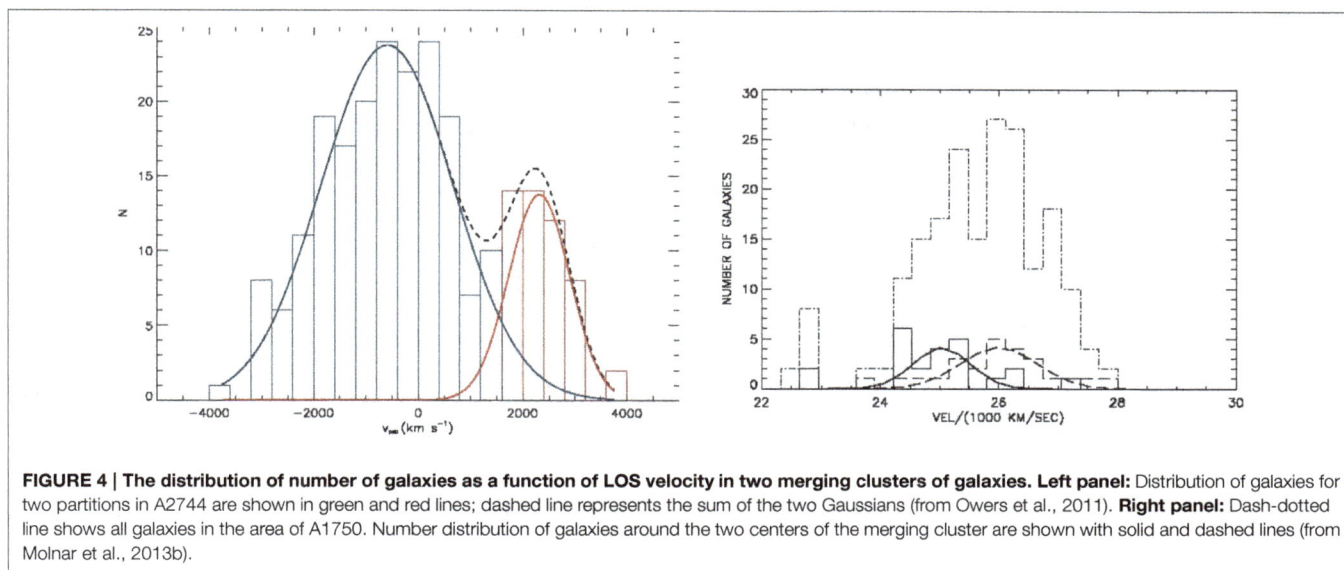

FIGURE 4 | The distribution of number of galaxies as a function of LOS velocity in two merging clusters of galaxies. Left panel: Distribution of galaxies for two partitions in A2744 are shown in green and red lines; dashed line represents the sum of the two Gaussians (from Owers et al., 2011). **Right panel:** Dash-dotted line shows all galaxies in the area of A1750. Number distribution of galaxies around the two centers of the merging cluster are shown with solid and dashed lines (from Molnar et al., 2013b).

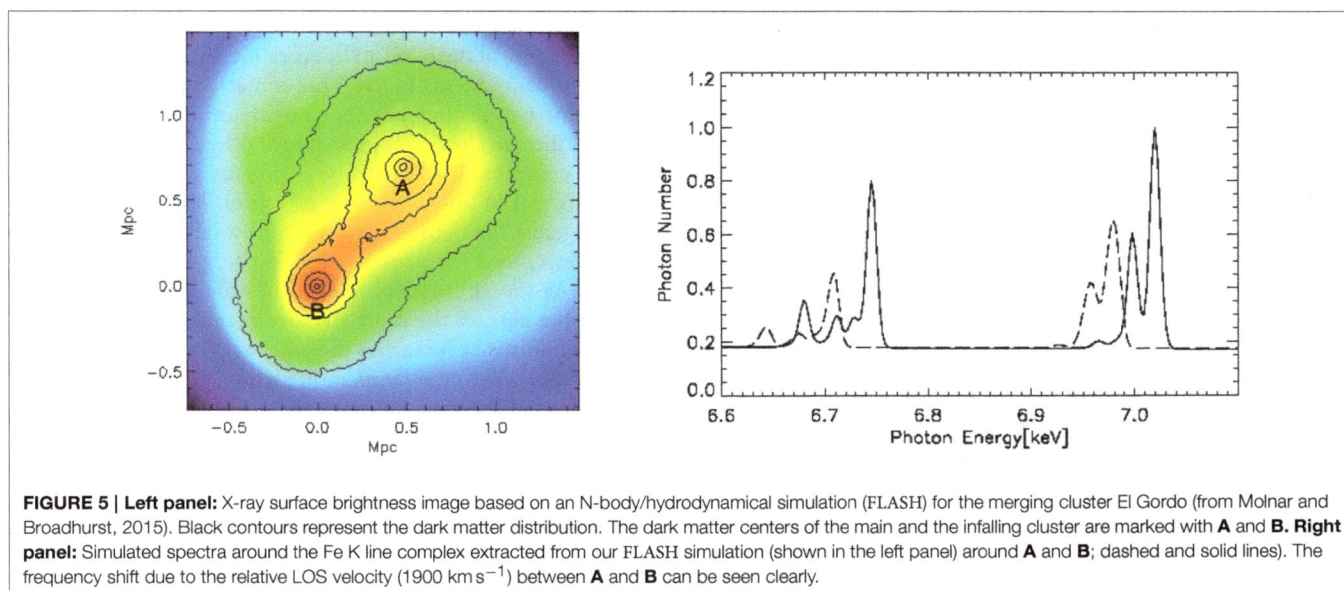

FIGURE 5 | Left panel: X-ray surface brightness image based on an N-body/hydrodynamical simulation (FLASH) for the merging cluster El Gordo (from Molnar and Broadhurst, 2015). Black contours represent the dark matter distribution. The dark matter centers of the main and the infalling cluster are marked with **A** and **B. Right panel:** Simulated spectra around the Fe K line complex extracted from our FLASH simulation (shown in the left panel) around **A** and **B**; dashed and solid lines). The frequency shift due to the relative LOS velocity (1900 km s^{-1}) between **A** and **B** can be seen clearly.

statistical and systematic errors) between these components. The future X-ray mission *Astro-H* will have a high spectral resolution of ~ 7 eV (compare the spectral resolution of 120 eV of the XIS on *Suzaku* at 6 keV photon energy), and reduced systematics, which will make it an ideal tool to measure radial velocity differences between the gas components of merging clusters, and study radial gas motion in clusters in general.

As an illustration of the capabilities of the *Astro-H* X-ray mission, we show a simulated X-ray surface brightness image of El Gordo based on FLASH simulations and simulated spectra associated with the two components of this massive merging cluster in **Figure 5** (Molnar and Broadhurst, 2015). In the left panel, the false color image is the X-ray emission, the black contours represent the dark matter distribution. The projection

angle for the simulation was chosen to reproduce the observed features this system the best. The infalling cluster is moving from north-west toward south-east, just passed the core of the main cluster. The dark matter centers of the main and the infalling clusters are marked with **A** and **B**. The LOS gas radial velocity difference between **A** and **B** derived from the best fit simulation was found to be $\Delta v_r = 1900$ km s^{-1}. This causes a large shift due to Doppler effect in the spectrum, as it is illustrated in the right panel in **Figure 5**. In this panel we show spectra extracted from simulation generated by APEC X-ray emission models including thermal broadening and the Doppler shift due to the line of sight motion of the ICG. The areas of the extraction regions were chosen to give about the same counts. The spectral resolution of ~ 7 eV of *Astro-H* should be sufficient to measure a relative velocity of few hundred km s^{-1} in merging clusters.

5.3.3. Radial Velocities of the ICG from the KSZ Effect
The change on the CMB due to the LOS motion of the galaxy cluster, the kinematic SZ (KSZ) effect (Sunyaev and Zel'dovich, 1980) can be expressed as a fractional radiation temperature change at frequency ν,

$$\frac{\Delta T_{KSZ}(\nu)}{T_{CMB}} = -\frac{x_\nu^2 \, e^x}{(e^{x_\nu} - 1)^2} \frac{v_r}{c} \tau, \qquad (54)$$

where the dimensionless frequency is $x_\nu = h_P \nu / (k_B T_{CMB})$, T_{CMB} is the temperature of the CMB, and τ is the LOS optical depth. In the Rayleigh-Jeans frequency regime, it becomes simply

$$\frac{\Delta T_{KSZ}^{RJ}(\nu)}{T_{CMB}} = -\frac{v_r}{c} \tau. \qquad (55)$$

Relativistic corrections for the KSZ effect were also calculated and found to be small (Nozawa et al., 1998b).

It is very difficult to measure of the KSZ effect due to its small amplitude. The KSZ signal was not detected statistically using *WMAP* and *Planck* observations, only upper limits were placed on the rms in the peculiar velocities of galaxy clusters (Osborne et al., 2011; Planck Collaboration et al., 2014). Hand et al. (2012) found evidence for the KSZ effect using statistical methods. They used maps derived from Atacama Cosmology Telescope (ALMA) observations and identified galaxy clusters based on a luminous galaxy catalog derived from the Baryon Oscillation Spectroscopic Survey, (part of the Sloan Digital Sky Survey III).

Measurements of the KSZ effect in individual massive clusters have been proven even more difficult. In general, the upper limits on radial peculiar velocities derived based on the KSZ effect were large, and no significant detections were reported (e.g., Mauskopf et al., 2012; Mroczkowski et al., 2012; Zemcov et al., 2012).

The first significant detection of the KSZ effect was performed by Sayers et al. (2013) using Caltech Submilimeter Observatory/Bolocam observations of MACS J0717.5+3745 in two frequency bands: 140 and 268 GHz (see also Mroczkowski et al., 2012). Sayers et al. derived a radial peculiar velocity of $v_r = +3450 \pm 900$ km s^{-1}. This result is in a good agreement with that based on optical spectroscopy: $v_r = +3238^{+252}_{-242}$ (Ma et al., 2009).

5.4. Relative Velocity of the Gas Component in Merging Clusters
When the main plane of the collision (the plane containing the two mass centers and the relative velocity vector) is close to the plane of the sky, shocks and contact discontinuities generated by the collision can be clearly identified and analyzed, and the shock velocity of the gas can be derived. Cold fronts have been found in many galaxy clusters (for review see Markevitch and Vikhlinin, 2007; Owers et al., 2009; Ghizzardi et al., 2010). Shock fronts have been detected only in a few merging clusters besides the Bullet cluster (e.g., A520: Markevitch et al., 2005; A754: Macario et al., 2011; A2744: Owers et al., 2011; A2146: Russell

et al., 2012; 1RXS J0603.3+4214, the "Toothbrush": Ogrean et al., 2013; RXJ0751.3+5012: Russell et al., 2014; CIZA J2242.8+5301: Ogrean et al., 2014).

5.4.1. Velocity of the Gas from Merging Shocks
Shocks in the intracluster plasma are expected to be collisionless because the mean free path of particles is large. A thin shock surface is developed where the kinetic energy of the colliding plasma is dissipated due to collective plasma instabilities. Fluctuations in the electron and ion distributions lead to electric currents which generate magnetic fields even in an unmagnetized plasma. As a consequence of these processes we expect the width of a shock in the ICG to be in the order of the ion inertial length, $\lambda_i = c/\omega_{pi}$, where ω_{pi} is the plasma frequency for ion i (e.g., Treumann, 2009). In addition, if external magnetic fields exist, the width of the shocks will be limited by the gyroradius perpendicular to the magnetic field. Cosmological simulations suggest that the shocked gas in clusters is heated up only to mildly relativistic temperatures with $T_{shock} \lesssim 100$ keV even for high infall velocity shocks, as merger simulations indicate (e.g., Molnar et al., 2012).

When two clusters collide, first a contact discontinuity forms where the edges of the ICG (at the accretion shock of each cluster) of the two components touch. This contact surface is generated because the two distinct gas components cannot penetrate each other. As the two clusters approach each other, the gas on the two sides get compressed and reach near pressure equilibrium (there are some contributions from ram pressure), The compressed gas pushes against the undistorted ICG, and when the velocities on both sides exceed the sound speed in the gas, two shocks are formed. As the collision proceeds, the infalling cluster passes through the core of the more massive main component, and a bow shock forms as it reaches the outer regions of the gas of the main cluster, where the temperature drops and thus the sound speed is lower, The most famous example of a merging cluster at this stage is the Bullet cluster (see left panel in **Figure 1**).

An upper limit for the Mach numbers we expect in merging clusters can be derived by assuming that the infall velocity of the smaller subcluster would be about the escape velocity, v_{esc}. Based on this argument, we obtain $M \lesssim v_{esc}/c_s$, where c_s is the sound speed in the gas of the main cluster, thus we expect the Mach number to be $M \lesssim 4$. However, this is an overestimate for two reasons: (1) massive clusters collect matter from a region of about 10 Mpc in radius, not from infinity; (2) the expansion of the Universe reduces the infall velocity derived applying Newtonian mechanics.

Assuming that no energy is lost at the shock due to radiation, particle acceleration, etc., the well-known Rankine–Hugoniot jump conditions across a shock front would determine the physical properties of the gas after the shock passed (Rankine, 1870; Hugoniot, 1889). Since the characteristic lengths of the physical processes that determine the properties of the shock are usually much smaller than the curvature of the shock, we can assume that the shock surface is flat. It is convenient to choose a coordinate system which is moving with the shock velocity (so $v_{sh} = 0$). In this coordinate system, the speed of the pre-shocked gas, $v_1 = -v_{sh}$, where v_{sh} is the speed of the shock as it is moving in the frame of the pre-shock gas. We use indices 1 and 2 for the gas properties in the pre- and post-shock gas hereafter. Integrating the mass, momentum and energy conservation equations, the jump conditions can be derived:

$$\rho_1 v_1 = \rho_2 v_2$$
$$P_1 + \rho_1 v_1^2 = P_2 + \rho_2 v_2$$
$$h_1 + \frac{1}{2} v_1^2 = h_2 + \rho_2 v_2, \qquad (56)$$

where $h_i = \epsilon_i + P_i/\rho_i$ is the enthalpy, and ϵ_i is the internal energy per unit mass of the pre- and post-shock gas ($i = 1$ or 2). Assuming that the gas is a perfect fluid, $\epsilon = [1/(\gamma - 1)]P/\rho$, where γ is the adiabatic index (the ratio of the specific heats). In the case of the almost fully ionized ICG, which can be regarded as a monatomic ideal gas, $\gamma = 5/3$. From Equation (56), we obtain the jump conditions:

$$\frac{P_2}{P_1} = \frac{2\gamma}{\gamma + 1} M^2 - \frac{\gamma - 1}{\gamma + 1}$$
$$\frac{\rho_1}{\rho_2} = \frac{1}{r} = \frac{2}{\gamma + 1} \frac{1}{M^2} - \frac{\gamma - 1}{\gamma + 1}, \qquad (57)$$

where r is the compression, $r = \rho_2/\rho_1 = v_1/v_2$, with a maximum value of 4, and M is the Mach number in the pre-shocked gas, $M = M_1$.

The temperatures and densities can be determined from X-ray spectroscopy. The compression can be derived from the measured X-ray temperature,

$$\frac{1}{r} = \left[\frac{1}{4}\left(\frac{\gamma + 1}{\gamma - 1}\right)^2 \left(\frac{T_2}{T_1} - 1\right)^2 + \frac{T_2}{T_1}\right]^{1/2} - \frac{1}{2}\left(\frac{T_2}{T_1} - 1\right), \quad (58)$$

as suggested by Markevitch et al. (1999). The compression can be used to derive the Mach number from the temperature jump using

$$v_{sh} = \frac{r}{(r-1)^{1/2}} \left[\frac{k_B T_1}{\mu m_p}\left(\frac{T_2}{T_1} - \frac{1}{r}\right)\right]^{1/2}, \qquad (59)$$

where μ is the mean molecular weight. This method was used to determine shock velocities in several clusters (e.g., Markevitch et al., 1999; Markevitch and Vikhlinin, 2001).

The Mach number can be derived from the shock compression, $r = \rho_2/\rho_1$, determined from X-ray observations using the Rankine–Hugoniot jump conditions as

$$M(r) = \left[\frac{2r}{\gamma + 1 - r(\gamma - 1)}\right]^{1/2}, \qquad (60)$$

and, using the Mach number, the shock velocity can be derived from the pre-shock sound speed obtained from X-ray observations as $v_{sh} = Mc_{s1}$. The relative velocity of the infalling cluster can be derived assuming that it is the same as the shock velocity, $v_2 = v_{sh}$, although this has been shown to be an overestimate (e.g., Springel and Farrar, 2007). The Mach number

can also be derived from the pressure jump based on the Rankine-Hugoniot jump conditions, as

$$M = \left[\frac{(\gamma + 1)P_2/P_1 + \gamma - 1}{2\gamma}\right]^{1/2}. \qquad (61)$$

In the case of strong shocks, the Mach number and the shock velocity are well constrained by the pressure jump, Equation (61), since these quantities grow without limit with increasing Mach number. The Mach number in weak shocks is more sensitive to small changes in the compression, thus, in this case, Equation (60) can be used. Using these techniques, the shock velocities have been found in the range of 2000–5000 km s^{-1} (e.g., Markevitch et al., 2002, 2005; Russell et al., 2012).

In principle, the opening angle, θ_M, of the Mach cone can also be used to derive the Mach number, if the shock is well visible: $M = 1/\sin\theta_M$. This method is not too reliable, however, since pressure gradients and substructure affect this simple scaling with angle. The distance to the stagnation point on the contact discontinuity (separating the two gas components of the colliding clusters) from the mid point of the bow shock, d_{st} (connecting the line of the stagnation point and the center of a spherical infalling cluster) and the radius of the curvature of the contact discontinuity, R_{cd}, can also be used to derive the Mach number, since the ratio d_{st}/R_{cd} is a function of $M^2 - 1$ for a spherical discontinuity (Schreier, 1982). Due to the fact that the subcluster is not a hard sphere, the existence of pressure gradients, substructure, ram pressure stripping, and substructure makes it difficult to use this method with accuracy.

5.4.2. Velocity of the Gas from Contact Discontinuities

Contact discontinuities (cold fronts) following the bow shock in merging clusters can also be used to derive the physical parameters of the gas (for a review see Markevitch and Vikhlinin, 2007). As the infalling subcluster gas is pushing the gas of the main cluster ahead, a shock forms if the Mach number, $M = M_1 = v_1/c_{s1} > 1$, where v_1 and c_{s1} are the speed and sound speed in the pre-shock gas. At the center of the contact discontinuity, where the flow of the gas of the main cluster is separated and starts flowing around the gas of the infalling cluster, there is a zero velocity point (relative to the moving cluster), which is called the stagnation point. The pressure ratio of at the stagnation point, P_{st}, and the pre-shock gas, P_1, can be expressed as

$$\frac{P_{st}}{P_1} = \begin{cases} [1 + 0.5(\gamma - 1)M^2]^{\gamma/(\gamma-1)} & M \leq 1 \\ M^2[0.5(\gamma + 1)]^{(\gamma+1)/(\gamma-1)}[\gamma - (\gamma - 1)/ & \\ (2M^2)]^{1/(1-\gamma)} & M > 1. \end{cases} \qquad (62)$$

where M is the Mach number in the pre-shocked gas.

In principle, determining the Mach number of strong shocks from the pressure ratio is better then using the compression, since it is increasing with Mach number without limit, so a small error in the ratio will not lead to a large error in the Mach number. Unfortunately the pressure ratio might not be determined if the gas on one side is too hot and low density, as it can be seen in cold fronts, where a colder gas is plunging through the hot gas of the main cluster (as in the Bullet cluster, because of the low photon counts ahead of the discontinuity; Markevitch et al., 2002). This problem can be solved by recognizing that at a contact discontinuity the pressure is continuous, so we can determine the pressure on the other side of the discontinuity, where the density is much higher, and thus more photons are available for spectroscopy. Using this approach, the Mach number for the upstream gas can be obtained, and from that, the velocity of the shock, $v_{sh} = v_1 = M_1 c_{s1}$. Merger velocities have been derived using this technique for several systems (e.g., Mazzotta et al., 2001; Vikhlinin et al., 2001; Kempner et al., 2002).

Systematic errors in the determination of the Mach number from these methods are due to non-uniform distribution of gas temperature and density, substructures, and projection effects. Particularly, the determination of the temperature is difficult, since we observe the integrated spectrum along the LOS. Attempts to deproject the temperature distribution in the LOS have been made using different weightings: spectroscopic, mass weighed, emission weighted (Mazzotta et al., 2004).

In the case of relaxed clusters, a simple geometry can be assumed (e.g., spherical or ellipsoidal), and a reliable deprojection can be performed. For example, a simultaneous spectral model can be fitted using LOS volume weighting assuming that the cluster is spherically symmetric and piecewise constant density and temperature distribution ("onion shell model") using an X-ray spectral fitting package (as it was done for ellipsoidal geometry in Chiu and Molnar, 2012). Merging clusters, however, are far from equilibrium, thus no simple geometry can be assumed to describe them.

It has been shown that, due to the unknown projection angle (the angle between the LOS and the main plane of collision), systematic errors in the temperature deprojection can be substantial (Molnar et al., 2013b). Errors in the temperature lead to bias in the derived Mach number. As it was demonstrated by Molnar et al. (2012) and Molnar et al. (2013b), systematic errors from deprojection can be minimized by using self-consistent N-body/hydrodynamical numerical simulations of merging galaxy clusters. The advantage of these numerical simulations is that, as long as substructures in the clusters before collision are not substantial, they will be washed out by the energetic merging event, and the features of the shock, which are used to determine the shock velocity and the Mach number of the collision, are not affected by them.

6. PHYSICAL PROPERTIES OF MERGING CLUSTERS FROM NUMERICAL SIMULATIONS

In principle, the best way to analyze merging clusters is to model the system using self-consistent N-body/hydrodynamical simulations constrained by multifrequency observations. Merging clusters are not in dynamical equilibrium. Their structure and evolution are determined by complex non-linear physical processes. The dark matter components react only to gravity, the ICG responsive to pressure forces and various feedback mechanisms (heating, cooling, etc.). As a consequence, it is not possible to use conventional methods, which assume

spherical or ellipsoidal geometry, to analyze observations of merging galaxy clusters.

Merging galaxy cluster simulations can be cast into two main categories: (i) simulations starting from cosmological initial conditions, and (ii) idealized cluster simulations, which assume simplified cluster models and follow their collisions. Both techniques have advantages and disadvantages (e.g., Molnar, 2015). Based on cosmological simulations, the general properties of merging clusters can be studied using the resimulation technique (e.g., Paul et al., 2011). Simulations based on initial conditions generated by controlled Gaussian statistic for the density field to simulate individual clusters is also possible (e.g., Mathis et al., 2005). Alternatively, multi-scale initial conditions can be used, in which case there is no need for resimulation (Hahn and Abel, 2011). These simulations are more realistic, but it is not possible to have full control over the initial conditions of the collision (infalling velocity, impact parameter, dark matter and gas distribution, etc...) to fit the resulting merging system to observations.

Idealized merging cluster simulations set up the initial conditions just before the collision and run the simulations in a Newtonian reference frame. The disadvantages of idealized merging simulations are that they adopt a symmetric cluster shape (mostly spherical), they lack of substructure and the large scale structure (filaments, infalling galaxies, etc...), and assume hydrostatic equilibrium (e.g., Ricker and Sarazin, 2001; Ritchie and Thomas, 2002; McCarthy et al., 2007; Poole et al., 2008; ZuHone, 2011; Molnar et al., 2012).

Self-consistent 3D merging cluster simulations, which took into account the mass of the gas and coupled gravity, adopting different implementation of N-body techniques, were first carried out using the method of smoothed particle hydrodynamics (SPH; Pearce, Thomas and Couchman, 1994; Roettiger et al., 1999a,b; Roettiger and Flores, 2000). The first N-body/magneto-hydrodynamical (MHD) simulations of merging clusters were carried out by Roettiger et al. (1999a). Roettiger et al. demonstrated that the magnetic field can be amplified as a consequence of merging. These simulations were used to analyze the merging cluster A3367, which contains double radio relics in the outskirts of the cluster, and concluded that the relics are a consequence of a 5:1 mass slightly off-axis merging nearly in the plane of the sky. The same method was used to model A3266 as a 2.5:1 off axis merger, and predicted that the rotation of the gas of the main component due to the angular momentum generated by merging might be observable in the future with a new generation of high-spectral resolution X-ray instruments (Roettiger and Flores, 2000).

Off-center collisions of equal mass clusters using different impact parameters were carried out by Ricker (1998), who studied the dependence of the virialization time, X-ray luminosity, and structure of the merger on the impact parameter. It was found that the X-ray luminosity increases due to merging as a function of the impact parameter. Ritchie and Thomas (2002) confirmed that merging can increase the X-ray luminosity and temperature substantially. They also showed that cool cores in clusters might be disrupted if the impact parameter is small enough and the mass ratio is not too large (e.g., less than 8:1). Merging cluster simulations with different mass ratios and impact parameters found that large scale turbulent motions are generated by merging with up to a few hundred kpc in sizes of eddies (Ricker and Sarazin, 2001).

SPH simulations were used to study electron-ion equilibration and non-thermal emission from off axis mergers (Takizawa, 2000). SPH merging cluster simulations assuming cool core clusters carried out by Poole et al. (2006) suggest that the offset between X-ray and mass surface density centroids is the best measure of how much the cluster deviates from dynamical equilibrium. These simulations demonstrate that a merging cluster at a redshift of 0.1 would exhibit no signs of dynamical activity after the second core passage on a 50 ks *Chandra* observation, although it takes another ~ 2 Gyr for the system to virialize.

N-body/MHD simulations of binary merging clusters found that merging amplifies the magnetic field perpendicular to the axis of collision (Takizawa, 2008). N-body/hydrodynamical simulations carried out to study electron-ion equilibration state in shocks due to cluster merging showed that there is a significant departure from equilibrium (Akahori and Yoshikawa, 2010). Detailed high resolution Eulerian AMR simulations (using FLASH) of binary cluster mergers were performed to study gas sloshing and entropy generation and the effect of different initial conditions by ZuHone et al. (2010) and ZuHone (2011). Binary merger simulations using SPH and Eulerian (FLASH) codes were carried out by Mitchell et al. (2009) to compare the results for mixing and turbulence for these two different methods. Mitchell et al. showed that SPH codes artificially suppress turbulence, while AMR codes treat them more realistically (see also Agertz et al., 2007).

Most merging cluster simulations have been carried out to study the astrophysics of merging systems and to understand how the dark matter and gas components interact. A large offset was found in the Bullet cluster, between the gravitational center of the bullet (derived from lensing observations) and the gas component derived from X-ray imaging (Clowe et al., 2006). In the Bullet cluster (left panel in **Figure 1**) the infalling subcluster seems to be plunging through the central region of the main cluster like a bullet showing an X-ray bright wedge shape morphology at the interface of the two ICG components. This merging cluster provided the first direct evidence for the existence of dark matter in large scales (Clowe et al., 2006). Detailed N-body/hydrodynamical simulations of binary mergers have been carried out to explain the morphology of the X-ray and mass surface density and the offset between the X-ray and mass peaks in the Bullet cluster (Springel and Farrar, 2007; Mastropietro and Burkert, 2008; Lage and Farrar, 2014).

The offsets between the dark matter peaks, the X-ray, and SZ centroids are generated by the different behavior of the collisionless component, the dark matter, and the collisional gas during the merging process: the dark matter responds only to gravity, and therefore the core of the infalling cluster passes through that of the main cluster relatively unchanged, but the gas is subject to hydrodynamical effects, which cause large disturbances, shocks and contact discontinuities. The amplitude of the offsets between the dark matter and the gas peaks is determined by the relative strength of the gravitational force,

which is trying to keep the gas locked into the potential well of the dark matter, and the ram pressure, which is acting to remove it.

A quantitative study of the offsets between the peaks of the mass surface density (dominated by dark matter, thus often just refer to as dark matter mass peak), X-ray emission, and SZ effect amplitude as a function of the masses of the components, the impact parameter, and the infall velocity using self-consistent N-body/hydrodynamical simulations of merging galaxy clusters we carried out by Molnar et al. (2012). They demonstrated that, in general, large infall velocities are necessary to produce offsets in the order of $\gtrsim 100$ kpc as long as the impact parameter is not close to zero. Mergers with small impact parameters may produce large offsets just because the gas of the infalling cluster cannot penetrate through that of the main cluster.

When interpreting multifrequency observations of merging clusters, it is possible to use results from simulations already performed, but the best way to analyze these systems is to carry out a series of simulations to reproduce the observations. This makes it possible to constrain quantitatively some important input parameters. As of today only a few detailed studies of merging clusters have been carried out to interpret multifrequency observations: Cl 0024+17 (ZuHone et al., 2009); CIZA J2242.8+5301 (van Weeren et al., 2011a); CL J0152-1347 (Molnar et al., 2012); 1RXS J0603.3+4214 (Brüggen, et al., 2012); using the AMR code FLASH; A3376 (Machado and Lima Neto, 2013) using the SPH code GADGET-2; El Gordo using GADGET-3 (Donnert, 2014), and the FLASH (Molnar and Broadhurst, 2015). All these merging clusters are in the phase soon after the first core passage, when the progenitors can be clearly identified and the morphology of the system is relatively simple. Detailed simulations were carried out to analyze the pre-merger cluster A1750, and it was demonstrated that they can also be used to constrain the initial parameters of the collision (Molnar et al., 2013b).

These numerical simulations assume the initial kinematics of the collision (velocities, impact parameters) and parameterized models for initial distribution of the DM and the gas. Since the parameter space is quite large, the question of uniqueness naturally arises. Assuming that the adopted model is correct, Molnar et al. (2012) demonstrated that observations in O/NIR (mass surface density), X-ray and radio (SZ effect) bands strongly constrain the initial parameters of the collision, and thus merging simulations can be used to uniquely determine them. It is possible, however, that we can obtain a viable solution assuming a different model. A detailed study of model parameterization has not been done yet.

6.1. Analyzing Merging Clusters after the First Core Passage

The best constraints on all initial parameters for a binary merger can be derived for a cluster observed just after the first core passage. The morphology of the X-ray emission and the offsets between the mass peaks, the X-ray and SZ centroids are sensitive to most initial parameters.

As an example of an analysis of a merging cluster after the first core passage using full N-body/hydrodynamical simulations we show the results for the massive merging cluster, El Gordo (Molnar and Broadhurst, 2015). The simulations were constrained by the X-ray morphology, the mass distribution (from gravitational lensing; Jee et al., 2014), and the relative LOS velocity of the two components (derived from optical spectroscopy; Menanteau et al., 2012). The detailed numerical modeling made it possible to constrain the infall velocity, the impact parameter, and the initial masses of the two merging clusters ($\simeq 2250$ km s^{-1}, $\simeq 300$ kpc, 1.4×10^{15} M$_\odot$ and 7.5×10^{14} M$_\odot$). The numerical simulations also suggested a physical explanation for the cometary structure with a "twin tail" feature in the X-ray morphology observed by *Chandra* (right panel in **Figure 1**): tidally stretched gas is responsible for the northern X-ray tail along the collision axis between the mass peaks, the southern tail, which lies off axis, marks compressed and shock heated gas due to the less massive infalling component (left panel in **Figure 5**).

Constraining physical parameters of the ICG using detailed numerical models for individual merging galaxy clusters has just started. We illustrate the power of this method on the extreme merging cluster, the Bullet cluster. The most recent N-body/MHD binary merger simulations to explain O/NIR and X-ray observations of the Bullet cluster were performed by Lage and Farrar (2014). The initial models for the clusters assumed triaxial geometry with a Navarro–Frenk–White (NFW; Navarro et al., 1997) distribution for the dark matter density and a triaxial β-model (Cavaliere and Fusco-Femiano, 1978) with three different exponents and core radii (three simple β model multiplied) in hydrostatic equilibrium with the dark matter. Parametric models were assumed for viscosity and the cluster magnetic field. Non-thermal pressure support was taken into account when calculating gas cooling assuming that it is a constant fraction of the thermal pressure. Images of X-ray emission and SZ effect were generated using the reduced temperature.

We show the results of Lage and Farrar in Figures 6, 7. In the left panel in **Figure 6** we show the mass surface density derived from observations and simulations based on the best fit model. The X-ray surface brightness in the 0.5–2 keV band is shown in the right panel. The effect of the magnetic field is illustrated in **Figure 7**. As a result of magnetic pressure, Lage and Farrar found that the amplitude and the position of the X-ray peak is shifted. As we can see from these figures, the simulations reproduced the observed mass surface density, and X-ray surface brightness well, but the resulting average metallicity, $Z_{ab} = 0.78 \, Z_\odot$, non-thermal pressure fraction, 0.52, and peak magnetic field, $B_0 = 63 \, \mu$G, seem to be too high for clusters. Note that the parametric forms of the distribution of viscosity and magnetic field are not well constrained by observations. As a consequence, a simultaneous fitting for all parameters may adjust the fitted parameters to compensate for modeling errors, and as a result, the derived parameters would be biased. This might be the cause of the large values of some fitted parameters. A study of the correlation between parameters would help to clarify the situation.

Numerical models of merging clusters can also be used to locate areas in the cluster where the LOS gas velocity can be measured with maximal S/N ratio for future high-spectral resolution missions such as *Astro-H* (see Section 5.3.3). For example, even without a map of the mass surface density, we can identify the dark matter centers of the main and the infalling clusters from where we expect the largest signal, the LOS relative velocity difference in the LOS between the two gas components (as marked with **A** and **B** in **Figure 5**).

6.2. Analyzing Pre-Merger Clusters

Perhaps it is the most straightforward to analyze merging clusters before the first core passage, because they are less disturbed due to the collision. Only a small volume of their gas component is affected by adiabatic gas compression and shocks, and the tidal distortions in their dark matter distributions are small, thus the initial conditions of the gas and dark matter distributions in the two merging clusters, in principle, can be straightforwardly derived. In this case, the numerical simulations are needed only to constrain the geometry of the collision, and the relative infall velocity.

We illustrate techniques applicable to analyze pre-merger clusters based on an analysis of A1750. Molnar et al. (2013b) carried out detailed N-body/hydrodynamical simulations using FLASH to model this system. A1750 contains two massive cluster components located at a mean redshift of $z = 0.086$. Optical and X-ray observations of this merging system were used to constrain the initial parameters of this system. The two components can be seen elongated in the northwest–southeast direction in the X-ray image of the cluster (left panel in **Figure 8**). The simulations were constrained by the projected mass centers of the two subclusters marked by their X-ray peaks (since we do not expect a large offset between mass centers and X-ray peaks before the first core passage), the projected X-ray temperature across the interaction region (with a maximum of ~ 6 keV; 3rd panel from the left in **Figure 8**), and the relative radial velocities derived from optical spectroscopic observations (left panel in **Figure 4**). The simulated X-ray surface brightness and temperature distributions based on the best fit model with $M_1 = 2.0 \times 10^{14}$ M$_\odot$ and $M_2 = 1.8 \times 10^{14}$ M$_\odot$, an impact parameter of $P = 150$ kpc, and an infall velocity of 1400 km s^{-1} are shown in the 2nd and 4th panel from the left in Figure 4; from Molnar et al., 2013b). The surface brightness is low in the interaction region. As a consequence, the temperature shows large variations in the map (3rd panel from the left in **Figure 8**), thus we used an average temperature in the interaction region as a function of the distance between the two X-ray peaks (a one dimensional distribution).

Molnar et al. (2013b) demonstrated that, in the case of mergers before the first core passage, the temperature of the compressed/shocked region can be used to put strong constraints on the infall velocity, and that the shape of the temperature distribution is sensitive to the impact parameter. They found that the temperature distribution derived from X-ray observations depends on the viewing angle of the system, and quantified the effect. Thus, due to projection effects, the Mach number derived directly from X-ray observations may also be biased since it depends on the temperature jump at the shock (as we described it in Section 5.4.1).

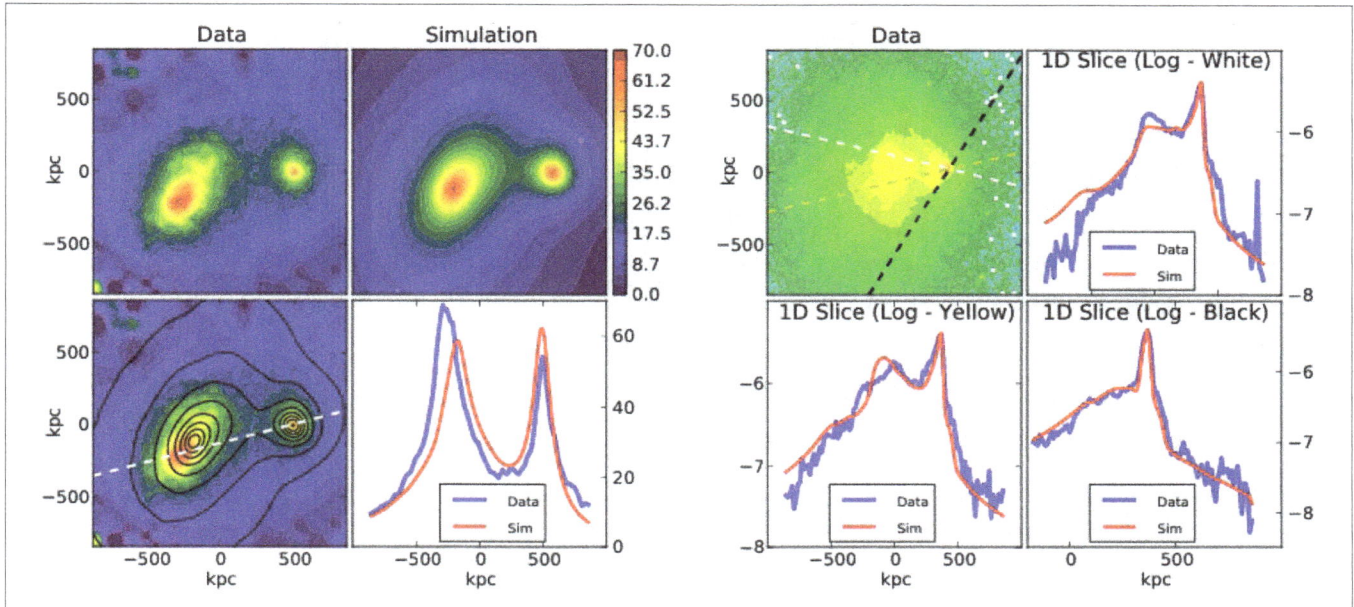

FIGURE 6 | Data and the results based on the best fit simulations for the Bullet cluster. Left Panel: The mass distribution reconstructed from weak lensing data and from simulations are shown in the first row, the data with contours from the best fit simulation shown in left panel in the second row. The right panel in the second row displays the mass distribution from the data and from simulations along the cut shown as dashed line in the left panel. **Right Panel:** The X-ray surface brightness distribution in the range of 0.5–2 keV is shown in the left panel in the first row. The other three panels show the data and the simulations along three cuts marked as white, yellow, and black dashed lines (from Lage and Farrar, 2014).

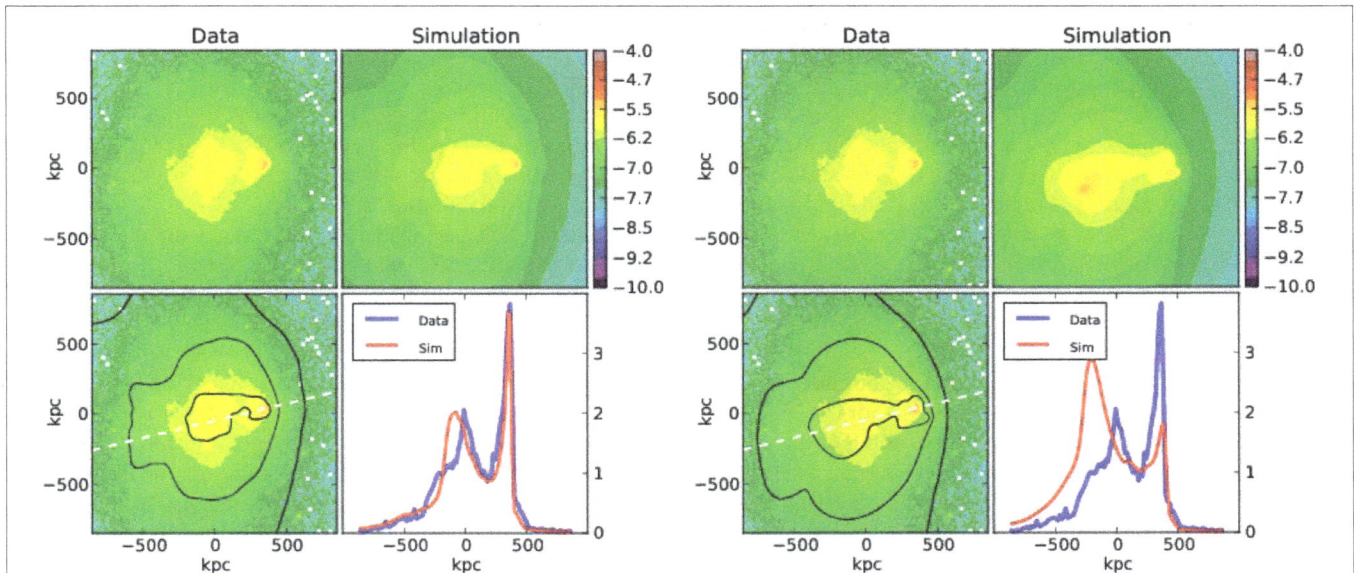

FIGURE 7 | The effect of the ICG magnetic field on the X-ray flux in the range of 0.5–2 keV. The peak of the X-ray emission is shifted as a result of magnetic pressure. **Left Panel:** First row (left and right panels): X-ray surface brightness from data and simulations assuming $B_0 = 61\,\mu G$ for the amplitude of the magnetic field (note that $B \propto \rho_g^{2/3}$ was assumed). Second row: left panel: data with contours from simulations over-plot; right panel: X-ray surface brightness along a cut marked as a dashed line on the left panel. **Right Panel:** Same as left panel but assuming $B_0 = 0.01\,\mu G$ (from Lage and Farrar, 2014).

7. SUMMARY AND FUTURE PROSPECTS

In this contribution, we presented a new, interesting interdisciplinary field, merging clusters of galaxies. In summary:

merging clusters are important for astrophysics and provide a unique tool for cosmology. The extreme physical environment in massive merging clusters amplifies the otherwise small effects due to, for example, viscosity, conduction, and magnetic field

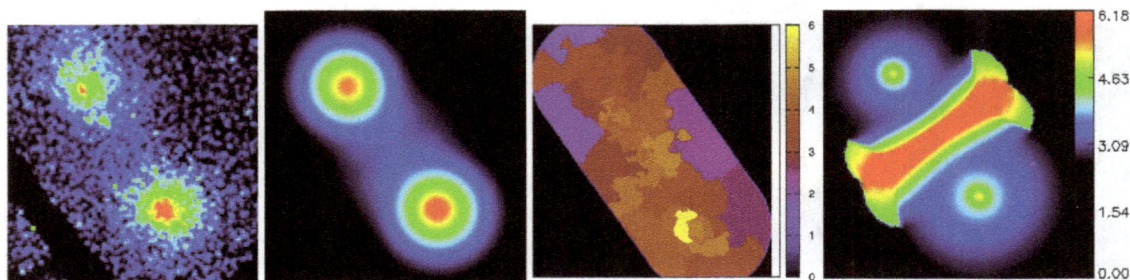

FIGURE 8 | Images of X–ray surface brightness and projected temperature (in units of keV) of A1750 from *Chandra* observations and from the FLASH **simulation with an impact velocity of 1400 km s⁻¹, which is the best match to observations (Molnar et al., 2013b).** From left to right: exposure corrected X–ray surface brightness maps from observations and simulation, projected temperature maps from observations and simulation.

in the ICG. Merging clusters trace the velocity distribution of massive halos, which can be used to test cosmological models.

We demonstrated that in order to interpret multifrequency observations of merging clusters, detailed N-body/hydrodynamical simulations are necessary. The mass surface density derived from O/NIR observations can be used to constrain the masses for the simulations and the phase of the collision. Optical spectroscopy is necessary to derive the LOS velocity of the components, and make it possible to use the phase space distribution of cluster galaxies to identify substructures. X-ray and SZ observations of the ICG can be used to derive the thermodynamical state of the ICG, analyze shocks and contact discontinuities, and constrain the geometry of the collision. High resolution X-ray spectroscopy will provide information on turbulence and the LOS motion of the gas. Observations of radio halos and relics along with X-ray observations can be used to study non-thermal particle population in clusters and constrain magnetic fields in the ICG. A detailed modeling of an extreme merging system, such as, for example, the Bullet cluster, using N-body/magneto-hydrodynamical simulations make it possible to constrain the initial parameters and the geometry of the merging clusters and derive constraints on viscosity, conduction, and magnetic fields in the ICG simultaneously.

In the next decade, we expect that a number of large observational projects will provide us a plethora of new data making it possible to improve our understanding of the physics of clusters of galaxies, and use them as a precision cosmological tool, complementary to other methods. Merging galaxy clusters will play an important role in this process. New powerful computational technology and code development will make it possible to model these new multifrequency data on merging clusters using more sophisticated N-body/magneto-hydrodynamical simulations including more realistic heating and cooling effects, transport processes, and, when necessary, radiative transfer tailored for each observed system.

In the O/NIR band, ground-based broad-band imaging surveys are already under way (e.g., Pan-STARRS, KIDS, VIKING, DES, and UKIDSS). In the near future (~ 2020), a large survey program, the Large Synoptic Survey Telescope (LSST)[1]

is planned using an 8-m-class telescope dedicated to a 10-year survey of the southern sky (Abell et al., 2009).

A 4-year all-sky survey will be carried out in the X-ray band by the extended ROentgen Survey with an Imaging Telescope Array (eROSITA[2]) scheduled for launch in 2016 (Merloni et al., 2012). Astro-H[3], the X-ray/soft Gamma-ray space-based pointing observatory developed by the Japan Aerospace Exploration Agency (JAXA) with participation from NASA and ESA, is planned for launch in 2016. The primary goal of this mission is to study clusters of galaxies and cluster physics using the high spectral-resolution X-ray detectors on board. The high spectral-resolution is achieved by using a micro calorimeter detector array, which provides a constant spectral resolution over the entire detected frequency range.

In the near future (~ 2021), a large FOV, and high angular (~ 5″), and spectral resolution (~ 2.5 eV) a large satellite mission, the International X-ray Observatory (IXO), is planned in an international collaboration including NASA, ESA, and JAXA. This mission will be capable of preforming, for the first time, X-ray polarimetry. IXO will carry four different instruments: (1) An X-ray Micro-calorimeter Spectrometer (XMS) with 2.5 eV spectral resolution; (2) A combined Wide Field Imager (WFI) and Hard X-ray Imager (HXI) with a FOV of 18′ sensitive in the photon energy range of 0.3–40 keV; (3) An X-ray grating spectrometer with a resolving power of ~ 3000 in the soft X-ray band, and (4) An X-ray polarimeter. One of the major goals of IXO is to map bulk motions and turbulence in galaxy clusters. The imaging instrument planned to operate in the hard X-ray energy band (10–40 keV) on board of IXO, with its high sensitivity and angular resolution, will make it possible to study non-thermal emission from clusters by detecting the expected inverse Compton emission due to high-energy electrons, and map its angular distribution. Combining IXO and future sensitive high-resolution radio interferometric observations, the properties and evolution of magnetic fields in the ICG can be studied.

Several large projects are planned to observe clusters of galaxies based on their SZ signal and the large scale structure

[1]http://www.lsst.org/lsst

[2]http://www.mpe.mpg.de/eROSITA
[3]http://heasarc.gsfc.nasa.gov/docs/astroh

in different wavebands. In the radio/submm band, the wide field bolometer array with broad frequency coverage and high sensitivity, Cerro Chajnantor Atacama Telescope (CCAT)[4], is planned to come online in 2017. CCAT is a large collaborative project involving universities from Canada, Germany, and the US. It will be the highest elevation large permanent ground-based telescope in the world. It has a large, 25 m aperture telescope with frequency coverage of 200–2200 μm (1499–1362.5 GHz) and a FOV of 1° with an angular resolution of 3.5 ". High spatial resolution millimeter-wave observations of the SZ effect in galaxy clusters coupled with submillimeter observations to remove contamination from cluster members and lensed galaxies will provide a unique opportunity to study the ICG.

The largest ground-based radio interferometer, the Atacama Large Millimeter Array (ALMA)[5], is already operational in its first phase since 2013. ALMA is a very expensive international project, a result of merging three large projects: the Millimeter Array of the USA, the Large Southern Array of the European Union, and the Large Millimeter Array of Japan. This interferometer will operate in the 30–1000 GHz band using 50 antennas with 12 m in diameter, with an extension of 4 12-m antennas and 12 smaller, 7-m, antennas, in a compact configuration to improve the capability to observe more extended sources, the Atacama Compact Array (ACA). Since June 2014, 34 out of the planned 50 12-m antennas of the full ALMA configuration, and two 12-m antennas and 9 smaller, 7-m antennas of ACA are operational. ALMA observations will provide maps of clusters at an angular resolution of ∼ 5" (Yamada et al., 2012). Although ALMA observations of accretion shocks in galaxy clusters and tangential motion of clusters will be a very difficult (Molnar et al., 2009, 2013a), shocks in merging clusters should be observable (Yamada et al., 2012). Thus, we expect that ALMA will provide us a unique opportunity to study shock fronts and relativistic electrons produced during cluster mergers even at high redshifts, which is not possible with X-ray observations due to the $1/(1 + z)^4$ dependence of the X-ray flux. The frequency dependence of the high-angular resolution maps generated by ALMA will allow us to study gas motion based on the kinematic Sunyaev-Zel'dovich (KSZ) effect, the electron temperature distribution based on the relativistic corrections to the thermal Sunyaev-Zel'dovich (TSZ) effect, and the effect of non-thermal electrons (e.g., Colafrancesco et al., 2011; Prokhorov et al., 2011). Mapping high-redshift galaxy clusters with ALMA's high spatial resolution and sensitivity is essential to asses the dynamical state of these clusters, which is necessary when using individual clusters for cosmology. However, these cluster observations at frequency bands higher than 100 GHz will become too time consuming due to the small effective beam size and higher system temperature (Yamada et al., 2012).

Our main goal to go beyond a simple review of past research on cluster physics with merging clusters of galaxies and discuss the opportunities the new data from instruments available currently and the near future was to show how important and interesting this field is, and inspire young scientists to consider working on this exciting new field.

---

[4]http://www.ccatobservatory.org/index.cfm/page/index.htm.

[5]http://www.almascience.org.

REFERENCES

Abell, G. O. (1958). The distribution of rich clusters of galaxies. *Astrophys. J. Suppl.* 3, 211–288. doi: 10.1086/190036

Abell, P. A., Allison, J., Anderson, S. F., Scott F., Andrew, J. R., Angel, J. R. P., et al., (2009). *The LSST Science Collaboration.* LSST Science Book, Version 2.0. arXiv:0912.0201, 1–596.

Ackermann, M., Ajello, M., Albert, A., Allafort, A., Atwood, W. B., Baldini, L., et al. (2014). Search for cosmic-ray-induced gamma-ray emission in galaxy clusters. *Astrophys. J.* 787, 1–26. doi: 10.1088/0004-637x/787/1/18

Agertz, O., Moore, B., Stadel, J., Potter, D., Miniati, F., Read, J., et al. (2007). Fundamental differences between SPH and grid methods. *Mon. Not. Roy. Astron. Soc.* 380, 963–978. doi: 10.1111/j.1365-2966.2007.12183.x

Akahori, T., and Yoshikawa, K. (2010). Non-equilibrium ionization state and two-temperature structure in the Bullet Cluster 1E 0657-56. *Publ. Astron. Soc. Jpn.* 62, 335–345. doi: 10.1093/pasj/62.2.335

Akimoto, F., Furuzawa, A., Tawara, Y., and Yamashita, K. (2000). Iron k-line analysis of clusters of galaxies with the resonance scattering effect. *Adv. Space Res.* 25, 603–606. doi: 10.1016/S0273-1177(99)00809-1

Andernach, H., Han, T., Sievers, A., Reuter, H.-P., Junkes, N., and Wielebinski, R. (1988). A radio survey of clusters of galaxies. VI - more observations of 34 Abell cluster areas at 11.1, 6.3 and 2.8 CM and a preliminary statistical review of data in papers I–VI. *Astron. Astrophys. Suppl.* 73, 265–324.

Asai, N., Fukuda, N., and Matsumoto, R. (2005). Three-dimensional MHD simulations of X-ray emitting subcluster plasmas in cluster of galaxies. *Adv. Space Res.* 36, 636–642. doi: 10.1016/j.asr.2005.04.041

Ascasibar, Y., and Markevitch, M. (2006). The origin of cold fronts in the cores of relaxed galaxy clusters. *Astrophys. J.* 650, 102–127. doi: 10.1086/506508

Bartelmann, M. (2010). TOPICAL REVIEW gravitational lensing. *Classical Quantum Grav.* 27, 1–72. doi: 10.1088/0264-9381/27/23/233001

Battaglia, N., Bond, J. R., Pfrommer, C., and Sievers, J. L. (2012). On the cluster physics of Sunyaev-Zel'dovich and X-ray surveys. I. The influence of feedback, non-thermal pressure, and cluster shapes on YM scaling relations. *Astrophys. J.* 758, 74–96. doi: 10.1088/0004-637X/758/2/74

Battaglia, N., Bond, J. R., Pfrommer, C., Sievers, J. L., and Sijacki, D. (2010). Simulations of the Sunyaev-Zel'dovich power spectrum with active galactic nucleus feedback. *Astrophys. J.* 725, 91–99. doi: 10.1088/0004-637X/725/1/91

Beers, T. C., Geller, M. J., and Huchra, J. P. (1982). Galaxy clusters with multiple components. I - the dynamics of Abell 98. *Astrophys. J.* 257, 23–32. doi: 10.1086/159958

Bershady, M. A., Jangren, A., and Conselice, C. J. (2000). Structural and photometric classification of galaxies. I. Calibration based on a nearby galaxy sample. *Astron. J.* 119, 2645. doi: 10.1086/301386

Bhattacharya, S., Nagai, D., Shaw, L., Crawford, T., and Holder, G. P. (2012). Bispectrum of the Sunyaev-Zel'dovich effect. *Astrophys. J.* 760, 5–15. doi: 10.1088/0004-637X/760/1/5

Birkinshaw, M. (1999). The Sunyaev-Zel'dovich effect. *Phys. Rep.* 310, 97–195. doi: 10.1016/S0370-1573(98)00080-5

Birkinshaw, M., and Gull, S. F. (1983). A test for transverse motions of clusters of galaxies. *Nature* 302, 315–317. doi: 10.1038/302315a0

Bonafede, A., Brüeggen, M., van Weeren, R., Vazza, F., Giovannini, G., Ebeling, H., et al. (2012). Discovery of radio haloes and double relics in distant MACS galaxy clusters: clues to the efficiency of particle acceleration. *Mon. Not. Roy. Astron. Soc.* 426, 40–56. doi: 10.1111/j.1365-2966.2012.21570.x

Bonafede, A., Feretti, L., Murgia, M., Govoni, F., Giovannini, G., Dallacasa, D., et al. (2010). The Coma cluster magnetic field from Faraday rotation measures. *Astron. Astrophys.* 513, 1–21. doi: 10.1051/0004-6361/200913696

Astronomy and Space Science

Bonafede, A., Vazza, F., Brüggen, M., Murgia, M., Govoni, F., Feretti, L., et al. (2013). Measurements and simulation of Faraday rotation across the Coma radio relic. *Mon. Not. Roy. Astron. Soc.* 433, 3208–3226. doi: 10.1093/mnras/stt960

Boschin, W., Girardi, M., and Barrena, R. (2013). The dynamical status of ZwCl 2341.1+0000: a very elongated galaxy structure with a complex radio emission. *Mon. Not. Roy. Astron. Soc.* 434, 772–783. doi: 10.1093/mnras/stt1070

Buote, D. A., and Tsai, J. C. (1995). The reliability of X-ray constraints of intrinsic cluster shapes. *Astrophys. J.* 439, 29–41. doi: 10.1086/175148

Bourdin, H., Arnaud, M., Mazzotta, P., Pratt, G. W., Sauvageot, J.-L., Martino, R. (2011). A2163: Merger events in the hottest Abell galaxy cluster. II. Subcluster accretion with galaxy-gas separation. *Astron. Astrophys.* 527, 1–13. doi: 10.1051/0004-6361/201014907

Boyarsky, A., Ruchayskiy, O., Iakubovskyi, D., and Franse, J. (2014). Unidentified line in X-ray spectra of the andromeda galaxy and Perseus galaxy cluster. 113, 1–11. doi: 10.1103/physrevlett.113.251301

Bradač, M., Allen, S. W., Treu, T., Ebeling, H., Massey, R., Glenn Morris, R., et al. (2008). Revealing the properties of dark matter in the merging cluster MACS J0025.4-1222. *Astrophys. J.* 687, 959–967. doi: 10.1086/591246

Broadhurst, T., Benítez, N., Coe, D., Sharon, K., Zekser, K., White, R., et al. (2005). Strong-lensing analysis of A1689 from deep advanced camera images. *Astrophys. J.* 621, 53–88. doi: 10.1086/426494

Brunetti, G., and Lazarian, A. (2011). Acceleration of primary and secondary particles in galaxy clusters by compressible MHD turbulence: from radio haloes to gamma-rays. *Mon. Not. Roy. Astron. Soc.* 410, 127–142. doi: 10.1111/j.1365-2966.2010.17457.x

Brunetti, G., Setti, G., Feretti, L., and Giovannini, G. (2001). Particle reacceleration in the Coma cluster: radio properties and hard X-ray emission. *Mon. Not. Roy. Astron. Soc.* 320, 365–378. doi: 10.1046/j.1365-8711.2001.03978.x

Brüggen, M., Hoeft, M., and Ruszkowski, M. (2005). X-Ray line tomography of AGN-induced motion in clusters of galaxies. *Astrophys. J.* 628, 153–159. doi: 10.1086/430732

Brüggen, M., van Weeren, R. J., and Röttgering, H. J. A. (2012). Simulating the toothbrush: evidence for a triple merger of galaxy clusters. *Mon. Not. Roy. Astron. Soc.* 425, L76–L80. doi: 10.1111/j.1745-3933.2012.01304.x

Bulbul, E., Markevitch, M., Foster, A., Smith, R. K., Loewenstein, M., and Randall, S. W. (2014). Detection of an unidentified emission line in the stacked X-ray spectrum of galaxy clusters. *Astrophys. J.* 789, 1–23. doi: 10.1088/0004-637x/789/1/13

Burbidge, G. R. (1959). Estimates of the total energy in particles and magnetic field in the non-thermal radio sources. *Astrophys. J.* 129, 849–852. doi: 10.1086/146680

Burns, J. O., Skillman, S. W., and O'Shea, B. W. (2010). Galaxy clusters at the edge: temperature, entropy, and gas dynamics near the virial radius. *Astrophys. J.* 721, 1105–1112. doi: 10.1088/0004-637X/721/2/1105

Bykov, A. M., Churazov, E. M., Ferrari, C., Forman, W. R., Kaastra, J. S., Klein, U., et al. (2015). Structures and components in galaxy clusters: observations and models. *Space Sci. Rev.* 188, 141–185. doi: 10.1007/s11214-014-0129-4

Carilli, C. L., and Taylor, G. B. (2002). Cluster magnetic fields. *Annu. Rev. Astron. Astrophys.* 40, 319–348. doi: 10.1146/annurev.astro.40.060401.093852

Carlstrom, J. E., Holder, G. P., and Reese, E. D. (2002). Cosmology with the Sunyaev-Zel'dovich effect. *Annu. Rev. Astron. Astrophys.* 40, 643–680. doi: 10.1146/annurev.astro.40.060401.093803

Cassano, R., and Brunetti, G. (2005). Cluster mergers and non-thermal phenomena: a statistical magneto-turbulent model. *Mon. Not. Roy. Astron. Soc.* 357, 1313–1329. doi: 10.1111/j.1365-2966.2005.08747.x

Cavaliere, A., and Fusco-Femiano, R. (1978). The distribution of hot gas in clusters of galaxies. *Astrophys. J.* 70, 677–684.

Chiu, I.-N. T., and Molnar, S. M. (2012). Testing hydrostatic equilibrium in galaxy cluster MS 2137. *Astrophys. J.* 756, 1–10. doi: 10.1088/0004-637X/756/1/1

Chluba, J., Switzer, E., Nelson, K., and Nagai, D. (2013). Sunyaev-Zeldovich signal processing and temperature-velocity moment method for individual clusters. *Mon. Not. Roy. Astron. Soc.* 430, 3054–3069. doi: 10.1093/mnras/stt110

Churazov, E., Forman, W., Jones, C., Sunyaev, R., and Böhringer, H. (2004). *XMM-Newton* observations of the Perseus cluster - II. Evidence for gas motions in the core. *Mon. Not. Roy. Astron. Soc.* 347, 29–35. doi: 10.1111/j.1365-2966.2004.07201.x

Churazov, E., Zhuravleva, I., Sazonov, S., and Sunyaev, R. (2010). Resonant scattering of X-ray emission lines in the hot intergalactic medium. *Space Sci. Rev.* 157, 193–209. doi: 10.1007/s11214-010-9685-4

Clowe, D., Bradač, M., Gonzalez, A. H., Markevitch, M., Randall, S. W., Jones, C., et al. (2006). A direct empirical proof of the existence of dark matter. *Astrophys. J. Lett.* 648, L109–L113. doi: 10.1086/508162

Clowe, D., Markevitch, M., Bradač, M., Gonzalez, A. H., Mi Chung, S., Massey, R., et al. (2012). On dark peaks and missing mass: a weak-lensing mass reconstruction of the merging cluster system A520. *Astrophys. J.* 758, 128–143. doi: 10.1088/0004-637X/758/2/128

Colafrancesco, S., Marchegiani, P., and Buonanno, R. (2011). Untangling the atmosphere of the Bullet cluster with Sunyaev-Zel'dovich effect observations. *Astron. Astrophys.* 527, L1–L5. doi: 10.1051/0004-6361/201016037

Crawford, T. M., Schaffer, K. K., Bhattacharya, S., Aird, K. A., Benson, B. A., Bleem, L. E., et al. (2014). A measurement of the secondary-CMB and millimeter-wave-foreground bispectrum using 800 deg2 of South Pole Telescope data. *Astrophys. J.* 784, 143–163. doi: 10.1088/0004-637X/784/2/143

Dahle, H., Sarazin, C. L., Lopez, L. A., Kouveliotou, C., Patel, S. K., Rol, E., et al. (2013). The Burst cluster: dark matter in a cluster merger associated with the short gamma-ray burst, GRB 050509B. *Astrophys. J.* 772, 23–37. doi: 10.1088/0004-637X/772/1/23

Dawson, W. A., Wittman, D., Jee, M., Gee, P., Hughes, J. P., Tyson, J. A., et al. (2012). Discovery of a dissociative galaxy cluster merger with large physical separation. *Astrophys. J. Lett.* 747, L42–L47. doi: 10.1088/2041-8205/747/2/L42

Diego, J. M., Broadhurst, T., Molnar, S. M., Lam, D., and Lim, J. (2015a). Free-form lensing implications for the collision of dark matter and gas in the Frontier Fields cluster MACS J0416.1-2403. *Mon. Not. Roy. Astron. Soc.* 447, 3130–3149. doi: 10.1093/mnras/stu2660

Diego, J. M., Broadhurst, T., Zitrin, A., Lam, D., Lim, J., Ford, H. C., et al.(2015b). Hubble Frontier Field free-form mass mapping of the massive multiple-merging cluster MACSJ0717.5+3745. *Mon. Not. Roy. Astron. Soc.* 451, 3920–3932. doi: 10.1093/mnras/stv1168

Diego, J. M., Protopapas, P., Sandvik, H. B., and Tegmark, M. (2005). Non-parametric inversion of strong lensing systems. *Mon. Not. Roy. Astron. Soc.* 360, 477–491. doi: 10.1111/j.1365-2966.2005.09021.x

Dodelson, S., and Widrow, L. M. (1994). Sterile neutrinos as dark matter. *Phys. Rev. Lett.* 72, 17–20. doi: 10.1103/PhysRevLett.72.17

Dolag, K., Bykov, A. M., and Diaferio, A. (2008). Non-thermal processes in cosmological simulations. *Space Sci. Rev.* 134, 311–335. doi: 10.1007/s11214-008-9319-2

Dolag, K., Vazza, F., Brunnetti, G., and Tormen, G. (2005). Turbulent gas motions in galaxy cluster simulations: the role of smoothed particle hydrodynamics viscosity. *Mon. Not. Roy. Astron. Soc.* 364, 753–772. doi: 10.1111/j.1365-2966.2005.09630.x

Donnert, J. M. F. (2014). Initial conditions for idealized clusters mergers, simulating 'El Gordo'. *Mon. Not. Roy. Astron. Soc.* 438, 1971–1984. doi: 10.1093/mnras/stt2291

Drury, L. O. (1983). An introduction to the theory of diffusive shock acceleration of energetic particles in tenuous plasmas. *Rep. Prog. Phys.* 46, 973–1027. doi: 10.1088/0034-4885/46/8/002

Eckert, D., Molendi, S., Owers, M., Gaspari, M., Venturi, T., Rudnick, L., et al. (2014). The stripping of a galaxy group diving into the massive cluster A2142. *Astron. Astrophys.* 570, 1–10. doi: 10.1051/0004-6361/201424259

Ettori, S., and Fabian, A. C. (2000). Chandra constraints on the thermal conduction in the intracluster plasma of A2142. *Mon. Not. Roy. Astron. Soc.* 317, L57–L59. doi: 10.1046/j.1365-8711.2000.03899.x

Evrard, A. E. (1990). Formation and evolution of X-ray clusters - a hydrodynamic simulation of the intracluster medium. *Astrophys. J.* 363, 349–366. doi: 10.1086/169350

Ferrari, C., Govoni, F., Schindler, S., Bykov, A. M., and Rephaeli, Y. (2008). Observations of extended radio emission in clusters. *Space Sci. Rev.* 134, 93–118. doi: 10.1007/s11214-008-9311-x

Feretti, L., Giovannini, G., Govoni, F., and Murgia, M. (2012). Clusters of galaxies: observational properties of the diffuse radio emission. *Astron. Astrophys. Rev.* 20, 1–60. doi: 10.1007/s00159-012-0054-z

Fixsen, D. J. (2009). The temperature of the cosmic microwave background. *Astrophys. J.* 707, 916–920. doi: 10.1088/0004-637X/707/2/916

Fujita, Y., Takizawa, M., and Sarazin, C. L. (2003). Nonthermal emissions from particles accelerated by turbulence in clusters of galaxies. *Astrophys. J.* 584, 190–202. doi: 10.1086/345599

Gaspari, M., and Churazov, E. (2013). Constraining turbulence and conduction in the hot ICM through density perturbations. *Astron. Astrophys.* 559, 1–18. doi: 10.1051/0004-6361/201322295

Gaspari, M., Churazov, E., Nagai, D., Lau, E. T., and Zhuravleva, I. (2014). The relation between gas density and velocity power spectra in galaxy clusters: high-resolution hydrodynamic simulations and the role of conduction. *Astron. Astrophys.* 569, 1–15. doi: 10.1051/0004-6361/201424043

Gastaldello, F., Limousin, M., Foëx, G., Muñoz, R. P., Verdugo, T., Motta, V., et al. (2014). Dark matter-baryons separation at the lowest mass scale: the Bullet Group. *Mon. Not. Roy. Astron. Soc.* 442, L76–L80. doi: 10.1093/mnrasl/slu058

Gastaldello, F., and Molendi, S. (2004). Ni Abundance in the core of the Perseus cluster: an answer to the significance of resonant scattering. *Astrophys. J.* 600, 670–680. doi: 10.1086/379970

Gastaldello, F., Wik, D. R., Molendi, S., Westergaard, N. J., Hornstrup, A., Madejski, G., et al. (2015). A NuSTAR observation of the center of the Coma cluster. *Astrophys. J.* 800, 139–146. doi: 10.1088/0004-637X/800/2/139

George, E. M., Reichardt, C. L., Aird, K. A., Benson, B. A., Bleem, L. E., Carlstrom, J. E., et al. (2015). A measurement of secondary cosmic microwave background anisotropies from the 2500 square-degree SPT-SZ survey. *Astrophys. J.* 799, 1–22. doi: 10.1088/0004-637X/799/2/177

Ghizzardi, S., Rossetti, M., and Molendi, S. (2010). Cold fronts in galaxy clusters. *Astron. Astrophys.* 516, 1–20. doi: 10.1051/0004-6361/200912496

Gil'fanov, M. R., Sunyaev, R. A., and Churazov, E. M. (1987). Radial brightness profiles of resonance X-ray lines in galaxy clusters. *Sov. Astron. Lett.* 13, 7–18.

Giovannini, G., Feretti, L., Venturi, T., Kim, K.-T., and Kronberg, P. P. (1993). The halo radio source Coma C and the origin of halo sources. *Astrophys. J.* 406, 399–406. doi: 10.1086/172451

Girardi, M., Bardelli, S., Barrena, R., Boschin, W., Gastaldello, F., and Nonino, M. (2011). Internal dynamics of Abell 2254: a merging galaxy cluster with a clumpy, diffuse radio emission. *Astron. Astrophys.* 536, 1–20. doi: 10.1051/0004-6361/201117332

Girardi, M., Mercurio, A., Balestra, I., Nonino, M., Biviano, A., Grillo, C., et al. (2015). CLASH-VLT: Substructure in the galaxy cluster MACS J1206.2-0847 from kinematics of galaxy populations. *Astron. Astrophys.* 579, 1–19. doi: 10.1051/0004-6361/201425599

Gregory, S. A., and Thompson, L. A. (1984). The A2197 and A2199 galaxy clusters. *Astron. Astrophys. J.* 286, 422–436. doi: 10.1086/162617

Griffin, R. D., Dai, X., and Kochanek, C. S. (2014). New limits on gamma-ray emission from galaxy clusters. *Astrophys. J. Lett.* 795, L21–L25. doi: 10.1088/2041-8205/795/1/L21

Gunn, J. E., and Gott, J. R. III. (1972). On the infall of matter into clusters of galaxies and some effects on their evolution. *Astrophys. J.* 176, 1–19. doi: 10.1086/151605

Hahn, O., and Abel, T. (2011). Multi-scale initial conditions for cosmological simulations. *Mon. Not. Roy. Astron. Soc.* 415, 2101–2121. doi: 10.1111/j.1365-2966.2011.18820.x

Hand, N., Addison, G. E., Aubourg, E., Battaglia, N., Battistelli, E. S., Bizyaev, D., et al. (2012). Evidence of galaxy cluster motions with the kinematic Sunyaev-Zel'dovich effect. *Phys. Rev. Lett.* 109, 1–6. doi: 10.1103/PhysRevLett.109.041101

Harvey, D., Massey, R., Kitching, T., Taylor, A., and Tittley, E. (2015). The nongravitational interactions of dark matter in colliding galaxy clusters. *Science* 347, 1462–1465. doi: 10.1126/science.1261381

Hill, J. C., and Sherwin, B. D. (2013). Cosmological constraints from moments of the thermal Sunyaev-Zel'dovich effect. *Phys. Rev. D* 87, 1–14. doi: 10.1103/PhysRevD.87.023527

Hinshaw, G., Weiland, J. L., Hill, R. S., Odegard, N., Larson, D., Bennett, C. L., et al. (2009). Five-year Wilkinson Microwave Anisotropy Probe observations: data processing, sky maps, and basic results. *Astrophys. J. Suppl.* 180, 225–245. doi: 10.1088/0067-0049/180/2/225

Hugoniot, A. (1889). Sur la propagation du mouvement dans les corps et spcialement dans les gaz parfaits (deuxiȒme partie). *J. Ecol. Polytech.* 58, 1–125.

Iapichino, L., and Niemeyer, J. C. (2008). Hydrodynamical adaptive mesh refinement simulations of turbulent flows - II. Cosmological simulations of galaxy clusters. *Mon. Not. Roy. Astron. Soc.* 388, 1089–1100. doi: 10.1111/j.1365-2966.2008.13518.x

Inogamov, N. A., and Sunyaev, R. A. (2003). Turbulence in clusters of galaxies and X-ray line profiles. *Astron. Lett.* 29, 791–824. doi: 10.1134/1.1631412

Itoh, N., Kawana, Y., Nozawa, S., and Kohyama, Y. (2001). Relativistic corrections to the multiple scattering effect on the Sunyaev-Zel'dovich effect in the isotropic approximation. *Mon. Not. Roy. Astron. Soc.* 327, 567–576. doi: 10.1046/j.1365-8711.2001.04740.x

Jaffe, W. J. (1980). On the morphology of the magnetic field in galaxy clusters. *Astrophys. J.* 241, 925–927. doi: 10.1086/158407

Jauzac, M., Richard, J., Jullo, E., Clement, B., Limousin, M., Kneib, J.-P., et al. (2015). Hubble Frontier fields: a high-precision strong-lensing analysis of the massive galaxy cluster Abell 2744 using ~180 multiple images. *Mon. Not. Roy. Astron. Soc.* 452, 1437–1446. doi: 10.1093/mnras/stv1402

Jee, M. J., Hoekstra, H., Mahdavi, A., and Babul, A. (2014). Hubble Space Telescope/Advanced Camera for Surveys confirmation of the dark substructure in A520. *Astrophys. J.* 783, 1–18. doi: 10.1088/0004-637x/783/2/78

Jee, M. J., Mahdavi, A., Hoekstra, H., Babul, A., Dalcanton, J. J., Carroll, P., et al. (2012). A Study of the dark core in A520 with the Hubble Space Telescope: the mystery deepens. *Astrophys. J.* 747, 1–8. doi: 10.1088/0004-637X/747/2/96

Kaiser, N., and Squires, G. (1993). Mapping the dark matter with weak gravitational lensing. *Astrophys. J.* 404, 441–450. doi: 10.1086/172297

Kahlhoefer, F., Schmidt-Hoberg, K., Kummer, J., and Sarkar, S. (2015). On the interpretation of dark matter self-interactions in Abell 3827. *Mon. Not. Roy. Astron. Soc.* 452, L54–L58. doi: 10.1093/mnrasl/slv088

Kempner, J. C., Sarazin, C. L., and Ricker, P. M. (2002). Chandra observations of A85: merger of the south subcluster. *Astrophys. J.* 579, 236–246. doi: 10.1086/342748

Keshet, U., Markevitch, M., Birnboim, Y., and Loeb, A. (2010). Dynamics and magnetization in galaxy cluster cores traced by X-ray cold fronts. *Astrophys. J. Lett.* 719, L74–L78. doi: 10.1088/2041-8205/719/1/L74

Kolb, E. W., and Turner, M. T. (1990). *The Early Universe.* Vol. 69. Boston, MA: Addison–Wesley.

Kolmogorov, A. (1941). The local structure of turbulence in incompressible viscous fluid for very large Reynolds' numbers. *Akad. Nauk. SSSR Dok.* 30, 301–305.

Korngut, P. M., Dicker, S. R., Reese, E. D., Mason, B. S., Devlin, M. J., Mroczkowski, T., et al. (2011). MUSTANG high angular resolution Sunyaev-Zel'dovich effect imaging of substructure in four galaxy clusters. *Astrophys. J.* 734, 10–17. doi: 10.1088/0004-637X/734/1/10

Kravtsov, A. V., Klypin, A., and Hoffman, Y. (2002). Constrained simulations of the real universe. II. Observational signatures of intergalactic gas in the local supercluster region. *Astrophys. J.* 571, 563–575. doi: 10.1086/340046

Laganá, T. F., de Souza, R. S., and Keller, G. R. (2010). On the influence of non-thermal pressure on the mass determination of galaxy clusters. *Astron. Astrophys.* 510, 1–10. doi: 10.1051/0004-6361/200911855

Lage, C., and Farrar, G. (2014). Constrained simulation of the Bullet cluster. *Astrophys. J.* 787, 144–162. doi: 10.1088/0004-637X/787/2/144

Lam, D., Broadhurst, T., Diego, J. M., Lim, J., Coe, D., Ford, H. C., et al. (2014). A rigorous free-form lens model of A2744 to meet the Hubble Frontier Fields challenge. *Astrophys. J.* 797, 98–125. doi: 10.1088/0004-637X/797/2/98

Lau, E. T., Kravtsov, A. V., and Nagai, D. (2009). Residual gas motions in the intracluster medium and bias in hydrostatic measurements of mass profiles of clusters. *Astrophys. J.* 705, 1129–1138. doi: 10.1088/0004-637X/705/2/1129

Limousin, M., Morandi, A., Sereno, M., Meneghetti, M., Ettori, S., Bartelmann, M., et al. (2013). The three-dimensional shapes of galaxy clusters. *Space Sci. Res.* 177, 155–194. doi: 10.1007/s11214-013-9980-y

Lindner, R. R., Baker, A. J., Hughes, J. P., Battaglia, N., Gupta, N., Knowles, K., et al. (2014). The radio relics and halo of El Gordo, a massive $z = 0.870$ cluster merger. *Astrophys. J.* 786, 1–12. doi: 10.1088/0004-637X/786/1/49

Lotz, J. M., Primack, J., and Madau, P. (2004). A new nonparametric approach to galaxy morphological classification. *Astron. J.* 128, 163-182. doi: 10.1086/421849

Ma, C.-J., Ebeling, H., and Barrett, E. (2009). An X-ray/optical study of the complex dynamics of the core of the massive intermediate-redshift cluster MACSJ0717.5+3745. *Astrophys. J. Lett.* 693, L56–L60. doi: 10.1088/0004-637X/693/2/L56

Macario, G., Markevitch, M., Giacintucci, S., Brunetti, G., Venturi, T., and Murray, S. (2011). A shock front in the merging galaxy cluster a754: X-ray and radio observations. *Astrophys. J.* 728, 82–90. doi: 10.1088/0004-637X/728/2/82

Machado, R. E. G., and Lima Neto, G. B. (2013). Simulations of the merging galaxy cluster Abell 3376. *Mon. Not. Roy. Astron. Soc.* 430, 3249–3260. doi: 10.1093/mnras/stt127

Mahdavi, A., Hoekstra, H., Babul, A., Balam, D. D., and Capak, P. L. (2007). A dark core in Abell 520. *Astrophys. J.* 668, 806–814. doi: 10.1086/521383

Markevitch, M. (2005). Chandra observation of the most interesting cluster in the universe. arXiv:astro-ph/0511345, 1–4.

Markevitch, M., Gonzalez, A. H., Clowe, D., Vikhlinin, A., Forman, W., Jones, C., et al. (2004). Direct constraints on the dark matter self-interaction cross section from the merging galaxy cluster 1E 0657-56. *Astrophys. J.* 606, 819–824. doi: 10.1086/383178

Markevitch, M., Gonzalez, A. H., David, L., Vikhlinin, A. Murray, S. Forman, W., et al. (2002). A textbook example of a bow shock in the merging galaxy cluster 1E 0657-56. *Astrophys. J. Lett.* 567, L27–L31. doi: 10.1086/339619

Markevitch, M., Govoni, F., Brunetti, G., and Jerius, D. (2005). Bow shock and radio halo in the merging cluster A520. *Astrophys. J.* 627, 733–738. doi: 10.1086/430695

Markevitch, M., Sarazin, C.L., and Vikhlinin, A. (1999). Physics of the merging clusters Cygnus A, A3667, and A2065. *Astrophys. J.* 521, 526–530. doi: 10.1086/307598

Markevitch, M., and Vikhlinin, A. (2001). Merger shocks in galaxy clusters A665 and A2163 and their relation to radio halos. *Astrophys. J.* 563, 95–102. doi: 10.1086/323831

Markevitch, M., and Vikhlinin, A. (2007). Shocks and cold fronts in galaxy clusters. *Phys. Rep.* 443, 1–57. doi: 10.1016/j.physrep.2007.01.001

Marriage, T. A., Acquaviva, V., Ade, P. A. R., Aguirre, P., Amiri, M., Appel, J. W., et al. (2011). The Atacama Cosmology Telescope: Sunyaev-Zel'dovich-selected galaxy clusters at 148 GHz in the 2008 survey. *Astrophys. J.* 737, 1–10. doi: 10.1088/0004-637X/737/2/61

Massardi, M., Ekers, R. D., Ellis, S. C., and Maughan, B. (2010). High angular resolution observation of the Sunyaev-Zel'dovich effect in the massive z 0.83 cluster CL J0152-1357. *Astrophys. J. Lett.* 718, L23–L26. doi: 10.1088/2041-8205/718/1/L23

Massey, R., Williams, L., Smit, R., Swinbank, M., Kitching, T. D., Harvey, D., et al. (2015). The behaviour of dark matter associated with four bright cluster galaxies in the 10 kpc core of Abell 3827. *Mon. Not. Roy. Astron. Soc.* 449, 3393–3406. doi: 10.1093/mnras/stv467

Mastropietro, C., and Burkert, A. (2008). Simulating the Bullet cluster. *Mon. Not. Roy. Astron. Soc.* 389, 967–688. doi: 10.1111/j.1365-2966.2008.13626.x

Mathis, H., Lavaux, G., Diego, J. M., and Silk, J. (2005). On the formation of cold fronts in massive mergers. *Mon. Not. Roy. Astron. Soc.* 357, 801–818. doi: 10.1111/j.1365-2966.2004.08589.x

Maurogordato, S., Sauvageot, J. L., Bourdin, H., Cappi, A., Benoist, C., Ferrari, C., et al. (2011). Merging history of three bimodal clusters. *Astron. Astrophys. J.* 525, 1–19. doi: 10.1051/0004-6361/201014415

Mauskopf, P. D., Horner, P. F., Aguirre, J., Bock, J. J., Egami, E., Glenn, J., et al. (2012). A high signal-to-noise ratio map of the Sunyaev-Zel'dovich increment at 1.1-mm wavelength in Abell 1835. *Mon. Not. Roy. Astron. Soc.* 421, 224–234. doi: 10.1111/j.1365-2966.2011.20295.x

Mazzotta, P., Markevitch, M., Vikhlinin, A., Forman, W. R., David, L. P., and van Speybroeck, L. (2001). Chandra observation of RX J1720.1+2638: a nearly relaxed cluster with a fast-moving core? *Astrophys. J.* 555, 205–214. doi: 10.1086/321484

Mazzotta, P., Rasia, E., Moscardini, L., and Tormen, G. (2004). Comparing the temperatures of galaxy clusters from hydrodynamical N-body simulations to Chandra and *XMM-Newton* observations. *Mon. Not. Roy. Astron. Soc.* 354, 10–24. doi: 10.1111/j.1365-2966.2004.08167.x

McCarthy, I. G., Bower, R. G., Balogh, M. L., Voit, G. M., Pearce, F. R., Theuns, T., et al. (2007). Modelling shock heating in cluster mergers - I. Moving beyond the spherical accretion model. *Mon. Not. Roy. Astron. Soc.* 376, 497–522. doi: 10.1111/j.1365-2966.2007.11465.x

Medezinski, E., Umetsu, K., Okabe, N., Nonino, M., Molnar, S., Massey, R., et al. (2016). Frontier fields: Subaru weak-lensing analysis of the merging galaxy cluster A2744. *Astrophys. J.* 748, 1–16. doi: 10.3847/0004-637X/817/1/24

Menanteau, F., Hughes, J. P., Sifón, C., Hilton, M., González, J. Infante, L., et al. (2012). The Atacama Cosmology Telescope: ACT-CL J0102-4915 el gordo, a massive merging cluster at redshift 0.87. *Astrophys. J.* 748, 7–24. doi: 10.1088/0004-637X/748/1/7

Merloni, A., Predehl, P., Becker, W., Böhringer, H., Boller, T., Brunner, H., et al. (2012). eROSITA science book: mapping the structure of the energetic universe. arXiv:1209.3114.

Merten, J., Coe, D., Dupke, R., Massey, R., Zitrin, A., Cypriano, E. S., et al. (2011). Creation of cosmic structure in the complex galaxy cluster merger Abell 2744. *Mon. Not. Roy. Astron. Soc.* 417, 333–347. doi: 10.1111/j.1365-2966.2011.19266.x

Merten, J., Meneghetti, M., Postman, M., Umetsu, K., Zitrin, A., Medezinski, E., et al. (2015). CLASH: The concentration-mass relation of galaxy clusters. *Astrophys. J.* 806, 1–26. doi: 10.1088/0004-637x/806/1/4

Miley, G. (1980). The structure of extended extragalactic radio sources. *Annu. Rev. Astron. Astrophys.* 18, 165–218. doi: 10.1146/annurev.aa.18.090180.001121

Mitchell, N. L., McCarthy, I. G., Bower, R. G., Theuns, T., and Crain, R. A. (2009). On the origin of cores in simulated galaxy clusters. *Mon. Not. Roy. Astron. Soc.* 395, 180–196. doi: 10.1111/j.1365-2966.2009.14550.x

Mohr, J. J., Fabricant, D. G., and Geller, M. J. (1993). An X-ray method for detecting substructure in galaxy clusters - application to Perseus, A2256, Centaurus, Coma, and Sersic 40/6. *Astrophys. J.* 413, 492–505. doi: 10.1086/173019

Molnar, S. M. (2015). *Cosmology with Clusters of Galaxies*. New York, NY: Nova Science Publishers.

Molnar, S. M., and Birkinshaw, M. (1999). Inverse Compton scattering in mildly relativistic plasma. *Astrophys. J.* 523, 78–86. doi: 10.1086/307718

Molnar, S. M., and Birkinshaw, M. (2000). Contributions to the power spectrum of cosmic microwave background from fluctuations caused by clusters of galaxies. *Astrophys. J.* 537, 542–554. doi: 10.1086/309042

Molnar, S. M., and Birkinshaw, M. (2003). Determining tangential peculiar velocities of clusters of galaxies using gravitational lensing. *Astrophys. J.* 586, 731–734. doi: 10.1086/346071

Molnar, S. M., Birkinshaw, M., and Mushotzky, R. F. (2006). Determining distances to clusters of galaxies using resonant X-ray emission lines. *Astrophys. J. Lett.* 643, L73–L76. doi: 10.1086/505301

Molnar, S. M., and Broadhurst, T. (2015). A hydrodynamical solution for the 'twin-tailed' colliding galaxy cluster 'El Gordo'. *Astrophys. J.* 800, 37–45. doi: 10.1088/0004-637X/800/1/37

Molnar, S. M., Broadhurst, T., Umetsu, K., Zitrin, A., Rephaeli, R., and Shimon, M. (2013a). Tangential velocity of the dark matter in the Bullet cluster from precise lensed image redshifts. *Astrophys. J.* 774, 70–80. doi: 10.1088/0004-637X/774/1/70

Molnar, S. M., Chiu, I.-N. T., Broadhurst, T., and Stadel, J. G. (2013b). The pre-merger impact velocity of the binary cluster A1750 from X-ray, lensing, and hydrodynamical simulations. *Astrophys. J.* 779, 63–73. doi: 10.1088/0004-637X/779/1/63

Molnar, S. M., Chiu, I.-N., Umetsu, K., Chen, P., Hearn, N., Broadhurst, T., et al. (2010). Testing strict hydrostatic equilibrium in simulated clusters of galaxies: implications for A1689. *Astrophys. J. Lett.* 724, L1–L4. doi: 10.1088/2041-8205/724/1/L1

Molnar, S. M., Hearn, N., Haiman, Z., Bryan, G., Evrard, A. E., and George, L. (2009). Accretion shocks in clusters of galaxies and their SZ signature from cosmological simulations. *Astrophys. J.* 696, 1640–1656. doi: 10.1088/0004-637X/696/2/1640

Molnar, S. M., Hearn, N. C., and Stadel, J. G. (2012). Merging galaxy clusters: offset between the Sunyaev-Zel'dovich effect and X-ray peaks. *Astrophys. J.* 748, 45–54. doi: 10.1088/0004-637X/748/1/45

Morandi, A., and Limousin, M. (2012). Triaxiality, principal axis orientation and non-thermal pressure in Abell 383. *Mon. Not. Roy. Astron. Soc.* 421, 3147–3158. doi: 10.1111/j.1365-2966.2012.20537.x

Morandi, A., Limousin, M., Rephaeli, Y., Umetsu, K., Barkana, R., Broadhurst, T., et al. (2011). Triaxiality and non-thermal gas pressure in Abell 1689. *Mon. Not. Roy. Astron. Soc.* 416, 2567–2573. doi: 10.1111/j.1365-2966.2011.19175.x

Morandi, A., Limousin, M., Sayers, J., Golwala, S. R., Czakon, N. G., Pierpaoli, E., et al. (2012). X-ray, lensing and Sunyaev-Zel'dovich triaxial analysis of Abell 1835 out to R_{200}. *Mon. Not. Roy. Astron. Soc.* 425, 2069–2082. doi: 10.1111/j.1365-2966.2012.21196.x

Mroczkowski, T., Dicker, S., Sayers, J., Reese, E. D., Mason, B., Czakon, N., et al. (2012). A multi-wavelength study of the Sunyaev-Zel'dovich effect in the triple-merger cluster MACS J0717.5+3745 with Mustang and Bolocam. *Astrophys. J.* 761, 47–61. doi: 10.1088/0004-637X/761/1/47

Muldrew, S. I., Hatch, N. A., and Cooke, E. A. (2015). What are protoclusters? - Defining high-redshift galaxy clusters and protoclusters. *Mon. Not. Roy. Astron. Soc.* 452, 2528–2539. doi: 10.1093/mnras/stv1449

Navarro, J. F., Frenk, C. S., and White, S. D. M. (1997). A universal density profile from hierarchical clustering. *Astrophys. J.* 490, 493–508. doi: 10.1086/304888

Nelson, K., Lau, E. T., and Nagai, D. (2014). Hydrodynamic simulation of non-thermal pressure profiles of galaxy clusters. *Astrophys. J.* 792, 25–32. doi: 10.1088/0004-637X/792/1/25

Nelson, K., Lau, E. T., Nagai, D., Rudd, D. H., and Yu, L. (2014). Weighing Galaxy clusters with gas. II. On the origin of hydrostatic mass bias in ΛCDM galaxy clusters. *Astrophys. J.* 782, 1–9. doi: 10.1088/0004-637X/782/2/107

Norman, M. L. and Bryan, G. L. (1999). "Cluster turbulence," in *The Radio Galaxy Messier 87, Proceedings of a Workshop Held at Ringberg Castle, Tegernsee, Germany. (Lecture Notes in Physics)*, Vol. 530, eds H.-J. Röser and K. Meisenheimer (Berlin: Springer), 106–115. doi: 10.1007/BFb0106425

Nozawa, S., Itoh, N., and Kohyama, Y. (1998a). Relativistic thermal bremsstrahlung Gaunt factor for the intracluster plasma. *Astrophys. J.* 507, 530–557. doi: 10.1086/306352

Nozawa, S., Itoh, N., and Kohyama, Y. (1998b). Relativistic corrections to the Sunyaev-Zeldovich effect for clusters of galaxies. II. Inclusion of peculiar velocities. *Astrophys. J.* 508, 17–24. doi: 10.1086/306401

Ogrean, G. A., Brüggen, M., van Weeren, R. J., Röttgering, H., Croston, J. H., and Hoeft, M. (2013). Challenges to our understanding of radio relics: X-ray observations of the Toothbrush cluster. *Mon. Not. Roy. Astron. Soc.* 433, 812–824. doi: 10.1093/mnras/stt776

Ogrean, G. A., Brüggen, M., van Weeren, R., Röttgering, H. J. A., Simionescu, A., Hoeft, M., et al. (2014). Multiple density discontinuities in the merging galaxy cluster CIZA J2242.8+5301. *Mon. Not. Roy. Astron. Soc.* 440, 3416–3425. doi: 10.1093/mnras/stu537

Osborne, S. J., Mak, D. S. Y., Church, S. E., and Pierpaoli, E. (2011). Measuring the galaxy cluster bulk flow from *WMAP* data. *Astrophys. J.* 737, 1–20. doi: 10.1088/0004-637X/737/2/98

Ota, N., Nagai, D., and Lau, E. T. (2015). Testing *Astro-H* measurements of bulk and turbulent gas motions in galaxy clusters. arXiv:1507.02730, 1–11.

Owers, M. S., Nulsen, P. E. J., Couch, W. J., and Markevitch, M. (2009). A high fidelity sample of cold front clusters from the Chandra archive. *Astrophys. J.* 704, 1349–1370. doi: 10.1088/0004-637X/704/2/1349

Owers, M. S., Randall, S. W., Nulsen, P. E. J., Couch, W. J., David, L. P., and Kempner, J. C. (2011). The dissection of Abell 2744: a rich cluster growing through major and minor mergers. *Astrophys. J.* 728, 27–53. doi: 10.1088/0004-637X/728/1/27

Parekh, V., van der Heyden, K., Ferrari, C., Angus, G., and Holwerda, B. (2015). Morphology parameters: substructure identification in X-ray galaxy clusters. *Astron. Astrophys. J.* 575, 1–28. doi: 10.1051/0004-6361/201424123

Parrish, I. J., McCourt, M., Quataert, E., and Sharma, P. (2012). Turbulent pressure support in the outer parts of galaxy clusters. *Mon. Not. Roy. Astron. Soc.* 419, L29–L33. doi: 10.1111/j.1745-3933.2011.01171.x

Paul, S., Iapichino, L., Miniati, F., Bagchi, J., and Mannheim, K. (2011). Evolution of shocks and turbulence in major cluster mergers. *Astrophys. J.* 726, 17–31. doi: 10.1088/0004-637X/726/1/17

Peebles, P. J. E. (1993). *Principles of Physical Cosmology*. Princeton, NJ: Princeton University Press.

Pearce, F. R., Thomas, P. A., and Couchman, H. M. P. (1994). Head-on mergers of systems containing gas. *Mon. Not. Roy. Astron. Soc.* 268, 953–965. doi: 10.1093/mnras/268.4.953

Pinto, C., Sanders, J. S., Werner, N., de Plaa, J., Fabian, A. C., Zhang, Y.-Y., et al. (2015). Chemical Enrichment RGS cluster Sample (CHEERS): constraints on turbulence. *Astron. Astrophys. J.* 575, 1–16. doi: 10.1051/0004-6361/201425278

Pinzke, A., Oh, S. P., and Pfrommer, C. (2013). Giant radio relics in galaxy clusters: reacceleration of fossil relativistic electrons? *Mon. Not. Roy. Astron. Soc.* 435, 1061–1082. doi: 10.1093/mnras/stt1308

Planck Collaboration, Ade, P. A. R., Aghanim, N., Arnaud, M., Ashdown, M., Aumont, J., et al. (2014). *Planck* intermediate results. XIII. Constraints on peculiar velocities. *Astron. Astrophys.* 561, 1–21. doi: 10.1051/0004-6361/201321299

Poole, G. B., Babul, A., McCarthy, I. G., Fardal, M. A., Bildfell, C. J., Quinn, T., et al. (2007). The impact of mergers on relaxed X-ray clusters II. Effects on global X-ray and Sunyaev-Zel'dovich properties and their scaling relations. *Mon. Not. Roy. Astron. Soc.* 380, 437–454. doi: 10.1111/j.1365-2966.2007.12107.x

Poole, G. B., Babul, A., McCarthy, I. G., Sanderson, A. J. R., and Fardal, M. A. (2008). The impact of mergers on relaxed X-ray clusters - III. Effects on compact cool cores. *Mon. Not. Roy. Astron. Soc.* 391, 1163–1175. doi: 10.1111/j.1365-2966.2008.14003.x

Poole, G. B., Fardal, M. A., Babul, A., McCarthy, I. G., Quinn, T., and Wadsley, J. (2006). The impact of mergers on relaxed X-ray clusters - I. Dynamical evolution and emergent transient structures. *Mon. Not. Roy. Astron. Soc.* 373, 881–905. doi: 10.1111/j.1365-2966.2006.10916.x

Postman, M., Coe, D., Benítez, N., Bradley, L., Broadhurst, T., Donahue, M., et al. (2012). The cluster lensing and supernova survey with Hubble: an overview. *Astrophys. J. Suppl.* 199, 1–23. doi: 10.1088/0067-0049/199/2/25

Prokhorov, D. A., Colafrancesco, S., Akahori, T., Million, E. T., Nagataki, S., and Yoshikawa K. (2011). A high-frequency study of the Sunyaev-Zel'dovich effect morphology in galaxy clusters. *Mon. Not. Roy. Astron. Soc.* 416, 302–310. doi: 10.1111/j.1365-2966.2011.19037.x

Randall, S. W., Markevitch, M., Clowe, D., Gonzalez, A. H., and Bradač, M. (2008). Constraints on the self-interaction cross section of dark matter from numerical simulations of the merging galaxy cluster 1E 0657-56. *Astrophys. J.* 679, 1173–1180. doi: 10.1086/587859

Rankine, W. J. M. (1870). On the thermodynamic theory of waves of finite longitudinal disturbance. *Philos. Trans. R. Soc.* 160, 277–288. doi: 10.1098/rstl.1870.0015

Rasia, E., Ettori, S., Moscardini, L., Mazzotta, P., Borgani, S., Dolag, K., et al. (2006). Systematics in the X-ray cluster mass estimators. *Mon. Not. Roy. Astron. Soc.* 369, 2013–2024. doi: 10.1111/j.1365-2966.2006.10466.x

Rasia, E., Tormen, G., and Moscardini, L. (2004). A dynamical model for the distribution of dark matter and gas in galaxy clusters. *Mon. Not. Roy. Astron. Soc.* 351, 237–252. doi: 10.1111/j.1365-2966.2004.07775.x

Reiprich, T. H., Basu, K., Ettori, S., Israel, H., Lovisari, L., Molendi, S., et al. (2013). Outskirts of galaxy clusterss. *Space Sci. Rep.* 177, 195–245. doi: 10.1007/s11214-013-9983-8

Rephaeli, Y. (1977). Spatial distribution of Compton-produced X-ray flux from rich and regular clusters of galaxies. *Astrophys. J.* 212, 608–615. doi: 10.1086/155083

Rephaeli, Y. (1995). Comptonization of the cosmic microwave background. *Annu. Rev. Astron. Astrophys.* 33, 541–579. doi: 10.1146/annurev.aa.33.090195.002545

Rephaeli, Y., Nevalainen, J., Ohashi, T., and Bykov, A. M. (2008). Nonthermal phenomena in clusters of galaxies. *Space Sci. Rep.* 134, 71–92. doi: 10.1007/s11214-008-9314-7

Ricker, P. M. (1998). Off-center collisions between clusters of galaxies. *Astrophys. J.* 496, 670–692. doi: 10.1086/305393

Ricker, P. M., and Sarazin, C. L. (2001). Off-axis cluster mergers: effects of a strongly peaked dark matter profile. *Astrophys. J.* 561, 621–644. doi: 10.1086/323365

Ritchie, B. W., and Thomas, P. A. (2002). Hydrodynamic simulations of merging clusters of galaxies. *Mon. Not. Roy. Astron. Soc.* 329, 675–688. doi: 10.1046/j.1365-8711.2002.05027.x

Roettiger, K., Burns, J., and Stone, J. M. (1999b). A cluster merger and the origin of the extended radio emission in Abell 3667. *Astrophys. J.* 518, 603–612. doi: 10.1086/307327

Roettiger, K., and Flores, R. (2000). A prediction of observable rotation in the intracluster medium of Abell 3266. *Astrophys. J.* 538, 92–97. doi: 10.1086/309132

Roettiger, K., Stone, J. M., and Burns, J. (1999a). Magnetic field evolution in merging clusters of galaxies. *Astrophys. J.* 518, 594–602. doi: 10.1086/307298

Russell, H. R., Fabian, A. C., McNamara, B. R., Edge, A. C., Sanders, J. S., Nulsen, P. E. J., et al. (2014). The bow shock, cold fronts and disintegrating cool core in the merging galaxy group RX J0751.3+5012. *Mon. Not. Roy. Astron. Soc.* 444, 629–641. doi: 10.1093/mnras/stu1469

Russell, H. R., McNamara, B. R., Sanders, J. S., Fabian, A. C., Nulsen, P. E. J., Canning, R. E. A., et al. (2012). Shock fronts, electron-ion equilibration and ICM transport processes in the merging cluster Abell 2146. *Mon. Not. Roy. Astron. Soc.* 423, 236–255. doi: 10.1111/j.1365-2966.2012.20808.x

Ruszkowski, M., Brüggen, M., Lee, D., and Shin, M.-S. (2014). Impact of magnetic fields on ram pressure stripping in disk galaxies. *Astrophys. J.* 784, 1–13. doi: 10.1088/0004-637x/784/1/75

Sakelliou, I., Peterson, J. R., Tamura, T., Paerels, F. B. S., Kaastra, J. S., Belsole, E., et al. (2002). High resolution soft X-ray spectroscopy of M 87 with the reflection grating spectrometers on *XMM-Newton*. *Astron. Astrophys.* 391, 903–909. doi: 10.1051/0004-6361:20020900

Sanders, J. S., Fabian, A. C., and Dunn, R. J. H., (2005). Non-thermal X-rays, a high-abundance ridge and fossil bubbles in the core of the Perseus cluster of galaxies. *Mon. Not. Roy. Astron. Soc.* 360, 133–140. doi: 10.1111/j.1365-2966.2005.09016.x

Sanders, J. S., Fabian, A. C., and Smith, R. K. (2011). Constraints on turbulent velocity broadening for a sample of clusters, groups and elliptical galaxies using *XMM-Newton*. *Mon. Not. Roy. Astron. Soc.* 410, 1797–1812. doi: 10.1111/j.1365-2966.2010.17561.x

Sanders, J. S., Fabian, A. C., Smith, R. K., and Peterson, J. R. (2010). A direct limit on the turbulent velocity of the intracluster medium in the core of Abell 1835 from *XMM-Newton*. *Mon. Not. Roy. Astron. Soc.* 402, L11–L15. doi: 10.1111/j.1745-3933.2009.00789.x

Sarazin, C. L. (1988). *X-Ray Emission from Clusters of Galaxies*. Cambridge: Cambridge University Press.

Sayers, J., Mroczkowski, T., Zemcov, M., Korngut, P. M., Bock, J., Bulbul, E., et al. (2013). A measurement of the kinetic Sunyaev-Zeldovich signal toward MACS J0717.5+3745. *Astrophys. J.* 778, 52–71. doi: 10.1088/0004-637X/778/1/52

Sazonov, S. Y., Churazov, E. M., and Sunyaev, R. A. (2002). Polarization of resonance X-ray lines from clusters of galaxies. *Mon. Not. Roy. Astron. Soc.* 333, 191–201. doi: 10.1046/j.1365-8711.2002.05390.x

Schneider, P., Ehlers, J., and Falco, E. E. (1992). *Gravitational Lenses*, Vol. XIV. Berlin; Heidelberg; New York, NY: Springer-Verlag.

Schreier, S. (1982). *Compressible Flow*. New York, NY: Wiley.

Schuecker, P., Finoguenov, A., Miniati, F., Böhringer, H., and Briel, U. G. (2004). Probing turbulence in the Coma galaxy cluster. *Astron. Astrophys. J.* 426, 387–397. doi: 10.1051/0004-6361:20041039

Shang, C., and Oh, S. P. (2013). Disentangling resonant scattering and gas motions in galaxy cluster emission line profiles. *Mon. Not. Roy. Astron. Soc.* 433, 1172–1184. doi: 10.1093/mnras/stt790

Shaw, L. D., Nagai, D., Bhattacharya, S., and Lau, E. T. (2010). Impact of cluster physics on the Sunyaev-Zel'dovich power spectrum. *Astrophys. J.* 725, 1452–1465. doi: 10.1088/0004-637X/725/2/1452

Shigeyama, T., (1998). Resonance line scattering modifies X-ray surface brightness of elliptical galaxies. *Astrophys. J.* 497, 587–593. doi: 10.1086/305498

Spitzer, L. (1962). *Physics of Fully Ionized Gases, 2nd Edn.* New York, NY: Wiley-Interscience.

Spitzer, L. (1978). *Physical Processes in the Interstellar Medium*. New York, NY: Wiley-Interscience.

Springel, V., and Farrar, G. R. (2007). The speed of the 'bullet' in the merging galaxy cluster 1E0657-56. *Mon. Not. Roy. Astron. Soc.* 380, 911–925. doi: 10.1111/j.1365-2966.2007.12159.x

Sunyaev, R. A. and Zel'dovich, Y. B. (1980). Microwave background radiation as a probe of the contemporary structure and history of the universe. *Annu. Rev. Astron. Astrophys.* 18, 537–560. doi: 10.1146/annurev.aa.18.090180.002541

Suto, D., Kawahara, H., Kitayama, T., Sasaki, S., Suto, Y., and Cen, R. (2013). Validity of hydrostatic equilibrium in galaxy clusters from cosmological hydrodynamical simulations. *Astrophys. J.* 767, 79–89. doi: 10.1088/0004-637X/767/1/79

Takizawa, M. (2000). Off-center mergers of clusters of galaxies and nonequipartition of electrons and ions in the intracluster medium. *Astrophys. J.* 532, 183–192. doi: 10.1086/308550

Takizawa, M. (2008). N-body + magnetohydrodynamical simulations of merging clusters of galaxies: characteristic magnetic field structures generated by bulk flow motion. *Astrophys. J.* 687, 951–958. doi: 10.1086/592059

Tamura, T., Hayashida, K., Ueda, S., and Nagai, M. (2011). Discovery of gas bulk motion in the galaxy cluster Abell 2256 with Suzaku. *Publ. Astron. Soc. Jpn.* 63, S1009–S1017. doi: 10.1093/pasj/63.sp3.s1009

Thompson, R., and Nagamine, K. (2012). Pairwise velocities of dark matter haloes: a test for the Λ Cold Dark Matter model using the Bullet cluster. *Mon. Not. Roy. Astron. Soc.* 419, 3560–3570. doi: 10.1111/j.1365-2966.2011.20000.x

Trac, H., Bode, P., and Ostriker, J. P. (2011). Templates for the Sunyaev-Zel'dovich angular power spectrum. *Astrophys. J.* 727, 1–14. doi: 10.1088/0004-637x/727/2/94

Treumann, R. A. (2009). Fundamentals of collisionless shocks for astrophysical application, 1. Non-relativistic shocks. *Astron. Astrophys. Rev.* 17, 409–535. doi: 10.1007/s00159-009-0024-2

Umetsu, K., Medezinski, E., Nonino, M., Merten, J., Postman, M., Meneghetti, M., et al. (2014). CLASH: Weak-lensing shear and magnification analysis of 20 galaxy clusters. *Astrophys. J.* 795, 1–25. doi: 10.1088/0004-637x/795/2/163

Umetsu, K., Sereno, M., Medezinski, E., Nonino, M., Mroczkowski, T., Diego, J. M., et al. (2015a). Three-dimensional multi-probe analysis of the galaxy cluster A1689. *Astrophys. J.* 806, 207. doi: 10.1088/0004-637X/806/2/207

Umetsu, K., Zitrin, A., Gruen, D., Merten, J., Donahue, M., and Postman, M. (2015b). CLASH: Joint analysis of strong-lensing, weak-lensing shear and magnification data for 20 galaxy clusters. arXiv:1507.04385, 1–26.

van Weeren, R. J., Brüggen, M., Röttgering, H. J. A., and Hoeft, M. (2011a). Using double radio relics to constrain galaxy cluster mergers: a model of double radio relics in CIZA J2242.8+5301. *Mon. Not. Roy. Astron. Soc.* 418, 230–243. doi: 10.1111/j.1365-2966.2011.19478.x

van Weeren, R. J., Hoeft, M., Röttgering, H. J. A., Brüggen, M., Intema, H. T., and van Velzen, S. (2011b). A double radio relic in the merging galaxy group RX J0751.3+5012. *Astron. Astrophys. J.* 528, 1–13. doi: 10.1051/0004-6361/201016185

van Weeren, R. J., Röttgering, H. J. A., Brüggen, M., and Hoeft, M. (2010). Particle acceleration on megaparsec scales in a merging galaxy cluster. *Science* 330, 347–357. doi: 10.1126/science.1194293

van Weeren, R. J., Röttgering, H. J. A., Intema, H. T., Rudnick, L., Brggen, M., Hoeft, M., et al. (2012). The "toothbrush-relic': evidence for a coherent linear 2-Mpc scale shock wave in a massive merging galaxy cluster? *Astron. Astrophys. J.* 546, 1–21. doi: 10.1051/0004-6361/201219000

Vazza, F., Brunetti, G., Kritsuk, A., Wagner, R., Gheller, C., and Norman, M. (2009). Turbulent motions and shocks waves in galaxy clusters simulated with adaptive mesh refinement. *Astron. Astrophys. J.* 504, 33–43. doi: 10.1051/0004-6361/2009 12535

Vazza, F., Eckert, D., Brüggen, M., and Huber, B. (2015). Electron and proton acceleration efficiency by merger shocks in galaxy clusters. *Mon. Not. Roy. Astron. Soc.* 451, 2198–2211. doi: 10.1093/mnras/stv1072

Vazza, F., Tormen, G., Cassano, R., Brunetti, G., and Dolag, K. (2006). Turbulent velocity fields in smoothed particle hydrodymanics simulated galaxy clusters: scaling laws for the turbulent energy. *Mon. Not. Roy. Astron. Soc.* 369, L14–L18. doi: 10.1111/j.1745-3933.2006.00164.x

Vikhlinin, A., Markevitch, M., and Murray, S. M. (2001). A moving cold front in the intergalactic medium of A3667. *Astrophys. J.* 551, 160–171. doi: 10.1086/320078

Wen, Z. L., and Han, J. L. (2013). Substructure and dynamical state of 2092 rich clusters of galaxies derived from photometric data. *Mon. Not. Roy. Astron. Soc.* 436, 275-293. doi: 10.1093/mnras/stt1581

Werner, N., Zhuravleva, I., Churazov, E., Simionescu, A., Allen, S. W., Forman, W., et al. (2009). Constraints on turbulent pressure in the X-ray haloes of giant elliptical galaxies from resonant scattering. *Mon. Not. Roy. Astron. Soc.* 398, 23–32. doi: 10.1111/j.1365-2966.2009.14860.x

Werner, N., ZuHone, J. A., Zhuravleva, I., Ichinohe, Y., Simionescu, A., Allen, S. W., et al. (2015). Deep Chandra observation and numerical studies of the nearest cluster cold front in the sky. *Mon. Not. Roy. Astron. Soc.* 455, 846-858. doi: 10.1093/mnras/stv2358

Wik, D. R., Hornstrup, A., Molendi, S., Madejski, G., Harrison, F. A., Zoglauer, A., et al. (2014). NuSTAR observations of the Bullet cluster: constraints on inverse compton emission. *Astrophys. J.* 792, 1–24. doi: 10.1088/0004-637X/792/1/48

Wik, D. R., Sarazin, C. L., Finoguenov, A., Finoguenov, A., Matsushita, K., Nakazawa, K., et al. (2009). A Suzaku search for nonthermal emission at hard X-ray energies in the Coma cluster. *Astrophys. J.* 696, 1700–1711. doi: 10.1088/0004-637X/696/2/1700

Wik, D. R., Sarazin, C. L., Zhang, Y.-Y., Baumgartner, W. H., Mushotzky, R. F., Tueller, J., et al. (2012). The Swift Burst Alert Telescope perspective on non-thermal emission in HIFLUGCS galaxy clusters. *Astrophys. J.* 748, 1–26. doi: 10.1088/0004-637X/748/1/67

Wucknitz, O., and Sperhake, U. (2004). Deflection of light and particles by moving gravitational lenses. *Phys. Rev. D* 69, 1–13. doi: 10.1103/physrevd.69.063001

Xu, H., Kahn, S. M., Peterson, J. R., Behar, E., Paerels, F. B. S., Mushotzky, R. F., et al. (2002). High-resolution observations of the elliptical galaxy NGC 4636 with the reflection grating spectrometer on board *XMM-Newton*. *Astrophys. J.* 579, 600–606. doi: 10.1086/342828

Yamada, K., Kitayama, T., Takakuwa, S., Iono, D., Tsutsumi, T., Kohno, K., et al. (2012). Imaging simulations of the Sunyaev-Zel'dovich effect for ALMA. *Publ. Astron. Soc. Jpn.* 64, 1–26. doi: 10.1093/pasj/64.5.102

Zandanel, F., and Ando, S. (2014). Constraints on diffuse gamma-ray emission from structure formation processes in the Coma cluster. *Mon. Not. Roy. Astron. Soc.* 440, 663–671. doi: 10.1093/mnras/stu324

Zemcov, M., Aguirre, J., Bock, J., Bradford, C. M., Czakon, N., Glenn, J., et al. (2012). High spectral resolution measurement of the Sunyaev-Zel'dovich effect null with Z-Spec. *Astrophys. J.* 749, 114–126. doi: 10.1088/0004-637X/749/2/114

Zhuravleva, I., Churazov, E., Sunyaev, R., Sazonov, S., Allen, S. W., Werner, N., et al. (2013). Resonant scattering in the Perseus cluster: spectral model for constraining gas motions with *Astro-H*. *Mon. Not. Roy. Astron. Soc.* 435, 3111–3121. doi: 10.1093/mnras/stt1506

ZuHone, J. A. (2011). A parameter space exploration of galaxy cluster mergers. I. Gas mixing and the generation of cluster entropy. *Astrophys. J.* 728, 54–77. doi: 10.1088/0004-637X/728/1/54

ZuHone, J. A., Markevitch, M., and Johnson, R. E. (2010). Stirring up the pot: can cooling flows in galaxy clusters be quenched by gas sloshing? *Astrophys. J.* 717, 908–928. doi: 10.1088/0004-637X/717/2/908

ZuHone, J. A., Ricker, P. M., Lamb, D. Q., and Karen Yang, H.-Y. (2009). Rings of dark matter in collisions between clusters of galaxies. *Astrophys. J.* 696, 694–700. doi: 10.1088/0004-637X/696/1/694

ZuHone, J. A., Markevitch, M., Ruszkowski, M., and Lee, D. (2013). Cold fronts and gas sloshing in galaxy clusters with anisotropic thermal conduction. *Astrophys. J.* 762, 69–85. doi: 10.1088/0004-637X/762/2/69

Conflict of Interest Statement: The author declares that the research was conducted in the absence of any commercial or financial relationships that could be construed as a potential conflict of interest.

2

Diagnostics of Coronal Magnetic Fields through the Hanle Effect in UV and IR Lines

Nour E. Raouafi[1]*, Pete Riley[2], Sarah Gibson[3], Silvano Fineschi[4] and Sami K. Solanki[5, 6]

[1] The John Hopkins University Applied Physics Laboratory, Laurel, MD, USA, [2] Predictive Science Inc., San Diego, CA, USA, [3] High Altitude Observatory, National Center for Atmospheric Research, Boulder, CO, USA, [4] The Astrophysical Observatory of Turin, National Institute for Astrophysics, Turin, Italy, [5] Max-Planck-Institut für Sonnensystemforschung, Göttingen, Germany, [6] School of Space Research, Kyung Hee University, Yongin, South Korea

Edited by:
Xueshang Feng,
National Space Science Center, China

Reviewed by:
Keiji Hayashi,
Naogya University, Japan
Meng Jin,
University Corporation for
Atmospheric Research and Lockheed
Martin Solar and Astrophysics
Laboratory, USA

*Correspondence:
Nour E. Raouafi
noureddine.raouafi@jhuapl.edu

Specialty section:
This article was submitted to
Stellar and Solar Physics,
a section of the journal
Frontiers in Astronomy and Space
Sciences

The plasma thermodynamics in the solar upper atmosphere, particularly in the corona, are dominated by the magnetic field, which controls the flow and dissipation of energy. The relative lack of knowledge of the coronal vector magnetic field is a major handicap for progress in coronal physics. This makes the development of measurement methods of coronal magnetic fields a high priority in solar physics. The Hanle effect in the UV and IR spectral lines is a largely unexplored diagnostic. We use magnetohydrodynamic (MHD) simulations to study the magnitude of the signal to be expected for typical coronal magnetic fields for selected spectral lines in the UV and IR wavelength ranges, namely the H I Ly-α and the He I 10,830 Å lines. We show that the selected lines are useful for reliable diagnosis of coronal magnetic fields. The results show that the combination of polarization measurements of spectral lines with different sensitivities to the Hanle effect may be most appropriate for deducing coronal magnetic properties from future observations.

Keywords: sun: corona, sun: magnetic fields, sun: UV radiation, sun: infrared, polarization, scattering, atomic processes, plasmas

1. INTRODUCTION

Our understanding of coronal phenomena, such as plasma heating and acceleration, particle energization, and explosive activity, faces major hurdles due to the lack of reliable measurements of key parameters such as densities, temperatures, velocities, and particularly magnetic fields. The knowledge of key plasma parameters, particularly the magnetic field and plasma velocity, in the solar corona is a prerequisite to advance our understanding of coronal manifestations that greatly affect and modulate the interplanetary medium, particularly the Earth's environment.

Spectroscopic diagnostics in the ultraviolet (UV) and extreme ultraviolet (EUV) wavelength regimes provide measurements of plasma densities, temperatures, and partial information on the velocity. But they cannot provide any insight into the coronal magnetic field, which is the key player in the structuring of the solar corona and in dominating most (if not all) physical processes underlying the multi-scaled solar activity. For instance, the abundant mechanical energy that is available in the convection zone is partially transferred to the corona, where it is stored in complex magnetic field structures and dissipated in the form of heat, acceleration, and energization of the plasma during activity events occurring at different spatial and temporal scales. The magnetic activity manifests itself in different forms such as Coronal Mass Ejections [CMEs], flares, jets, waves and instabilities, magnetic reconnection, and turbulence.

Magnetic fields measurements in the photosphere and, to a lesser degree, in the chromosphere, which are based mainly on the Zeeman effect, have been a routine exercise for decades. The Zeeman effect in the higher layers of the solar atmosphere, particularly the corona, are limited to regions of relatively strong magnetic fields (i.e., above active regions) and to infrared lines due to the wavelength squared scaling of this effect and the relatively large widths of coronal lines (Lin et al., 2004). Other methods based on radio emissions could provide constraints on the magnetic field in the corona (White, 1999; Gibson et al., 2016). Coronal magnetic fields are usually approximated through MHD modeling and extrapolation of photospheric measurements (e.g., Wiegelmann et al., 2014). These approaches have, however, their limitations. For instance, extrapolation models are based on the assumption that the magnetic field is force-free at the lower boundary of the calculation, which is not the case in the photosphere. Additionally, large-scale MHD models of the solar corona are based on synoptic maps of the photospheric magnetic fields, which are built up from images taken by near-Earth observatories recorded over a whole solar rotation. Finally, the MHD models may underestimate the magnetic field strength in the corona (Riley et al., 2012). Moreover, coronal plasma parameters obtained through the models cannot be constrained without direct measurements. For more details on methods for the measurements of magnetic fields in upper solar atmosphere, see reviews by Fineschi (2001) and Raouafi (2005, 2011).

In this paper, we focus on the diagnostics of coronal magnetic fields through the linear polarization of selected spectral lines (i.e., H I Ly-α and He I 10,830 Å) that are sensitive to the "Hanle effect." Other spectral lines are also of interest, but the analysis of their polarization is left for future publications.

2. THE HANLE EFFECT

The Hanle effect (Hanle, 1924), which is the modification of the linear polarization of a spectral line by a local magnetic field, may provide strong diagnostics of regions of weak magnetic fields such as the solar corona, where a number of spectral lines with different but complementary sensitivity ranges are present. Unlike the Zeeman effect, the Hanle effect does not create polarization but requires its presence through other physical processes such as radiation scattering. The Hanle effect is a purely quantum phenomenon and has no classical equivalent. However, for brevity, to provide a simplified illustration of such a complex effect, it can be explained by approximating the excited atom/ion to a damped oscillator with Larmor frequency that is scattering incident non-polarized radiation (see **Figure 1**). The Larmor frequency, ω_L, of the precession motion around the magnetic field vector is directly related to the magnetic field strength. The damping is proportional to the finite lifetime, τ, of the upper level of the atomic transition. We note that this classical description could explain only the case of the normal Zeeman triplet (i.e., two-level atom $J_u = 1; J_l = 0$). Significant advances in the theory of radiation scattering in the presence of magnetic fields have been achieved in the last four decades (Sahal-Bréchot et al., 1977; Bommier, 1980; Landi Degl'Innocenti, 1982; Casini and Judge, 1999; López Ariste and Casini, 2002; Raouafi, 2002; Trujillo Bueno et al., 2002a; Landi Degl'Innocenti and Landolfi, 2004).

The sensitivity of a given spectral line to the Hanle effect is a function of the lifetime of the atomic transition and $|B|$. Ideally, a spectral line is sensitive to magnetic field strengths satisfying the relation

$$\gamma B \tau \approx 1, \qquad (1)$$

where $\gamma = g_{J_u} \mu_B / \hbar$, g_{J_u} is the Landé factor of the upper atomic level, μ_B is the Bohr magneton, and $\hbar$ is the reduced Planck constant. Practically, the Hanle effect is measurable for $0.2 \leq \omega_L \tau \leq 10$ (Bommier and Sahal-Bréchot, 1978, 1982). **Table 1** provides the magnetic field strengths corresponding to the ideal sensitivity of the different spectral lines to the Hanle effect.

Theoretically, direct determination of the magnetic field in the solar corona could be achieved through linear polarization

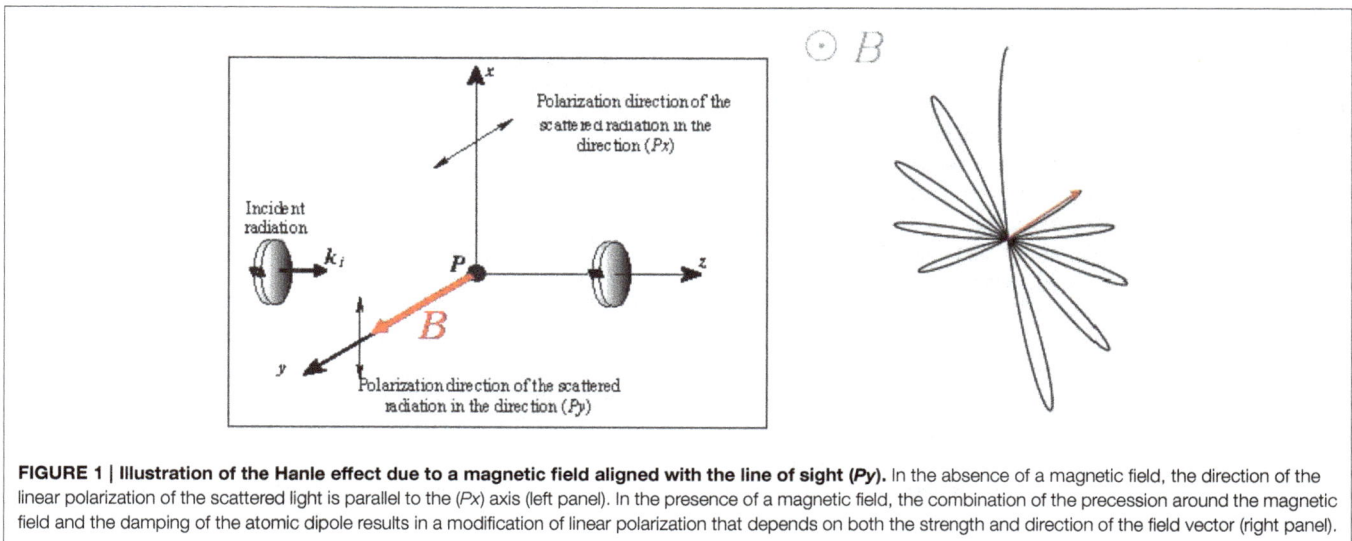

FIGURE 1 | Illustration of the Hanle effect due to a magnetic field aligned with the line of sight (Py). In the absence of a magnetic field, the direction of the linear polarization of the scattered light is parallel to the (Px) axis (left panel). In the presence of a magnetic field, the combination of the precession around the magnetic field and the damping of the atomic dipole results in a modification of linear polarization that depends on both the strength and direction of the field vector (right panel).

TABLE 1 | Magnetic field strengths corresponding to the ideal Hanle effect for a number of spectral lines (column 4).

Spectral line	Wavelength (Å)	A_{ul} (10^8 s^{-1})	B (Gauss)
H I Ly-α	1215.16	6.265	53.43
H I Ly-β	1025.72	1.672	14.26
H I Ly-γ	972.53	0.682	5.81
H I Ly-δ	949.74	0.344	2.93
O VI	1031.91	4.16	35.48
He I	10,830.0	0.344	0.82

A_{ul} is the Einstein coefficient of spontaneous emission of the upper level of the corresponding transitions.

of spectral lines with suitable sensitivity to the Hanle effect. The Hanle effect in selected spectral lines yields a powerful diagnostic tool for magnetic fields typically ranging from a few milli-Gauss to several hundred Gauss (depending strongly on the chosen line and the strength and direction of the magnetic field). Unlike the Zeeman effect, the depolarization of spectral lines by turbulent magnetic fields can be detected in the Hanle regime allowing the determination of the strength of the field in mixed-polarity regions (e.g., Stenflo, 1982; Trujillo Bueno et al., 2004).

In the solar corona, the Hanle effect manifests itself primarily through a depolarization and a rotation of the plane of linear polarization, with respect to the zero-field case where the plane of polarization is parallel to the local solar limb. Bommier et al. (1981) studied various measurement scenarios allowing for the complete diagnostic of the coronal magnetic field vector. The Hanle effect diagnostic of magnetic fields has been successful in solar prominences (Leroy et al., 1977; Sahal-Bréchot et al., 1977; Bommier, 1980; Landi Degl'Innocenti, 1982; Querfeld et al., 1985; López Ariste and Casini, 2002; Trujillo Bueno et al., 2002b), as well as in arch filament systems (Solanki et al., 2003; Lagg et al., 2004; Xu et al., 2010; Merenda et al., 2011).

2.1. Prominence Magnetic Fields

Bommier et al. (1994) and previous related papers (Sahal-Bréchot et al., 1977; Bommier and Sahal-Bréchot, 1978) have successfully demonstrated the power of the Hanle effect method for measuring the magnetic fields in solar prominences. Leroy et al. (1983, 1984) used observations obtained with the coronagraph polarimeter at the Pic du Midi observatory (France) to study the magnetic field of several hundreds of prominences based on the Hanle effect of spectral lines such as H-α, H-β, and He I 5876 Å. Leroy et al. (1983) found that magnetic field strengths increased with the rise of the solar cycle. They reported an average field strength of $\sim$ 6 Gauss at the beginning of the cycle and about twice this value near solar maximum. Furthermore, Leroy et al. (1984) found that the magnetic field strength and direction depend also on the prominence height: prominences with heights lower than 30 Mm have $\sim$ 20 Gauss fields with $\alpha \sim 20°$ and prominences higher than 30 Mm have 5 − 10 Gauss fields with $\alpha \sim 25°$ (α is the angle between the magnetic field vector and the prominence spine).

More recently, the He I 10,830 Å triplet has provided additional, very detailed diagnostics of the magnetic field in filaments. Thus, Kuckein et al. (2009, 2012), Sasso et al. (2011), and Xu et al. (2012) found that active region filaments have hectoGauss field strengths, i.e., an order of magnitude larger than the quiet filaments and prominences studied earlier. The spectropolarimetry of this set of lines even revealed the complex multicomponent structure of an activated filament, with the different components displaying magnetic vectors with different field strengths and directions and gas flowing at different speeds and in different directions (Sasso et al., 2014).

2.2. Polarization of Coronal Forbidden Lines

Charvin (1965) has shown that the direction of polarization of some forbidden lines is expected to be either parallel or perpendicular to the local magnetic field projected onto the plane of the sky. This provides a useful approach to study the orientation (direction) of coronal magnetic fields. No information on the field strength can, however, be obtained from such diagnostics.

The polarization of forbidden lines such as Fe XIV 530.3 nm and Fe XIII 1074.7 nm have been studied for more than three decades during solar eclipses and using coronagraph observations (Querfeld, 1974, 1977; Querfeld and Elmore, 1976; Arnaud and Newkirk, 1987, etc.). The relatively low resolution observations show a striking evidence of a predominant radial orientation of the polarization, found everywhere independently of the phase of the solar cycle, which depicts the direction of the coronal field projected on the plane of the sky (see Arnaud, 1982a,b; Arnaud and Newkirk, 1987). This may, however, be attributed to the low resolution of the instruments used in the above studies. In addition due to the van Vleck ambiguity, the magnetic field can also be perpendicular to the direction of the linear polarization. This is likely the case at tops of large coronal loops where the magnetic field is nearly horizontal.

Habbal et al. (2001) analyzed intensity and polarization maps with better resolution of the Fe XIII 1074.7 nm line. They found evidence for two magnetic components in the corona: a non-radial field associated with the large-scale structures known as streamers (with loop-like structures at their base) and a more pervasive radial magnetic field, which corresponds to the open coronal magnetic field. More recent observations from the CoMP telescope (Tomczyk et al., 2008) with higher resolutions show significant non-radiality of the coronal magnetic field projected on the plane of the sky. For examples of CoMP observations, see Gibson et al. (2016).

2.3. Polarization of FUV and EUV Coronal Lines

Several lines in the far UV (FUV) and EUV wavelength ranges have suitable sensitivity to determine the coronal magnetic field via the Hanle effect. The coronal Hanle effect in the FUV and EUV wavelength ranges is largely unexplored despite the high potential of this diagnostic. Li-like ion lines (O VI, N V, C IV, ...)

are presumed to be observed high in the corona due to their broad abundance curves (Sahal-Bréchot et al., 1986). Li-like ion lines are very intense lines of the chromosphere-corona transition region (the O VI 103.2 nm line is one of the most intense lines after H I Ly-α). For instance, the observed emission of the O VI ion by Vial et al. (1980) at $30''$ above the limb (as well as Reeves and Parkinson, 1970) has shown that the O VI emission extends out into the corona to a few *arcmin* above the limb. Observations from the Ultraviolet Coronagraph Spectrometer (UVCS; Kohl et al., 1995) on the The Solar and Heliospheric Observatory (SOHO; Domingo et al., 1995) show that the O VI emission extends several solar radii above the limb, with a line ratio and line widths sensitive to Doppler dimming and anisotropic velocity distributions (Kohl et al., 1998; Li et al., 1998). Such lines have small natural widths and short lifetimes of the upper levels of the corresponding atomic transitions. The magnetic field strength corresponding to their sensitivity to the Hanle effect ranges from a few Gauss to more than 300 Gauss. This interval contains the expected magnitude of the magnetic field strength in the solar corona.

In a series of papers, Fineschi et al. (1991, 1993) and Fineschi and Habbal (1995) studied both theoretically and from an instrumental point of view the feasibility of coronal magnetic field diagnostics through the Hanle effect. In particular, they considered the case of the strongest UV coronal line, H I Ly-α. One of the advantages of using H I Ly-α is the very broad line profile of transition region incident radiation, which makes it insensitive to effects of the solar wind velocity at low coronal heights. Additionally, H I Ly-α has a negligible collisional component compared to that of the other H I Lyman series. This results in a H I Ly-α zero-field polarization larger than that of the other H I Lyman lines, increasing the overall line sensitivity to the Hanle effect (see Fineschi et al., 1999).

Raouafi et al. (1999a) used spectroscopic observations from the SOHO Solar Ultraviolet Measurements of Emitted Radiation spectrometer (SUMER; Wilhelm et al., 1995) to measure the linear polarization of the O VI 103.2 nm line. SUMER calibration before launch shows that the instrument is sensitive to the linear polarization of the observed light (Hassler et al., 1997). The observations were made during the roll manoeuver of the SOHO spacecraft on March 19, 1996, in the southern coronal polar hole at 1.3 $R_\odot$. For more details on the observations see Raouafi et al. (1999a,b). The data show in particular that the plane of polarization has an angle of $\sim 9°$ with respect to the solar limb. In contrast, the polarization direction is expected to be tangent to the local solar limb in the absence of the magnetic field effect. Raouafi et al. (1999a,b, 2002) interpreted these measurements in terms of the Hanle effect due to the coronal magnetic field. They developed models to simulate the observational results and inferred a field strength of ~ 3 Gauss at 1.3 $R_\odot$ above the solar pole. The main result from this work is a clear evidence for the Hanle effect in the strong O VI 103.2 nm coronal lines. This opens a window for direct diagnostic of the coronal magnetic field by using different FUV-EUV lines with complementary sensitivities to the magnetic field.

Manso Sainz and Trujillo Bueno (2009) discussed the possibility of mapping the on-disk coronal magnetic fields using forward scattering in permitted lines at EUV wavelengths (e.g., the Fe X 17.4 nm line).

3. SIMULATION DATA: MAGNETIC FIELD CONFIGURATION AND PLASMA PARAMETERS

To develop useful model solutions for the solar corona, we use the magnetohydrodynamic (MHD) approximation, which is appropriate for long-scale, low-frequency phenomena in magnetized plasmas. In the past, we employed a "*polytropic approximation*" for treating the heating of the coronal plasma and the acceleration of the solar wind (Riley et al., 2001; Riley and Luhmann, 2012). While this approach produces remarkably good solutions for the structure of the coronal magnetic field, this is at the expense of poorer velocity and density profiles. In this study, however, we use our "*thermodynamic*" model, which relies on coronal heating functions that are guided by observational constraints (Lionello et al., 2009; Riley et al., 2015). Detailed comparisons with EUV and X-ray observations from several spacecraft have allowed us to constrain the likely functional forms for this heating, such that they reproduce the observed emission.

Most (if not all) of the Hanle effect studies in the literature were based on well-defined magnetic field configurations (e.g., theoretical models or extrapolated photospheric magnetic fields), which were lacking well-defined plasma parameters such as densities, temperatures, and velocities. These quantities enter directly into the definition of the Stokes parameters encompassing the magnetic field signature that is the Hanle effect. *Ad-hoc* approximations may provide valuable results and order of magnitude estimates of different quantities as well as estimates of linear polarization of a given spectral line, but they cannot yield physically meaningful estimates of the coronal magnetic field through the Hanle effect. These assumptions can be improved upon by considering self-consistent magnetic fields and plasma parameters obtained through MHD simulations.

To obtain realistic estimates of the polarization parameters of the UV H I Ly-α (1216 Å), we utilize high-resolution MHD simulations using Predictive Science's state-of-the-art MAS code. The simulation includes a full thermodynamic description of the plasma. All parameters needed for the calculation of the polarization of the spectral line are obtained in a self-consistent fashion, thus removing any need for heuristic assumptions. All quantities are provided on the nodes of spherical grids whose resolution changes with the heliodistance. **Figure 2** displays the magnetic field configuration of the solar corona corresponding to Carrington rotation 2130. The synthetic white-light coronal emission is shown in the right-hand-side panel.

For the coronal Hanle effect of the different spectral lines, we use the magnetic field, density, temperature, and velocity data cube. Line-of-sight (LOS) integration is taken into account. All quantities at any given point on the LOS are obtained by interpolation of the simulation data.

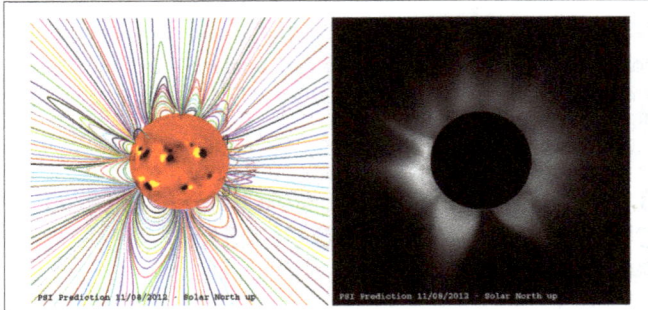

FIGURE 2 | Predictions of the coronal magnetic field configuration (left) and white-light synthetic image (right) for the solar eclipse of November 13, 2012. The images shown below are aligned such that solar north is vertically upward and includes the tilt due to the solar B_0 angle. The date on the panels is when the prediction was made. PSI have been providing predictions for solar eclipses for many years, which can be found at: http://www.predsci.com/corona/.

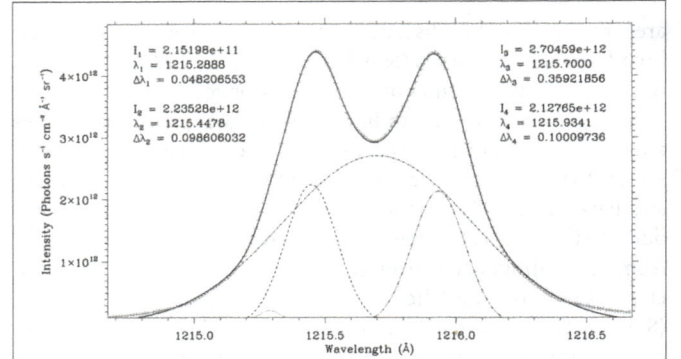

FIGURE 3 | Average transition region profile (Solid profile) of the H I Ly-α profile as observed by SOHO/SUMER (Lemaire et al., 1998). The "+" signs are the best 4-Gaussian fit of the observed profile. The individual Gaussians are given by the dotted, dashed, dot-dashed, and triple-dot-dashed profiles. The incident H I Ly-α radiation is assumed to be composed of four individual Gaussian profiles. The parameters of the individual profiles (i.e., amplitude [in photons cm^{-2} s^{-1} Å^{-1} sr^{-1}], central wavelength [in Å] and width [in Å]) are also displayed.

4. H I LY-α SOLAR DISK RADIATION

The H I Ly-α line is formed in the transition region at a temperature $\sim$ 1 MK. The coronal counterpart is formed by scattering of this incident radiation, resulting in the most intense UV coronal line. The solar disk radiation of this line is characterized by very little to no center-to-limb variations (Bonnet et al., 1980). The line profile is substantially wider than other lines and is typically complex. **Figure 3** shows an average H I Ly-α line profile as observed by SOHO/SUMER (Lemaire et al., 1998). It is characterized by an inversed peak at the center of the line, making a single Gaussian fit meaningless. For the present study, this profile is fitted with four Gaussians whose parameters are shown in the same figure. Numerically, we assume four individual Gaussian spectral lines (**Figure 3**) at different frequencies and with different widths and intensities. This assumption is, we believe, the best way to realistically mimic the incident radiation from the solar transition region.

We consider a two level atomic model for the H I Ly-α line. This assumption is sufficient to describe the light scattering by hydrogen atoms in the solar corona. The spectral line has two components, 2p ^{2}P$^o_{3/2}$ $\rightarrow$ 1s ^{2}S$_{1/2}$ (polarizable) and 2p ^{2}P$^o_{1/2}$ $\rightarrow$ 1s ^{2}S$_{1/2}$ (non-polarizable), with virtually the same Einstein coefficients of spontaneous emission, $A_{ul} \approx 6.2648 \times 10^8$ s^{-1}. The coronal electron collisional component represents less than 1% of the total intensity (Raymond et al., 1997). We neglect this component and assume that the H I Ly-α coronal line results only from the scattering of incident radiation from the transition region. Since we are interested in the magnetic field, we also neglect the effect of the solar wind velocity. This assumption is justified by the fact that the polarization calculations presented in this paper are achieved at coronal heights lower than 1 $R_\odot$, where the solar wind speed is lower than 100 km s^{-1}. Considering the line width of the incident line, the Doppler distribution effects are neglected. The incident radiation from the solar transition region is unpolarized and the radiation field is assumed to be cylindrically symmetric around the solar vertical, with half-cone

angle α_r (i.e., solar disk radiation inhomogeneities [e.g., active regions] are also neglected).

4.1. The Atomic Model and Polarization

We consider the case of a two-level atom $(\alpha_l J_l, \alpha_u J_u)$ in the presence of a magnetic field B. We also assume a non-polarizable lower level, such as that of the H I Ly-α line whose spherically symmetric lower level ^{2}S$_{1/2}$ (that is not polarizable). Within the frame of the density matrix formalism, the lower level is described only by its population represented by $^{\alpha_l J_l}\rho_0^0$ within the frame of density matrix formalism.

We assume that the incident radiation is characterized by a Gaussian spectral profile

$$\mathcal{I}(\Omega, \nu) = \frac{\mathcal{I}_c f(\Omega)}{\sqrt{\pi}\, \sigma_i} e^{-\left(\frac{\nu - \nu_0}{\sigma_i}\right)^2}, \quad (2)$$

where $\mathcal{I}_c$ is the solar disk center radiance (in erg cm^{-2} s^{-1} sr^{-1} Hz^{-1}), ν_0 and σ_i are the line center frequency and width of the incident profile, and $f(\Omega)$ describes the center-to-limb variation of the incoming radiation field. In the case of Ly-α, $f(\Omega) \approx 1$ (see Bonnet et al., 1980). In the solar frame, the properties of the incident radiation field are given by

$$\begin{aligned}\mathfrak{J}_0^0(\nu) &= \oint \frac{d\Omega}{4\pi} \mathcal{I}(\Omega, \nu) \\ \mathfrak{J}_0^2(\nu) &= \oint \frac{d\Omega}{4\pi} \frac{1}{2\sqrt{2}}(3\cos^2\alpha_r - 1)\mathcal{I}(\Omega, \nu)\end{aligned} \quad (3)$$

All other multipoles (i.e., $\mathfrak{J}_{0,\pm1}^1$ and $\mathfrak{J}_{\pm1,\pm2}^2$) are zero since the incident radiation is not polarized. The density matrix multipoles of the incident radiation have to be re-written in the magnetic field reference frame, which is obtained from the solar frame by a rotation $\mathcal{R}(\psi, \eta, 0)$, where ψ and η are, respectively, the azimuth and co-latitude of the vector magnetic field with respect to the solar vertical reference frame.

For a two-level atom with unpolarized lower level, the atomic polarization properties of the upper level are given by the atomic density matrix:

$$
\begin{aligned}
{}^{\alpha_u J_u}\rho_Q^K(\nu, \Omega) &= \frac{\alpha_l J_l \rho_0^0(\nu)}{A_{ul} + i\,Q\,\omega_L} \sqrt{3(2J_l+1)}\, B_{lu}\,(-1)^{1+J_l+J_u+Q} \\
&\quad \begin{Bmatrix} 1 & 1 & K \\ J_u & J_u & J_l \end{Bmatrix} {}^{B}\mathfrak{J}_{-Q}^K(\nu, \Omega) \\
&= \sqrt{\frac{2J_l+1}{2J_u+1}}\, \frac{B_{lu}\,{}^{\alpha_l J_l}\rho_0^0(\nu)}{A_{ul} + i\,Q\,\omega_L}\, w_{ul}^{(K)}\,(-1)^Q\,{}^{B}\mathfrak{J}_{-Q}^K(\nu, \Omega)
\end{aligned}
\tag{4}
$$

where $K = 0, \cdots, 2J_u$ and $Q = -K, \cdots, K$. $\omega_L = 2\pi\, g_u\, \nu_L$ is the Larmor angular frequency. For details on the derivation of Equation (4), see Landi Degl'Innocenti and Landolfi (2004). ν_L is the Larmor frequency and reflects the Hanle effect due to the presence of the magnetic field, g_u is the Landé factor of the upper level, and A_{ul} and B_{lu} are the Einstein coefficients for the spontaneous emission and absorption from the lower to upper levels. The symbol between the brackets is the Wigner 6j-symbol.

Equation (4) is the solution of the statistical system of linear equations of the atomic system in the steady-state case.

5. HANLE EFFECT OF THE H I Ly-α CORONAL LINES

The results of the forward modeling of the linear polarization of the coronal H I Ly-α line are shown in **Figure 4**. In the absence of the effect of the coronal magnetic field (i.e., Hanle effect), the direction of polarization is parallel to the local solar limb regardless of coronal altitude. The fractional linear polarization increases as a function of altitude because of the increased anisotropy of the incident solar disk radiation. The LOS integration is also more important with increasing altitude above the solar limb. This is due to the increasingly shallower density gradient. For the present calculations, the LOS integration is done for a range of 5 $R_\odot$ centered on the plane of the sky.

The top panels of **Figure 4** illustrated the altitude variation of the Stokes parameters ($\log_{10} I$, Q/I, and U/I, respectively). The lowest coronal altitude of the calculations is 1.015 $R_\odot$. The bottom panels show the degree of linear polarization (in %, left),

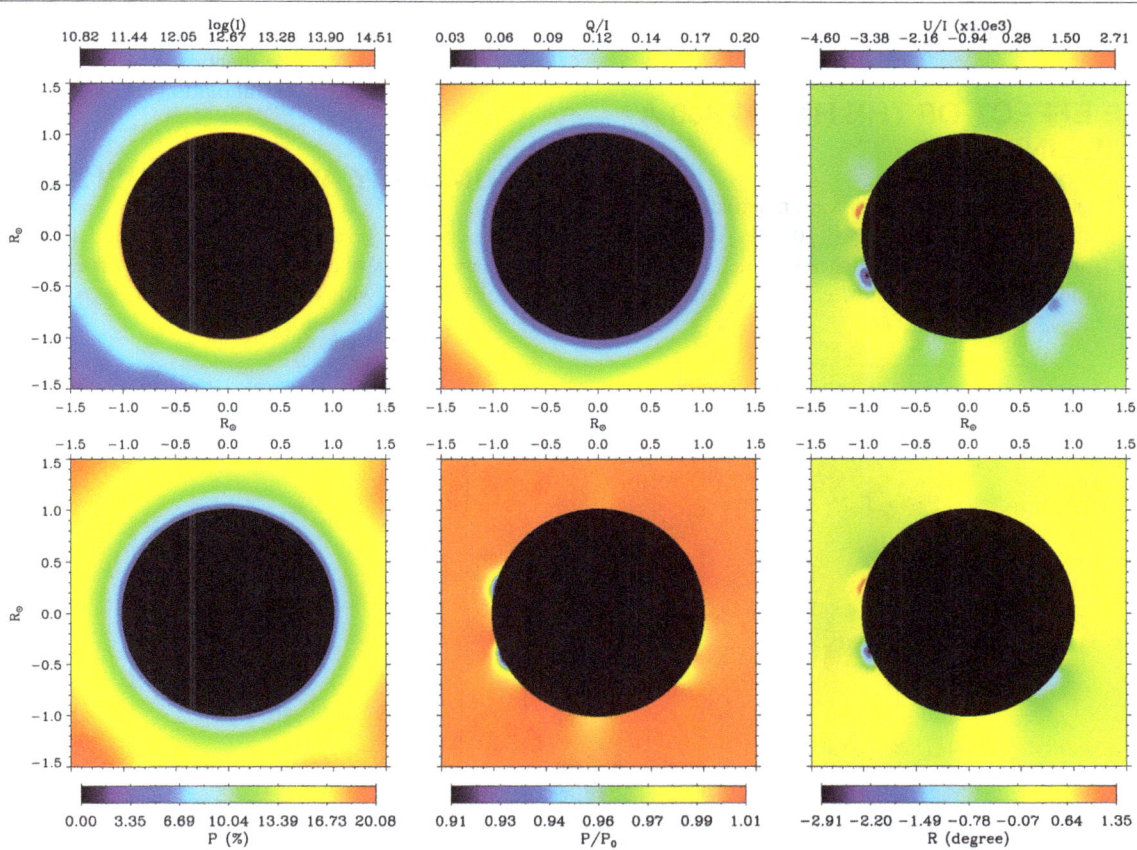

FIGURE 4 | Line-profile-integrated Stokes parameters (top) and linear polarization (bottom) of the coronal H I Ly-α line. P, P_0, and R are the polarization degree, the polarization degree in zero magnetic field, and the rotation of the plane of polarization with respect to the local solar limb, respectively. Above regions of relatively strong coronal magnetic fields the Hanle depolarization attains about 10% and the rotation of the plane of polarization is about 3°. This illustrates that although this line is not the best in terms of sensitivity to the Hanle effect, the effects of the magnetic field on the linear polarization are significant.

depolarization (i.e., the ratio of the degree of polarization to that in the absence of magnetic field, middle), and the rotation of the plane of linear polarization with respect to the local solar limb (in degrees). It is clear that the signature of the Hanle effect, which is given primarily by the depolarization and the rotation of plane of polarization, is limited to low-altitude regions where the magnetic field is relatively strong. This is expected because of the sensitivity of this line to the Hanle effect.

Although these results are still preliminary, they suggest that the H I Ly-α line could be very useful for measuring the coronal magnetic field, particularly at low latitude, and in strong field regions (i.e., above active regions). Theoretical polarization rates (not shown here) in these coronal line are reasonably high and could be easily measured. The depolarization with respect to the zero magnetic field case attains $\sim$ 10% in some areas. The main parameter that limits the measurability of the Hanle effect in this line is the rotation of the plane of polarization. Above active regions, a rotation of about 3° is obtained, and rotations of more than 1° are obtained in larger areas. These results show that this line is promising in terms of constraining the coronal magnetic field, despite the fact that it is not the most suitable line in terms of sensitivity to the coronal Hanle effect. In combination with other spectral lines with complementary sensitivities to the effect of the coronal magnetic field, the Hanle effect could provide reliable constraints on the coronal magnetic field.

6. HANLE EFFECT OF THE He I 10,830 Å CORONAL LINE

Kuhn et al. (2007) showed evidence for an extended diffuse surface brightness flux at the He I 10,830 Å line using observations from the SOLARC coronagraph. The observations show that emissions result from cold helium (i.e., narrow line profiles), which is unlikely to be scattered by the solar wind helium that is presumably significantly hotter. The authors argue that cold helium atoms form on dust grains, which provide an atomic population different to that of the solar wind. The importance of these observations stems from the sensitivity of the He I 10,830 Å line to the Hanle effect, which corresponds to magnetic field strength ranging from $\sim$ 0.2 Gauss to <10 Gauss. This line may be the most suitable line for the diagnostic of coronal magnetic fields through the Hanle effect.

Figure 5 shows forward modeling results of the linear polarization of the He I 10,830 Å. The LOS-integration scheme is the same as for H I Ly-α. Unlike H I Ly-α where the Hanle effect is limited to low-height, strong field regions, the linear polarization of He I 10,830 Å shows more variations as it depicts coronal structures, such as streamers, closed field regions. The variation of the polarization parameters spreads over larger intervals, which make their measurement easier. This is expected because of the higher sensitivity of this line to the relatively weak coronal magnetic fields. Dima et al. (2016) present a complementary analysis of the polarization of the He I 10,830 Å line.

We believe that the linear polarization of the He I 10,830 Å line could provide valuable constraints on the coronal magnetic field. In combination with UV lines such as H I Ly-α, -β, and O VI 103.2 nm, as well as IR forbidden lines, the coronal Hanle effect could provide reliable diagnostic of the coronal magnetic field and consequently to extrapolation and MHD models.

7. SUMMARY AND CONCLUSIONS

The preliminary results of the forward modeling of the linear polarization of two coronal spectral lines (i.e., H I Ly-α and He I 10,830 Å) with different sensitivities to the Hanle effect

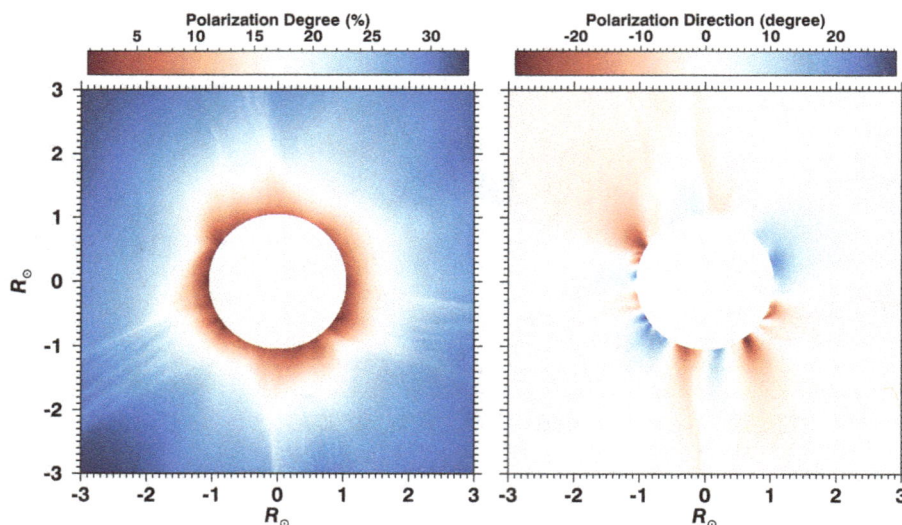

FIGURE 5 | Polarization parameters of the He I 10,830 line. (Left) polarization degree (in %) and (Right) rotation of the plane of polarization (in degrees) with respect to the local tangent to the solar limb.

are promising. The UV H I Ly-α line is mainly sensitive to low-height, strong magnetic fields above active regions. The polarization of the He I 10,830 Å line, which is sensitive to magnetic fields ranging roughly from ~ 0.2 Gauss up to <10 Gauss, shows more variations with coronal height and traces different coronal structures with different magnetic topologies (e.g., streamers and closed field regions).

Forward modeling of the Hanle effect is an important step in our quest for direct measurements of the magnetic field in the solar corona, which is a very difficult problem that includes different issues that observations will be subject to (e.g., $180°$ and Van Vleck $90°$ ambiguities and LOS-integration). Forward modeling will allow us to fully understand these problems and develop the necessary tools to analyze the observations. It also shows the potential of the Hanle effect in different wavelength regimes, which can be utilized for future space solar missions with UV and IR polarimeters (e.g., Peter et al., 2012).

We believe that the combination of the linear polarization of coronal lines with complementary sensitivities to the Hanle effect is promising and could provide long-sought measurements of the coronal magnetic field. The Hanle effect is a powerful tool that may provide the most reliable diagnostics of the magnetic field at relatively low coronal heights. Radio observations along with polarimetric measurements in the Zeeman and saturated Hanle regimes may provide complementary constrains that could help help piece together the coronal magnetic field structure.

AUTHOR CONTRIBUTIONS

All authors listed, have made substantial, direct and intellectual contribution to the work, and approved it for publication.

ACKNOWLEDGMENTS

We, the authors, would like to thank the reviewers for the constructive comments and criticism that helped improving the quality of the manuscript. This work was enabled by discussions with members of the International Space Science Institute (ISSI) working group on coronal magnetism (2013–2014), particularly J. Kuhn. SG acknowledges support from the Air Force Office of Space Research, FA9550-15-1-0030; NCAR is supported by the National Science Foundation.

REFERENCES

Arnaud, J. (1982a). Observed polarization of the Fe xiv 5303 coronal emission line. *Astron. Astrophys.* 112, 350–354.

Arnaud, J. (1982b). The analysis of Fe xiv 5303 coronal emission-line polarization measurements. *Astron. Astrophys.* 116, 248–254.

Arnaud, J., and Newkirk, G. Jr. (1987). Mean properties of the polarization of the Fe XIII 10747 Å coronal emission line. *Astron. Astrophys.* 178, 263–268.

Bommier, V. (1980). Quantum theory of the hanle effect. II - effect of level-crossings and anti-level-crossings on the polarization of the D$_3$ helium line of solar prominences. *Astron. Astrophys.* 87, 109–120.

Bommier, V., and Sahal-Bréchot, S. (1978). Quantum theory of the hanle effect - calculations of the stokes parameters of the D$_3$ helium line for quiescent prominences. *Astron. Astrophys.* 69, 57–64.

Bommier, V., and Sahal-Bréchot, S. (1982). The Hanle effect of the coronal L-α line of hydrogen: theoretical investigation. *Sol. Phys.* 78, 157–178. doi: 10.1007/BF00151151

Bommier, V., Sahal-Bréchot, S., and Leroy, J. L. (1981). Determination of the complete vector magnetic field in solar prominences, using the Hanle effect. *Astron. Astrophys.* 100, 231–240.

Bommier, V., Landi Degl'Innocenti, E., Leroy, J.-L., and Sahal-Bréchot, S. (1994). Complete determination of the magnetic field vector and of the electron density in 14 prominences from linear polarization measurements in the He I D$_3$ and H-α lines. *Sol. Phys.* 154, 231–260.

Bonnet, R. M., Decaudin, M., Bruner, E. C. Jr., Acton, L. W., and Brown, W. A. (1980). High-resolution Lyman-α filtergrams of the sun. *Astrophys. J. Lett.* 237, L47–L50. doi: 10.1086/183232

Casini, R., and Judge, P. G. (1999). Spectral lines for polarization measurements of the coronal magnetic field. II. Consistent treatment of the stokes vector for magnetic-dipole transitions. *Astrophys. J.* 522, 524–539. doi: 10.1086/307629

Charvin, P. (1965). Étude de la polarisation des raies interdites de la couronne solaire. Application au cas de la raie verte λ5303. *Ann. d'Astrophys.* 28:877.

Dima, G. I., Kuhn, J. R., and Berdyugina, S. V. (2016). Infrared dual-line hanle diagnostic of the coronal vector magnetic field. *Front. Astron. Space Sci.* 3:13. doi: 10.3389/fspas.2016.00013

Domingo, V., Fleck, B., and Poland, A. I. (1995). The SOHO mission: an overview. *Sol. Phys.* 162, 1–37. doi: 10.1007/BF00733425

Fineschi, S. (2001). Space-based instrumentation for magnetic field studies of solar and stellar atmospheres, in magnetic fields across the Hertzsprung-Russell diagram. *ASP Conf. Ser.* 248:597.

Fineschi, S., Chiuderi, C., Poletto, G., Hoover, R. B., and Walker, A. B. C. Jr. (1993). LY-A-CO-PO (LY-α coronagraph/polarimeter): an instrument to measure coronal magnetic fields. *Memorie della Societ Astronomia Italiana* 64, 441.

Fineschi, S., Gardner, L. D., Kohl, J. L., Romoli, M., Pace, E., Corti, G., et al. (1999). "Polarimetry of the UV Solar Corona with ASCE", in *Ultraviolet and X-Ray Detection, Spectroscopy, and Polarimetry III*, Vol. 3764, eds S. Fineschi, B. E. Woodgate, and R. A. Kimble (Denver, CO: Proceedings of SPIE), 147–160. doi: 10.1117/12.371079

Fineschi, S., and Habbal, S. R. (1995). Coronal magnetic field diagnostics via the Hanle effect of Lyman series lines. *Proc. Solar Wind* 8:68.

Fineschi, S., Hoover, R. B., Fontenla, J. M., and Walker, A. B. C. Jr. (1991). Polarimetry of extreme ultraviolet lines in solar astronomy. *Opt. Eng.* 30, 1161–1168. doi: 10.1117/12.55922

Gibson, S. E., Kucera, T. A., White, S. M., Dove, J. B., Fan, Y., Forland, B. C., et al. (2016). FORWARD: a toolset for multiwavelength coronal magnetometry. *Front. Astron. Space Sci.* 3:8. doi: 10.3389/fspas.2016.00008

Habbal, S. R., Woo, R., and Arnaud, J. (2001). On the predominance of the radial component of the magnetic field in the solar corona. *Astrophys. J.* 558, 852–858. doi: 10.1086/322308

Hanle, W. (1924). Über magnetische Beeinflussung der Polarisation der Resonanzfluoreszenz. *Zeitschrift für Physik* 30, 93. doi: 10.1007/BF01331827

Hassler, D. M., Lemaire, P., Longval, Y. (1997). Polarization sensitivity of the SUMER instrument on SOHO. *Appl. Opt.* 36, 353–359. doi: 10.1364/AO.36.000353

Kohl, J. L., Esser, R., Gardner, L. D., Habbal, S., Daigneau, P. S., Dennis, E. F., et al. (1995). The ultraviolet coronagraph spectrometer for the solar and heliospheric observatory. *Sol. Phys.* 162, 313–356. doi: 10.1007/BF00733433

Kohl, J. L., Noci, G., Antonucci, E., Tondello, G., Huber, M. C. E., Cranmer, S. R., et al. (1998). UVCS/*SOHO* empirical determinations of anisotropic velocity distributions in the solar corona. *Astrophys. J. Lett.* 501:L127. doi: 10.1086/311434

Kuckein, C., Centeno, R., Martínez Pillet, V., Casini, R., Manso Sainz, R., and Shimizu, T. (2009). Magnetic field strength of active region filaments. *Astron. Astrophys.* 501, 1113–1121. doi: 10.1051/0004-6361/200911800

Kuckein, C., Martínez Pillet, V., and Centeno, R. (2012). An active region filament studied simultaneously in the chromosphere and photosphere. I. Magnetic structure. *Astron. Astrophys.* 539:A131. doi: 10.1051/0004-6361/2011 17675

Kuhn, J. R., Arnaud, J., Jaeggli, S., Lin, H., and Moise, E. (2007). Detection of an extended near-sun neutral helium cloud from ground-based infrared coronagraph spectropolarimetry. *Astrophys. J. Lett.* 667:L203. doi: 10.1086/522370

Lagg, A., Woch, J., Krupp, N., Solanki, S. K. (2004). Retrieval of the full magnetic vector with the He I multiplet at 1083 nm. Maps of an emerging flux region. *Astron. Astrophys.* 414, 1109–1120. doi: 10.1086/522370

Landi Degl'Innocenti, E. (1982). The determination of vector magnetic fields in prominences from the observations of the Stokes profiles in the D_3 line of helium. *Sol. Phys.* 79, 291–322.

Landi Degl'Innocenti, E., and Landolfi, M. (2004). "Polarization in spectral lines," in *Astrophysics and Space Science Library*, Vol. 307 (Dordrecht: Springer).

Lemaire, P., Emerich, C., Curdt, W., Schuehle, U., and Wilhelm, K. (1998). Solar H I Lyman-α full disk profile obtained with the SUMER/SOHO spectrometer. *Astron. Astrophys.* 334, 1095–1098.

Leroy, J. L., Ratier, G., and Bommier, V. (1977). The polarization of the D_3 emission line in prominences. *Astron. Astrophys.* 54, 811–816.

Leroy, J. L., Bommier, V., and Sahal-Bréchot, S. (1983). The magnetic field in the prominences of the polar crown. *Sol. Phys.* 83, 135–142. doi: 10.1007/BF00148248

Leroy, J. L., Bommier, V., and Sahal-Bréchot, S. (1984). New data on the magnetic structure of quiescent prominences. *Astron. Astrophys.* 131, 33–44.

Li, X., Habbal, S. R., Kohl, J. L., and Noci, G. (1998). The effect of temperature anisotropy on observations of doppler dimming and pumping in the inner corona. *Astrophys. J. Lett.* 501, L133–L137. doi: 10.1086/311428

Lionello, R., Linker, J. A., and Mikić, Z. (2009). Multispectral emission of the sun during the first whole sun month: magnetohydrodynamic simulations. *Astrophys. J.* 690, 902–912. doi: 10.1088/0004-637X/690/1/902

Lin, H., Kuhn, J. R., and Coulter, R. (2004). Coronal magnetic field measurements. *Astrophys. J. Lett.* 613:L177. doi: 10.1086/425217

López Ariste, A., and Casini, R. (2002). Magnetic fields in prominences: inversion techniques for spectropolarimetric data of the He I D_3 line. *Astrophys. J.* 575, 529–541. doi: 10.1086/341260

Manso Sainz, R., and Trujillo Bueno, J. (2009). A possible polarization mechanism of EUV coronal lines, in solar polarization 5. *ASP Conf. Ser.* 405:423.

Merenda, L., Lagg, A., and Solanki, S. K. (2011). The height of chromospheric loops in an emerging flux region. *Astron. Astrophys.* 532:A63. doi: 10.1051/0004-6361/201014988

Peter, H., Abbo, L., Andretta, V., Auchère, F., Bemporad, A., Berrilli, F., et al. (2012). Solar magnetism eXplorer (SolmeX). Exploring the magnetic field in the upper atmosphere of our closest star. *Exp. Astron.* 33, 271–303. doi: 10.1007/s10686-011-9271-0

Querfeld, C. W. (1974). The high altitude observatory coronal-emmission polarimeter, in planets, stars, and nebulae: studied with photopolarimetry. *IAU Colloq.* 23:254.

Querfeld, C. W., and Elmore, D. E. (1976). Observation of polarization in Fe XIII 10747 Å coronal emission line. *Bull. Am. Astron. Soc.* 8, 368.

Querfeld, C. W. (1977). Observations of Fe XIII 10747 Å coronal emission-line polarization. *Rep. Lund Observ.* 12:109.

Querfeld, C. W., Smartt, R. N., Bommier, V., Landi Degl'Innocenti, E., and House, L. L. (1985). Vector magnetic fields in prominences. II - He I D_3 stokes profiles analysis for two quiescent prominences. *Sol. Phys.* 96, 277–292.

Raouafi, N.-E., Lemaire, P., and Sahal-Bréchot, S. (1999a). Detection of the O VI 103.2 nm line polarization by the SUMER spectrometer on the SOHO spacecraft. *Astron. Astrophys.* 345, 999–1005.

Raouafi, N.-E., Sahal-Bréchot, S., Lemaire, P., and Bommier, V. (1999b). Doppler redistribution of resonance polarization of the O VI 103. 2 nm line observed above a polar hole, Proc. of Solar Polarization 2. *Astrophy. Space Sci. Library* 243:349.

Raouafi, N.-E. (2002). Stokes parameters of resonance lines scattered by a moving, magnetic medium. Theory of the two-level atom. *Astron. Astrophys.* 386, 721–731. doi: 10.1051/0004-6361:20020113

Raouafi, N.-E., Sahal-Bréchot, S., and Lemaire, P. (2002). Linear polarization of the O VI lambda 1031.92 coronal line. II. Constraints on the magnetic field and the solar wind velocity field vectors in the coronal polar holes. *Astron. Astrophys.* 396, 1019–1028.

Raouafi, N.-E. (2005). Measurement methods for chromospheric and coronal magnetic fields, in chromospheric and coronal magnetic fields. *ESA Spec. Publ.* 596:3.

Raouafi, N.-E. (2011). Coronal polarization, proceedings of solar polarization 6. *ASP Conf. Ser.* 437, 99.

Raymond, J. C., Kohl, J. L., Noci, G., Antonucci, E., Tondello, G., Huber, M. C. E., et al. (1997). Composition of coronal streamers from the SOHO ultraviolet coronagraph spectrometer. *Sol. Phys.* 175, 645–665. doi: 10.1023/A:1004948423169

Reeves, E. M., and Parkinson, W. H. (1970). An atlas of extreme-ultraviolet spectroheliograms from OSO-IV. *Astrophys. J. Suppl.* 21, 1. doi: 10.1086/190217

Riley, P., Linker, J. A., and Mikić, Z. (2001). An empirically–driven global MHD model of the solar corona and inner heliosphere. *J. Geophys. Res.* 106, 15889–15901. doi: 10.1029/2000JA000121

Riley, P., and Luhmann, J. G. (2012). Interplanetary signatures of unipolar streamers and the origin of the slow solar wind. *Sol. Phys.* 277, 355–373. doi: 10.1007/s11207-011-9909-0

Riley, P., Linker, J. A., Lionello, R., and Mikić, Z. (2012). Corotating interaction regions during the recent solar minimum: the power and limitations of global MHD modeling. *J. Atmos. Solar Terrestr. Phys.* 83, 1–10. doi: 10.1016/j.jastp.2011.12.013

Riley, P., Lionello, R., Linker, J. A., Cliver, E., Balogh, A., Beer, J., et al. (2015). Inferring the structure of the solar corona and inner heliosphere during the maunder minimum using global thermodynamic magnetohydrodynamic simulations. *Astrophys. J.* 802:105. doi: 10.1088/0004-637X/802/2/105

Sahal-Bréchot, S., Bommier, V., and Leroy, J. L. (1977). The hanle effect and the determination of magnetic fields in solar prominences. *Astron. Astrophys.* 59, 223–231.

Sahal-Bréchot, S., Malinovsky, M., and Bommier, V. (1986). The polarization of the O VI 1032 Å line as a probe for measuring the coronal vector magnetic field via the Hanle effect. *Astron. Astrophys.* 168, 284–300.

Sasso, C., Lagg, A., and Solanki, S. K. (2011). Multicomponent He I 10830 Å profiles in an active filament. *Astron. Astrophys.* 526, A42. doi: 10.1051/0004-6361/200912956

Sasso, C., Lagg, A., and Solanki, S. K. (2014). Magnetic structure of an activated filament in a flaring active region. *Astron. Astrophys.* 561:A98. doi: 10.1051/0004-6361/201322481

Solanki, S. K., Lagg, A., Woch, J., Krupp, N., and Collados, M. (2003). Three-dimensional magnetic field topology in a region of solar coronal heating. *Nature* 425, 692–695. doi: 10.1038/nature02035

Stenflo, J. O. (1982). The Hanle effect and the diagnostics of turbulent magnetic fields in the solar atmosphere. *Sol. Phys.* 80, 209–226. doi: 10.1007/BF00147969

Tomczyk, S., Card, G. L., Darnell, T., Elmore, D. F., Lull, R., Nelson, P. G., et al. (2008). An instrument to measure coronal emission line polarization. *Sol. Phys.* 247, 411–428. doi: 10.1007/s11207-007-9103-6

Trujillo Bueno, J., Casini, R., Landolfi, M., and Landi Degl'Innocenti, E. (2002a). The physical origin of the scattering polarization of the Na I D lines in the presence of weak magnetic fields. *Astrophys. J. Lett.* 566:L53.

Trujillo Bueno, J., Landi Degl'Innocenti, E., Collados, M., Merenda, L., and Manso Sainz, R. (2002b). Selective absorption processes as the origin of puzzling spectral line polarization from the sun. *Nature* 415, 403–406.

Trujillo Bueno, J., Shchukina, N., and Asensio Ramos, A. (2004). A substantial amount of hidden magnetic energy in the quiet sun. *Nature* 430, 326–329. doi: 10.1038/nature02669

Vial, J. C., Lemaire, P., Artzner, G., and Gouttebroze, P. (1980). O VI ($\lambda = 1032$ Å) profiles in and above an active region prominence, compared to quiet sun center and limb profiles. *Sol. Phys.* 68, 187–206. doi: 10.1007/BF00153276

White, S. M. (1999). Radio versus EUV/X-ray observations of the solar atmosphere. *Sol. Phys.* 190, 309–330. doi: 10.1023/A:1005253501584

Wiegelmann, T., Thalmann, J. K., and Solanki, S. K. (2014). The magnetic field in the solar atmosphere. *Astron. Astrophys. Rev.* 22:78. doi: 10.1007/s00159-014-0078-7

Wilhelm, K., Curdt, W., Marsch, E., Schühle, U., Lemaire, P., Gabriel, A., et al. (1995). SUMER – solar ultraviolet measurements of emitted radiation. *Sol. Phys.* 162, 189–231. doi: 10.1007/BF00733430

Xu, Z., Lagg, A., and Solanki, S. K. (2010). Magnetic structures of an emerging flux region in the solar photosphere and chromosphere. *Astron. Astrophys.* 520:13. doi: 10.1051/0004-6361/200913227

Xu, Z., Lagg, A., Solanki, S. K., and Liu, Y. (2012). Magnetic fields of an active region filament from full stokes analysis of Si I 1082.7 nm and He I 1083.0 nm. *Astrophys. J.* 749:138. doi: 10.1088/0004-637X/749/2/138

Conflict of Interest Statement: The authors declare that the research was conducted in the absence of any commercial or financial relationships that could be construed as a potential conflict of interest.

3

Line-of-Sight Velocity As a Tracer of Coronal Cavity Magnetic Structure

Urszula Bąk-Stęślicka[1]*, Sarah E. Gibson[2] and Ewa Chmielewska[1]

[1] Astronomical Institute, University of Wrocław, Wrocław, Poland, [2] High Altitude Observatory, National Center for Atmospheric Research, Boulder, CO, USA

We present a statistical analysis of 66 days of observations of quiescent (non-erupting) coronal cavities and associated velocity and thermal structures. We find that nested rings of LOS-oriented velocity are common in occurrence and spatially well correlated with cavities observed in emission. We find that the majority of cavities possess multiple rings, and a range in velocity on the order of several *km/sec*. We find that the tops of prominences lie systematically below the cavity center and location of largest Doppler velocity. Finally, we use DEM analysis to consider the temperature structure of two cavities in relation to cavity, prominence, and flows. These observations yield new constraints on the magnetic structure of cavities, and on the conditions leading up to solar eruptions.

Keywords: sun: corona, sun: filaments, prominence, sun: infrared, sun: magnetic field

Edited by:
Xueshang Feng,
National Space Science Center, China

Reviewed by:
Gordon James Duncan Petrie,
National Solar Observatory, USA
Chaowei Jiang,
Chinese Academy of Sciences, China

*Correspondence:
Urszula Bąk-Stęślicka
bak@astro.uni.wroc.pl

Specialty section:
This article was submitted to
Stellar and Solar Physics,
a section of the journal
Frontiers in Astronomy and Space
Sciences

1. INTRODUCTION

Solar coronal cavities are regions of rarefied density and elliptical cross-section (Fuller and Gibson, 2009; Gibson et al., 2010; Forland et al., 2013). They are often observed in eruption as a component of the classic three-part structure of a Coronal Mass Ejections (CME), as they surround the CME's bright core formed by the dense material from the eruptive prominence (Illing and Hundhausen, 1986; Tandberg-Hanssen, 1995). Cavities exist not only as eruptive phenomena, however. Non-erupting, or quiescent cavities may exist in equilibrium for many days or weeks (Gibson et al., 2006). They usually surround quiescent prominences (Tandberg-Hanssen, 1995), especially in the polar crown regions, but in some cases cavities are observed in the absence of prominences. Cavities are mostly observed when a filament or filament channel is large and oriented along the line-of-sight, and where there are no bright neighboring structures (Gibson, 2015; McCauley et al., 2015).

The first recorded observations of cavities were made in white light (WL) during the solar eclipse of January 22, 1898 (Wesley, 1927) and since then cavities have been studied many times (Waldmeier, 1941; von Kluber, 1961; Williamson et al., 1961; Waldmeier, 1970; Gibson et al., 2006). The first explanation of the cavity phenomena as an area of reduced electron density was proposed by Waldmeier (1941). Cavities have been observed in a wide wavelength range, not only in WL, but also in radio, EUV and SXR (Vaiana et al., 1973a,b; Hudson et al., 1999; Hudson and Schwenn, 2000; Marqué et al., 2002; Marqué, 2004; Heinzel et al., 2008; Berger et al., 2012; Reeves et al., 2012). Observations in WL are particularly useful for larger cavities; smaller cavities are better viewed in EUV (Gibson, 2015). *Yohkoh*/Soft X-ray Telescope observations revealed hot central cores within cavities (Hudson et al., 1999; Hudson and Schwenn, 2000). Other analyses also suggest the existence of hotter plasma inside cavities (Fuller et al., 2008; Habbal et al., 2010; Reeves et al., 2012).

Cavities have been modeled as a flux rope (Low, 1994; Low and Hundhausen, 1995), but the physical nature of cavities is still under investigation. Although quiescent cavities are long-lived and their structure evolves slowly with time, they have been observed to bodily erupt as a CME

(Maričič et al., 2004; Vršnak et al., 2004; Gibson et al., 2006; Régnier et al., 2011; Forland et al., 2013; Su et al., 2015). Understanding the magnetic structure of such cavities may be essential for establish the pre-CME configurations and for choosing between models for CME eruptive drivers.

The Coronal Multi-channel Polarimeter (CoMP) observes the full-Sun lower coronal magnetic field via spectropolarimetric measurements of the the forbidden lines of Fe XIII. These observations give us information about line intensity (polarized and unpolarized), Doppler shift, and line width. These in turn constrain coronal density, temperature, velocity, and through the polarization measurements, magnetic field. Observations of linear polarization in particular diagnose the direction of the magnetic field in the plane of sky (POS), and are well-suited for analysis of the magnetic configuration in polar-crown prominence cavities. CoMP observations have revealed that polar-crown cavities showed characteristic structures in linear polarization (**Figure 1**) which we termed "lagomorphic," due to their resemblance to rabbit heads seen in silhouette. This characteristic structure may be explained by a magnetic flux-rope model (Bąk-Stęślicka et al., 2013). Lagomorphic structures are very common, and they may be observed in most of the cavities oriented along the line of sight (LOS). The size of the CoMP lagomorphic signature generally scales with the cavity size seen in EUV (Bąk-Stęślicka et al., 2014).

The first CoMP observations of cavities, taken while it was placed at the National Solar Observatory in 2005, revealed interesting results. Schmit et al. (2009) found, for the first time, Doppler velocities in the range of $5 - 10\,\mathrm{km\,s^{-1}}$ within a coronal cavity. In our previous paper (Bąk-Stęślicka et al., 2013) we showed another example of LOS flows within cavities. What was the most interesting about these flows was that they occurred in the form of nested ring-like structures with apparently counterstreaming velocities.

In the present work we present a statistical analysis of Doppler velocities within cavities. In Section 2 we describe the data used in our study of 66 days of cavity flow observations. In Section 3 we present our results, and discuss relations between cavities, flows,

associated prominences, and hot cores. In Section 4 we give our conclusions.

2. INSTRUMENTS AND DATA ANALYSIS

2.1. Instruments

The Atmospheric Imaging Assembly (AIA, Lemen et al., 2012) on board *Solar Dynamics Observatory* (*SDO*, Pesnell et al., 2012) continuously makes full-disk images of the Sun through ten passbands with a spatial resolution $\sim$ 1 arcsecond, temporal cadence of 12 s, and FOV of $1.3\,\mathrm{R_\odot}$. AIA consists of four telescopes and provides narrow-band imaging of seven extreme ultraviolet (EUV) band passes centered on lines: Fe XVIII (94 Å), Fe VIII, XXI (131 Å), Fe IX (171 Å), Fe XII, XXIV (193 Å), Fe XIV (211 Å), He II (304 Å), and Fe XVI (335 Å). For effective temperature diagnostics of EUV emissions these cover the range from 0.6 to 20 MK.

The CoMP was installed in 2010 at the Mauna Loa Solar Observatory (MLSO) in Hawaii. Since 2010 October CoMP has made daily (subject to weather conditions) observations of the coronal magnetic field in the lower corona with a field of view (FOV) of about 1.04–1.4 solar radii and a spatial resolution of 4.46"/pixel. CoMP measures a polarimetric signal (Stokes I, Q, U, V) of the forbidden lines of Fe XIII at 1074.7 and at 1079.8 nm (Tomczyk et al., 2008). The line-of-sight (LOS) directed strength of the magnetic field can be obtained from the circular polarization (Stokes V) although such observations require long integration times on the order of multiple hours due to the very low intensity of circular polarization. As discussed above, the direction of the magnetic field in the POS – subject to a (resolvable) 90° ambiguity – can be determined from the observations in the linear polarization. LOS velocity is obtained via Doppler line shift measurements. All measurements are integrated along the LOS since the corona is optically thin; however forward modeling and CoMP observations have demonstrated that for extended structures such as polar-crown cavities critical information is preserved (see Gibson et al., 2016).

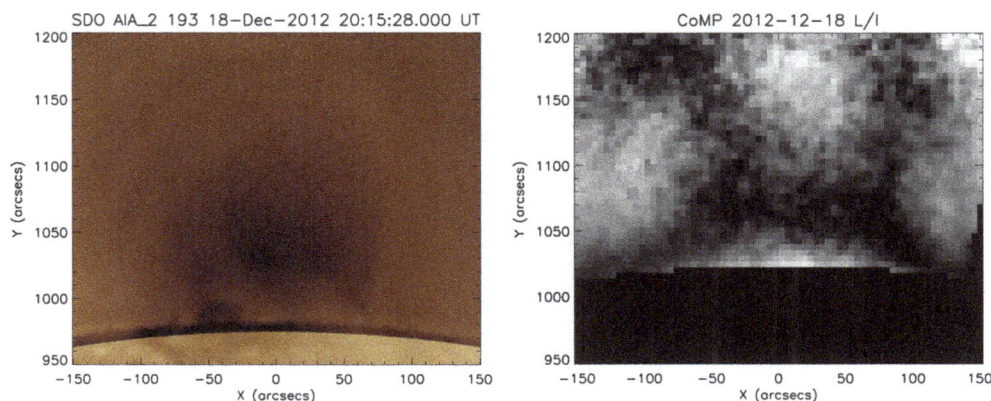

FIGURE 1 | Left: Cavity observed on 2012 December 18 by SDO/AIA 193 Å. **Right:** LOS-integrated linear polarization fraction (*L/I*) from CoMP showing lagomorphic, or rabbit-head shaped, structure. The occulting disk of CoMP extends to 1.05 R⊙.

2.2. Data Analysis

Using daily AIA 193 Å images and CoMP data we examined multiple years of polar-crown cavities and analyzed their properties. In particular, we investigated Doppler velocity patterns, quiescent prominences, and hot cores in relation to cavities.

In the set of CoMP observations between January 2012 and October 2015, we analyzed more than 70 days for which a coherent Doppler velocity pattern was observed in the cavity. For the purpose of our analysis we excluded those for which the cavity center was at a height lower than the CoMP occulter, and cavities positioned too close to the telescope occulter arm where there was no possibility of obtaining complete intensity and velocity profiles. This left us with 66 days of cavity flow observations, of which 46 appear to be independent cavity systems. We note that it is difficult to determine whether cavities are truly "independent," since they are extended along the line of sight and may go in and out of view with a curve of the the neutral line (see Gibson et al., 2006 for further discussion). For the purposes of this paper we consider distinct days of observation as data points.

From CoMP observations we used intensity (Stokes-I) images and Doppler shift maps. We mainly used the Level 2 three-point data of the Fe XIII 1074.7 nm line taken between January 2012 and October 2015 available on the MLSO web page (http://mlso.hao.ucar.edu). The zero point for CoMP Doppler velocity is not currently well established (*G. de Toma., private communication*); we therefore used the median value of Doppler shift in each map as our zero point in the scale (Tian et al., 2013). In addition, both intensity and Doppler shifts images were averaged over tens of minutes to hours, to improve the signal-to-noise ratio. Such averaging may affect our obtained velocity gradient (average values are likely to be smaller than for individual moments of time). Doppler velocity for Level 2 is also partially corrected for solar rotation through a model described by Tian et al. (2013);

this model assumption may be source of uncertainty. Since cavities are relativity small compared to the velocity gradients of the Tian model, and since our analysis depends on relative velocity values rather than absolute ones, our conclusions will be generally robust to these uncertainties.

AIA 193 Å images were used to measure the height of the cavity center [a detailed description of this method can be found in Gibson et al. (2010)]. For each cavity we extracted polar-angle cuts in the averaged CoMP intensity image at the same height as the AIA cavity center height (**Figure 2**, left). Using such CoMP intensity profiles we measured an average signal at the cavity rim on both sides of the cavity, and fit a straight line between those points. We defined a cavity width using an area where the signal decreased more than 3σ in relation to the fitted line. We repeated this analysis three times. The width, presented in this paper, is the average value from those measurements and the error is their standard deviation. We note that the cavity width obtained in this manner is somewhat larger than for widths in previously reported results based on AIA data (Forland et al., 2013).

We applied the same polar-angle cuts to the Doppler velocity images and used the same averaging as for the CoMP intensity images (**Figure 2**, right). Using these cavity-center velocity profiles, we measured the velocity range (minimum to maximum) within the cavity and the position of the strongest flow. Since many cavities show a nested-rings pattern in Doppler velocity images, we also calculated the slope of the velocity profiles and number of the velocity gradient changes. We also smoothed velocity profiles (using 2, 3, and 4 points), then measured the number of the velocity gradient changes. The values presented in this paper are the averaged value (from smoothed and unsmoothed profiles), error is their standard deviation.

For two cavities we used six of the AIA filters (94, 131, 171, 193, 211, and 335 Å) to calculate Differential Emission

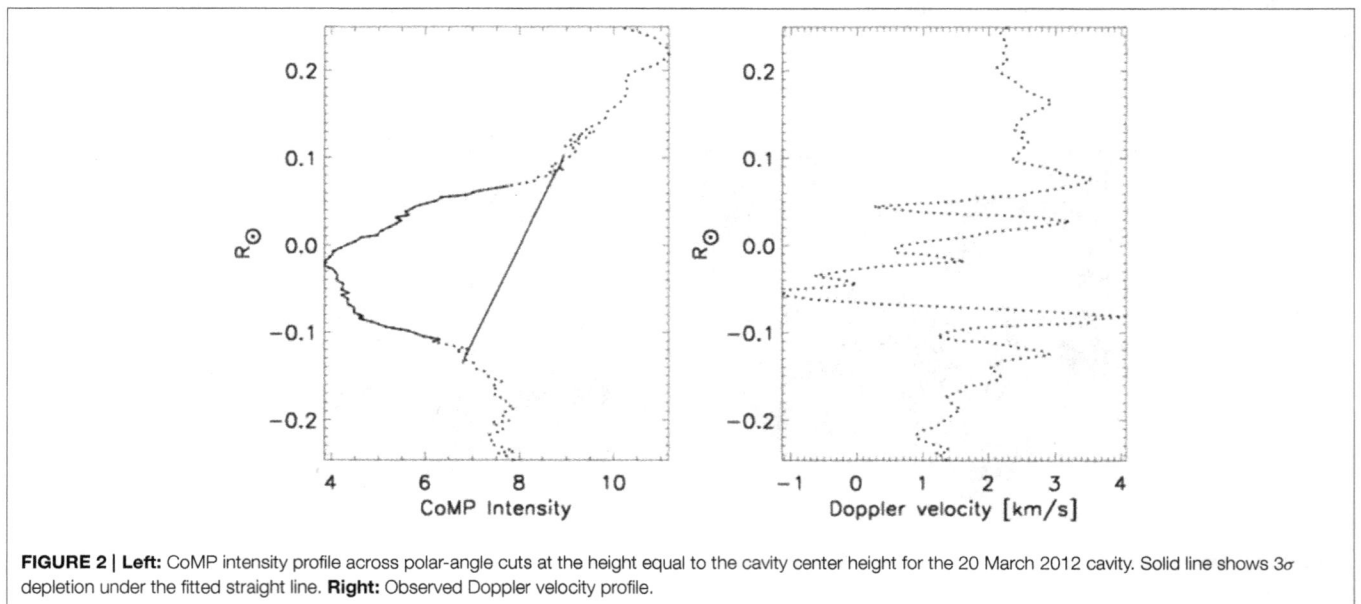

FIGURE 2 | **Left:** CoMP intensity profile across polar-angle cuts at the height equal to the cavity center height for the 20 March 2012 cavity. Solid line shows 3σ depletion under the fitted straight line. **Right:** Observed Doppler velocity profile.

Measure (DEM). We calculated response functions for all six filters using the CHIANTI database (Dere et al., 1997). In order to calculate DEM profiles and maps we used the iterative forward-fitting method originally developed for *HINODE*/XRT data (Weber et al., 2004). This XRT method of DEM calculation is available through SSW and was slightly modified by Cheng et al. (2012) to work with the AIA filters (see Appendix of Cheng et al., 2012). In this forward fitting method the differences between fitted and the observed intensities in six EUV AIA filters are minimized. For each pixel in the map we calculated the DEM-weighted average temperature. This parameter characterizing the overall temperature was introduced by Cheng et al. (2012):

$$\overline{T} = \frac{\int DEM(T) \times T dT}{\int DEM(T) dT} \qquad (1)$$

Finally, for the full set of cavities with flows, we used AIA 304 Å images to calculate the height of any associated prominence (at the top as measured in an averaged image).

3. RESULTS

First of all, we find that coherent – often ring-shaped – flows in Doppler velocity are almost as common within cavities as the "lagomorphic" signature in linear polarization discussed in our previous paper (Bąk-Stęślicka et al., 2013), although the visibility of both depends upon the orientation and extent of the cavity; see Jibben et al. (2016) for an example of a cavity which does not have a clear lagomorph/quiescent velocity structure. The flows in the 66 days we analyzed are usually in the form of the characteristic concentric rings within the cavity (**Figures 3**, **4**), and may be observed to persist for a few days (as is also true for lagomorphic

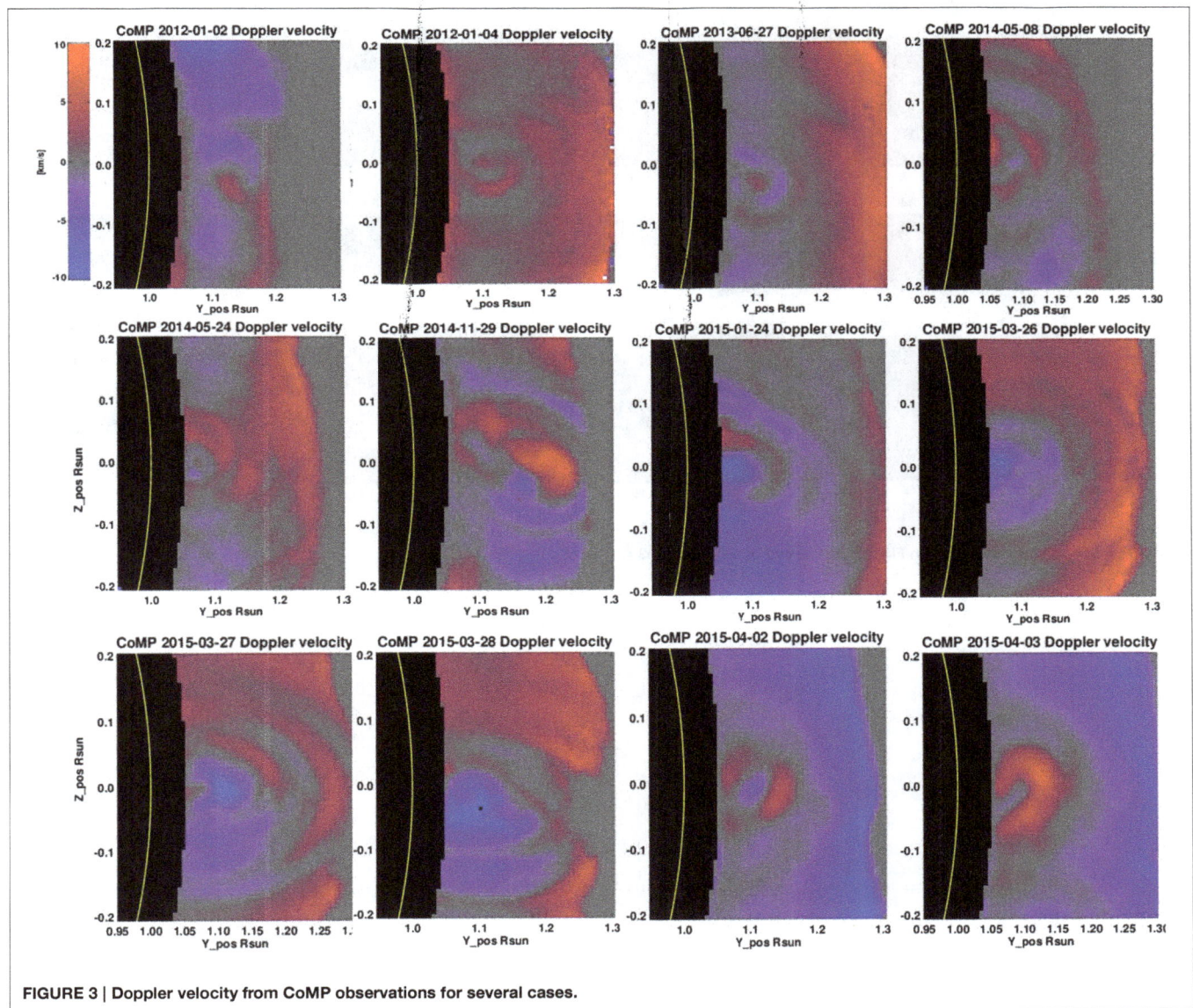

FIGURE 3 | Doppler velocity from CoMP observations for several cases.

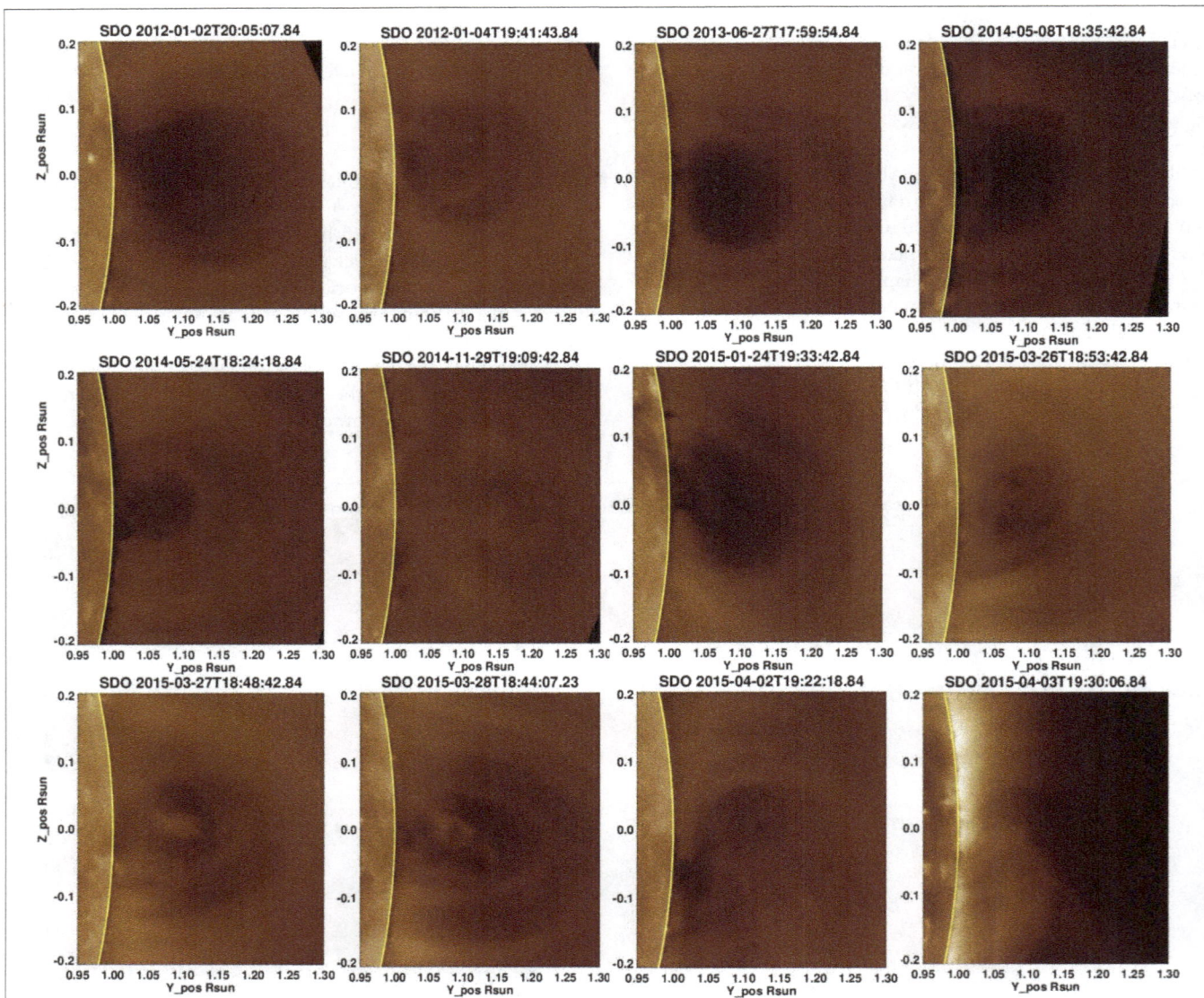

FIGURE 4 | The same cavities as in Figure 3 observed by SDO/AIA 193 Å.

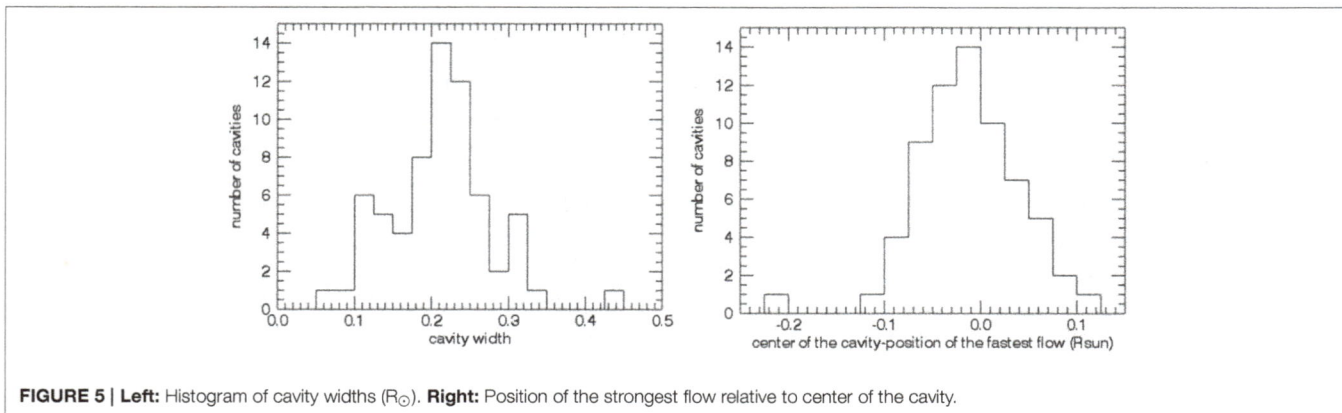

FIGURE 5 | Left: Histogram of cavity widths ($R_\odot$). **Right:** Position of the strongest flow relative to center of the cavity.

structures). Eventually, the change of the cavity's orientation as it rotates past the limb – and possibly also evolution of the structure – leads to the loss of a clearly viewed cavity, lagomorph, and velocity rings.

We found that almost half of the analyzed cavities have a width of about 0.2–0.25 $R_\odot$ (**Figure 5**, left). Detailed inspection of CoMP images and the intensity and velocity profiles indicate that flows are localized within the cavity (**Figures 2–4**). A histogram comparing the position of the strongest flow to the position of the cavity center is presented in **Figure 5** (right). In most cases this difference is within 0.05 $R_\odot$ of the cavity center which suggest that central flows are the strongest ones. In cases where the strongest flows are not observed in the center, a weaker central flow is often still present. We note that the errors in our method for finding the cavity center from AIA data are on the order .01 $R_\odot$, and that there are errors in alignment between AIA and CoMP of similar or somewhat larger size. These alignment errors may have a systematic component, which could also contribute to the skew in the distribution.

In most cases the range of velocities observed within a cavity is between 4 and 8 km/s (see **Figure 6**, left), which is consistent with previous reports (Schmit et al., 2009), but larger ranges (> 10 km/s) are also sometimes observed. **Figure 6** (right) shows the relation between velocity range and cavity width. The strongest ranges in velocities are observed in the widest and largest cavities.

Because of the uncertainty in the zero point of the velocity, it is still not clear if flows are truly counterstreaming, however, velocity gradients are clearly observed. Even if concentric rings in velocity (**Figure 3**) do not imply a change in flow direction, each ring may be an indicator of a discontinuity in LOS-directed velocity. The number of loops (or whole rings for higher cavities) is thus of interest. Taking the derivative of the velocity profile we obtain the number of velocity gradient changes (if the structure were symmetric about the center, the number of loops would be half the number of gradient changes). Most cavities indicate 6 − 12 such gradient changes (see **Figure 7**, left), but this number changes with smoothing of the velocity profile (see error bars in the **Figure 7**, right). This number of changes is well correlated with the cavity width (**Figure 7**, right).

In all but three cases, we observed prominences in AIA 304 Å associated with the cavities. We analyzed the relation between cavity center height and the heights of these prominences. A histogram of the difference between the two values is presented in **Figure 8**, left. In all but two cases, the prominence top was below the cavity center and central flows (in the two cases, the prominence was only slightly higher). The prominence top heights are also correlated with the cavity height: for higher cavities, higher prominences were observed. Sample images for two cavities are presented in **Figure 9**.

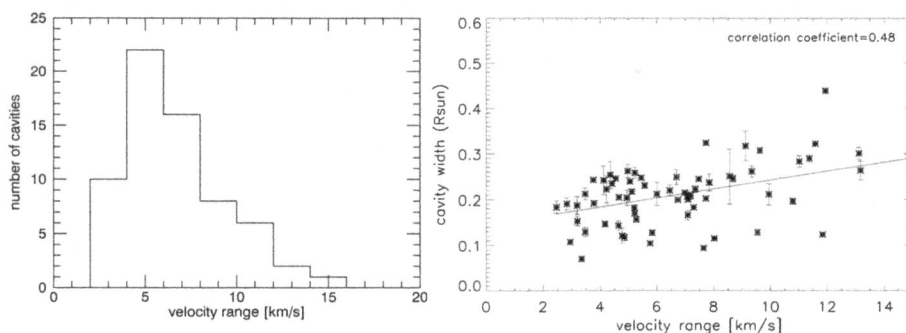

FIGURE 6 | Left: Histogram of velocity ranges, **Right:** Relation between velocity range and cavity width. Correlation coefficient is equal to 0.48.

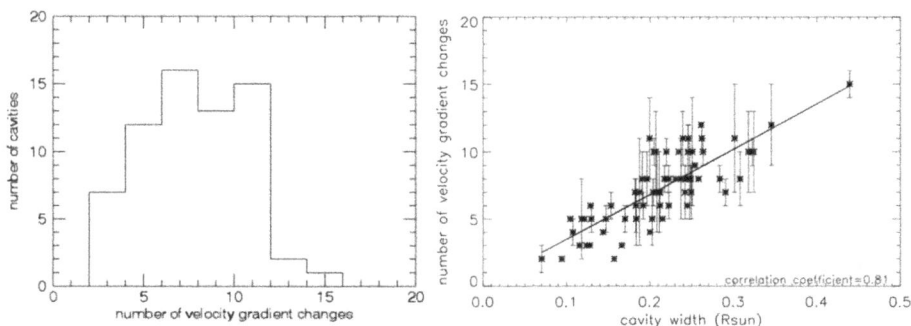

FIGURE 7 | Left: Histogram of number of velocity gradient changes. **Right:** Cavity width vs. number of velocity gradient changes. Correlation coefficient is equal to 0.81.

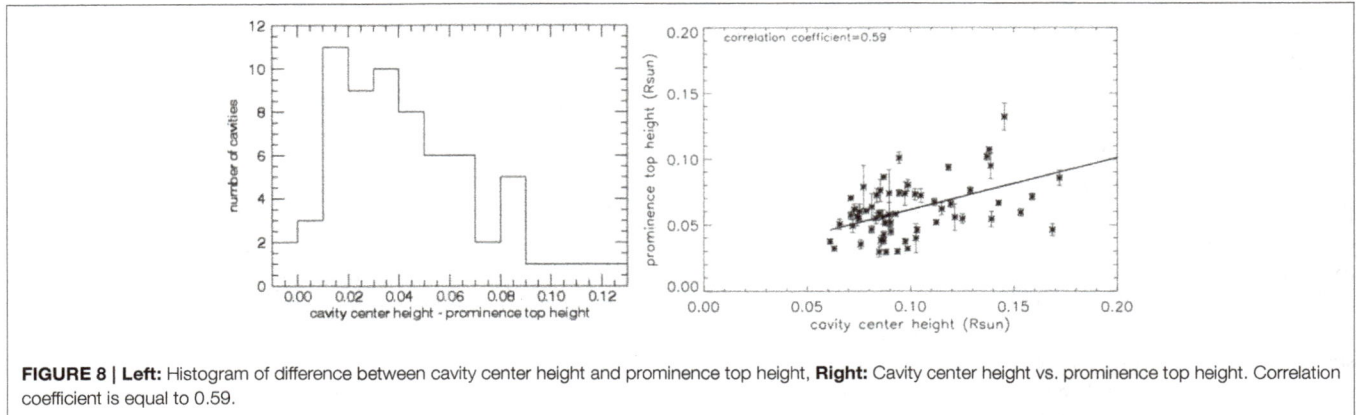

FIGURE 8 | Left: Histogram of difference between cavity center height and prominence top height, **Right:** Cavity center height vs. prominence top height. Correlation coefficient is equal to 0.59.

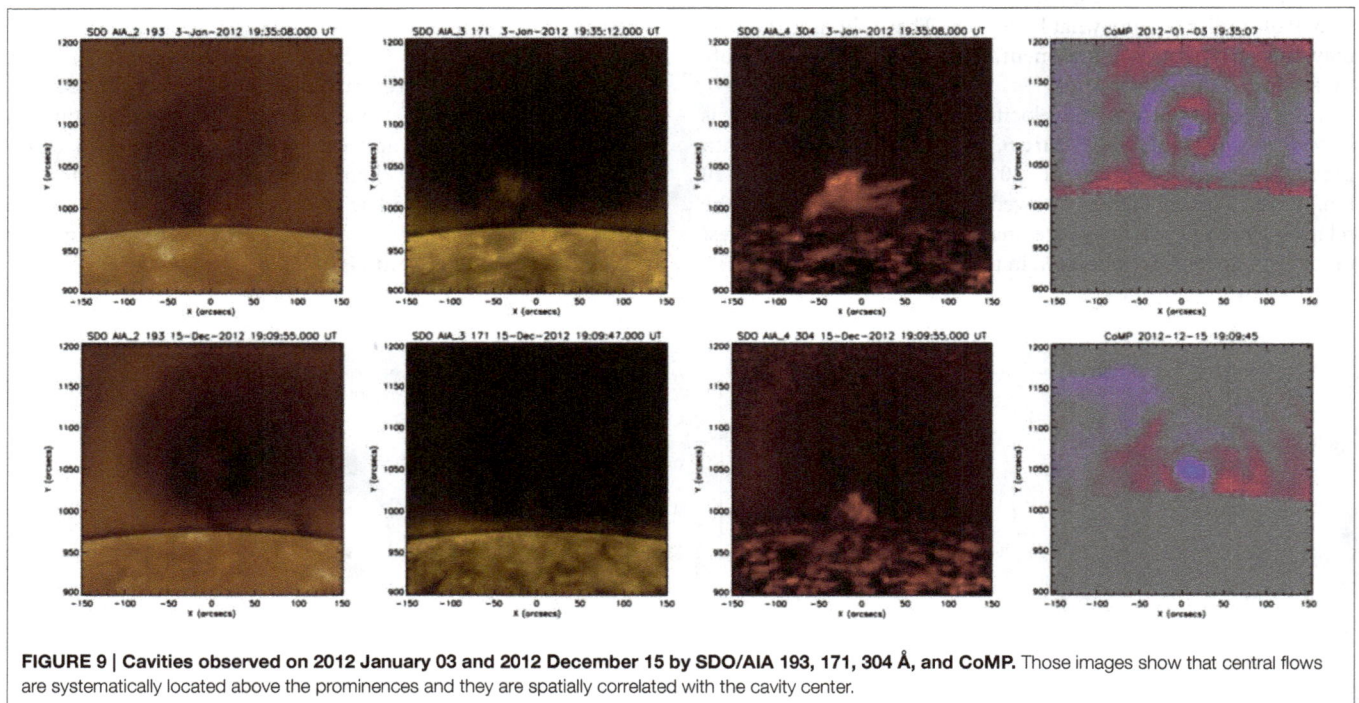

FIGURE 9 | Cavities observed on 2012 January 03 and 2012 December 15 by SDO/AIA 193, 171, 304 Å, and CoMP. Those images show that central flows are systematically located above the prominences and they are spatially correlated with the cavity center.

For two of the cavities we calculated DEM maps and profiles and estimated the average temperature using Equation 1. Maps characterizing the overall temperature are presented in **Figures 10, 11** (middle). In both cases, at least the central portion of the cavity seems to be filled with hotter material than its surroundings, which is consistent with results of Habbal et al. (2010). We found that the temperature in the center of these cavities were 2.3 and 2.1 MK for 2012 March 19 and 2015 June 05, respectively. The position of these hot cores is spatially correlated within the cavity seen in AIA 193 Å (**Figures 10, 11**, top panels, left) and flows (**Figures 10, 11**, top panels, right). In the second example, the position of the hot core is strongly correlated with the position of the part of the cavity with the lowest density and the strongest central flow observed by CoMP. DEM maps are presented in **Figures 10, 11** (bottom panels).

4. CONCLUSIONS

We have found that LOS-flow structure within coronal cavities is clearly related both to cavity morphology and to cool (prominence) and hot (corona) plasma distribution. Because the corona is magnetically dominated, these relations must ultimately derive from the magnetic field. The fact that prominences systematically lie below the center of cavities and the peak LOS coronal flow, indicates an association of the cavity center with a LOS-oriented, axial magnetic field. This, in combination with the concentric rings of flows surrounding this axis implies toroidal flux surfaces consistent with a magnetic flux rope topology. These observations thus provide complementary evidence toward conclusions based on previously discovered linear-polarization lagomorph signatures

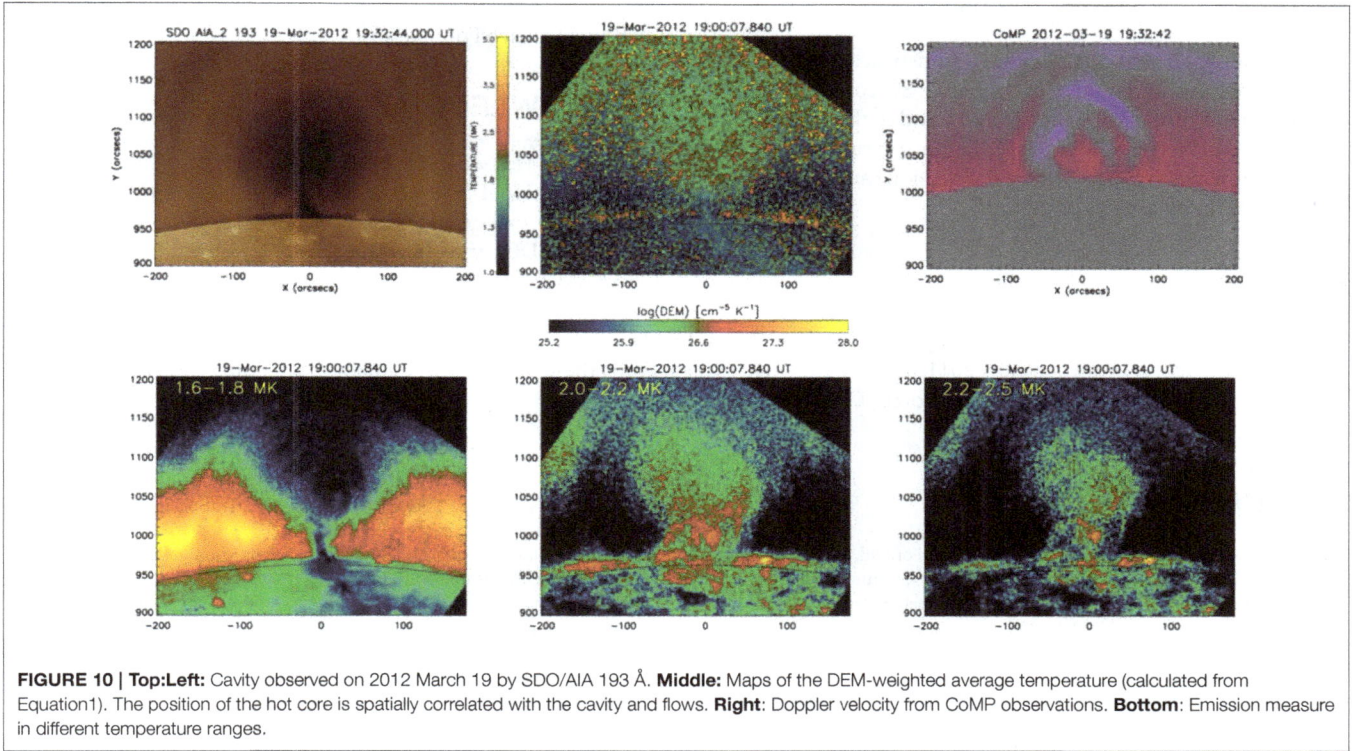

FIGURE 10 | Top:Left: Cavity observed on 2012 March 19 by SDO/AIA 193 Å. **Middle:** Maps of the DEM-weighted average temperature (calculated from Equation1). The position of the hot core is spatially correlated with the cavity and flows. **Right**: Doppler velocity from CoMP observations. **Bottom**: Emission measure in different temperature ranges.

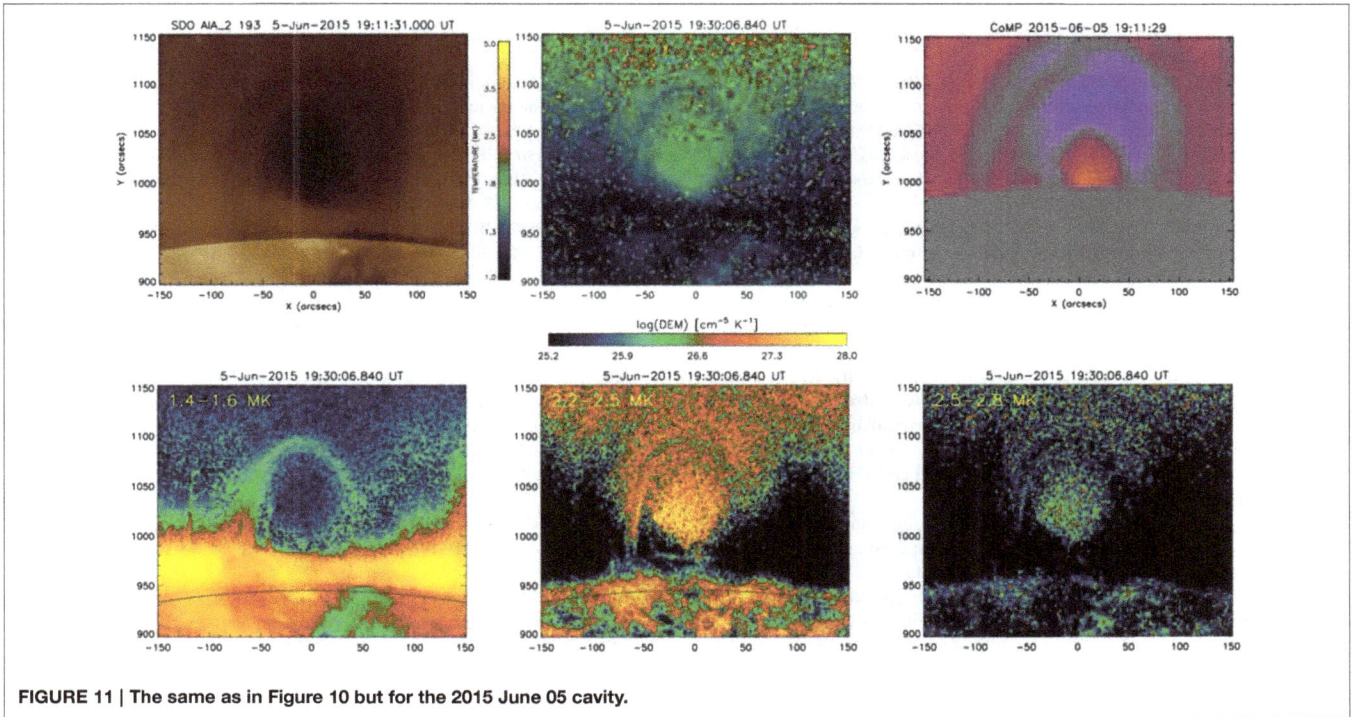

FIGURE 11 | The same as in Figure 10 but for the 2015 June 05 cavity.

within cavities. They also motivate efforts to obtain large-aperture telescope measurements of circular polarization [see further discussion in Gibson (2015)]. The details of how these field-aligned flows originate remain uncertain, however, and represent a challenge to magnetohydrodynamic models of prominence and cavity formation.

AUTHOR CONTRIBUTIONS

UBS led the design and carried out the statistical analysis, and led the interpretation and writing of the paper. SG contributed to the design and interpretation of the analysis, and to the writing of the paper. EC did the DEM analysis. All authors read and critically revised the paper, approved the final version, and agreed to be accountable for all aspects of the work.

FUNDING

UBS and EC acknowledge financial support from the Polish National Science Centre grant 2011/03/B/ST9/00104. SG acknowledges support from the Air Force Office of Space Research, FA9550-15-1-0030. NCAR is supported by the National Science Foundation.

ACKNOWLEDGMENTS

This work was enabled by participation of UBS and SG in the International Space Science Institute (ISSI) working group on coronal magnetism (2013–2014). AIA data were courtesy of NASA/SDO and the AIA, EVE, and HMI science teams. The CoMP data was provided courtesy of the MLSO, operated by the HAO, as part of the NCAR. We thank Giuliana de Toma and Steve Tomczyk in particular for assistance with these data, and Giuliana de Toma for internal HAO review.

REFERENCES

Bąk-Stęślicka, U., Gibson, S. E., Fan, Y., Bethge, C., Forland, B., and Rachmeler, L. A. (2013). The magnetic structure of solar prominence cavities: new observational signature revealed by coronal magnetometry. *Astrophys. J.* 770, 28. doi: 10.1088/2041-8205/770/2/L28

Bąk-Stęślicka, U., Gibson, S. E., Fan, Y., Bethge, C., Forland, B., and Rachmeler, L. A. (2014). "The spatial relation between EUV cavities and linear polarization signatures," in *Nature of Prominences and their role in Space Weather. Proceedings of the International Astronomical Union, IAU Symposium*, Vol. 300, eds B. Schmieder, J.-M. Malherbe, and S. T. Wu (Cambridge, UK: Cambridge University Press), 395–396.

Berger, T. E., Liu, W., and Low, B. C. (2012). SDO/AIA detection of solar prominence formation within a coronal cavity. *Astrophys. J.* 758, 37. doi: 10.1088/2041-8205/758/2/L37

Cheng, X., Zhang, J., Saar, S. H., and Ding, M. D. (2012). Differential emission measure analysis of multiple structural components of coronal mass ejections in the inner corona. *Astrophys. J.* 761, 62. doi: 10.1088/0004-637X/761/1/62

Dere, K. P., Landi, E., Mason, H. E., Monsignori Fossi, B. C., and Young, P. R. (1997). CHIANTI - an atomic database for emission lines. *Astron. Astrophys. Supp.* 125, 149. doi: 10.1051/aas:1997368

Forland, B. C., Gibson, S. E., Dove, J. B., Rachmeler L. A., and Fan, Y., (2013). Coronal cavity survey: morphological clues to eruptive magnetic topologies. *Sol. Phys.* 288, 603. doi: 10.1007/s11207-013-0361-1

Fuller, J., Gibson, S. E., de Toma, G., and Fan, Y. (2008). Observing the unobservable? Modeling coronal cavity densities. *Astrophys. J.* 678, 515. doi: 10.1086/533527

Fuller, J., and Gibson, S. E. (2009). A survey of coronal cavity density profiles. *Astrophys. J.* 700, 1205. doi: 10.1088/0004-637X/700/2/1205

Gibson, S. E. (2015). "Coronal cavities: observations and implications for the magnetic environment of prominences," in *Solar Prominences, Astrophysics and Space Science Library*, Vol. 415, eds J.-C. Vial and O. Engvold (Springer International Publishing Switzerland), 323.

Gibson, S. E., Foster, D., Burkepile, J., de Toma, G., and Stanger, A. (2006). The calm before the storm: the link between quiescent cavities and coronal mass ejections. *Astrophys. J.* 641, 590. doi: 10.1086/500446

Gibson, S. E., Kucera, T. A., Rastawicki, D., Dove, J., and de Toma, G. (2010). Three-dimensional morphology of a coronal prominence cavity. *Astrophys. J.* 724, 1133. doi: 10.1088/0004-637X/724/2/1133

Gibson, S. E., Kucera, T, White, S. M., Dove, J., Fan, Y., Forland, B. et al. (2016). FORWARD: A toolset for multiwavelength coronal magnetometry. *Front. Astron. Space Sci.* 3:8. doi: 10.3389/fspas.2016.00008

Habbal, S. R., Druckmueller, M., Morgan, H., Scholl, I., Rusin, V., Daw, A., et al. (2010). Total solar eclipse observations of hot prominence shrouds. *Astrophys. J.* 719, 1362. doi: 10.1088/0004-637X/719/2/1362

Heizel, P., Schmieder, B., Fárnik, F., Schwartz, P., Labrosse, N., Kotrč, P. et al. (2008). Hinode, TRACE, SOHO, and ground-based observations of a quiescent prominence. *Astrophys. J.* 686, 1383. doi: 10.1086/591018

Hudson, H. S., Acton, L. W., Harvey, K. A., and McKenzie, D. M. (1999). A stable filament cavity with a hot core. *Astrophys. J.* 513, 83. doi: 10.1086/311892

Hudson, H. S., and Schwenn, R. (2000). Hot cores in coronal filament cavities. *Adv. Space Res.* 25, 1859. doi: 10.1016/S0273-1177(99)00618-3

Illing, R. M., and Hundhausen, J. R. (1986). Disruption of a coronal streamer by an eruptive prominence and coronal mass ejection. *J. Geophys. Res.* 91, 10951.

Jibben, P. R., Reeves, K. K., and Su, Y. (2016). Evidence for a magnetic flux rope in observations of a solar prominence-cavity system. *Front. Astron. Space Sci.* 3:10. doi: 10.3389/fspas.2016.00010

Lemen, J. R., Title, A. M., Akin, D. J., Boerner, P. F., Chou, C., Drake, J. F., et al. (2012). The Atmospheric Imaging Assembly (AIA) on the Solar Dynamics Observatory (SDO). *Sol. Phys.* 275, 17. doi: 10.1007/s11207-011-9776-8

Low, B. C. (1994). Magnetohydrodynamic processes in the solar corona: flares, coronal mass ejections, and magnetic helicity. *Phys. Plasmas* 1, 1684. doi: 10.1063/1.870671

Low, B. C., and Hundhausen, J. R. (1995). Magnetostatic structures of the solar corona. 2: the magnetic topology of quiescent prominences. *Astrophys. J.* 443, 818.

Maričič, D., Vršnak, B., Stanger, A. L., and Veronig, A. M. (2004). Coronal mass ejection of 15 may 2001: I. evolution of morphological features of the eruption. *Sol. Phys.* 225, 337. doi: 10.1007/s11207-004-3748-1

Marqué, C. (2004). Radio metric observations of quiescent filament cavities. *Astrophys. J.* 602, 1037. doi: 10.1086/381085

Marqué, C., Lantos, P., and Delaboudinère, J.-P. (2002). Multi wavelength investigation of the eruption of a sigmoidal quiescent filament. *Astron. Astrophys.* 387, 317. doi: 10.1051/0004-6361:20020309

McCauley, P. I., Su, Y. N., Schanche, N., Evans, K. E., Su, C., McKillop, S., et al. (2015). Prominence and filament eruptions observed by the solar dynamics observatory: statistical properties, kinematics, and online catalog. *Sol. Phys.* 290, 1703. doi: 10.1007/s11207-015-0699-7

Pesnell, W. D., Thompson, B. J., and Chamberlin, P. C. (2012). The Solar Dynamics Observatory (SDO). *Sol. Phys.* 275, 3. doi: 10.1007/s11207-011-9841-3

Reeves, K. K., Gibson, S. E., Kucera, T. A., Hudson, H. S., and Kano, R. (2012). Thermal properties of a solar coronal cavity observed with the X-ray telescope on Hinode. *Astrophys. J.* 746, 146. doi: 10.1088/0004-637X/746/2/146

Régnier, S., Walsh, R. W., and Alexander, C. E. (2011). A new look at a polar crown cavity as observed by SDO/AIA. Structure and dynamics. *Astron. Astrophys.* 533, L1. doi: 10.1051/0004-6361/201117381

Schmit, D. J., Gibson, S. E., Tomczyk, S., Reeves, K. K., Sterling, A. C., Brooks, D. H., et al (2009). Large-scale flows in prominence cavities. *Astrophys. J.* 700:L96. doi: 10.1088/0004-637X/700/2/L96

Su, Y., van Ballegooijen, A., McCauley, P., Ji, H., Reeves, K. K., and DeLuca, E. E. (2015). Magnetic structure and dynamics of the erupting solar polar crown prominence on 2012 march 12. *Astrophys. J.* 807, 144. doi: 10.1088/0004-637X/807/2/144

Tandberg-Hanssen, E. (1995). *The Nature of Solar Prominences, 2nd Edn.* Dordrecht: Kluwer.

Tian, H., Tomczyk, S., McIntosh, S. W., Bethge, C., de Toma, G., and Gibson, S. E. (2013). Observations of coronal mass ejections with the coronal multichannel polarimeter. *Sol. Phys.* 288, 637. doi: 10.1007/s11207-013-0317-5

Tomczyk, S., Card, G. L., Darnell, T., Elmore, D. F., Lull, R., Nelson, P. G. et al (2008). An instrument to measure coronal emission line polarization. *Sol. Phys.* 247, 411. doi: 10.1007/s11207-007-9103-6

Vaiana, G. S., Davis, J. M., Giacconi, R., Krieger, A. S., Silk, J. K., Timothy, A. F., et al. (1973a). Identification and analysis of structures in the corona from X-ray photography. *Sol. Phys.* 32, 81. doi: 10.1007/BF00152731

Vaiana, G. S., Krieger, A. S., and Timothy, A. F. (1973b). X-Ray observations of characteristic structures and time variations from the solar corona: preliminary results from SKYLAB. *Astrophys. J.* 185, 47. doi: 10.1086/181318

Vršnak, B., Maričič, D., Stanger, A. L., and Veronig, A. M. (2004). Coronal mass ejection of 15 may 2001: II. Coupling of the Cme acceleration and the flare energy release. *Sol. Phys.* 225, 355. doi: 10.1007/s11207-004-4995-x

von Kluber, H. (1961). Photometric investigation of the inner solar corona using an eclipse plate of 1927 June 29. *Monthly Notices R. Astron. Soc.* 123, 61. doi: 10.1093/mnras/123.1.61

Waldmeier, M. (1941). *Ergebnisse und Probleme der Sonnenforschung.* Leipzig: Becker & Erler kom.-ges.

Waldmeier, M. (1970). The structure of the monochromatic corona in the surroundings of prominences. *Sol. Phys.* 15, 167.

Weber, M. A., DeLuca, E. E., Golub, L., and Sette, A. L. (2004). "Temperature diagnostics with multichannel imaging telescopes," in *Multi-Wavelength Investigations of Solar Activity, IAU Symposium,* Vol. 223, eds A. V. Stepanov, E. E. Benevolenskaya, and A. G. Kosovichev (Cambridge, UK: Cambridge University Press), 312–328.

Wesley, W. H. (1927). *Memoirs of the British Astronomical Association, Vol. 64, Appendix.*

Williamson, N. K., Fullerton, C. M., and Billings, D. E. (1961). Coronal emission in the vicinity of quiescent prominences. *Astrophys. J.* 133, 973. doi: 10.1086/147102

Conflict of Interest Statement: The authors declare that the research was conducted in the absence of any commercial or financial relationships that could be construed as a potential conflict of interest.

The reviewer CJ and handling Editor declared a past collaboration and the handling Editor states that the process nevertheless met the standards of a fair and objective review.

4

Evidence for a Magnetic Flux Rope in Observations of a Solar Prominence-Cavity System

Patricia R. Jibben[1]*, Katharine K. Reeves[1] and Yingna Su[2]

[1] Harvard-Smithsonian Center for Astrophysics, Cambridge, MA, USA, [2] Key Laboratory for Dark Matter and Space Science, Purple Mountain Observatory, Chinese Academy of Sciences, Nanjing, China

Edited by:
Stephen M. White,
Air Force Research Laboratory, USA

Reviewed by:
Gordon James Duncan Petrie,
National Solar Observatory, USA
Meng Jin,
University Corporation for
Atmospheric Research/Lockheed
Martin Solar and Astrophysics
Laboratory, USA

*Correspondence:
Patricia R. Jibben
pjibben@cfa.harvard.edu

Specialty section:
This article was submitted to
Stellar and Solar Physics,
a section of the journal
Frontiers in Astronomy and Space
Sciences

Coronal cavities are regions of low coronal emission that usually sit above solar prominences. These systems can exist for days or months before erupting. The magnetic structure of the prominence-cavity system during the quiescent period is important to understanding the pre-eruption phase. We describe observations of a coronal cavity situated above a solar prominence observed on the western limb as part of an Interface Region Imaging Spectrograph (IRIS) and Hinode coordinated Observation Program (IHOP 264). During the observation run, an inflow of hot plasma observed by the Hinode X-Ray Telescope (XRT) envelopes the coronal cavity and triggers an eruption of chromospheric plasma near the base of the prominence. During and after the eruption, bright X-ray emission forms within the cavity and above the prominence. IRIS and the Hinode EUV Imaging Spectrometer (EIS) show strong blue shifts in both chromospheric and coronal lines during the eruption. The Hinode Solar Optical Telescope (SOT) Ca II H-line data show bright emission during the ejection with complex, turbulent, flows near the prominence and along the cavity wall. These observations suggest a cylindrical flux rope best represents the cavity structure with the ejected material flowing along magnetic field lines supporting the cavity. We also find evidence for heating of the plasma inside the cavity after the flows. A model of the magnetic structure of the cavity comprised of a weakly twisted flux rope can explain the observed loops in the X-ray and EUV data. Observations from the Coronal Multichannel Polarimeter (CoMP) are compared to predicted models and are inconclusive. We find that more sensitive measurements of the magnetic field strength along the line-of-sight are needed to verify this configuration.

Keywords: sun, prominence, coronal cavity, magnetic field modeling, magnetic field

1. INTRODUCTION

Solar prominences are sheets of cool dense plasma suspended in the solar corona observed on the limb. Their formation and stability require several mechanisms working in tandem. It is widely accepted that the magnetic field provides the structural support of the prominence but direct observations of the magnetic field in the corona are not currently available. Comprehensive reviews of prominence systems and their dynamics are provided by Martin (1990), Mackay et al. (2010), Parenti (2014), Priest (2014), and Vial and Engvold (2015) and we will provide a brief review here. Prominences only form between regions of opposite magnetic field polarity. In other words, along the polarity inversion line (PIL; Smith, 1968; Martin, 1973). But not all PILs will exhibit prominence formation; another condition required is a predominant transverse magnetic field aligned with the long axis of the prominence. The path in the chromosphere where this happens is referred to as a filament channel (Martin, 1990). Along filament channels, there is a significant decrease in the number of observed spicules compared to the surrounding area (Martin, 1990). Reduced spicule activity indicates weak radial magnetic fields and quiescent prominences may form along giant cell boundaries that separate unipolar magnetic field regions (Malherbe and Priest, 1983; Schröter et al., 1987). Bipoles within the filament channel are often characterized by a bald-patch topology (Titov and Démoulin, 1999) where the magnetic field at the photosphere is largely horizontal and points from negative to positive polarity (López Ariste et al., 2006). The long-term converging patches of opposite polarity flow into juxtaposition along the PIL and as they encounter one another, they disappear concurrently at their boundaries (Martin, 1990). Finally, prominences will quickly dissipate unless there is a closed arcade of magnetic field lines overlying and connecting regions of opposite polarity. The closed loops not only hold down the prominence material but they also create a magnetically stable system in which the cool prominence material interacts with the hot plasmas in the corona.

Under the arcade, a region of reduced coronal emission (the coronal cavity) can develop above quiescent prominences (Vaiana et al., 1973). Despite their reduced coronal emissions, these cavities are filled with complex, twisted structures when observed in white light eclipse data and are distinct from the magnetic structures defining the rest of the overlying arcade that forms the base of streamers as well as the boundaries of streamers (Habbal et al., 2010). This suggests the overlying arcade and underlying prominence are independent magnetic structures that can interact via magnetic reconnection at their boundaries. Therefore, the magnetic structure of the cavity and prominence system prior to an eruption is important to understanding how instabilities form.

A magnetic flux rope is often used to model the prominence and cavity system (Priest et al., 1989; Rust and Kumar, 1994; van Ballegooijen, 2004), with much of the prominence material sitting in the dips of the magnetic field lines (Kuperus and Raadu, 1974; Pneuman, 1983; Priest et al., 1989; van Ballegooijen and Martens, 1989; Rust and Kumar, 1994; Low and Hundhausen, 1995; Aulanier et al., 1998; Chae et al., 2001; van Ballegooijen, 2004; Gibson et al., 2006; Dudík et al., 2008). Fan and Gibson (2006) modeled a prominence as a twisted flux rope and found that a current sheet forms within the flux rope cavity along a bald-patch separatrix surface (BPSS), composed of the field lines that graze the anchoring lower boundary, enclosing the detached helical field that supports the prominence. They further show that resistive dissipation of the current sheet would produce a hot sheath surrounding the prominence material in the cavity, which could provide an explanation for the observed development of X-ray bright cores within a coronal cavity (Hudson et al., 1999; Hudson and Schwenn, 2000; Reeves et al., 2012). Su et al. (2015) constructed a series of magnetic field models with different configurations based on the observed photospheric magnetogram for a polar crown prominence, and they found that the model with a twisted flux rope best matches the observations.

Coronal magnetic fields provide the structure and support for the coronal cavity, but measuring coronal magnetic fields is difficult to do (Lin et al., 2004). Fortunately, some information about the coronal magnetic fields making up coronal cavities has been achieved with the Coronal Multi-Channel Polarimeter (Tomczyk et al., 2008; CoMP). A recent statistical study by Bąk-Stęślicka et al. (2013) found that quiescent prominence cavities consistently posses a "lagomorphic" signature in linear polarization indicating twist or shear extending up into the cavity above the PIL. They also compared the CoMP observations with synthetic CoMP-like data created using a forward magnetohydrodynamics (MHD) model and concluded that a cylindrical magnetic flux rope better represents polar-crown prominence cavities.

In this paper, we present chromospheric and coronal observations of a prominence-cavity system observed on the west limb. We develop a magnetic field model of the system based on these observations. We utilize data from Hinode (Kosugi et al., 2007), Interface Region Imaging Spectrograph (IRIS; De Pontieu et al., 2014), and the Solar Dynamics Observatory's Atmospheric Imaging Assembly (AIA; Lemen et al., 2012) and Helioseismic and Magnetic Imager (HMI; Schou et al., 2012). Together, these instruments provide near simultaneous multithermal observations of the prominence-cavity system and its surroundings. The X-Ray Telescope (XRT; Golub et al., 2007) observes some of the hottest coronal temperatures between 2 and 10 MK. The EUV Imaging Spectrometer (EIS; Culhane et al., 2007) takes spectral data from the transition region to coronal temperatures. Finally, the Solar Optical Telescope (SOT; Tsuneta et al., 2008) and IRIS image chromospheric and transition region plasmas. We use photospheric line-of-sight (LOS) magnetic field data to derive a model of the structure of the system with the assumption it is a magnetic flux rope under an arcade. We vary the axial and poloidal fluxes of the model to best fit the observations and then we compare the coronal features predicted by the model with CoMP observations. We find evidence of heating within the cavity during an eruption and we find evidence that magnetic bipoles within the filament channel exhibit a bald-patch topology. Therefore, we conclude that a weakly twisted magnetic flux rope best represents the prominence-cavity system but further instrumentation is needed to resolve coronal magnetic signatures of a quiescent flux rope within the corona.

2. MATERIALS AND METHODS

The observations we use were part of an IRIS and Hinode Joint Observation Program (IHOP 264[1]) that included observations from all three Hinode instruments. The IHOP was run three times pointing at the same prominence on the west limb between 9 and 10 October 2014. We present the data taken between 18 and 22 UT on 9 October because CoMP was also observing at this time. We do not use the other two data sets because they either do not have corresponding CoMP observations or in the case of the 10 October data, a non-related filament erupted.

---

[1]http://www.isas.jaxa.jp/home/solar/hinode_op/hop.php?hop=0264.

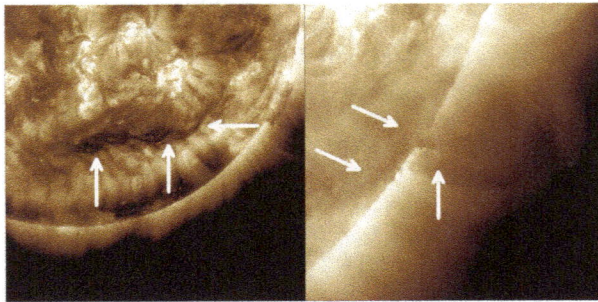

FIGURE 1 | AIA 193 Å observations of the filament. **Left**: 4 October 2014 at 18:18:06 UT. **Right**: 9 October 2014 at 18:24:42 UT. The arrows point to the prominence location.

2.1. Hinode XRT, EIS, and SOT Data Reduction

The XRT observations used in this study include 8 s thin-Be exposures at 60 s cadence. The field of view is $\approx 790'' \times 790''$ and the images are binned 2×2 giving a resolution of $2''0572$ per pixel. Observations were paused during times when Hinode passed through the South Atlantic Anomaly (SAA) causing 20–30 min gaps in the data. The data is processed using standard data reduction routines provided by the XRT team (Kobelski et al., 2014) and aligned using the database developed by Yoshimura and McKenzie (2015) distributed in SolarSoft (Freeland and Handy, 1998). Either individual or 5 min averaged data were spatially enhanced using the *à trous* wavelet transform with a cubic spline scaling function. See page 29, and Appendix A in Stark and Murtagh (2002) for a complete description of the routine. This method separates the image into different spatial scales based on pixel size along with a residual image containing the portion of the image outside of the spatial scales. We display the data without the residual image using a log-like scale so that only bright features with intensity gradients that vary over 1–3 pixels remain. Regions that do not change rapidly are threshold to white or are set to be transparent.

The EIS data utilize the $2''$ slit with 50 s exposures, 75 raster positions with binning along the x-direction giving a $300'' \times 512''$ FOV. Two raster scans are used in this study. The two scans were taken between 18:16–19:21 UT and 19:21–20:26 UT. The scans are processed in IDL using software provided by the EIS team and a thorough discussion of the routines is provided in the EIS data analysis guide[2]. A brief overview is given here.

The data are calibrated using EIS_PREP with the default parameters outlined in the analysis guide. We use three coronal lines for this study, Fe XII 195.12 Å, Fe XIII 202.04 Å, and Fe XV 284.16 Å. The Fe XV 284.16 Å spectra is imaged on a different camera resulting in a slightly different field of view. The lines are fit with a Gaussian profile using EIS_AUTO_FIT routine (Young, 2013). From this, intensity, LOS Doppler velocity maps and line width maps are created.

The default velocity scale used in the data reduction software is derived using the Kamio method (Kamio et al., 2010). We

[2]http://solarb.mssl.ucl.ac.uk:8080/eiswiki/Wiki.jsp?page=EISAnalysisGuide.

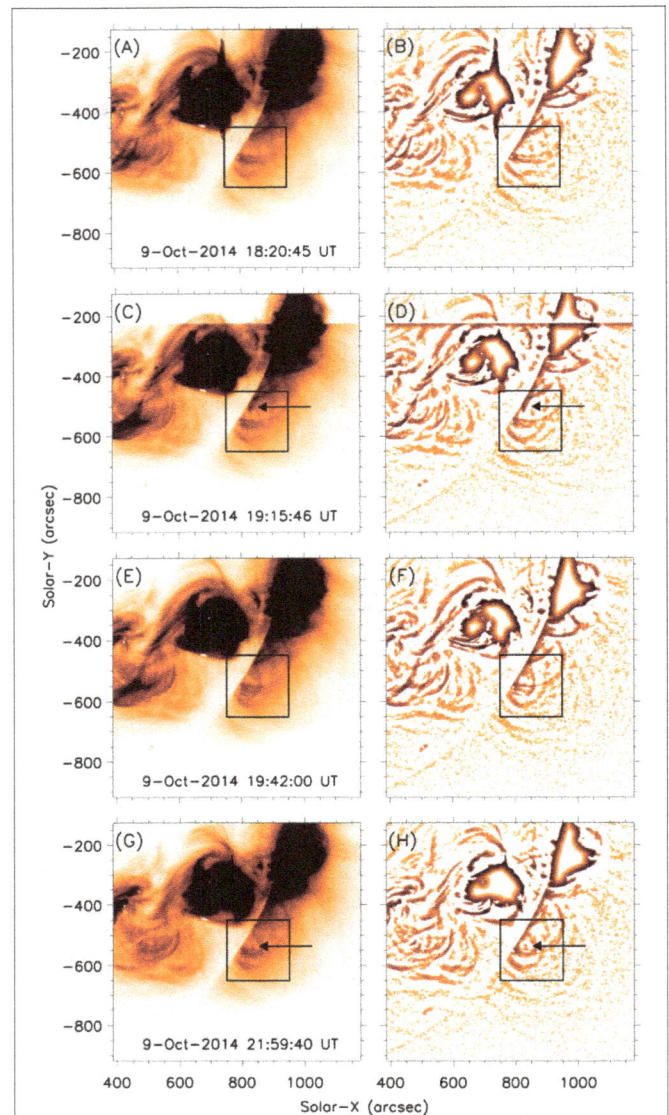

FIGURE 2 | **Left column:** Hinode XRT thin-Be (inverse log) intensity (5 min average) observations of the coronal cavity. **Right column**: Spatially enhanced image with the background emission threshold to white. Panels **(A,B)** shows the cavity prior to the eruption. Panels **(C,D)** shows the initial phase of the eruption with an arrow pointing to an increase in X-ray emission. Panels **(E–H)** show the cavity during and after the eruption with the arrows pointing to an increase in X-ray emission within the cavity.

update the velocity scale using a patch of quiet sun. We assume that the LOS velocities will average to zero in this region for each of the spectral lines (Warren et al., 2011). The off limb patch contains 120 pixels in the y-direction and we bin the data by 20 pixels at each raster position, a Gaussian profile is fit to the average of the bottom 6 binned pixels, thus defining the reference wavelength. We perform this procedure for each line separately because an absolute velocity scale cannot be derived for the Fe XV 284.16 Å line with respect to the other two since it is on a separate camera. The error in the

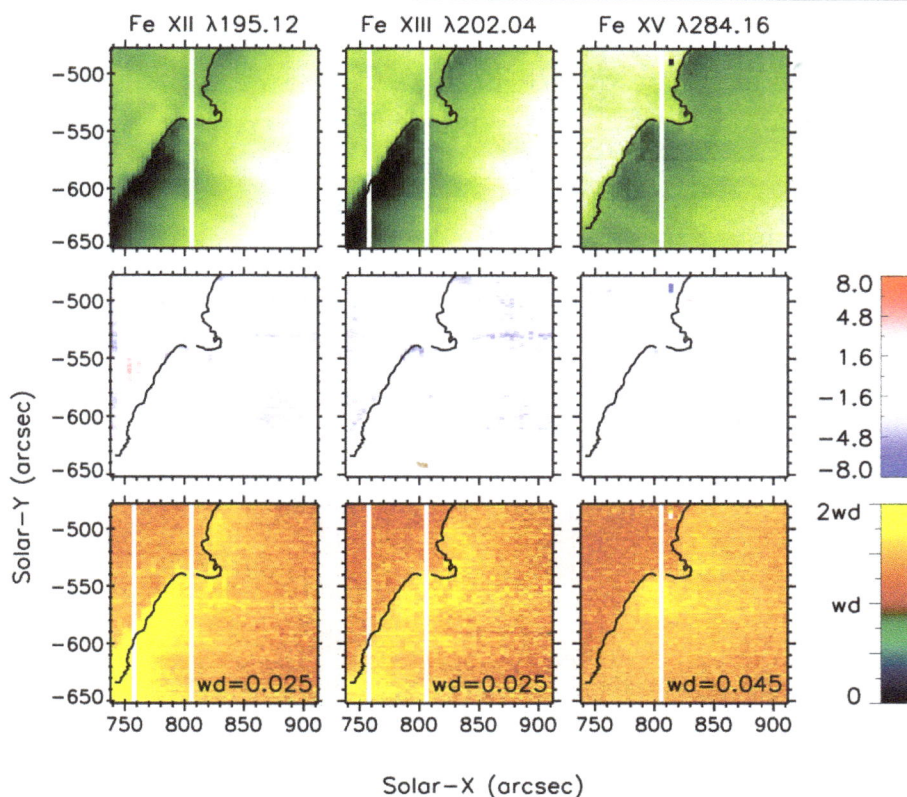

FIGURE 3 | A subfield of the EIS raster scan starting from the left at 18:16 UT and finishing at 19:21 UT. Top row: EIS intensity (inverse log) for Fe XII 195.12 Å, Fe XIII 202.04 Å, and Fe XV 284.16 Å. **Middle row**: Doppler velocity maps for the three lines. **Bottom row**: Line widths of the given lines with the median line width (wd) given in each window. The white stripes are regions of missing data.

velocity is provided from the routines used to generate the velocities.

The SOT data consists of $\approx 112'' \times 112''$ Ca II H-line images taken at 30 s cadence between 18 and 20 UT and then 60 s cadence from 20 to 22 UT. The data are calibrated using routines provided in SolarSoft. In addition, the images are spatially enhanced using the same method used on the XRT data except the residual image is preserved. Small scale features are enhanced before recombining the image, which acts to sharpen the image while preserving information about the enhancement. After the images are spatially enhanced, a radial density filter is applied to reduce the intensity of the disk and spicule regions. The radial density filter applied is similar to the one described in Berger et al. (2010). After the images are scaled and sharpened they are aligned using the SolarSoft routine `fg_rigidalign.pro`. There was a shift in the SOT pointing between the 21:07 and 21:08 frames. The images after 21:08 UT are aligned manually by aligning features visible in both images. Features in SOT are compared with features in AIA 211 Å to coalign SOT with other instruments.

2.2. IRIS Data Reduction

IRIS performed a 16-step coarse raster from 18:24 UT to 21:58 UT. The telescope was pointed on the west limb at $803''$, $-546''$

capturing most of the prominence. The raster field of view was $30'' \times 119''$ with a raster step cadence of 9.4 s making a raster cadence of 150 s with 8 s exposures. Two broadband filter (2796 and 1400 Å) slit-jaw images (SJI) were taken at a cadence of 19 s with a $119'' \times 119''$ field of view. The calibrated level 2 data were used in this study and downloaded from the IRIS website[3]. The prominence material is significantly dimmer than on-disk regions for the chosen lines. To simultaneously observe both regions, we apply an intensity filter to the Si IV 1400 Å images decreasing the on-disk and spicule intensities. The on-disk intensity is decreased by 90% of its original intensity. The intensity of the spicule region is linearly increased from 10 to 100% at the edge of the spicule region. The intensity of the prominence and off limb features do not have their intensity altered. The resultant images are displayed using a square-root inverse intensity scaling.

We use the Mg II 2796 Å and Si IV 1394 Å spectra for this study. The Mg II h and k lines are formed at chromospheric temperature (10^4 K). Emission from the Si IV 1394 Å line form in the prominence transition region (PCTR). The UV continuum at 1400 Å formed in the lower chromosphere is not present for a

---

[3]https://iris.lmsal.com/index.html.

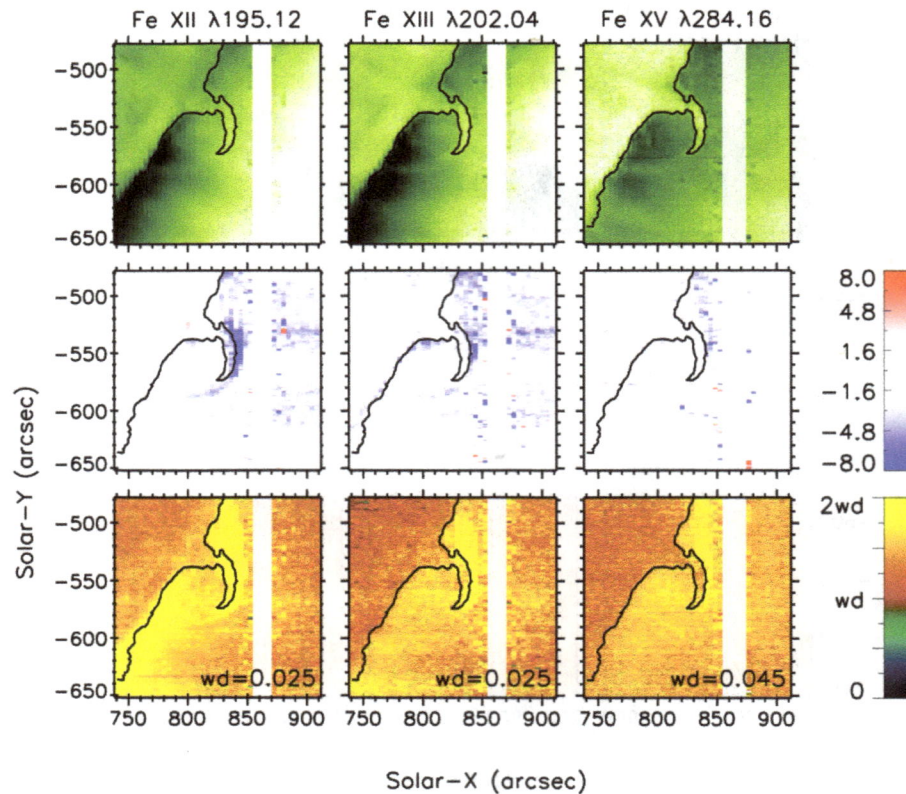

FIGURE 4 | A subfield of the EIS raster scan starting from the left at 19:22 UT and finishing at 20:27 UT. Top row: EIS intensity (inverse log) for Fe XII 195.12 Å, Fe XIII 202.04 Å, and Fe XV 284.16 Å. **Middle row**: Doppler velocity maps for the three lines. **Bottom row**: Line widths of the given lines with the median line width (wd) given in each window. The white stripes are regions of missing data.

prominence observed at the limb contrary to observations on the disk (Schmieder et al., 2014).

We perform a relative wavelength calibration to measure the Doppler velocities of the eruption. The reference wavelength for each spectral line window is selected at the centroid of the line profile averaged over the on-disk scan positions for each raster scan. Therefore, the relative Doppler velocities are measured with respect to the quiet Sun regions. The absolute uncertainty of the relative wavelength calibration is estimated to be 4 km s^{-1} by Liu et al. (2015). Their estimates include a wavelength shift, 20 mÅ from disk center to the limb and the IRIS orbital thermal variation of 3 km s^{-1}. In that paper, portions of the IRIS slit were on the disk throughout the observations.

2.3. CoMP Data Reduction

CoMP makes daily polarimetric (Stokes I, Q, U, V) of the forbidden lines of Fe XIII at 1074.4 nm and 1078.9 nm with a FOV of 1.4–2 $R_\odot$. The degree of linear polarization (L/I) constrains the direction of the plane-of-sky (POS) magnetic field. The amount of circular polarization (V/I) provides information about the strength of the magnetic field along the LOS. The CoMP data consists of QuickInvert data of the Fe XIII 1074.7 nm coronal emission line. The data were downloaded from the High

Altitude Observatory/Mauna Loa Solar Observatory website[4]. The Quick Invert files contain five images: Stokes I, Q, U, linear polarization (L), and magnetic field azimuth. The file is read into the FORWARD (Gibson et al., 2016) toolset where L/I is calculated and used for this study.

2.4. Magnetic Model of Prominence-Cavity System

A three-dimensional magnetic model of the prominence-cavity system is constructed using the Coronal Modeling System (CMS) developed by van Ballegooijen (2004). The CMS model assumes the prominence material is supported against gravity by a helical flux rope. The model utilizes SDO/HMI magnetograms to establish the magnetic field strength and topology at the photosphere where it is assumed to be radial. The model is constructed by inserting a flux rope along the PIL under a potential field representing the overlying coronal arcade. The axial flux (in Mx) and the poloidal flux per unit length (in Mx/cm) along the filament are set to an initial value and then magnetofrictional relaxation is used to drive the system to a nonlinear force free state. Detailed description of the methodology can be found in the

---

[4]http://www2.hao.ucar.edu/mlso/mlso-home-page.

literature and references therein (Su et al., 2009, 2011; Su and van Ballegooijen, 2012) and we describe the method briefly below.

First, the potential field is computed from the observed magnetic maps. Then, by appropriate modifications of the vector potentials a "cavity" is created above the selected path, and a thin flux bundle, representing the axial flux of the flux rope, is inserted into the cavity with the footpoints of the flux rope embedded in regions near the PIL. The footpoints are chosen so that the flux rope begins and ends in a patch of positive and negative polarity, respectively. To preserve the radial component of the inner boundary, the patch representing the footpoints is removed/added from the photospheric flux distribution and is equal to the axial flux of the inserted flux rope. The poloidal flux is inserted by adding circular loops around the flux bundle (Su et al., 2015). The inserted flux rope is not in force-free equilibrium. We use magneto-frictional relaxation to drive the field toward a force-free state. This method is an iterative relaxation method (van Ballegooijen et al., 2000) specifically designed for use with vector potentials. Magnetofriction has the effect of expanding the flux rope until its magnetic pressure balances the magnetic tension applied by the surrounding potential arcade. Significant magnetic reconnection between the inserted flux rope and the ambient flux may occur during the relaxation process.

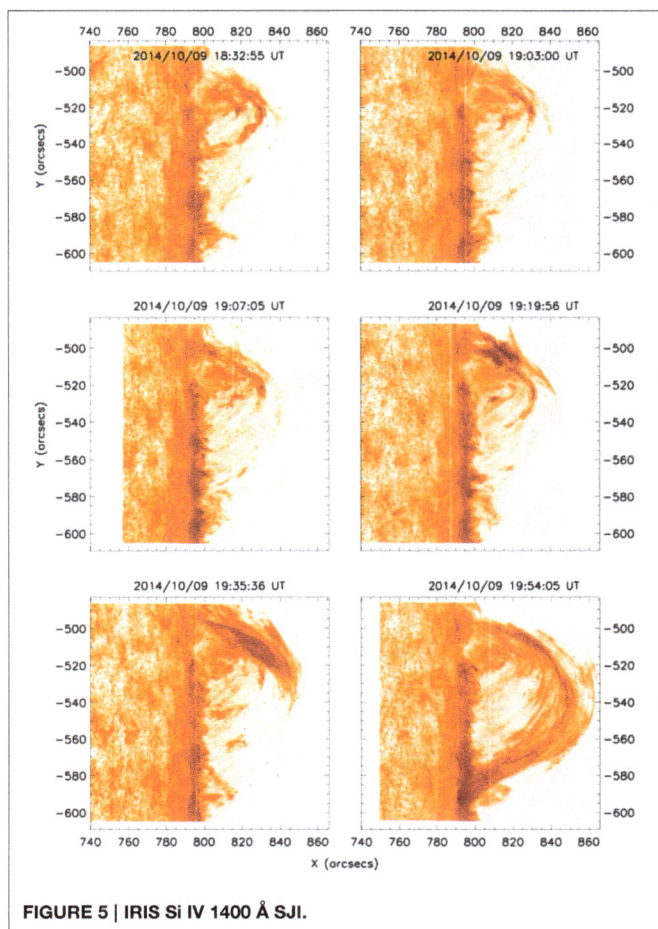

FIGURE 5 | IRIS Si IV 1400 Å SJI.

Therefore, the end points of the flux rope in the relaxed model may be different from that in the original model (Su et al., 2015).

The region around the filament is modeled with high spatial resolution (HIRES) on a variable grid, while more distant regions have a lower resolution on a uniform grid. The HIRES region may contain electric currents, whereas the global model is a current-free potential field. The lower boundary condition for the HIRES region is derived by combining several LOS photospheric magnetograms obtained with the SDO/HMI as these provide better signal to noise than vector magnetograms in quiet sun regions. Since the prominence is observed near the west limb, we use magnetograms that are taken several days before the prominence reaches the limb. We combine four magnetograms, each taken at 19:00 UT, between 2014 October 2–5 to construct a high-resolution map of the radial component B_r of the magnetic field as a function of longitude and latitude at the lower boundary of the HIRES region (0.002 $R_\odot$) (Su et al., 2015). The high-resolution computational domain extends about 117° in longitude, 36° in latitude, and up to 2.05 $R_\odot$ from the Sun. We use the corresponding HMI synoptic map of B_r to compute a low-resolution (1°) global potential field, which provides the side boundary conditions for the HIRES domain, and allows us to trace field lines that pass through the side boundaries of the HIRES region (Su et al., 2015).

We construct a series of models with different combinations of axial and poloidal fluxes of the inserted flux rope. We compare each model with the size, location, and shape of the filament channel and cavity, including the emission structure on the two sides of the filament channel as well as the trajectory of plasma motions. We require the best-fit model to have an overall structure consistent with the observed LOS velocities observed by IRIS and EIS.

2.4.1. Forward Modeling of Stokes Profiles

To compare the models with the CoMP we calculate what the expected L/I would be for our models. To calculate the Stokes vector produced along a given LOS for the magnetic field models, we use the forward models developed by Judge and Casini (2001) and implemented in the FORWARD (Gibson et al., 2016) suite of IDL codes. The FORWARD database is available to the public[5] and details are provided at the website and in the literature including Rachmeler et al. (2013). A brief summary is provided here.

The forward code uses the magnetic field, temperature, density, and velocity along the LOS to calculate the level population and emitted polarization profile for the Fe XIII 1074.7 nm transition. For our model, we assume an exponential isothermal atmosphere with a temperature of 1.5 MK and use `HYDROCALC.pro` to calculate the remaining parameters required for the forward calculations. It outputs Stokes I, Q, U, V and for the purpose of this study we use the relative linear polarization (L/I) and relative circular polarization (V/I). The models were based on SDO/HMI magnetograms and were

---

[5]http://www.hao.ucar.edu/FORWARD/.

initially rotated to the 2014-10-04 23:59 UT. To compare with the CoMP observations, the models were rotated to the limb so that they match the observation time of the CoMP QuickInvert data, at 21:13 UT.

3. RESULTS

3.1. Observations

The prominence is composed of two linear structures with a N-S oriented component and an E-W component with a southern pitch. **Figure 1** shows what the prominence looked like in AIA 193 Å on 4 October 2014 at 18:18 UT (left) and at the beginning of the observation campaign at 18:24 UT on October 9 (right). It is sandwiched between several active regions to the north and the polar coronal hole to the south. The Sun is active during this period with small scale flares and coronal mass ejections associated with the active regions and numerous filaments on the disk.

The X-ray data displayed in **Figure 2** shows a small coronal cavity associated with the prominence. **Figures 2A,B** demonstrate how the region (inside black box) looked near the beginning of the observation run. There are several bright formations and it is not apparent which structures, if any, are associated with the prominence. The region remains stable until at 19:09 UT when there is an increase in X-ray emission just above the limb. The black arrows in **Figures 2C,D** point to this

region of increased X-ray emission in the 19:15 UT image. This is the last image XRT took until 19:42 UT, **Figures 2E,F**. At this time, the X-ray emission has increased around a circular structure which we identify as the cavity. Furthermore, there is now X-ray emission near the center of the cavity that persists throughout the remainder of the observations. **Figures 2G,H** shows the cavity near the end of the observation run. Black arrows point to the bright X-ray emission near the center of the cavity.

The coronal cavity and overlying arcade are also well sampled with EIS. **Figures 3**, **4** compare raster scans before and during the eruption. The top row of **Figure 3** relates the intensity map (inverse log) for the three Fe lines of the prominence system prior to the eruption. The vertical white stripes represent missing data. The outline of the prominence as seen in Fe XV is overlaid for each image. The southern edge of the cavity is seen in the Fe XII 195.12 Å and Fe XIII 202.04 Å lines as a sharp decrease in intensity just south of the prominence starting from the limb at $(770'', -570'')$ extending radially out to the edge of the field of view. Interestingly, the Fe XV 284.16 Å line does not show this trend. The Doppler velocity map (middle row) shows a quiet region lacking large-scale LOS flows. Velocities that fall within the error for the measurements are scaled to white. The images showing line widths (bottom row) exhibit some regions around the prominence with higher than average widths, especially in the Fe XV 284.16 Å line. These elevated line widths could indicate that turbulent motions are present.

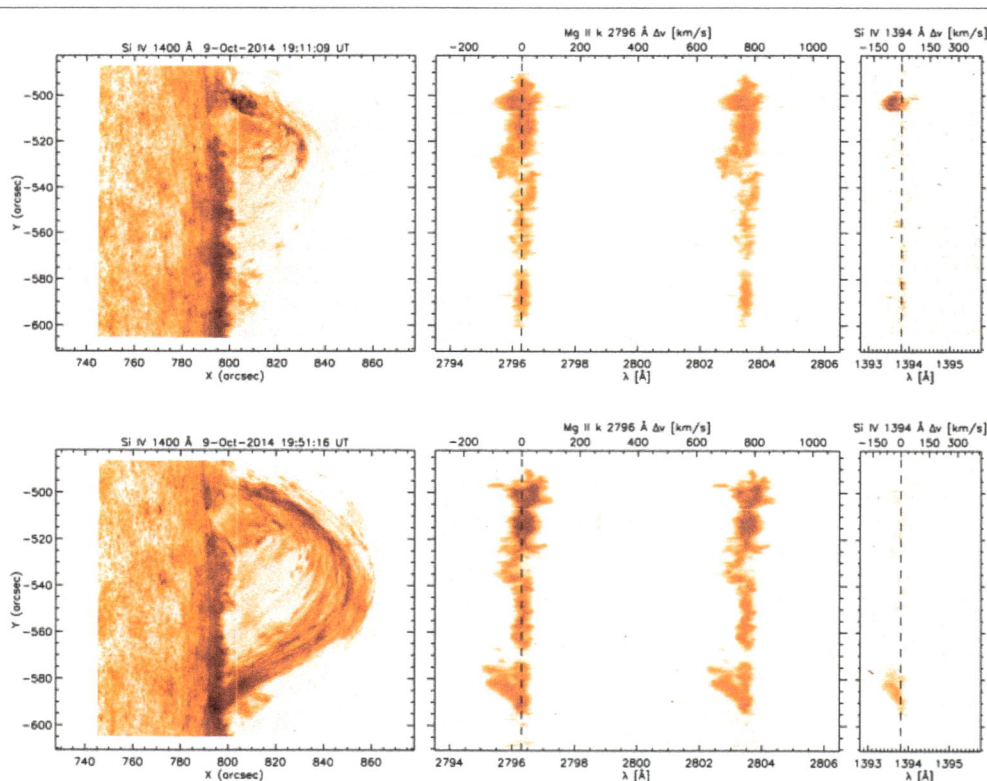

FIGURE 6 | Left: IRIS Si IV 1400 Å. **Middle:** Mg II 2796 Å spectra. **Right:** Si IV 1394 Å spectra. In each spectra panel, the top axis x-axis is the Doppler velocity whose reference wavelength is marked by a dashed line.

The top row of **Figure 4** shows the region during the eruption. Cool plasma is now present along an arc as an absorption feature in the Fe XV 284.16 Å . The Doppler velocity maps show the eruption is strongly blue shifted for all three Fe lines throughout the eruption site as well as the region just north of the prominence. Additionally, the line widths for these regions are large compared to the pre-eruptive state. There is flowing material around the cavity but the structure of the cavity and prominence remain stable. During and after the eruption there is evidence for turbulence and heating within the cavity.

The chromospheric observations provide clues about the structure of the coronal cavity and overlying arcade when an eruption forces chromospheric plasma to flow over the cavity. **Figure 5** provides an overview of the evolution of the eruption as observed in the IRIS Si IV 1400 Å SJI. The observations start with a prominence that appears in a stable configuration exhibiting minor plasma flows (top row). At 19:07 UT, the northern edge of the prominence brightens (middle row) and the bright plasma travels up and out along the outer edge of the prominence. Once the plasma reaches a certain height it cascades back toward the limb (bottom row). The cool plasma flows along an arc that mimics the shape of the prominence. The eruption is over by 21 UT when the system returns to its original state. Doppler velocity measurements of the chromospheric plasmas also show a predominantly blue-shifted flow. **Figure 6** presents Si IV 1400 Å SJI with simultaneous spectra of Mg II 2796 Å and Si IV 1394 Å when the slit is just above the spicule region near the beginning of the eruption (top row) and near the end of the eruption (bottom row).

Figure 7 and Supplementary Video 1 present high resolution SOT data showing striking details of the prominence and eruption. Initially, the prominence appears in a stable configuration with bi-directional plasma flows along the northern edge. This part of the prominence is highly stratified with the flows divided by regions with scant emission. **Figures 7A,B** show the prominence prior to the eruption with arrows pointing in the directions of plasma flow and a circle around the region where an intensity enhancement is seen in the minutes before the eruption. **Figure 7C** shows the time when the eruption starts. At this time, the bulk motion is in the direction of the arrow. As the eruption evolves, two bright ridges are prominent with regions of decreased intensity on either side. **Figures 7D,E** show that despite the upward bulk motion, the plasma does not move beyond the linear extrusion at the top of the prominence. In fact, the plasma flow is stalled as it encounters this barrier and the prominence experiences oscillatory motions in the regions around the two bright ridges. Eventually, the barrier is breached **Figure 7F** and plasma flows up along an arc, over the spine, exiting the FOV. Motions in the central regions of the prominence do not significantly change during or after the eruption. The motion slows and the plasma falls back down along the original trajectory path with some of the plasma flowing northward leaving the upper FOV **Figure 7G**. By the end of the observation run, the prominence is noticeably smaller **Figure 7H**.

Composite images of the X-ray emission with the SOT data are shown in **Figure 8** and Supplementary Video 2. The X-ray data is scaled using an orange color table while the SOT image is a grayscale image. The time differences between the XRT images and nearest SOT image range from 1 to 30 s. Regions with low X-ray emission are set to be transparent with respect to the SOT data. Prior to the eruption, the prominence sits in a region with little X-ray emission. An arrow points to the eruption site where the X-ray emission increases just off the limb and the chromospheric plasma is ejected. After the pause in observations, the X-ray emission is strongest just outside the chromospheric plasma flows. A circle encompasses the bright X-ray emission that forms near the top of the prominence in

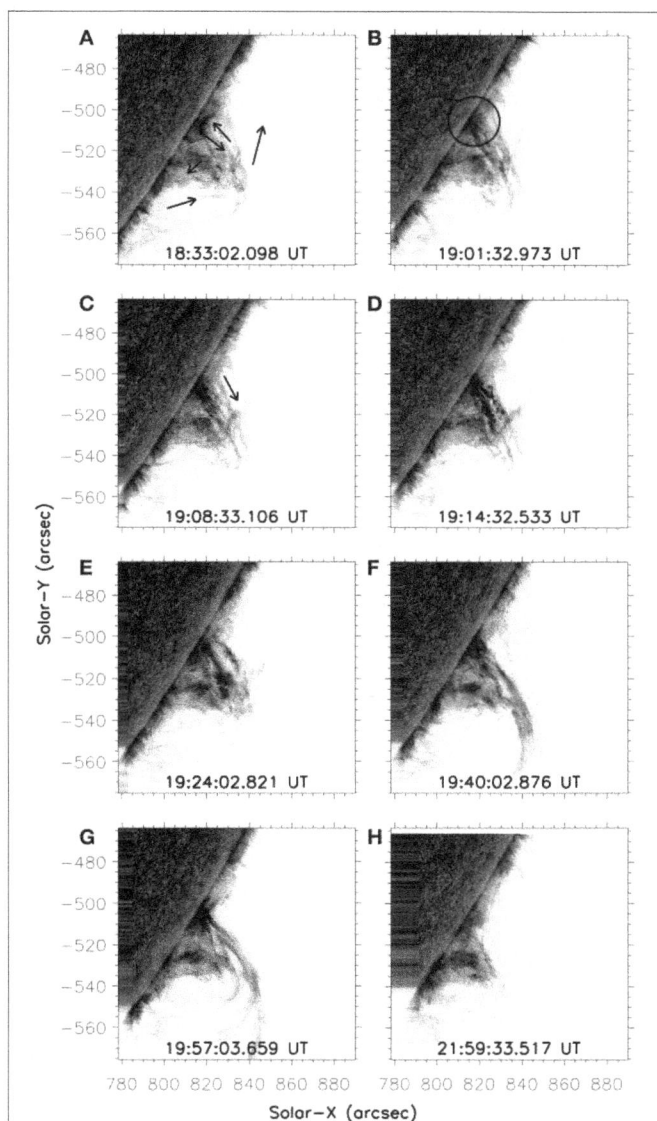

FIGURE 7 | Hinode SOT Ca II H-line observations (inverse intensity) of the prominence during the eruption. Panels **(A,B)** shows the prominence prior to the eruption with the arrows **(A)** indicating plasma motions and the base of the eruption is circled in **(B)**. Panels **(C–E)** show the initial phase of the eruption. The arrow points in the direction of the eruption. Panels **(F,G)** show the trajectory of the eruption and Panel **(H)** shows the prominence after the eruption.

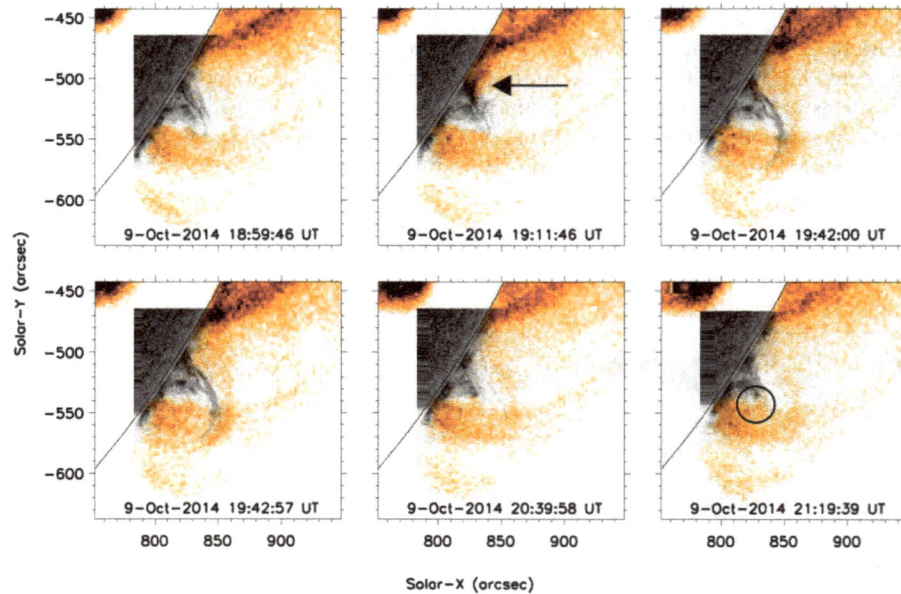

FIGURE 8 | SOT inverse intensity (black and white) with XRT inverse intensity overlay (orange) of the prominence and cavity system. The arrow points to the increased X-ray emission at the start of the eruption. The circle outlines a region of increased X-ray emission after the eruption.

Figure 8. These observations indicate that the eruption of the cool plasma observed by SOT was initiated by the incursion of hot plasma observed by XRT.

3.2. Model Results and Comparison to Observations

We construct several models with varying axial and poloidal fields and compare them with the observations. The model that best fit the observations has the correct magnetic field orientation to account for the observed plasma motions, Doppler velocities, and structures seen in XRT and the EUV data. **Figure 9** presents the best model (Model 1) along with the potential field model and a highly twisted flux rope model (Model 2). The initial inserted flux rope for Model 1 has axial and poloidal fluxes of 2e20 Mx and 0 Mx/cm, respectively. Model 2 has the same initial axial flux and −2e10 Mx/cm poloidal flux. Both flux rope models have left helical twist and dextral chirality. After a 30,000-iteration relaxation, the models relax toward a force-free state.

Figure 9A shows a grayscale map of the LOS magnetic field with positive fields scaled white and negative fields black. The blue line shows the path of the inserted flux rope and the circles at the ends represent the footpoints of that flux rope. The path is selected to be along the PIL and the footpoints of the flux rope are embedded within patches of strong magnetic fields near the PIL. The same path is utilized for all of the models and a comparison of selected magnetic field lines (colored lines) for the three models are shown in **Figures 9 D,G,J** with an AIA 193 Å background image taken on 4 October 2014 at 23:59 UT. White arrows point to field lines that represent the orientation of the magnetic field.

The middle column of **Figure 9** compares the three models rotated to the limb. The background image is an AIA 171 Å taken at 20:00 UT on 9 October 2014. **Figure 9B** shows the prominence with white arrows pointing to regions of plasma flow around the cavity. The right column compares the models with the background image as XRT thin-Be taken at 19:42 UT. **Figure 9C** shows the cavity with white arrows pointing to the regions of increased X-ray intensity around the coronal cavity. The bottom three rows of **Figure 9** show selected field lines from the models in comparison to the Hinode/XRT and SDO/AIA observations.

Figure 9 shows that the observed arc-like filament structure is corresponding to the overlying magnetic field lines in the models, which are more sheared in Model 1 (**Figure 9G**) and nearly perpendicular to the filament channel for Model 2 (**Figure 9J**). Model 1 exhibits a weakly twisted flux rope structure after the relaxation, although the initial inserted flux bundle has no twist. This twist may be produced during the relaxation due to reconnection between the inserted sheared flux bundle and the overlying arcade. Model 1, shows magnetic field lines oriented in a way that could produce the observed Doppler velocities but the highly twisted flux rope, has magnetic field lines in the wrong orientation to account for the observed LOS Doppler velocities. The sheared overlying field lines can account for the aforementioned observed blue-shift flow in the overlying arcade. Therefore, we think that the weakly twisted flux rope fit the observations better. In comparison to the potential field model, and Model 2, the weakly twisted flux rope clearly shows a much better match to both the on-disk filament channel and the cavity observed on the limb by XRT.

One feature in the IRIS Si IV 1394 Å spectra is a persistent region with no emission on a portion of the on-disk scans. This

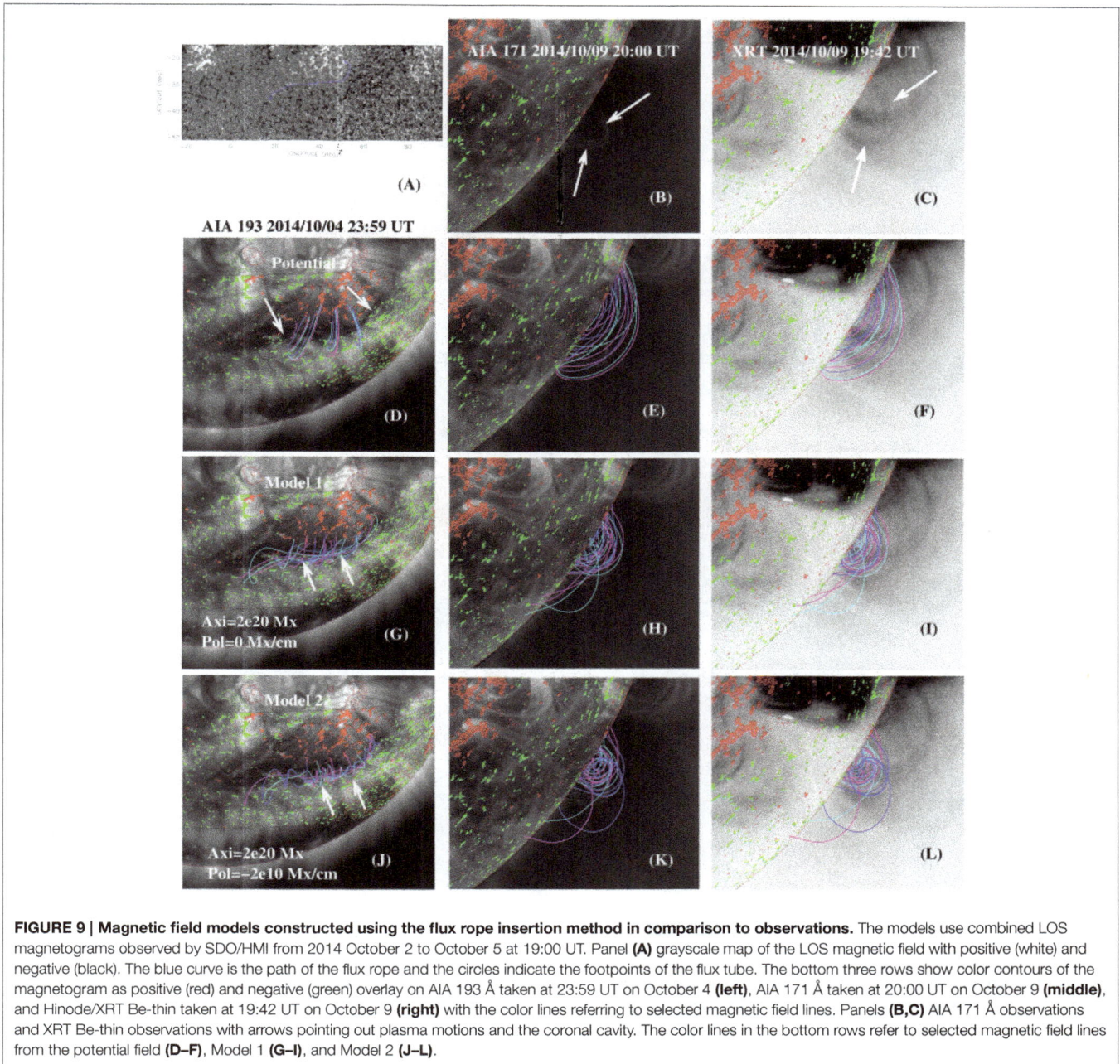

FIGURE 9 | Magnetic field models constructed using the flux rope insertion method in comparison to observations. The models use combined LOS magnetograms observed by SDO/HMI from 2014 October 2 to October 5 at 19:00 UT. Panel **(A)** grayscale map of the LOS magnetic field with positive (white) and negative (black). The blue curve is the path of the flux rope and the circles indicate the footpoints of the flux tube. The bottom three rows show color contours of the magnetogram as positive (red) and negative (green) overlay on AIA 193 Å taken at 23:59 UT on October 4 **(left)**, AIA 171 Å taken at 20:00 UT on October 9 **(middle)**, and Hinode/XRT Be-thin taken at 19:42 UT on October 9 **(right)** with the color lines referring to selected magnetic field lines. Panels **(B,C)** AIA 171 Å observations and XRT Be-thin observations with arrows pointing out plasma motions and the coronal cavity. The color lines in the bottom rows refer to selected magnetic field lines from the potential field **(D–F)**, Model 1 **(G–I)**, and Model 2 **(J–L)**.

region is outlined by two dotted lines in the top and bottom rows of **Figure 10**. This region persists throughout the observations but its location depends on the slit position. As the slit moves toward the limb, the region with no emission shifts northward until the slit reaches the limb. The Si IV emission just above the horizontal lines is also red-shifted relative to the average line centroid. The middle row of **Figure 10** shows one of two slit positions that are almost exactly on the solar limb. The emission is strong in this region throughout the entire slit length. Once the slit clears the limb, the region of reduced emission is located over the prominence (bottom row) and the region continues to

exhibit a redshift relative to the line average. There is also a noticeable decrease in the number of spicules near the base of the prominence. The orientation of the spicules in the peripheral regions of the prominence suggest they are curved away from the prominence.

We compare the location of the depleted region observed in the IRIS Si IV 1394 Å spectra with the location of the PIL in Model 1. **Figure 11** shows an IRIS Si IV 1400 Å SJ image, with the correct prominence orientation, along with contours (green/negative; red/positive) of the SDO/HMI LOS magnetic field model data. The region of reduced emission is located

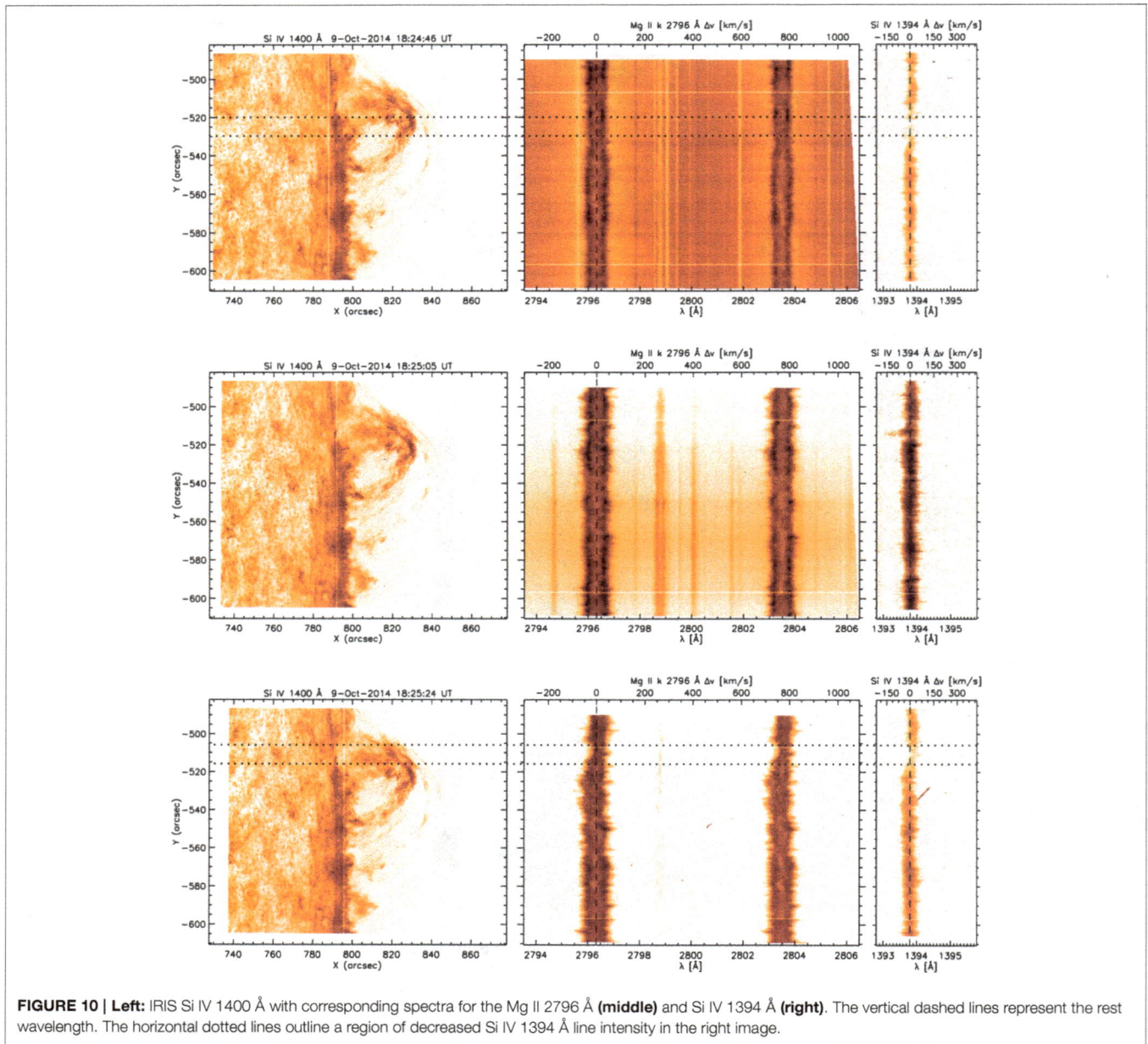

FIGURE 10 | Left: IRIS Si IV 1400 Å with corresponding spectra for the Mg II 2796 Å **(middle)** and Si IV 1394 Å **(right)**. The vertical dashed lines represent the rest wavelength. The horizontal dotted lines outline a region of decreased Si IV 1394 Å line intensity in the right image.

between the two white dotted lines. These are at the same location of the horizontal lines in the top row of **Figure 10**. A bipole sits near the region of reduced emission. The pink line is a small field line from Model 1 that crosses the PIL indicating its location. The right image shows the same location on the model data before it was rotated to the limb.

To potentially constrain the model parameters we compare theoretical L/I measurements of Model 1 and the potential arcade with L/I CoMP measurements in **Figure 12**. The left image is CoMP L/I (log scale) observations of the prominence region. The prominence sits just above the limb (below 1.3 $R_\odot$) so we cannot directly observe the prominence-cavity system in the CoMP data. However, some elements of the structure could still

be present. The CoMP data does exhibit a linear decrease in intensity in a similar location to the potential arcade (middle panel) and Model 1 (right panel). The bright feature in the CoMP data is not seen in either of the models. The models do not contain information about other structures near the region so it is possible that the bright feature is not associated with the prominence. There are also minor differences between the potential arcade model and Model 1 far away from the disk. However, the CoMP data alone is not different enough to truly distinguish between the two models. Model 1 is a small flux rope embedded in a potential arcade, so at distances far from the flux rope, the L/I signatures are very similar to those of the potential arcade.

4. DISCUSSION

We present observations of a prominence and cavity system with an ensuing ancillary eruption that serves to highlight some of the topological features of the system. We find the prominence-cavity system maintains its structure during the event but heating is observed as an increase in X-ray emission around the coronal cavity and just above the prominence. Previous observational studies of bright X-ray emission within coronal cavities observed long-lived polar crown prominences where the bright core had already formed (Hudson et al., 1999; Reeves et al., 2012). The X-ray bright core always sits directly above the prominence although temperature structures found using EUV data (Schmit et al., 2009; Kucera et al., 2012) and white light studies find dynamic structures throughout the cavity (Habbal et al., 2010). The longevity of polar crown prominences, sometimes lasting several solar rotations, suggest a continuous heating process is needed to maintain the bright central emissions. Our observations suggest the heating inside the cavity is from a current sheet formed at a BPSS (Fan and

Gibson, 2006). The BPSS forms a sheath or tunnel enclosing the dipped prominence field lines extending from the prominence footpoints in the photosphere, up into the cavity and would appear to be central to the cavity when viewed edge on. The BPSS can explain the steady-state X-ray emissions observed in long-lived polar crown prominences and it can explain the rapid increase in X-ray emission when a stable prominence system is disturbed.

The eruption causes oscillatory motions in the prominence near the eruption site but plasma motions within the central regions of the prominence do not change suggesting the inner prominence is structurally isolated from the eruption site. We model the prominence-cavity system as a flux rope situated under a coronal arcade. After testing several combinations of axial and poloidal fluxes we found the model that fit the observations best was that of a weakly twisted flux rope with dextral chirality.

The flux rope has opposite chirality than we would expect for a southern prominence Martin et al. (1994). The active regions north of the prominence have a positive (red) leading polarity whereas the prominence has the opposite (**Figure 9**). The dextral chirality is based on the comparison with the AIA emissions on the two sides of the filament channel. Through a statistical study Su et al. (2010) found that the emission on the two sides of the filament channel are asymmetric with one side showing bright and curved loops and the other side faint and straight emissions. They proposed that the bright curve features (on the southern side of filament channel for our case) are corresponding to the field lines that turn into the flux rope, and the straight faint features to the north, are the legs of the large overlying arcade. This idea was also confirmed by the magnetic field modeling in Su and van Ballegooijen (2012). The dextral flux rope model matches the direction of the observed bright curved feature on the southern side for our prominence, **Figure 9G**. The configuration also explains the trajectory of the erupting plasma as it flowed along magnetic field lines within the overlying arcade. The orientation of the arcade is such that any plasma flowing within the arcade would exhibit a predominantly blue-shifted LOS velocity when viewed on the limb.

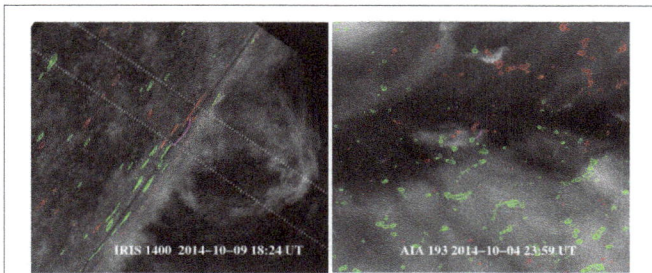

FIGURE 11 | Left: IRIS Si IV 1400 Å rotated to its orientation on the limb taken at 18:24 UT on 9 October 2014. The white dashed lines indicate the region of reduced emission in the Si IV 1394 Å spectra shown in **Figure 10**. The green (negative) and red (positive) contours are the LOS photospheric magnetic field taken by SDO/HMI. The purple line crosses the PIL for this region. **Right:** AIA 193 Å with HMI magnetic field contours with the PIL at 23:59 UT on 4 October 2014.

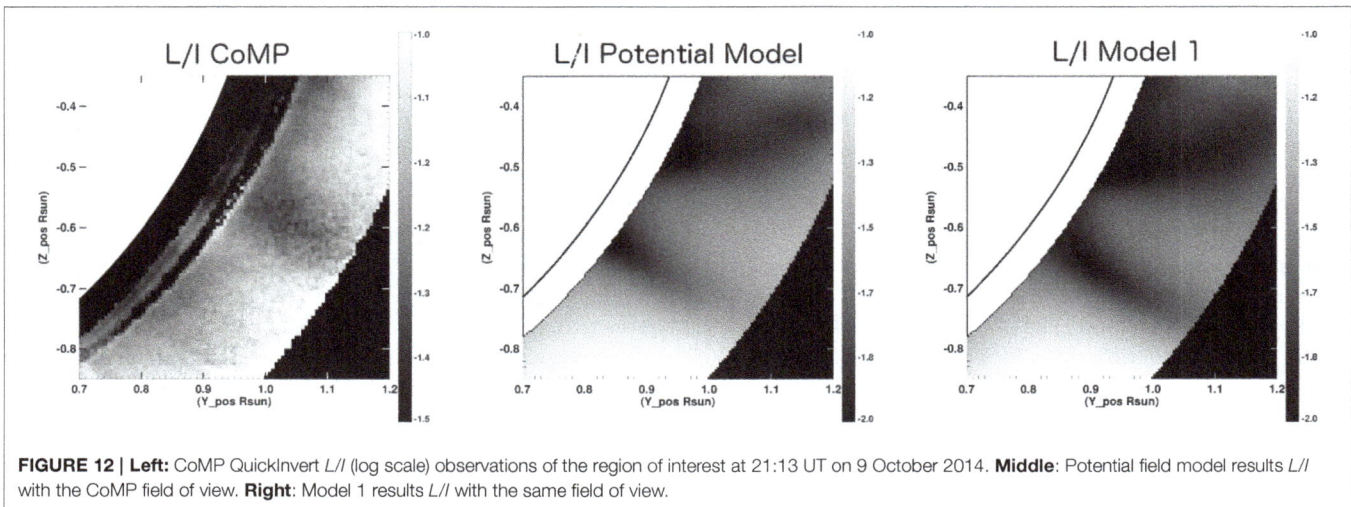

FIGURE 12 | Left: CoMP QuickInvert *L/I* (log scale) observations of the region of interest at 21:13 UT on 9 October 2014. **Middle:** Potential field model results *L/I* with the CoMP field of view. **Right:** Model 1 results *L/I* with the same field of view.

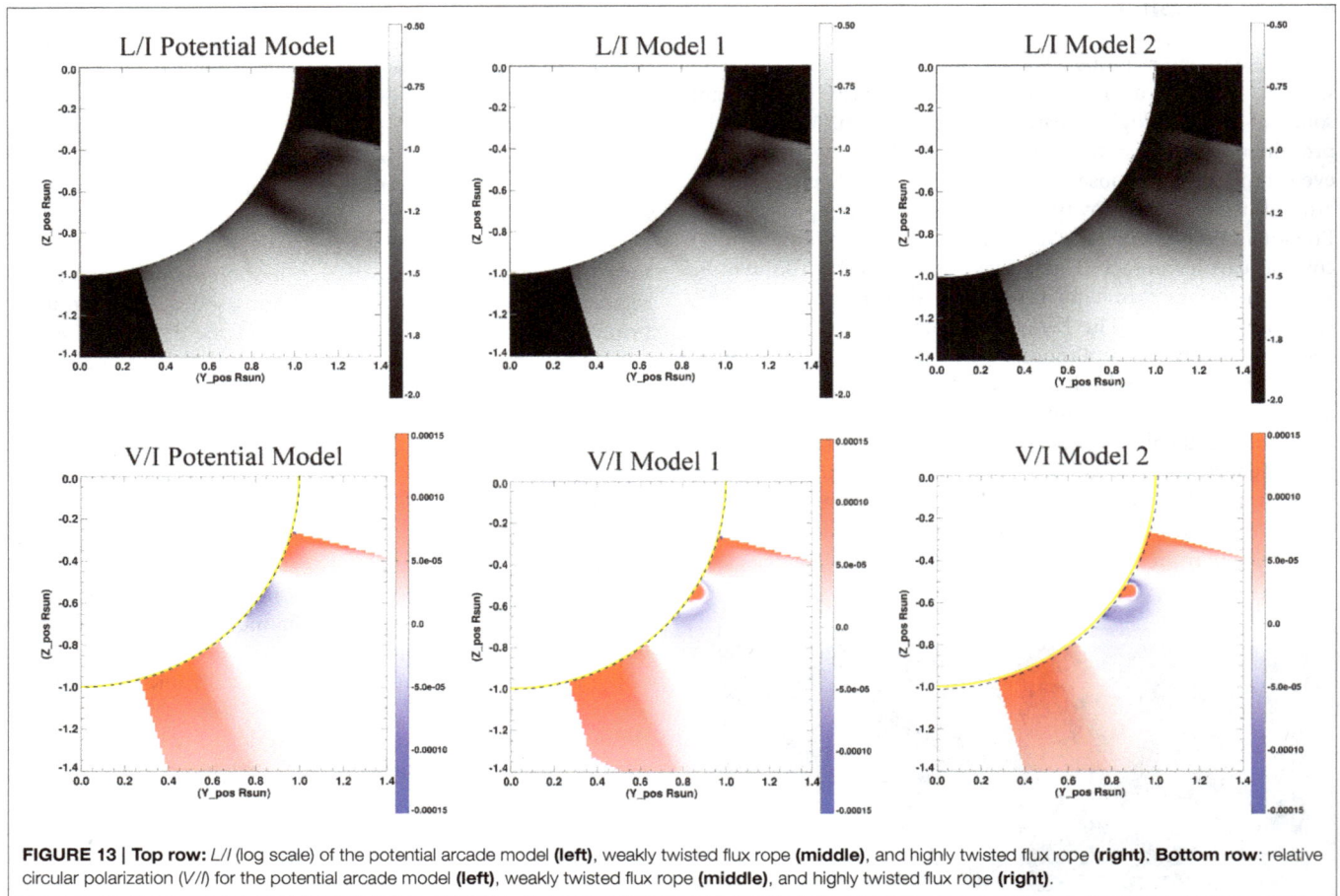

FIGURE 13 | Top row: *L/I* (log scale) of the potential arcade model **(left)**, weakly twisted flux rope **(middle)**, and highly twisted flux rope **(right)**. **Bottom row**: relative circular polarization (*V/I*) for the potential arcade model **(left)**, weakly twisted flux rope **(middle)**, and highly twisted flux rope **(right)**.

The decreased emission of the Si IV 1394 Å spectra on the disk and in close proximity to the prominence coincides with the location of a bipole within the PIL of the model and thus we interpret it as evidence for a bald-patch underneath the prominence. A study by López Ariste et al. (2006) used vector magnetic fields to analyze bipolar regions within a filament channel. They found that at least four of the six bipolar regions exhibited a bald-patch topology forming photospheric dips where the horizontal component of the magnetic field points from a negative toward positive polarity. They concluded the observed magnetic field topology in the photosphere tends to support models of prominence based on magnetic dips located within weakly twisted flux tubes. Their underlying and lateral extensions form photospheric dips both within the channel and below barbs.

A comparison of the model with CoMP *L/I* observations were inconclusive as the prominence structure lies just above the limb but below the CoMP FOV. The flux rope for this model is small and embedded in a potential arcade. Far from the flux rope, the linear polarization will be similar to that of the overlying arcade. Bąk-Stęślicka et al. (2016) performed a statistical study of quiescent coronal cavities observed with CoMP and found that coherent, often, ring-shaped, LOS Doppler velocity flows are common within cavities that possess a "lagomorphic" signature in the *L/I* polarization. The portion of the prominence we are studying is not oriented in the E-W direction and may not be in the best orientation to observe these signatures. Another reason that could account for the differences between the CoMP data and models is that our model only considers the local fields around the prominence. Differences observed in CoMP could be from other coronal structures.

Even if we could make linear polarization measurements up to the solar disk we would still have a difficult time distinguishing the *L/I* signatures of small flux ropes from those of the overlying potential arcade. The top row of **Figure 13** compares the *L/I* signatures of a potential arcade model, Model 1 and Model 2. The linear polarization signatures are similar with varying differences in intensity. To observe the differences between the models we need to have *V/I* polarization measurements closer to the limb. The bottom row **Figure 13** shows the *V/I* polarization measurements for the three models. The *V/I* can distinguish between the three models. Currently, *V/I* measurements are not practical as they require hours long integration times. However, an observatory that would be capable of making high resolution polarization measurements close to the solar limb is the proposed COronal Solar Magnetism Observatory (COSMO) (de Wijn et al., 2014). Future measurements from COSMO would clearly be very useful in determining magnetic structures of prominence-cavity systems.

AUTHOR CONTRIBUTIONS

PJ performed the data analysis and document preparation. KR analyzed and interpreted the data. YS provided the NLFF models and documentation thereof.

FUNDING

PJ and KR are supported by under contract 80111112705 from Lockheed-Martin to SAO, contract NNM07AB07C from NASA to SAO, grant number NNX12AI30G from NASA to SAO, and contract Z15-12504 from HAO to SAO under a grant from AFOSR. YS is supported by the Youth Fund of Jiangsu No. BK20141043, the NSFC No. 11473071, and the "One Hundred Talent Program" of the Chinese Academy of Sciences.

ACKNOWLEDGMENTS

IRIS is a NASA small explorer mission developed and operated by LMSAL with mission operations executed at NASA Ames Research center and major contributions to downlink communications funded by ESA and the Norwegian Space Centre. *Hinode* is a Japanese mission developed and launched by ISAS/JAXA, with NAOJ as domestic partner and NASA and STFC (UK) as international partners. It is operated by these agencies in co-operation with ESA and NSC (Norway). The authors acknowledge Dr. Adriaan van Ballegooijen for his valuable suggestions on the magnetic field modeling.

REFERENCES

Aulanier, G., Demoulin, P., van Driel-Gesztelyi, L., Mein, P., and Deforest, C. (1998). 3-D magnetic configurations supporting prominences. II. The lateral feet as a perturbation of a twisted flux-tube. *Astron. Astrophys.* 335, 309–322.

Bąk-Stęślicka, U., Gibson, S. E., and Chmielewska, E. (2016). Line-of-sight velocity as a tracer of coronal cavity magnetic structure. *Front. Astron. Space Sci.* 3:7. doi: 10.3389/fspas.2016.00007

Bąk-Stęślicka, U., Gibson, S. E., Fan, Y., Bethge, C., Forland, B., and Rachmeler, L. A. (2013). The magnetic structure of solar prominence cavities: new observational signature revealed by coronal magnetometry. *Astrophys. J. Lett.* 770:L28. doi: 10.1088/2041-8205/770/2/L28

Berger, T. E., Slater, G., Hurlburt, N., Shine, R., Tarbell, T., Title, A., et al. (2010). Quiescent prominence dynamics observed with the hinode solar optical telescope. I. Turbulent upflow plumes. *Astrophys. J.* 716, 1288–1307. doi: 10.1088/0004-637X/716/2/1288

Chae, J., Wang, H., Qiu, J., Goode, P. R., Strous, L., and Yun, H. S. (2001). The formation of a prominence in active region NOAA 8668. I. SOHO/MDI observations of magnetic field evolution. *Astrophys. J.* 560, 476–489. doi: 10.1086/322491

Culhane, J. L., Harra, L. K., James, A. M., Al-Janabi, K., Bradley, L. J., Chaudry, R. A., et al. (2007). The EUV imaging spectrometer for hinode. *Solar Phys.* 243, 19–61. doi: 10.1007/s01007-007-0293-1

De Pontieu, B., Title, A. M., Lemen, J. R., Kushner, G. D., Akin, D. J., Allard, B., et al. (2014). The Interface Region Imaging Spectrograph (IRIS). *Solar Phys.* 289, 2733–2779. doi: 10.1007/s11207-014-0485-y

de Wijn, A. G., Tomczyk, S., and Burkepile, J. (2014). "A progress update for the coronal solar magnetism observatory for coronal and chromospheric polarimetry," in *Solar Polarization 7, Vol. 489 of Astronomical Society of the Pacific Conference Series*, eds K. N. Nagendra, J. O. Stenflo, Q. Qu, and M. Samooprna (San Francisco), 323.

Dudík, J., Aulanier, G., Schmieder, B., Bommier, V., and Roudier, T. (2008). Topological departures from translational invariance along a filament observed by THEMIS. *Solar Phys.* 248, 29–50. doi: 10.1007/s11207-008-9155-2

Fan, Y., and Gibson, S. E. (2006). On the nature of the X-ray bright core in a stable filament channel. *Astrophys. J. Lett.* 641, L149–L152. doi: 10.1086/504107

Freeland, S. L., and Handy, B. N. (1998). Data analysis with the solarsoft system. *Solar Phys.* 182, 497–500. doi: 10.1023/A:1005038224881

Gibson, S. E., Foster, D., Burkepile, J., de Toma, G., and Stanger, A. (2006). The calm before the storm: the link between quiescent cavities and coronal mass ejections. *Astrophys. J.* 641, 590–605. doi: 10.1086/500446

Gibson, S. E., Kucera, T. A., White, S. M., Dove, J. B., Fan, Y., Forland, B. C., et al. (2016). FORWARD: a toolset for multiwavelength coronal magnetometry. *Front. Astron. Space Sci.* 3:8. doi: 10.3389/fspas.2016.00008

Golub, L., Deluca, E., Austin, G., Bookbinder, J., Caldwell, D., Cheimets, P., et al. (2007). The X-Ray Telescope (XRT) for the hinode mission. *Solar Phys.* 243, 63–86. doi: 10.1007/s11207-007-0182-1

Habbal, S. R., Druckmüller, M., Morgan, H., Scholl, I., Rušin, V., Daw, A., et al. (2010). Total solar eclipse observations of hot prominence shrouds. *Astrophys. J.* 719, 1362–1369. doi: 10.1088/0004-637X/719/2/1362

Hudson, H., and Schwenn, R. (2000). Hot cores in coronal filament cavities. *Adv. Space Res.* 25, 1859–1861. doi: 10.1016/S0273-1177(99)00618-3

Hudson, H. S., Acton, L. W., Harvey, K. L., and McKenzie, D. E. (1999). A stable filament cavity with a hot core. *Astrophys. J. Lett.* 513, L83–L86. doi: 10.1086/311892

Judge, P. G., and Casini, R. (2001). "A synthesis code for forbidden coronal lines," in *Advanced Solar Polarimetry – Theory, Observation, and Instrumentation, volume 236 of Astronomical Society of the Pacific Conference Series*, ed M. Sigwarth (San Francisco), 503.

Kamio, S., Hara, H., Watanabe, T., Fredvik, T., and Hansteen, V. H. (2010). Modeling of EIS spectrum drift from instrumental temperatures. *Solar Phys.* 266, 209–223. doi: 10.1007/s11207-010-9603-7

Kobelski, A. R., Saar, S. H., Weber, M. A., McKenzie, D. E., and Reeves, K. K. (2014). Calibrating data from the hinode/X-ray telescope and associated uncertainties. *Solar Phys.* 289, 2781–2802. doi: 10.1007/s11207-014-0487-9

Kosugi, T., Matsuzaki, K., Sakao, T., Shimizu, T., Sone, Y., Tachikawa, S., et al. (2007). The hinode (solar-B) mission: an overview. *Solar Phys.* 243, 3–17. doi: 10.1007/s11207-007-9014-6

Kucera, T. A., Gibson, S. E., Schmit, D. J., Landi, E., and Tripathi, D. (2012). Temperature and extreme-ultraviolet intensity in a coronal prominence cavity and streamer. *Astrophys. J.* 757, 73. doi: 10.1088/0004-637X/757/1/73

Kuperus, M., and Raadu, M. A. (1974). The support of prominences formed in neutral sheets. *Astron. Astrophys.* 31:189.

Lemen, J. R., Title, A. M., Akin, D. J., Boerner, P. F., Chou, C., Drake, J. F., et al. (2012). The Atmospheric Imaging Assembly (AIA) on the Solar Dynamics Observatory (SDO). *Solar Phys.* 275, 17–40. doi: 10.1007/s11207-011-9776-8

Lin, H., Kuhn, J. R., and Coulter, R. (2004). Coronal magnetic field measurements. *Astrophys. J.* 613, L177–L180. doi: 10.1086/425217

Liu, W., De Pontieu, B., Vial, J.-C., Title, A. M., Carlsson, M., Uitenbroek, H., et al. (2015). First high-resolution spectroscopic observations of an erupting prominence within a coronal mass ejection by the Interface Region Imaging Spectrograph (IRIS). *Astrophys. Lett.* 803, 85. doi: 10.1088/0004-637X/803/2/85

López Ariste, A., Aulanier, G., Schmieder, B., and Sainz Dalda, A. (2006). First observation of bald patches in a filament channel and at a barb endpoint. *Astron. Astrophys.* 456, 725–735. doi: 10.1051/0004-6361:20064923

Low, B. C., and Hundhausen, J. R. (1995). Magnetostatic structures of the solar corona. 2: The magnetic topology of quiescent prominences. *Astrophys. Lett.* 443, 818–836. doi: 10.1086/175572

Mackay, D. H., Karpen, J. T., Ballester, J. L., Schmieder, B., and Aulanier, G. (2010). Physics of solar prominences: II-magnetic structure and dynamics. *Space Sci. Rev.* 151, 333–399. doi: 10.1007/s11214-010-9628-0

Malherbe, J. M., and Priest, E. R. (1983). Current sheet models for solar prominences. I Magnetohydrostatics of support and evolution through quasistatic models. *Astron. Astrophys.* 123, 80–88.

Martin, S. F. (1973). The evolution of prominences and their relationship to active centers (a review). *Solar Phys.* 31, 3–21. doi: 10.1007/BF00156070

Martin, S. F. (1990). "Conditions for the formation of prominences as inferred from optical observations," in *IAU Colloq. 117: Dynamics of Quiescent Prominences, Vol. 363 of Lecture Notes in Physics*, eds V. Ruzdjak and E. Tandberg-Hanssen (Berlin: Springer Verlag), 1–44.

Martin, S. F., Bilimoria, R., and Tracadas, P. W. (1994). "Magnetic field configurations basic to filament channels and filaments," in *NATO Advanced Science Institutes (ASI) Series C, Vol. 433 of NATO Advanced Science Institutes (ASI) Series C*, eds R. J. Rutten and C. J. Schrijver (Dordrecht), 303.

Parenti, S. (2014). Solar prominences: observations. *Living Rev. Solar Phys.* 11, 1. doi: 10.12942/lrsp-2014-1

Pneuman, G. W. (1983). The formation of solar prominences by magnetic reconnection and condensation. *Solar Phys.* 88, 219–239. doi: 10.1007/BF00196189

Priest, E. (2014). *Magnetohydrodynamics of the Sun*. Cambridge, MA: Cambridge University Press.

Priest, E. R., Hood, A. W., and Anzer, U. (1989). A twisted flux-tube model for solar prominences. I - General properties. *Astrophys. Lett.* 344, 1010–1025. doi: 10.1086/167868

Rachmeler, L. A., Gibson, S. E., Dove, J. B., DeVore, C. R., and Fan, Y. (2013). Polarimetric properties of flux ropes and sheared arcades in coronal prominence cavities. *Solar Phys.* 288, 617–636. doi: 10.1007/s11207-013-0325-5

Reeves, K. K., Gibson, S. E., Kucera, T. A., Hudson, H. S., and Kano, R. (2012). Thermal properties of a solar coronal cavity observed with the X-ray telescope on hinode. *Astrophys. J.* 746, 146. doi: 10.1088/0004-637X/746/2/146

Rust, D. M., and Kumar, A. (1994). Helical magnetic fields in filaments. *Solar Phys.* 155, 69–97. doi: 10.1007/BF00670732

Schmieder, B., Tian, H., Kucera, T., López Ariste, A., Mein, N., Mein, P., et al. (2014). Open questions on prominences from coordinated observations by IRIS, Hinode, SDO/AIA, THEMIS, and the Meudon/MSDP. *Astron. Astrophys.* 569:A85. doi: 10.1051/0004-6361/201423922

Schmit, D. J., Gibson, S. E., Tomczyk, S., Reeves, K. K., Sterling, A. C., Brooks, D. H., et al. (2009). Large-scale flows in prominence cavities. *Astrophys. J.* 700, L96–L98. doi: 10.1088/0004-637X/700/2/L96

Schou, J., Scherrer, P. H., Bush, R. I., Wachter, R., Couvidat, S., Rabello-Soares, M. C., et al. (2012). Design and ground calibration of the Helioseismic and Magnetic Imager (HMI) instrument on the Solar Dynamics Observatory (SDO). *Solar Phys.* 275, 229–259. doi: 10.1007/s11207-011-9842-2

Schröter, E.-H., Vázquez, M., and Wyller, A. A. (eds.). (1987). *The Role of Fine-Scale Magnetic Fields on the Structure of the Solar Atmosphere*. Cambridge: Cambridge University Press.

Smith, S. F. (1968). "The formation, structure and changes in filaments in active regions," in *Structure and Development of Solar Active Regions, Vol. 35 of IAU Symposium*, ed K. O. Kiepenheuer, 267. doi: 10.1007/978-94-011-6815-1_42

Stark, J.-L., and Murtagh, F. (2002). *Astronomical Image and Data Analysis*. Berlin: Springer.

Su, Y., Surges, V., van Ballegooijen, A., DeLuca, E., and Golub, L. (2011). Observations and magnetic field modeling of the flare/coronal mass ejection event on 2010 april 8. *Astrophys. J.* 734, 53. doi: 10.1088/0004-637X/734/1/53

Su, Y., and van Ballegooijen, A. (2012). Observations and magnetic field modeling of a solar polar crown prominence. *Astrophys. J.* 757, 168. doi: 10.1088/0004-637X/757/2/168

Su, Y., van Ballegooijen, A., and Golub, L. (2010). Structure and dynamics of quiescent filament channels observed by hinode/XRT and STEREO/EUVI. *Astrophys. J.* 721, 901–910. doi: 10.1088/0004-637X/721/1/901

Su, Y., van Ballegooijen, A., Lites, B. W., Deluca, E. E., Golub, L., Grigis, P. C., et al. (2009). Observations and nonlinear force-free field modeling of active region 10953. *Astrophys. J.* 691, 105–114. doi: 10.1088/0004-637X/691/1/105

Su, Y., van Ballegooijen, A., McCauley, P., Ji, H., Reeves, K. K., and DeLuca, E. E. (2015). Magnetic structure and dynamics of the erupting solar polar crown prominence on 2012 march 12. *Astrophys. J.* 807, 144. doi: 10.1088/0004-637X/807/2/144

Titov, V. S., and Démoulin, P. (1999). Basic topology of twisted magnetic configurations in solar flares. *Astron. Astrophys.* 351, 707–720.

Tomczyk, S., Card, G. L., Darnell, T., Elmore, D. F., Lull, R., Nelson, P. G., et al. (2008). An instrument to measure coronal emission line polarization. *Solar Phys.* 247, 411–428. doi: 10.1007/s11207-007-9103-6

Tsuneta, S., Ichimoto, K., Katsukawa, Y., Nagata, S., Otsubo, M., Shimizu, T., et al. (2008). The solar optical telescope for the hinode mission: an overview. *Solar Phys.* 249, 167–196. doi: 10.1007/s11207-008-9174-z

Vaiana, G. S., Krieger, A. S., and Timothy, A. F. (1973). Identification and analysis of structures in the corona from X-Ray photography. *Solar Phys.* 32, 81–116. doi: 10.1007/BF00152731

van Ballegooijen, A. A. (2004). Observations and modeling of a filament on the sun. *Astrophys. J.* 612, 519–529. doi: 10.1086/422512

van Ballegooijen, A. A., and Martens, P. C. H. (1989). Formation and eruption of solar prominences. *Atrophys. J.* 343, 971–984. doi: 10.1086/167766

van Ballegooijen, A. A., Priest, E. R., and Mackay, D. H. (2000). Mean field model for the formation of filament channels on the sun. *Astrophys. J.* 539, 983–994. doi: 10.1086/309265

Vial, J.-C., and Engvold, O. (eds.). (2015). *Solar Prominences, Vol. 415 of Astrophysics and Space Science Library*. Springer International Publishing Switzerland.

Warren, H. P., Ugarte-Urra, I., Young, P. R., and Stenborg, G. (2011). The temperature dependence of solar active region outflows. *Astrophys. J.* 727, 58. doi: 10.1088/0004-637X/727/1/58

Yoshimura, K., and McKenzie, D. E. (2015). Calibration of hinode/XRT for coalignment. *Solar Phys.* 290, 2355–2372. doi: 10.1007/s11207-015-0746-4

Young, P. (2013). *EIS Software Note No. 16, ver. 2.5*. Available online at: http://solarb.mssl.ucl.ac.uk:8080/eiswiki/Wiki.jsp?page=EISAnalysisGuide

Conflict of Interest Statement: The authors declare that the research was conducted in the absence of any commercial or financial relationships that could be construed as a potential conflict of interest.

5

Infrared Dual-Line Hanle Diagnostic of the Coronal Vector Magnetic Field

Gabriel I. Dima[1]*, Jeffrey R. Kuhn[1] and Svetlana V. Berdyugina[1, 2, 3]

[1] *Institute for Astronomy, University of Hawaii, Pukalani, HI, USA, [2] Kiepenheuer Institut fuer Sonnenphysik, Freiburg, Germany, [3] Predictive Science Inc., San Diego, CA, USA*

Measuring the coronal vector magnetic field is still a major challenge in solar physics. This is due to the intrinsic weakness of the field (e.g., $\sim$ 4G at a height of $0.1 R_\odot$ above an active region) and the large thermal broadening of coronal emission lines. We propose using concurrent linear polarization measurements of near-infrared forbidden and permitted lines together with Hanle effect models to calculate the coronal vector magnetic field. In the unsaturated Hanle regime both the direction and strength of the magnetic field affect the linear polarization, while in the saturated regime the polarization is insensitive to the strength of the field. The relatively long radiative lifetimes of coronal forbidden atomic transitions implies that the emission lines are formed in the saturated Hanle regime and the linear polarization is insensitive to the strength of the field. By combining measurements of both forbidden and permitted lines, the direction and strength of the field can be obtained. For example, the SiX 1.4301 μm line shows strong linear polarization and has been observed in emission over a large field-of-view (out to elongations of 0.5 $R_\odot$). Here we describe an algorithm that combines linear polarization measurements of the SiX 1.4301 μm forbidden line with linear polarization observations of the HeI 1.0830 μm permitted *coronal* line to obtain the vector magnetic field. To illustrate the concept we assume that the emitting gas for both atomic transitions is located in the plane of the sky. The further development of this method and associated tools will be a critical step toward interpreting the high spectral, spatial and temporal infrared spectro-polarimetric measurements that will be possible when the Daniel K. Inouye Solar Telescope (DKIST) is completed in 2019.

Keywords: corona, magnetic fields, spectro-polarimetry, magnetometry, infrared, Hanle effect

Edited by:
Sarah Gibson,
National Center for Atmospheric
Research/High Altitude Observatory,
USA

Reviewed by:
Veronique Bommier,
Observatoire de Paris, France
Roberto Casini,
High Altitude Observatory-National
Center for Atmospheric Research,
USA

*Correspondence:
Gabriel I. Dima
gdima@hawaii.edu

Specialty section:
This article was submitted to
Stellar and Solar Physics,
a section of the journal
Frontiers in Astronomy and Space
Sciences

1. INTRODUCTION

Magnetometry using optical spectropolarimetry has yielded some of the most precise direct measurements of coronal magnetic fields (Kuhn, 1995; Lin et al., 2000, 2004; Tomczyk et al., 2008). Earlier infrared (IR) coronal Zeeman observations (e.g., Arnaud and Newkirk, 1987; Kuhn, 1995) have used forbidden FeXIII transitions near 1 micron. The larger context of all coronal magnetometry techniques has been reviewed elsewhere (e.g., Penn, 2014), but the great promise of the Daniel K. Inouye Solar Telescope (DKIST) will be to use near-IR coronal lines to routinely observe the so far seldom measured weak solar coronal magnetic field. Up until now attempts from the ground to measure the magnetic field strength have depended on the ability to detect very weak Zeeman splitting through Stokes-V (circular) polarization observations. A Gauss-scale coronal magnetic field creates very weak Stokes-V signals (typically 10^{-4}) in spectral lines that are dominated by much stronger linear scattering polarization amplitudes (e.g., Stokes-Q and U of order 10^{-2} and sometimes up to 10^{-1}, Lin et al., 2004).

Most recently linear polarization observations of permitted lines combined with forward calculations of field configurations have been productive tools for understanding solar prominence magnetic fields (Bommier et al., 1981; López Ariste and Casini, 2003; Merenda et al., 2006). A powerful coronal field diagnostic follows from simultaneous measurements of the optical scattering linear polarization of combined forbidden and permitted spectral lines. Early work on the possibility of using lines with different Hanle sensitivity used the HeI 0.5875 μm and HeI 1.0830 μm (hereafter HeI1083) lines for measuring the magnetic field in a prominence located in the plane of the sky (Bommier et al., 1981). Recently space spectropolarimetric observations of the permitted coronal Lyα line have been attempted (Ishikawa et al., 2011). The discovery of HeI1083 line far into the corona (Kuhn et al., 1996, 2007) has now made it feasible to measure coronal fields in the 0.1 − 10G range using only linear polarimetry of the HeI1083 line and another forbidden coronal line—such as the newly characterized SiX 1.4301 μm (hereafter SiX1430) line.

For practical reasons the IR spectrum is particularly useful for ground-based studies of the corona because spurious background noise from both the atmosphere and optical scattering in telescopes and instruments decreases with increasing wavelength (Kuhn et al., 2003). Terrestrial thermal emission below 1.8 μm is also inconsequential. Observations (Kuhn et al., 1996) and calculations (Judge, 1998) have described new IR forbidden lines that could be useful as spectropolarimetry diagnostics. Only the HeI1083 line has been observed as a promising IR permitted line for Hanle magnetometry. Some earlier measurements revealed diffuse coronal neutral triplet-state Helium associated with streamers (Kuhn et al., 1996). This initial measurement was eventually confirmed to have solar origin through ground-based spectro-polarimetric observations using the Scatter-free Observatory for Limb, Active Regions, and Coronae (SOLARC) telescope on Haleakala (Kuhn et al., 2007; Moise et al., 2010). The diffuse HeI emission is generated by scattering of photospheric radiation by the triplet state of HeI. The narrow line-width observed for this emission is consistent with the triplet states being produced primarily through electron collisional excitation of singlet-state neutral He in the higher density K-corona, rather than collisional recombination of He$^+$ ions (Moise et al., 2010).

2. DUAL-LINE HANLE MAGNETIC DIAGNOSTICS

The Hanle effect causes a change in the polarization of atomically scattered optical radiation due to the presence of a magnetic field. The magnetic field splits atomic levels into 2J+1 magnetic sublevels (J is the total angular momentum) via the Zeeman effect. If sublevels of the upper level are unevenly populated through their coupling to an anisotropic solar radiation field, then the emission line can be polarized. When the Zeeman splitting is comparable to the energy spread of the upper level

(i.e., the Larmor frequency is smaller than or comparable to the total line emission transition rate), quantum mechanically induced wavefunction interferences will modify the scattering polarization magnitude and rotate the polarization plane by an amount that depends on the field—this is the unsaturated Hanle effect.

The coronal vector magnetic field at a point in the corona is uniquely described by the magnetic flux density $|\mathbf{B}| \equiv B$, the inclination angle θ_B (with respect to the local outward solar radial direction) and the azimuth angle χ_B in a plane perpendicular to the radial direction (**Figure 1**). For a scattering geometry where the emission takes place in the plane-of-sky (POS) we can freely choose the reference axis for the χ_B angle to coincide with the line of sight axis. In the unsaturated Hanle regime, when the atomic Larmor frequency is comparable to the inverse upper-level lifetime, the linear polarization of an emission line is sensitive to all three B-vector parameters, while in the saturated Hanle regime (when the Larmor frequency is much larger than the inverse lifetime) only the angles (θ_B, χ_B) influence the linear polarization. The B value at which the transition between the two regimes takes places is not a sharp value. In fact, a gradual loss of sensitivity takes place above the critical field strength B_H, which depends on the Lande factor g$'$ and the lifetime τ' of the upper level:

$$B_H = \frac{\hbar}{\mu_B g' \tau'} \qquad (1)$$

where μ_B is the Bohr magneton.

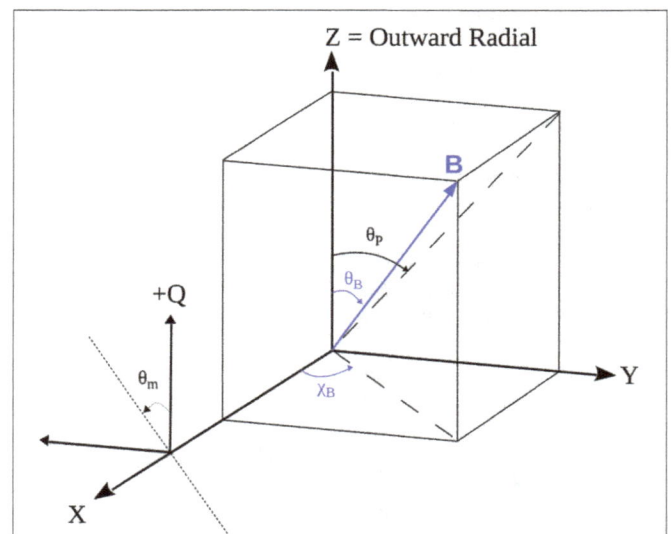

FIGURE 1 | Observing geometry for a magnetic field located in the plane of the sky (corresponding to the ZY plane). The +Z direction indicates the local outward radial direction so moving around the solar limb corresponds to a rotation about the X axis which is taken to coincide with the line of sight. The projected angle of magnetic field on the plane sky θ_P is measured clockwise, while the angle of polarization θ_m is measured counter-clockwise adhering to the common polarimetric convention. The reference direction for the polarization measurement is oriented along the outward radial direction.

The dual-line vector magnetometry technique we propose here relies on simultaneous observations of both permitted and forbidden coronal lines. Near-IR observable coronal lines such as SiX1430, FeXIII 1.0747 μm (hereafter FeXIII1075) and HeI1083 have good polarized atomic modeling available (e.g., House, 1974; Sahal-Brechot, 1977; Casini and Judge, 1999; Asensio Ramos et al., 2008). The critical field strength B_H for the HeI1083 transition is 0.77G (Bommier et al., 1981) while the forbidden lines have critical field strengths in the 10^{-5}G range (House, 1974). The two forbidden lines are firmly in the saturated Hanle regime, while the permitted HeI line maintains Hanle sensitivity up to $\sim$ 8G. In their analysis, Bommier et al. (1981) found the unsaturated Hanle magnetic sensitivity of the HeI1083 line to be significant between $0.1B_H < B < 10B_H$.

The only known visible or IR coronal permitted line is HeI1083. Using current observatories like SOLARC, it is possible to combine near-IR observations of HeI1083 with the FeXIII1075 or SiX1430 lines. When DKIST comes on-line, potentially longer wavelength IR spectropolarimetry in the near-thermal IR will be possible. To date, emphasis has been placed on FeXIII1075 observations for coronal spectro-polarimetry (Tomczyk et al., 2008), although observations during the total solar eclipse on March 29, 2006 (Dima et al., 2016, in preparation) show that SiX1430 emission can be significantly brighter than FeXIII1075. The experiment for that eclipse used a wide-field fiber fed spectropolarimeter. **Figure 2** gives a comparative view of the line signal/noise in each of the fibers. During the same eclipse HeI1083 emission was also observed, although that spectropolarimeter did not have the sensitivity to demonstrate Hanle magnetometry. Nevertheless, these IR measurements clearly point to the importance of the SiX1430 line. Since the FeXIII and SiX ion abundances peak at different temperatures this result highlights the need to have multiple coronal lines accessible for polarimetry that sample different temperature regimes of the corona. While the analysis and examples presented below discuss the SiX1430 line, they can apply equally well to FeXIII1075 observations since the two lines have very similar polarization properties (Judge et al., 2006).

3. ALGORITHM DESCRIPTION

Forbidden lines like FeXIII1075 and SiX1430 have radiative decay rates that are not so different from the electron collision rate at coronal densities. Thus, isotropic collisions can depolarize the Zeeman substate populations in the upper levels of the lines. Mixing occurs through both electron collisions and indirectly through cascades from excited higher levels that can have substantially higher downward transition rates (Sahal-Brechot, 1977; Judge et al., 2006). This collisional depolarization has a density dependence which is difficult to accurately model, but only affects the amplitude of the forbidden line polarization (Judge and Casini, 2001). Consequently, our method in its current form only employs the polarization angle in the forbidden lines which is independent of isotropic collisional effects.

Lines in the saturated Hanle regime maintain a fixed angular relationship between the linear polarization plane (characterized by the polarization angle θ_m and the projected magnetic field orientation on the plane of the sky (characterized by the projected angle θ_P) as shown in **Figure 1**. The magnetic field orientation angles (θ_B, χ_B) are related to the projected angle θ_P by

$$\tan\theta_P = \tan\theta_B \sin\chi_B \qquad (2)$$

For magnetic dipole transitions like SiX1430 the polarization plane is parallel to the magnetic field when $\theta_B < \theta_{VV}$ or $\theta_B > 180° - \theta_{VV}$ and perpendicular when $\theta_{VV} < \theta_B < 180° - \theta_{VV}$, where $\theta_{VV} = 54.7°$ is the Van Vleck angle. This effect leads to the Van Vleck ambiguity (e.g., House, 1974): one measured pair of Stokes Q, U corresponds to at least two pairs of possible magnetic field orientation angles. This ambiguity only applies to a subset of possible field inclinations: all linearly polarized emission from

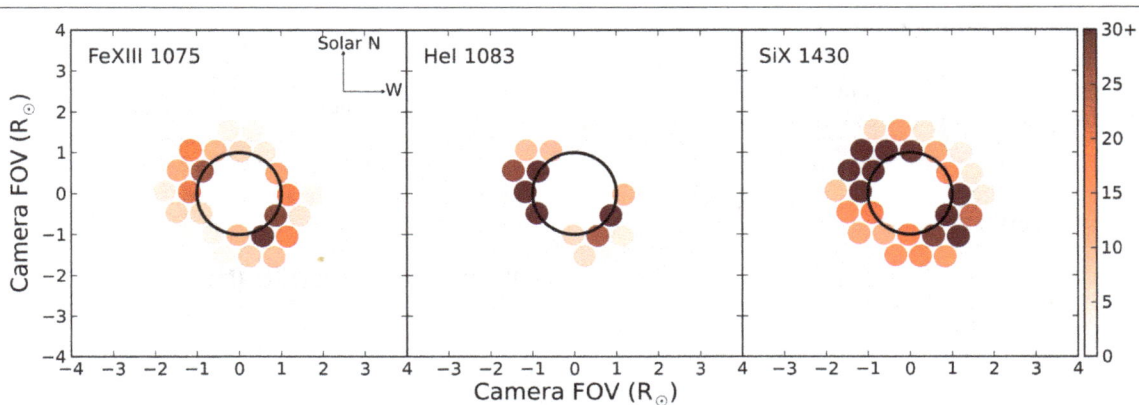

FIGURE 2 | Spatial sampling of the corona during the March 29, 2006 total eclipse. A hexagonal array of 127 fibers sparsely sampled the coronal image plane. Each plot shows the signal/noise measured in each fiber for the lines indicated. One key result from these measurements is the large spatial extent of bright SiX1430 emission compared to FeXIII1075 emission.

TABLE 1 | Summary of parameters and algorithm solutions for two example magnetic field cases.

	Assumed values (B, θ_B, χ_B) Height($R_\odot$) Density(cm^{-3})	SiX1430 polarization (angle[b], amplitude[c])	HeI1083 polarization (angle[b], amplitude)	Solutions[a] (B, θ_B, χ_B)
Field I	(0.65G, 156°, −90°)	(−24 ± 2°, 0.097 ± 0.005)	(78 ± 1°, 0.128 ± 0.005)	$(0.65^{+0.15G}_{-0.15G}, 156^{+1°}_{-1°}, -90^{+10°}_{-15°})$
	0.26			$(0.65^{+0.15G}_{-0.15G}, 24^{+1°}_{-1°}, 90^{+15°}_{-10°})$
	0.2×10^8			$(0.35^{+0.15G}_{-0.10G}, 152^{+3°}_{-2°}, -125^{+10°}_{-5°})$
				$(0.35^{+0.15G}_{-0.10G}, 27^{+3°}_{-2°}, 125^{+5°}_{-10°})$
Field II	(1.3G, 76°, −63°)	(−16 ± 14°, 0.01 ± 0.005)	(−90 ± 6°, 0.024 ± 0.005)	$(1.3^{+0.4G}_{-0.3G}, 76^{+1°}_{-1°}, -63^{+8°}_{-7°})$
	0.08			$(1.3^{+0.4G}_{-0.3G}, 104^{+1°}_{-1°}, 63^{+7°}_{-8°})$
	2×10^8			$(3.2^{+1.8G}_{-0.9G}, 141^{+1°}_{-1°}, -22^{+3°}_{-3°})$
				$(3.2^{+1.8G}_{-0.9G}, 39^{+1°}_{-1°}, 22^{+3°}_{-3°})$

[a] Parameter errors only account for polarization errors in the HeI1083 line. The error contributions from SiX1430 polarization angle uncertainty is discussed in the text.
[b] Polarization angles are given in the [−90°, 90°] domain and the reference direction is along the local solar radial.
[c] Polarization amplitude for SiX1430 are given to show amplitude/noise for each case, but the values themselves are not part of the algorithm.

fields with $\theta_{VV} < \theta_B < 180° - \theta_{VV}$ is ambiguous with respect to a set of field inclinations outside this inclination domain.

In contrast, the HeI1083 permitted line has an upper level lifetime six orders of magnitude shorter. Collisions have a negligible effect on polarization amplitudes permitted lines at coronal densities. Thus, both the polarization angle and amplitude can be modeled without detailed knowledge of the coronal electron density. In our analysis synthetic Stokes I, Q, U profiles for the HeI1083 line are created using the Hanle and Zeeman Light (HAZEL)[1] code (Asensio Ramos et al., 2008). The HeI1083 line is a multiplet between the $2p^3S$ and $2s3S$ terms of the triplet system of HeI. The upper term has three levels with $J = 0, 1, 2$ while the lower term has one level with $J = 1$ with corresponding transition wavelengths: 10829.09Å, 10830.25Å and 10830.34Å. The blue component is not polarizable in emission because the upper level with $J = 0$ has only one magnetic sublevel and is intrinsically unpolarizable. The final Stokes parameters are obtained from integrating the synthetic line profiles over the two red components which typically appear blended due to the small wavelength separation. For the analysis we choose to work in terms of the concepts of linear polarization angle and amplitude (degree) which are related to the line-profile integrated Stokes I, Q, U by the simple relations:

$$\text{Polarization amplitude} = \frac{\sqrt{Q^2 + U^2}}{I} \quad (3)$$

$$\text{Polarization angle} = 0.5 \tan^{-1}\left(\frac{U}{Q}\right) \quad (4)$$

To ensure the polarization angle is correctly calculated an "arctan2"-type function should be applied. This function accounts for the signs of the U an Q values and correctly maps the polarization angle over the domain [−90°, 90°].

[1] http://www.iac.es/proyecto/magnetismo/pages/codes/hazel.php

The algorithm steps for co-spatial sources in the plane of the sky proceed as follows:

1. From the measured forbidden line linear polarization angle θ_m we generate two sets of angle pairs (θ_B, χ_B) satisfying (Equation 2) with $\theta_P = -\theta_m$ or $\theta_P = -(\theta_m + 90°)$. The two sets correspond to the situations where the plane of polarization is respectively parallel or perpendicular to the projected magnetic field direction.
2. HAZEL is used to generate two model Stokes profile grids for each set of angle pairs together with a suitably chosen value range for the magnetic field strength ($0 < B < 8G$). Thus, each point on the grid corresponds to one or more (B, θ_B, χ_B) magnetic fields. The two dimensional grids are expressed in terms of polarization angles and amplitudes calculated using Equations (3) and (4).
3. The measured HeI1083 polarization angle and amplitude are now compared to each of the model grids to find the magnetic field solution grid points consistent with the measurements and errors. If the measured linear polarization parameters only intersects the parallel model grid and lie outside the perpendicular model grid then the deduced magnetic field solution is not affected by the Van Vleck uncertainty. Alternatively if the measured value intersects both grids the deduced magnetic field has at least two degenerate solutions due to the Van Vleck uncertainty.

3.1. Example Application

To demonstrate the method we use as examples two magnetic fields with different (B, θ_B, χ_B) parameters that are typical of coronal fields (**Table 1**). The fields, named Fields I and II are influencing scattering points located in the plane of the sky at different heights, $0.26R_\odot$ and $0.08R_\odot$ respectively. We synthesize "measurements" using the assumed magnetic field parameters and height. HeI1083 measurements are calculated using the HAZEL code, while SiX1430 measurements are

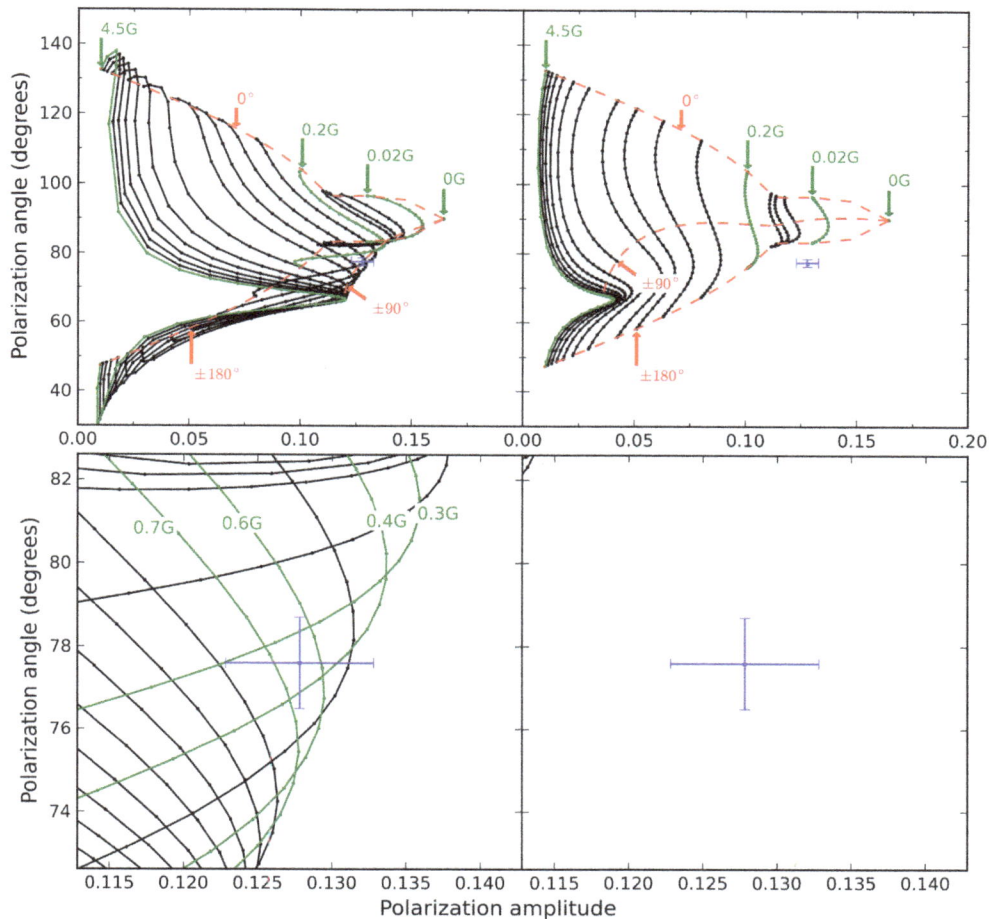

FIGURE 3 | Field I model grids for HeI1083 linear polarization with polarization angle drawn against polarization amplitude. B-isocontours are drawn as solid black lines. **The top panel** shows the entire solution space with some B-isocontours highlighted and labeled in green three χ_B-isocontours drawn with red dashed. **The bottom panel** shows an enhanced region around the measured value for Field I with some B-isocontours highlighted in green. The B-isoncotours in the bottom panel are all separated by 0.1G. For both the top and bottom panels the left plot shows the model grids for plane of polarization parallel to the field projection, while the right plot shows the grid for the plane of polarization perpendicular to the field projection. The measured HeI1083 polarization value for Field I is drawn in blue with errors bars corresponding in size to intensity errors ~0.5%. For Field I the measurement intersects only the parallel grid. This is consistent with an inclination measurement outside the Van Vleck uncertainty region.

calculated using the FORWARD[2] code (Gibson et al., 2010) which generates polarized emission from a multi-level SiX atomic model (Judge and Casini, 2001). To synthesize the SiX1430 polarized emission we also assumed coronal electron densities typical of the heights at which the two fields are located: 0.2×10^8 cm^{-3} for Field I and 2×10^8 cm^{-3} for Field II. The larger exciting radiation anisotropy and lower densities found at larger heights leads to an increase in the amplitude of the SiX1430 polarization. For observations that are not photon limited this leads to improved accuracy for measurements higher above the solar limb.

Following our algorithm two angle/amplitude grids are generated separately for Field I and II from the SiX1430 polarization angle measurement. **Figures 3**, **4** show the model grids generated for Field I and II respectively. By convention the polarization angle is defined over $[-90°, 90°]$, but we redefine it for display purposes over the interval $[0°, 180°]$ without any loss of information. This is done because the model grids shown below are easier to interpret over the modified domain. While the algorithm grid are arbitrarily dense, only some of the grid points are shown to avoid overcrowding the plot space. To visualize the variation with magnetic field strength B-isocontours are highlighted. The errors in the HeI1083 measurement are typical measurement errors of ~0.5% in the line intensity, although more accurate measurements are possible. The solution grids are not uniform so the same measurement error translates differently into inverted magnetic field errors depending on the strength of the field. Visually this is evident in the way the B-isocontours become closer together as the field strength increases. The top panel in each figure shows the full model domain while the lower

[2]http://www.hao.ucar.edu/FORWARD/

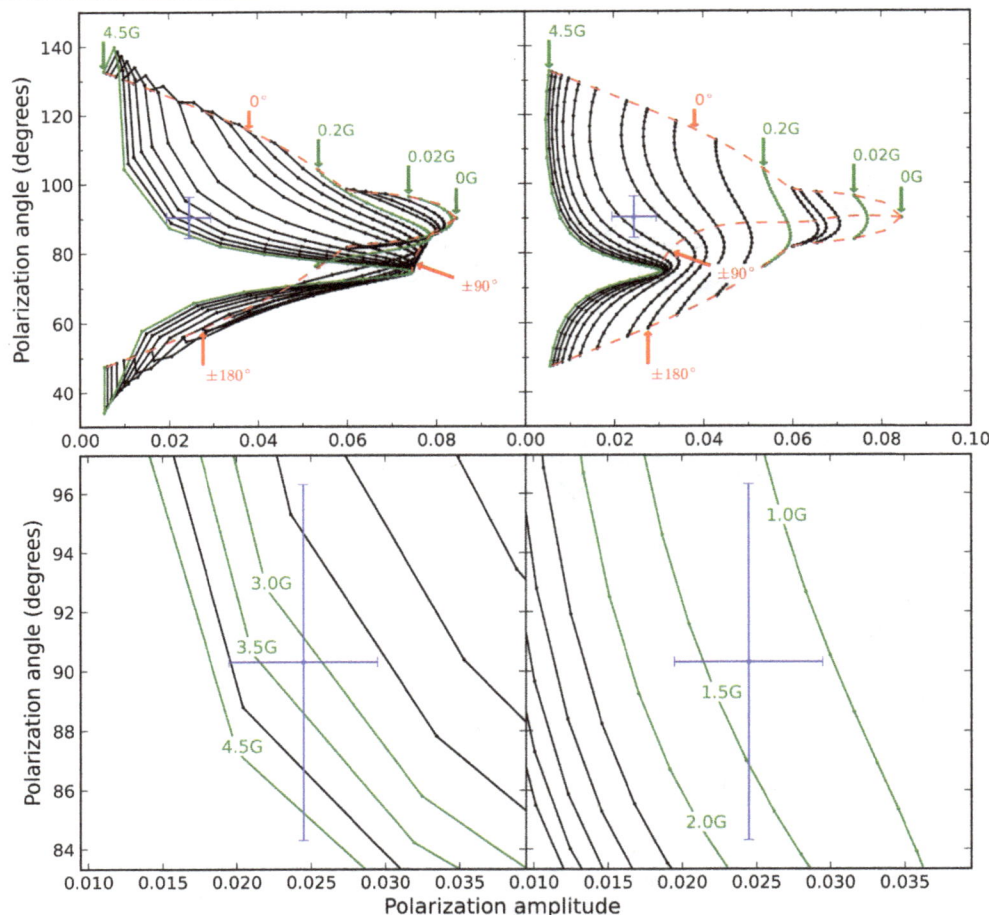

FIGURE 4 | The same as Figure 3 for the Field II model grids. For the bottom panel only the B-isocontours spaced by 0.5G are drawn. For Field II the HeI1083 polarization measurement intersects both model grids which is consistent with inclination solutions inside the Van Vleck uncertainty region.

panels show an enhanced view of each grid near the measured values.

For Field I four independent solutions are obtained as shown in **Table 1**. The four solutions can be divided into two solution pairs with unique values of the magnetic field strength B. For each pair with a unique B there are two degenerate solutions for the angle variables θ_B and χ_B. This "classical degeneracy" is independent of the Van Vleck degeneracy and is inherent in the matter-radiation interaction problem and plane of sky scattering geometry (Bommier, 1980). The two ambiguous solutions can be obtained from each other by reflection of the B vector through the line of sight. For Field I it is evident that the measured polarization value does not intersect the solution grid for the case where the polarization plane is perpendicular to the magnetic field vector. This shows that the magnetic field is not in the Van Vleck degeneracy region.

The recovered solution space for Field II also consists of four independent solutions which can be broken down into two pairs of solutions with unique B values. Same as for Field I each pair with a unique B has two degenerate solutions for the angle variables due to the classical degeneracy. However, for Field II the

origin of the different solutions for the magnetic field strength B lies in the Van Vleck degeneracy. This is seen from the fact that the measured polarization value intersects both model grids.

An important source of error in the analysis is the uncertainty in measuring the SiX1430 polarization angle. This uncertainty can be quite large as is the case for Field II due to the low radiation anisotropy and higher electron density. This uncertainty changes the parallel and perpendicular sets of (θ_B, χ_B) angles that satisfy Equation (2). The effect this uncertainty has on the model grids is shown in **Figures 5, 6** for Field I and II respectively. To produce the variation shown the maximum uncertainty is added and subtracted to the measured SiX1430 polarization angle and new model grids are created using HAZEL. For Field I solutions the errors given in **Table 1** are roughly one and a half time larger for all the parameters. Field II has a much larger uncertainty in measured SiX1430 polarization angle so the effect is larger but mostly concentrated in the angle determination with the variation in the angles increasing to $\pm 25°$ while the magnetic field strength B uncertainty increases by one and a half times the values given in **Table 1**.

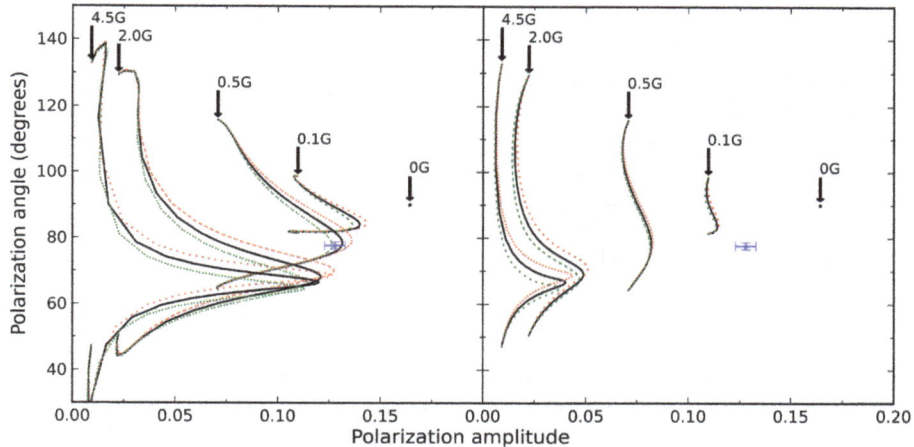

FIGURE 5 | Shown in black are the parallel(left) and perpendicular(right) model grids for Field I as they also appear in Figure 3. The dotted lines represent added (red) and subtracted (green) uncertainties in the SiX1430 polarization angle. Only a few B-isocontours are shown and labeled to avoid overcrowding due to intersecting contour lines. The measured HeI1083 polarization parameters are shown with corresponding measurement uncertainties. Propagating the SiX1430 uncertainty requires new grids to be computed since the shapes of the grid changes as seen by the bending and crossing of model contours.

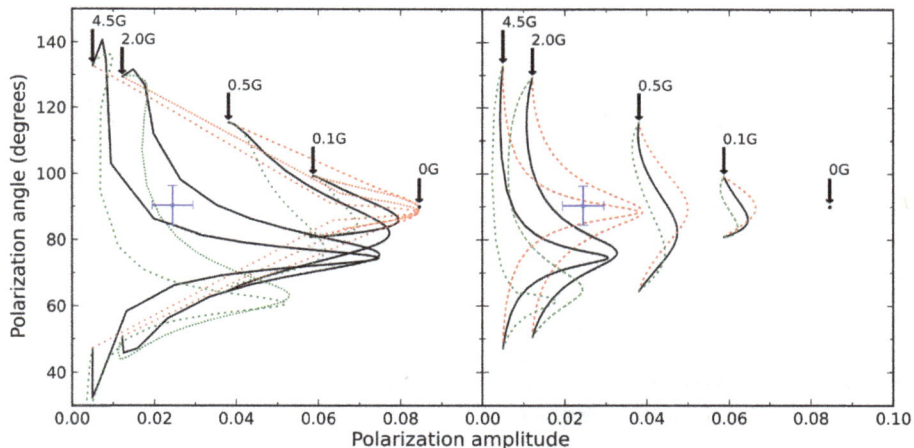

FIGURE 6 | The same as Figure 5 but for the Field II model grids and corresponding errors. The SiX1430 linear polarization signal for Field II is weak and thus relatively uncertain for the selected realistic measurement accuracy. This uncertainty leads to the large distortions in the model grids. High accuracy forbidden line polarization directions will be required in this regime.

For these test cases we assume the line intensity measurement to be ~0.5% for both situation. Since the polarized signal for Field II is ten times weaker than the signal for Field I this translates into a significant increase in the error of the calculated magnetic field strength. However, there is nothing fundamentally limiting about the uncertainty we adopted since the source of the uncertainty is random rather than systematic. For weak polarimetric signals we can increase the integration time to improve the the uncertainty to acceptable levels for errors in the calculated parameters. To achieve the quoted 0.5% accuracy using the current spectrograph on the 0.45 m SOLARC telescope around 12 min of integration time is needed assuming a SiX1430 line brightness of $5 \times 10^{-6} B_{\odot}$ and a spatial resolution element $7''$ in diameter. The larger 4 m telescope DKIST will have improved light collecting power as well

as improved signal throughput. It will make this type of accuracy possible for an observation region $1''$ in diameter in less than 1 s of integration time.

4. DISCUSSION AND CONCLUSIONS

The method proposed provides important constraints on the coronal magnetic field and shows promise as a detailed magnetic field diagnostic, since it drastically constrains the coronal source region local magnetic field to four independent solutions using potentially high signal-to-noise IR linear polarization measurements. This is achieved without knowledge of polarization amplitudes for the forbidden lines that depends on the coronal electron density. It is interesting to note that the method obtains four degenerate solutions for magnetic fields located inside or outside the Van Vleck degeneracy region. Merenda et al. (2006) proposed a chromospheric algorithm that uses measured HeI1083 linear and circular polarization to determine the vector magnetic field for prominences located in the POS. Their method recovered two degenerate solutions for a magnetic field outside the Van Vleck region and four degenerate solutions for a field inside the Van Vleck region. However, all the examples analyzed by them were for field strengths in excess of 10G, which means the HeI1083 emission is in the saturated Hanle regime. From our solutions for Field I we conclude that the extra degeneracy (not related to the classical degeneracy) that appears even for fields outside the Van Vleck regions is due to the unsaturated Hanle effect. Independent knowledge of the electron density (or the forbidden line polarization amplitude) can reduce the degeneracy outside the Van Vleck region from four to two and uniquely recover the magnetic field strength B. For future work we are testing how accurate the density estimate needs to be in order to reliably distinguish between the degenerate solutions. It may be possible to exclude one pair of solutions even with an average electron model consistent with coronal white light observations.

In principle, the information on the electron density is contained in the polarization amplitude of the forbidden line which we excluded from the present algorithm. If it is possible to distinguish between the two solution pairs, we can then recover information about the electron density from the measured polarization amplitude.

Resolving the final ambiguity from the radiation field geometry requires more information. One solution to this problem is through forward modeling, using 3D coronal MHD models, perhaps constrained by photospheric magnetic field measurements. It is noticeable that these degenerate solutions have complementary values for the inclination angle, so constraints just from the photospheric magnetic polarity changes may provide the key to breaking this degeneracy. Note that similar degeneracies are encountered when measuring vector magnetic fields near the photosphere. Leka et al. (2009) summarizes the types of algorithms used to break the ambiguities in photospheric vector magnetograms. Another possibility involves using tomographic inferences from observing the same region over a few days of solar rotation. Bommier et al. (1981) successfully distinguished between ambigous solutions by observing a prominence as it rotates through the plane of the sky.

While detections of coronal HeI1083 emission shows strong correlation with streamers (Moise et al., 2010) more polarimetric observations of this line are needed to determine the exact geometry and line formation mechanisms of the emitting region. One of our principal assumptions is that the HeI1083 and forbidden line emission is co-spatial but a relaxed version of this assumption is that the emitters experience the same magnetic field. Since the magnetic field expands to fill the coronal volume it is not unreasonable to assume that some large volumes of the corona will experience the same magnetic field. However, obtaining a better understanding and characterization of the HeI1083 coronal signal in the context of simultaneous forbidden line emission measurements will provide more information toward understanding the validity of this assumption. Currently we are pursuing a dedicated campaign to obtain co-spatial and quasi-simultaneous spectropolarimetric observations of the FeXIII1075, SiX1430 and HeI1083 lines in the solar corona using the SOLARC telescope on Haleakala. These observations will form the data set needed to test the proposed method.

AUTHOR CONTRIBUTIONS

The bulk of the effort in testing and implementing this algorithm has been done by GD in the course of developing his PhD thesis. JK conceived and advised the primary concept. SB provided support in the Hanle effect theory. SB and JK developed the funding and work plan and contributed ideas and material to the manuscript.

FUNDING

We gratefully acknowledge support from the NSF through grant number ATM-1358270.

ACKNOWLEDGMENTS

We are grateful to Tom Schad for useful clarifying discussion on the HeI spectropolarimetry. SB thanks Predictive Science Inc. and IfA, University of Hawaii for the opportunity to carry out this project as a visiting scientist. The algorithm has benefited from discussion at NSF SHINE program meetings and at the International Space Science Institute (ISSI) International Team Coronal Magnetism meetings in 2013 and 2014.

REFERENCES

Arnaud, J., and Newkirk, Jr. G. (1987). Mean properties of the polarization of the Fe XIII 10747 A coronal emission line. *Astron. Astrophys.* 178, 263–268.

Asensio Ramos, A., Trujillo Bueno, J., and Landi Degl'Innocenti, E. (2008). Advanced forward modeling and inversion of stokes profiles resulting from the joint action of the Hanle and Zeeman effects. *Astrophys. J.* 683, 542–565. doi: 10.1086/589433

Bommier, V. (1980). Quantum theory of the Hanle effect. II - Effect of level-crossings and anti-level-crossings on the polarization of the D3 helium line of solar prominences. *Astron. Astrophys.* 87, 109–120.

Bommier, V., Sahal-Brechot, S., and Leroy, J. L. (1981). Determination of the complete vector magnetic field in solar prominences, using the Hanle effect. *Astron. Astrophys.* 100, 231–240.

Casini, R., and Judge, P. G. (1999). Spectral lines for polarization measurements of the coronal magnetic field. II. Consistent treatment of the stokes vector for magnetic-dipole transitions. *Astrophys. J.* 522, 524–539. doi: 10.1086/307629

Gibson, S. E., Kucera, T. A., Rastawicki, D., Dove, J., de Toma, G., Hao, J., et al. (2010). Three-dimensional morphology of a coronal prominence cavity. *Astrophys. J.* 724, 1133–1146. doi: 10.1088/0004-637X/724/2/1133

House, L. L. (1974). The theory of the polarization of coronal forbidden lines. *Pub. Astron. Soc. Pacific* 86, 490. doi: 10.1086/129637

Ishikawa, R., Bando, T., Fujimura, D., Hara, H., Kano, R., Kobiki, T., et al. (2011). "A sounding rocket experiment for spectropolarimetric observations with the Ly line at 121.6 nm (CLASP)," in *Solar Polarization 6, Vol. 437 of Astronomical Society of the Pacific Conference Series*, eds J. R. Kuhn, D. M. Harrington, H. Lin, S. V. Berdyugina, J. Trujillo-Bueno, and S. L. Keil (San Francisco, CA: Astronomical Society of the Pacific), 287.

Judge, P. G. (1998). Spectral lines for polarization measurements of the coronal magnetic field. I. Theoretical intensities. *Astrophys. J.* 500, 1009–1022. doi: 10.1086/305775

Judge, P. G., and Casini, R. (2001). "A synthesis code for forbidden coronal lines," in *Advanced Solar Polarimetry – Theory, Observation, and Instrumentation, Vol. 236 of Astronomical Society of the Pacific Conference Series*, ed M. Sigwarth (San Francisco, CA: Astronomical Society of the Pacific), 503.

Judge, P. G., Low, B. C., and Casini, R. (2006). Spectral lines for polarization measurements of the coronal magnetic field. IV. Stokes signals in current-carrying fields. *Astrophys. J.* 651, 1229–1237. doi: 10.1086/507982

Kuhn, J. R. (1995). "Infrared coronal magnetic field measurements," in *Infrared Tools for Solar Astrophysics: What's Next?*, eds J. R. Kuhn and M. J. Penn (Singapore: World Scientific), 89–94.

Kuhn, J. R., Arnaud, J., Jaeggli, S., Lin, H., and Moise, E. (2007). Detection of an extended near-sun neutral helium cloud from ground-based infrared coronagraph spectropolarimetry. *Astrophys. J.* 667, L203–L205. doi: 10.1086/522370

Kuhn, J. R., Coulter, R., Lin, H., and Mickey, D. L. (2003). "The SOLARC off-axis coronagraph," in *Innovative Telescopes and Instrumentation for Solar Astrophysics, Vol. 4853 of Society of Photo-Optical Instrumentation Engineers (SPIE) Conference Series*, eds S. L. Keil and S. V. Avakyan (Bellingham, WA), 318–326.

Kuhn, J. R., Penn, M. J., and Mann, I. (1996). The near-infrared coronal spectrum. *Astrophys. J.* 456:L67. doi: 10.1086/309864

Leka, K. D., Barnes, G., Crouch, A. D., Metcalf, T. R., Gary, G. A., Jing, J., et al. (2009). Resolving the 180° ambiguity in solar vector magnetic field data: evaluating the effects of noise, spatial resolution, and method assumptions. *Sol. Phys.* 260, 83–108. doi: 10.1007/s11207-009-9440-8

Lin, H., Kuhn, J. R., and Coulter, R. (2004). Coronal magnetic field measurements. *Astrophys. J.* 613, L177–L180. doi: 10.1086/425217

Lin, H., Penn, M. J., and Tomczyk, S. (2000). A new precise measurement of the coronal magnetic field strength. *Astrophys. J.* 541, L83–L86. doi: 10.1086/312900

López Ariste, A., and Casini, R. (2003). Improved estimate of the magnetic field in a prominence. *Astrophys. J.* 582, L51–L54. doi: 10.1086/367600

Merenda, L., Trujillo Bueno, J., Landi Degl'Innocenti, E., and Collados, M. (2006). Determination of the magnetic field vector via the Hanle and Zeeman effects in the He I λ10830 multiplet: evidence for nearly vertical magnetic fields in a polar crown prominence. *Astrophys. J.* 642, 554–561. doi: 10.1086/501038

Moise, E., Raymond, J., and Kuhn, J. R. (2010). Properties of the diffuse neutral helium in the inner heliosphere. *Astrophys. J.* 722, 1411–1415. doi: 10.1088/0004-637X/722/2/1411

Penn, M. J. (2014). Infrared Solar Physics. *Living Rev. Solar Phys.* 11:2. doi: 10.12942/lrsp-2014-2

Sahal-Brechot, S. (1977). Calculation of the polarization degree of the infrared lines of Fe XIII of the solar corona. *Astrophys. J.* 213, 887–899. doi: 10.1086/155221

Tomczyk, S., Card, G. L., Darnell, T., Elmore, D. F., Lull, R., Nelson, P. G., et al. (2008). An instrument to measure coronal emission line polarization. *Sol. Phys.* 247, 411–428. doi: 10.1007/s11207-007-9103-6

Conflict of Interest Statement: The authors declare that the research was conducted in the absence of any commercial or financial relationships that could be construed as a potential conflict of interest.

The reviewer RC declared a past co-authorship with one of the authors JK to the handling Editor, who ensured that the process met the standards of a fair and objective review.

6

The Effects of Clinorotation on the Host Plant, *Medicago truncatula,* and Its Microbial Symbionts

Ariel J. C. Dauzart, Joshua P. Vandenbrink and John Z. Kiss *

Department of Biology, Graduate School, University of Mississippi, University, MS, USA

Edited by:
Jack Van Loon,
VU University, Netherlands

Reviewed by:
Zurab Silagadze,
Budker Institute of Nuclear Physics,
Russia
Ruediger Hampp,
University of Tuebingen, Germany
Hideyuki Takahashi,
Tohoku University, Japan

***Correspondence:**
John Z. Kiss
jzkiss@olemiss.edu

Specialty section:
This article was submitted to
Cosmology,
a section of the journal
Frontiers in Astronomy and Space
Sciences

Understanding the outcome of the plant-microbe symbiosis in reduced or altered is vital to developing life support systems for long-distance space travel and colonization of other planets. Thus, the aim of this research was to understand mutualistic relationships between plants and endophytic microbes under the influence of altered gravity. This project utilized the model tripartite relationship among *Medicago truncatula—Sinorhizobium meliloti—Rhizophagus irregularis*. Plants were inoculated with rhizobial bacteria (*S. meliloti),* arbuscular mycorrhizal fungi (*R. irregularis*), or both microbes, and placed on a rotating clinostat. Vertical and horizontal static controls were also performed. Clinorotation significantly reduced *M. truncatula* dry mass and fresh mass compared to the static controls. The addition of rhizobia treatments under clinorotation also altered total root length and root-to-shoot fresh mass ratio. Nodule size decreased under rhizobia + clinorotation treatment, and nodule density was significantly decreased compared to the vertical treatment. However, inoculation with arbuscular mycorrhizal fungi was shown to increase biomass accumulation and nodule size. Thus, clinorotation significantly affected *M. truncatula* and its symbiotic relationships with *S. meliloti* and *R. irregularis.* In the long term, the results observed in this clinostat study on the changes of plant-microbe mutualism need to be investigated in spaceflight experiments. Thus, careful consideration of the symbiotic microbes of plants should be included in the design of bioregenerative life support systems needed for space travel.

Keywords: Clinorotation, *Medicago truncatula,* **nodulation, plant biology, plant symbiosis,** *Rhizophagus irregularis, Sinorhizobium meliloti,* **space biology**

INTRODUCTION

To achieve the goal of long-distance space travel, researchers need to first develop a biological self-sustained life support system (Ferl et al., 2002). Plants constitute the cornerstone of this system as they have the ability to recycle waste water for drinking and carbon dioxide into breathable oxygen (National Research Council, 2011). The National Research Council also states a need for "basic plant and microbial research to define how these organisms sense and respond to the varied environments presented in space." Using the legume *Medicago truncatula* as a model for plant space biology research, new information can be obtained on the effects of reduced gravity on a valuable crop species and two of its associated microbial symbionts, rhizobial bacteria, and arbuscular mycorrhizal fungi (AMF). To date, the relationship among legumes, rhizobia and AMF has not previously been studied under reduced gravity. However, studying these symbioses under an altered gravity should provide insight into the role that gravity plays in this symbiotic tripartite relationship on Earth and beyond.

M. truncatula is a plant that is a fast growing, herbaceous legume, and it forms symbiotic relationships with both rhizobia and AMF. Interest in *M. truncatula* is also due to the agriculturally significant close relative, *Medicago sativa*, or alfalfa (Kouchi et al., 2010). *M. truncatula* is a forage crop but is used as a model legume due to its relatively small diploid genome ($2n = 16$) and autogamous reproduction, whereas alfalfa contains a larger tetraploid genome and obligate outbreeding reproduction (Connor et al., 2011). More importantly, *M. truncatula* forms symbiotic relationships with multiple arbuscular mycorrhizal species, including *Rhizophagus irregularis* (formerly *Glomus intraradices*; Hogekamp and Kuster, 2013) and with rhizobial bacteria, including *Sinorhizobium meliloti*.

Research to date suggests that plants can survive and produce a viable food source on the ISS; however, these plants may be stressed growing in microgravity conditions and may not be reaching their full potential as a food source (Colla et al., 2007; Wolverton and Kiss, 2009; Vandenbrink and Kiss, 2016). On Earth, plants have microbial symbionts that are beneficial under stressful conditions such as increased salinity, drought, and heavy metal toxicity (Nadeem et al., 2014). However, altered gravity is a stressor that plants have not evolved to handle. Thus, to date, there is a lack of published information on how colonization of symbiotic microbes such as AMF and rhizobia changes under altered gravity. We cannot be sure that a symbiotic relationship will remain unchanged because there are changes within a plant's growth and physiological responses under an altered gravity.

AMF colonize between 80 and 90% of plant species, including our most valuable food crops (Smith and Read, 2008). Legumes colonized by AMF benefit from a mutualistic relationship with the fungi, leading to an increase in phosphorus (P) and nitrogen (N) uptake by the plant (Wang and Dong, 2011; Larimer and Bever, 2014). Legumes colonized with AMF can produce more protein rich foods and tolerate abiotic stressors more vigorously as a result of an increase in nutrient availability (Vazquez et al., 2001; Ashrafi et al., 2014). Rhizobia colonize 88% of legumes and fix atmospheric nitrogen into a useable form for plants to assimilate into proteins and nucleic acids (Graham and Vance, 2003). Legumes benefit from forming a mutualistic tripartite symbiosis with both rhizobia and AMF through nutrient uptake greater than the individual microbe alone could provide the host (Larimer et al., 2010). The dual colonized host also is more resistant to pathogens and abiotic stressors such as drought, increased soil salinity and metal toxicity (Nadeem et al., 2014). Thus, plants grown with symbiotic microbes in an Advanced Life Support (ALS) system would be more efficient at assimilating the vital elements, P and N, within foods essential to astronauts.

Ground-based methods of altering gravity provide a practical method to studying organisms under such conditions (Herranz et al., 2013). There are many instruments that can simulate the effects of an altered gravity vector, and these include diamagnetic levitation (Qian et al., 2013), random positioning machines (RPMs; Kraft et al., 2000; Herranz et al., 2013) and 2-dimensional (2D) clinostats, the latter which we employ here. This device will rotate organisms at a programmed speed on a single axis. In addition, these instruments can provide valuable, preliminary information as to how organisms will react to microgravity

conditions (Kraft et al., 2000). Many studies with plants under altered gravity that have implemented clinostats have found similar results (depending on the parameter) to microgravity conditions in space (Herranz et al., 2013). Gravity-sensing plant cells in the columella contain dense starch granules that cannot settle under clinorotation due to the constant alteration of the gravity vector. Because these amyloplasts within the columella do not settle, the plant is unable to detect the direction of gravity (Kiss, 2000). In many cases, clinostats and RPMs provide similar results in experiments in true microgravity. For instance, Matia et al. (2010) studied *Arabidopsis* meristematic growth on an RPM and true microgravity on the ISS. The two experimental groups produced similar results that were significantly different from the 1 g controls on the ISS and Earth. This research team concluded that, for the parameters measured, an RPM was comparable to spaceflight experiments on *Arabidopsis* meristematic tissue. In addition, Kraft et al. (2000) used an RPM to study *Arabidopsis* plastid (amyloplast) position in columella cells. The results indicated that the location of the plastids within the cell on the RPM was similar to those from a true microgravity study in spaceflight.

In the future, humans that travel long distances beyond the Earth's atmosphere for extended periods of time will be reliant on an ALS system. This system will be a small scale replication of a quintessential ecosystem pieced together from organisms found in ecosystems on Earth. At the base of any cyclical system that humans rely on for survival, plants will serve as cornerstone organisms. Although plants can produce food, recycle wastes into water and oxygen autonomously, host plants are more efficient with microbial symbionts assisting in nutrient uptake. Beneficial plant-microbe symbioses have the potential for many advantages in a self-sustainable system, but this relationship has not yet been studied in an agriculturally important crop plant in a ground-based, gravity-vector-altering apparatus nor has it been studied on the ISS. Thus, we report on the *M. truncatula—S. meliloti—R. irregularis* symbiotic model system in order to identify the effects of altered gravity on the plant-microbe mutualistic symbiosis. The major questions addressed in this research are: (1) does *M. truncatula* growth change under altered gravity; (2) does altered gravity affect the tripartite symbiosis among *M. truncatula—S. meliloti—R. irregularis*?

MATERIALS AND METHODS

M. truncatula seeds (accession A17) were scarified using 98% (v/v) sulfuric acid for 10 min followed by a rinse with sterile water. Next, seeds were surface sterilized in 30% (v/v) bleach solution for 10 min followed by a second rinse with water. Scarified seeds were shaken for 4 h in sterile water. Wet seeds were placed in a 100 mm sterile Petri dish, turned over so the seeds adhered to the top of the Petri dish, and then were left to germinate overnight. After 24 h, the seedlings were planted in autoclaved Metro Mix 366 soil (Sungro Horticulture, Massachusetts, USA). Seedlings were grown in a growth chamber for 2 weeks (18–22°C, 16 h light, 8 h dark) before transplanting to cone-tainers (#SC10R, Stuewe and Sons, Oregon, USA).

Kanamycin resistant *S. meliloti* 1021 was incubated in TY/Ca$^+$ broth at 30°C for 46–48 h (Journet et al., 2006). At this point, the rhizobia are in late logarithmic to early stationary stage. The absorbance reading at 600 nm was between 0.9 and 1.0 OD as assayed with a spectrophotometer. The culture was then rinsed with sterile water three times to remove TY/Ca$^+$ broth, and then re-suspended in sterile water. The rinsed culture's absorbance was measured again at 600 nm and ranged between 0.8 and 0.9 OD.

R. irregularis was grown in sterile conditions in M media (Bécard and Fortin, 1988) on a split plate. One side (1) of the plate had sucrose added to the media for mutated carrot root to grow on, and the other (side 2) was without sucrose. The AMF was initially placed on side 1 with the carrot root but over time the hyphae spread to side 2. Spores were harvested from the gel matrix by blending in a Hamilton Beach Single Server Blender (item# 1568444, Walmart, Arkansas, USA) with 10 mM sodium citrate. The mixture was blended for five, 5 s intervals. The solution was filtered through 250 μm mesh cloth to remove carrot root debris. Only spores and liquid passed through this mesh. To then isolate the spores, we filtered the solution through a 50 μm mesh. Spores remained on the mesh and were rinsed with sterile water into a Falcon tube. To quantify the amount of spores collected, 25 μL of spore solution was placed onto a microscope slide and examined under a dissecting microscope. The spores from each 5 μL drop were counted then averaged. From this average, the concentration of the spore solution was extrapolated.

Each cone-tainer was layered from the bottom of the cone-tainer to the top as follows: a foam plug (#98140-960, VWR, Georgia, USA), Profile® Greens Grade porous ceramic as the lowermost layer followed by a 2:1 sand to pebble mixture (125 ml), a centimeter of packed sand, and the 2 weeks old seedling on top of the sand layer. The total volume of the cone-tainer was 164 ml. A ¼ inch polyethylene tube (#HSVEB20, Watts, Home Depot, USA) was placed into the soil to allow efficient watering of rotating plants via syringe (#301604, Becton Dickenson Disposable Syringe). The roots of an individual seedling were inoculated with ∼800 spores of *R. irregularis*, 8–9 ×108 cells of *S. meliloti*, or both, and subsequently covered with a thin layer of sand (5–10 ml). The cone-tainer was capped by an Erlenmeyer flask foam plug (#98140-960, VWR, Georgia, USA). Inoculated and non-inoculated (microbial control) 14-day-old plants are rotated for 28 days on a 1 rpm rotating clinostat (**Figure 1A**). We also inoculated two sets of control plants with the same concentrations of spores and bacteria, and these gravitational control plants were place in a horizontal or vertical position for 28 days (**Figure 1B**). In addition, non-inoculated (microbial control) and non-rotated (gravitational control) plants were set in a vertical and horizontal position as well. A fifth run that only tested AMF and microbial control plants was performed. In total, 186 plants were analyzed for statistical differences.

During each experiment, plants were grown under a 16 h day/8 h night cycle at ∼23°C. The plants were given 10 ml of distilled water via syringe twice a week through a watering tube. In addition, Hoagland's nutrient solution was diluted to 1/8 concentration and delivered in the same manner as water once a

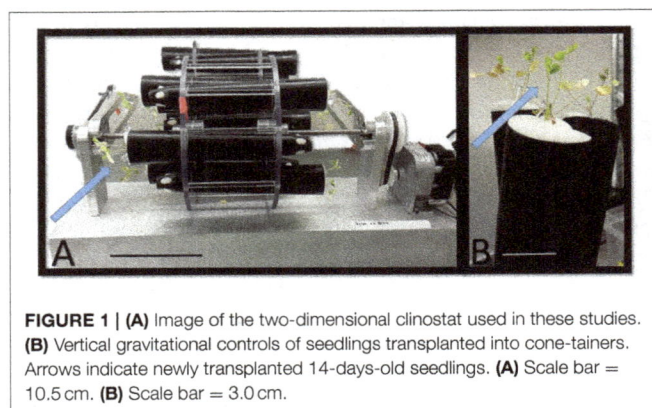

FIGURE 1 | (A) Image of the two-dimensional clinostat used in these studies. **(B)** Vertical gravitational controls of seedlings transplanted into cone-tainers. Arrows indicate newly transplanted 14-days-old seedlings. **(A)** Scale bar = 10.5 cm. **(B)** Scale bar = 3.0 cm.

week. After 28 days post inoculation (dpi), plants were removed from the rotating clinostat. Stem length, total root length, and total biomass were measured as well as nodule number and size. Root tissue was cleaned of soil, photographed, and placed into 50% (v/v) ethanol. Aerial tissue was dried at 60°C in an oven for 48 h. The total root length was measured using the images taken at the end of each experiment in program GIA Roots (Galkovskyi et al., 2012). The same images were used to measure nodule size using the program ImageJ (Rasband, 2014).

Data were analyzed using two- and three-way analysis of variance (ANOVA) and post hoc tests in the program R (version 3.2.1) with the "car" and "multcomp" packages. The response variables tested in the three-way ANOVA models were dry mass of aerial tissue, change in total fresh biomass after 28 dpi, total root length, and root:shoot fresh mass ratio. Each three-way ANOVA model was run using the two microbial (+/− AMF and +/− Rhizobia) predictor variables and gravitational predictor variables (GrT: vertical, horizontal, and clinorotated), as well as with the interactions among those three variables. Two-way ANOVA tests were performed for the response variables—nodule size, nodule density, and AMF percent colonization. For analysis of nodule size and nodule density, we utilized plants that were inoculated with rhizobia. This data subset was analyzed with two-way ANOVA models containing the predictor variables—AMF, GrT, and the AMF × GrT interaction. For analyses of nodule size and nodule density, we analyzed only plants that were inoculated with AMF and this subset was used for a two-way ANOVA model that included the predictor variables rhizobia, GrT, and the rhizobia × GrT interaction. After the ANOVA tests were performed and significance among treatments was found, means were compared using Tukey's Honestly Significant Difference test ($p < 0.05$). The relationship between dry mass of above-ground tissue and nodule size was analyzed in Pearson's Correlation Test.

RESULTS

The goal of this study was to use clinorotation to provide altered gravity conditions in order to analyze any potential changes in the symbiotic relationships between the host plant and its associated symbiotic fungi and bacteria. When *M. truncatula* was grown under altered gravity on a clinostat for 28 days after being

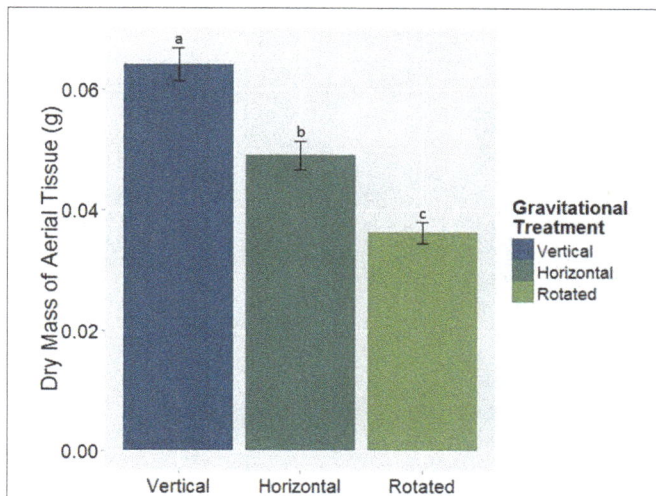

FIGURE 2 | The main effect of the gravitational treatments on the above-ground dry biomass in *M. truncatula* (*n* = 186, *p* < 0.05). The letters above each standard error bar denote significance between the gravitational treatments. Treatments with the same different letters are significantly different. Clinorotation significantly reduces the above-ground mass of *M. truncatula* ($p < 0.05$).

TABLE 1 | Three-way ANOVA results for dry mass of the above-ground tissue in microbial, gravitational treatments (GrT), and microbial × gravitational treatment groups.

Three-way ANOVA of dry mass in aerial tissue

Treatment	df	F-value	p-value
AMF	1	3.54	0.061
Rhizobia	1	0.75	0.387
*GrT	2	6.04	0.003
AMF × Rhizobia	1	1.87	0.172
AMF × GrT	2	0.39	0.677
Rhizobia × GrT	2	0.80	0.451
AMFx Rhiz × GrT	2	0.08	0.920

Significant differences are indicated by an asterisk ($p < 0.05$).

TABLE 2 | Three-way ANOVA results indicating the significance in the change in fresh biomass of *M. truncatula* 28 dpi and gravitational treatment (GrT).

Three-way ANOVA of change in fresh mass

Treatment	df	F-value	p-value
*AMF	1	8.38	4.29×10^{-3}
*Rhizobia	1	32.34	5.38×10^{-8}
*GrT	2	43.78	3.96×10^{-6}
AMF × Rhizobia	1	2.23	0.137
AMF × GrT	2	1.21	0.300
Rhizobia × GrT	2	1.33	0.266
AMFx Rhiz × GrT	2	0.04	0.961

AM fungi and rhizobia treatments and GrT created a main effect in the host plant's ability to accumulate fresh biomass. Significant differences are indicated by asterisks ($p < 0.05$).

TABLE 3 | Three-way ANOVA results indicating the significance of total root length with a combination of microbial treatments, gravitational treatments (GrT) and a combination of both.

Three-way ANOVA results of total root length

Treatment	df	F-value	p-value
AMF	1	2.21	0.139
Rhizobia	1	0.02	0.896
*GrT	2	18.17	6.84×10^{-8}
AMF × Rhizobia	1	0.52	0.471
AMF × GrT	2	1.43	0.243
*Rhizobia × GrT	2	6.22	2.46×10^{-3}
AMFx Rhiz × GrT	2	0.07	0.934

Significant differences are indicated by asterisks ($p < 0.05$).

inoculated with AMF, rhizobia, or both symbionts, it exhibited significant changes in distribution of above- and below-ground biomass. Thus, clinorotation produces a significant reduction of dry aerial tissue relative to the vertical control and the horizontal control, which also differed from each other (**Figure 2**, **Table 1**). At no time did any plants exhibit signs of drought response in any of the three orientation conditions, suggesting adequate access to water in these experiments. Interestingly, microbial treatments appear to have no significant effect ($p < 0.05$) on dry mass of aerial tissue under altered gravity.

Total fresh biomass accumulation after 28 dpi was influenced by significant main effects of AMF, rhizobia, and gravitational treatment (GrT; **Table 2**). Plants inoculated with AMF gained significantly more fresh mass compared to plants without AMF (**Figure 3A**, $p < 0.05$). The inoculation with rhizobia had the opposite effect on plants. Thus, nodulated plants gained significantly less fresh mass than groups without nodules (**Figure 3B**, $p < 0.05$). Clinorotation reduced the fresh mass accumulated over 28 days, and the vertical treatment had the largest fresh mass gain (**Figure 3C**).

The total root length was measured to assess below-ground morphological changes. There was a significant interaction effect between the rhizobial and gravitational treatments (**Figure 4**, **Table 3**) resulting in total root length being much more reduced in nodulated plants compared to non-nodulated plants ($p < 0.05$, **Figure 4**). Rhizobia-treated plants in the vertical (**Figure 5A**) and horizontal (**Figure 5B**) control groups developed significantly greater roots lengths than the rhizobia + clinorotated group respectively (**Figure 5C**).

The root:shoot fresh biomass ratio was measured as an indication of the plants' allocation of resources within the plants. There was a significant interaction between the rhizobia

inoculum and GrT on the root:shoot ratio (**Figure 6**, **Table 4**). Non-nodulated plants gained more below-ground biomass compared to rhizobia-inoculated plants, as indicated by the overall higher root:shoot ratio., but differences among the gravitational treatments were much more dramatic in the non-nodulated plants compared to the nodulated plants.

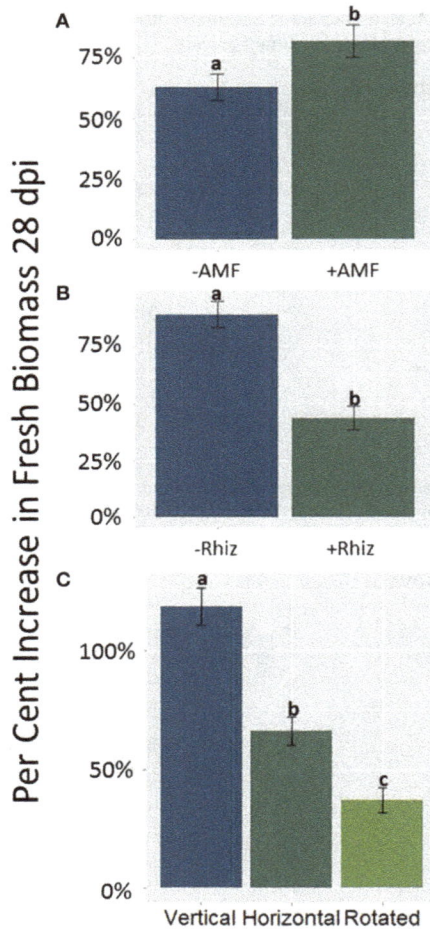

FIGURE 3 | (A–C) The main effects of AMF (A), rhizobia (B), and gravitational treatment (C) on the difference in fresh mass of seedlings 28 days post inoculation (dpi) in the host, *M. truncatula*. (A) AMF treated plants from all three orientations ($n = 94$) gained significantly more fresh mass compared to plants not treated with AMF ($p < 0.05$). (B) Rhizobia treated plants from all three orientations ($n = 69$) gained significantly less fresh mass compared to plants not treated with rhizobia ($p < 0.05$). (C) The clinorotated plants ($n = 66$) gained less fresh mass compared to vertically and horizontally grown plants ($p < 0.05$; $n = 58$ and $n = 62$, respectively). The letters above each standard error bar denote significance between treatments.

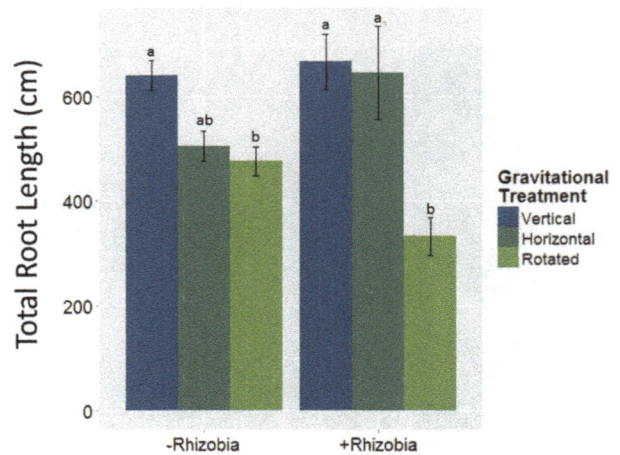

FIGURE 4 | The interaction between rhizobia × gravitational treatment in total root length. Letters indicate significance among nodulated treated plant (+Rhizobia) and non-nodulated plants (-Rhizobia). Plants treated with rhizobia ($n = 69$) had significantly less total root length ($p < 0.05$). Total root length is significantly reduced in clinorotated plants inoculated with rhizobia ($n = 30$) compared to the vertically and horizontally grown plants ($p < 0.05$; $n = 19$ and $n = 20$). The letters above each standard error bar denote significance between rhizobia treatments.

TABLE 4 | Three-way ANOVA results of root:shoot fresh mass of *M. truncatula* 28 dpi after microbial inoculation and gravitational treatments (GrT).

Three-way ANOVA results of root:shoot fresh mass

Treatment	df	F-value	p-value
AMF	1	0.40	0.529
*Rhizobia	1	154.61	2.20×10^{-16}
*GrT	2	37.45	2.97×10^{-14}
AMF × Rhizobia	1	1.20	0.274
AMF × GrT	2	2.05	0.132
*Rhizobia × GrT	2	4.29	0.015
AMF× Rhiz × GrT	2	0.14	0.872

Significant differences are indicated by asterisks ($p < 0.05$).

TABLE 5 | Two-way ANOVA results of nodule size in rhizobia and AMF +rhizobia treated host plants under gravitational treatments (GrT).

Two-way ANOVA results of nodule size

Treatment	df	F-value	p-value
*AMF	1	5.39	0.024
GrT	2	1.93	0.153
*AMF × GrT	2	3.41	0.039

Significant differences are indicated by an asterisk ($p < 0.05$).

Alterations in the gravity treatment interacted with AMF inoculation to influence nodule size (**Figure 7**, **Table 5**) and had a significant main effect on nodule density (**Figure 8**, **Table 6**). Plants inoculated with AMF did not have significantly different nodule sizes among gravitational treatments, whereas plants not inoculated with AMF exhibited negative effects of clinorotation on nodule size. Rhizobia-inoculated plants produced a reduced number of nodules per centimeter under clinorotation compared to the vertical control (**Figure 8**, $p < 0.05$).

In addition, gravity did not affect the percent colonization of *R. irregularis* and neither did the addition of rhizobia in the inoculum (data not shown). The mean number ± SE of nodules for vertical plants inoculated with rhizobia only was 53.5 ± 5.6

($N = 10$) and 73.9 ± 4.1 ($N = 10$) for vertical plants inoculated with both microbes. For horizontal plants inoculated, the mean ± SE nodule count was 45 ± 4.2 ($N = 10$) for plants inoculated

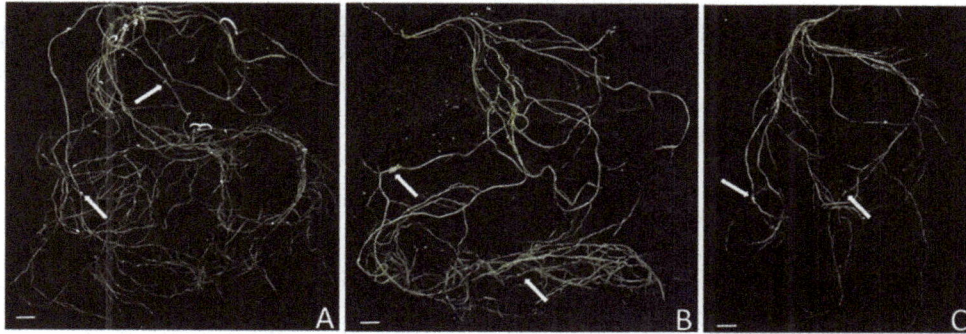

FIGURE 5 | Images are of the root system of 42-days-old *M. truncatula* plants inoculated with *S. meliloti* then grown for 28 dpi. The arrows indicate the position of a nodule. Note the differences in the total root length and size of nodules in each treatment. **(A)** Roots of vertical gravitational control seedlings. **(B)** Roots of horizontal control seedlings. **(C)** Roots from clinorotated seedlings. Scale bars = 1 cm.

with rhizobia only, and 47.5 ± 6.7 ($N = 10$) for those inoculated with both microbes. Finally, for clinorotated plants, the mean number $\pm$ SE of nodules for plants inoculated with rhizobia only was 53.5 ± 5.6 ($N = 15$) while those plants inoculated with both microbes averaged 73.9 ± 4.15 ($N = 15$) nodules.

There was a significant correlation ($r = 0.35$, $p < 0.05$) between nodule size and biomass in plants that are only inoculated with rhizobia (**Figure 9A**). As nodules increase in size, above-ground biomass increases as well. However, this trend was no longer significant when AMF was added to rhizobia inoculated plants (**Figure 9B**), suggesting AMF may indirectly influence nodule size.

DISCUSSION

The first goal of this research was to determine if the development of *M. truncatula* is changed under clinorotation. The second objective of the research was to determine if the symbiotic relationships among the host plant, rhizobia, and AMF were altered due to clinorotation. Parameters used to measure changes in *M. truncatula* development and the effects of symbioses were dry mass of aerial tissue, change in fresh mass of the host plant, total root length, and root:shoot fresh mass ratio. To determine changes between the host plant and microsymbionts under clinorotation, we measured nodule size, nodule density, and AMF percent colonization.

Effects of Clinorotation on *M. truncatula* Growth

It is well known that microgravity can have profound effects on the growth and development of plants. For example, the growth rate and biomass of various plant organs changes significantly in microgravity compared to 1 g controls (Hilaire et al., 1996; Musgrave et al., 2000; Wolverton and Kiss, 2009; Paul et al., 2012). Physiological processes such as cellular division and other parameters are altered under microgravity (Matia et al., 2010). Previous research has shown that simulated microgravity affects auxin regulation (Oka et al., 1995), which has also been shown

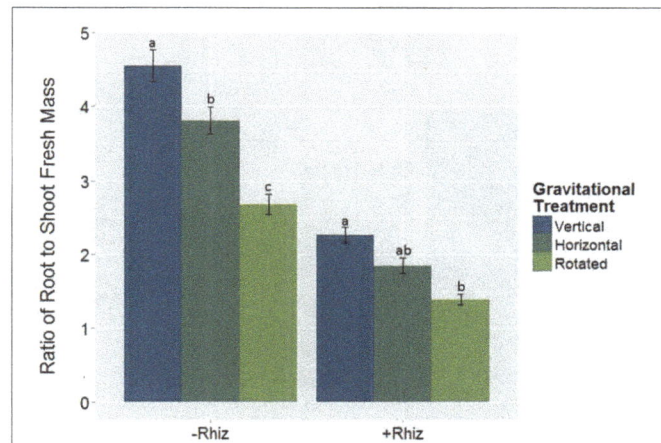

FIGURE 6 | The root:shoot fresh mass ratio responding to the significant rhizobia × gravitational treatment interaction ($p < 0.05$). The root:shoot fresh mass reduces with clinorotation and the addition of rhizobia (+Rhizobia). Rhizobia + clinorotated treatments ($n = 30$) have a significantly smaller root to shoot mass ratio than vertical controls ($p < 0.05$; $n = 39$). Plants without nodules (−Rhizobia; $n = 117$) have a higher root:shoot fresh mass ratio than +Rhizobia ($p < 0.05$; $n = 69$). The letters above each standard error bar denote significance between rhizobia treatments.

to be involved in nodule development in *Medicago* (Wasson et al., 2006). In *Arabidopsis* seedlings grown on the ISS, these latter researchers found that cortical meristematic tissue had increased individual cell numbers, but growth decreased in those cells compared to 1 g controls. The changes in cellular processes due to true microgravity led to increased hypocotyl and total root length. In a different spaceflight experiment, seedlings of mung bean, oat, and sunflower were exposed to microgravity (Krikorian and O'Connor, 1984). In this study, the root cells showed a decreased number of cells dividing in the meristem, chromosome breakage, deletions, and translocations compared to ground controls.

The results of our research indicate that clinorotation had significant effects on *M. truncatula* compared to vertical

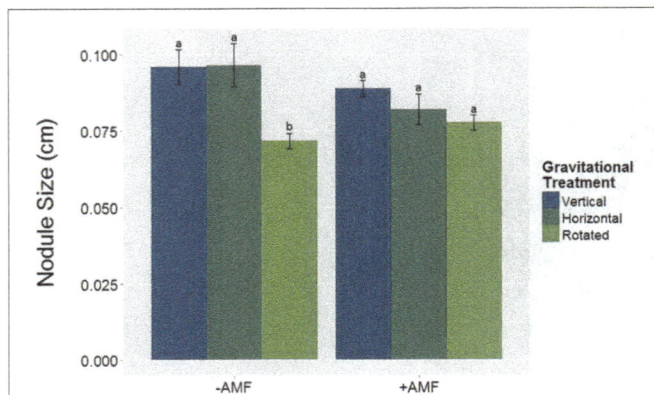

FIGURE 7 | The interaction between AMF x gravitational treatment in nodulated plants (*n* = 71). Nodule size is significantly smaller (*p* < 0.05) under rhizobia + clinorotated (*n* = 15) treatments compared to the static controls. The letters above each standard error bar denote significance between AMF treatments.

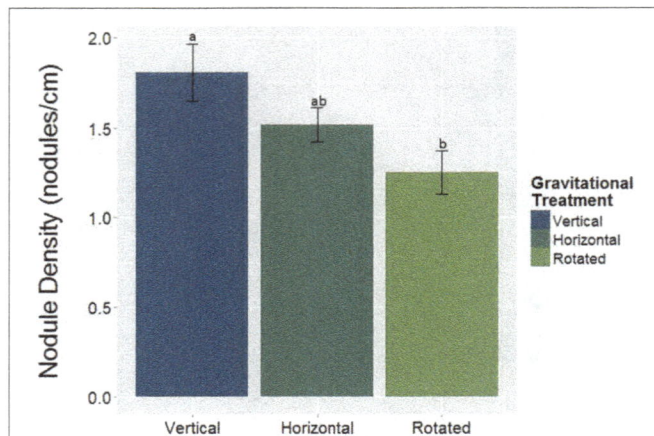

FIGURE 8 | The main effect of gravitational treatments on nodule density (nodules/cm) in *M. truncatula* (*n* = 71). Nodule density was calculated to compensate for the differences in total root length among gravitational treatments. The number of nodules per centimeter of root is significantly reduced in nodulated, clinorotated (*n* = 30) groups compared to the nodulated, vertical controls (*p* < 0.05; *n* = 19). The letters above each standard error bar denote significance between gravity treatments.

TABLE 6 | Two-way ANOVA results of nodule density in rhizobia and AMF + rhizobia treated plants under gravitational treatments (GrT).

Two-way ANOVA results of nodule density

Treatment	df	F-value	p-value
AMF	1	1.55	0.217
*GrT	2	6.44	2.8×10^{-3}
AMF × GrT	2	4.48	0.159

Significant differences are indicated by an asterisk (p < 0.05).

and Piastuch (2005) studied soybean seedlings (*Glycine max*) under clinorotation and found that roots were shorter in the soybean plants clinorotated compared to 1 g ground controls and plants grown in microgravity on the ISS. Other clinostat and spaceflight studies have shown the opposite effect (Hilaire et al., 1996; Aarrouf et al., 1999; Mortley et al., 2008). Thus, in contrast, these researchers found that in reduced gravity, plants will grow faster and are elongated compared to 1 g controls. Aarrouf et al. (1999) germinated plants on nutrient rich agar plates while rotating at 1 RPM on a clinostat. In contrast, after 5 days, these researchers found that horizontally clinorotated plants had increased total root length and increased fresh biomass similar to spaceflight seedlings in other experiments.

These experiments implemented by other researchers were performed on seedlings that germinated under altered gravity and for shorter periods of time in comparison to our research. Thus, young seedlings are using resources initially from cotyledons then from nutrient rich agar. Hilaire et al. (1996) grew plants in Biological Research in Canisters (BRIC), which is a sealed system that may have increased ethylene content in the surrounding environment. The host plants in this study were 14-days-old when inoculated and grown in an open environment in which ethylene could be released. Thus, these differences among experimental conditions potentially can be attributed to the deviation from morphological obtained from other studies utilizing clinorotation and space flight.

In addition, in our studies, plants grown in a horizontal orientation experienced a reduction in biomass when compared to the vertical control. We hypothesize that this slight reduction in overall plant biomass was the result of reduced physical capacity for proper root growth of the plants. As opposed to the clinorotated samples, horizontal plants had a constant gravity vector, resulting in roots growing a short distance before being impeded by the vessel wall, restricting the ability of the roots to grow and proliferate evenly throughout the soil. Due to the changing effective gravity vector of clinorotation, this same impediment was not present in clinorotated samples.

The Effects of Clinorotation on *M. truncatula–S. meliloti*

Our results show that over the course of these experiments, nodulated plants accumulated less fresh mass compared to the microbial control and AMF groups. Rhizobia require photosynthate from the host to fix nitrogen. Conditions and species compatibility may not have been conducive to a beneficial

and horizontal growth within the microbial control group. Clinorotated plants had accumulated less total fresh and above-ground dry mass compared to the static controls (**Figures 2, 3C**). Similarly, Tripathy et al. (1996) found a reduction in fresh biomass of *Tritium aestivium* in a space experiment. Hilaire et al. (1996) also found that clinorotation reduces the fresh mass in plants compared to vertically grown plants, and attributed this result to increased ethylene production found in clinorotated plants.

In our study, total root length was reduced in clinorotated treatments. Published reports have shown an increase, decrease, or no effect (Johnson et al., 2015) in total root length after simulated or actual microgravity treatment. For instance, Levine

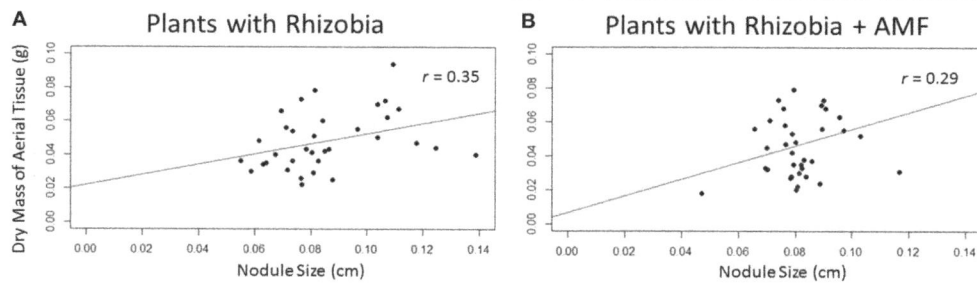

FIGURE 9 | Scatterplots depict the relationship between nodule size and dry mass of aerial tissue. (A) Plants inoculated with only rhizobia ($n = 34$) showed a significant ($p < 0.05$, $p = 0.35$) positive correlation between nodule size and biomass. (B) Rhizobia + AMF ($n = 35$) inoculated plants do not have a significant correlation between nodule size and biomass.

legume-rhizobium symbiosis (Heath and Tiffin, 2007; Heath et al., 2010).

In a report by another group, Mhadhbi et al. (2005) used above-ground biomass as an indicator of nitrogen-fixing efficiency in the *M. truncatula–S. meliloti* relationship. Their study showed that above-ground biomass was positively correlated to acetylene reduction activity, an indicator of nitrogenase activity. In our study, **Figure 9A** depicts the positive correlation ($r = 0.35$) between the nodule size and dry mass of aerial tissue. Plants treated with clinorotation + rhizobia have reduced nodule size causing significantly less dry above-ground biomass to accumulate. This negative trend is attenuated in clinorotation + rhizobia + AMF as the correlation is not significant between nodule size and dry above-ground mass. Furthermore, the correlation between biomass and nodule size could be due to the stress effect that clinorotation has on the symbiotic relationship, which potentially may include changes in water availability and vibrations among other things. Additional stresses such as drought and salinity stress have been shown to reduce the nitrogen-fixing ability of colonized nodules (Serraj and Drevon, 1998; Serraj et al., 1999). Previous studies on host-pathogen relationships and animal-bacteria symbiosis under altered gravity indicate that colonization of the host by the microbe is significantly altered (Bishop et al., 1997; Ryba-White et al., 2001; McFall-Ngai et al., 2010). Thus, the stress of altered gravity may inhibit rhizobia benefits by reducing the efficiency of nitrogen fixation.

Clinorotation Influences Relationships among Host and Microbes

In previous studies, host-microbe interactions are known to be affected in various ways by microgravity. Ryba-White et al. (2001) studied changes in virulence of the soybean fungal pathogen, *Phytophthora sojea* in a spaceflight experiment. Plants inoculated with the fungus had significantly more colonized roots while exposed to microgravity compared to ground controls. The researchers attributed the increased virulence of the pathogen to the stress of microgravity weakening the plants immune response. Similarly, Bishop et al. (1997) sent seeds of wheat cv. Super Dwarf to the ISS for a study on peroxidase activity but found that a fungal pathogen, *Neotyphodium* sp., had infected the

germplasm of the seeds. Once germination took place, the fungus began to infect 50% of the plants grown in space while only 30% were affected in the ground controls.

Another mutualistic symbiosis studied in reduced gravity was the water fern *Azolla* and nitrogen-fixing cyanobacteria, *Anabeana azollea*, along with the concomitant bacteria associated with the symbiotic partners (Kordyum et al., 1983). In this study, the *Azolla-Anabeana* were sent into space for 8 days along with concomitant bacteria to test the effects of microgravity on the microcosm within the *Azolla* system. The cyanobacteria, *Anabeana*, was reported to have increased in number in the space flight experiment along with the concomitant bacteria living within the *Azolla*, although only data from the unknown bacteria were present in the paper. Currently, parasitic relationships among plants and microbes under microgravity are better understood than the mutualistic relationships, which are the focus of the present study.

Effects of Both AMF and Rhizobia on *M. truncatula* under Clinorotation

AMF and rhizobia often cohabitate the same host roots. Larimer and Bever (2014) used the perennial prairie legume, *Amorpha canescens*, to study the effects of the symbionts on the host. Their studies showed that in this legume, AMF-rhizobia symbionts benefit the plant as well as each other. The AMF-rhizobia symbionts increased above-ground biomass and provided more nitrogen and phosphorus to the above- and below-ground tissue compared to plants colonized by only one of the symbionts. AMF-rhizobia compatibility dictates the extent of nutritional benefits received by the host plant (Lisette et al., 2003). In plants inoculated with compatible AMF and rhizobia, nodule count and mass increased leading to higher nitrogen and phosphorus in shoots, above-ground biomass and increased fruit output (Lisette et al., 2003; Larimer and Bever, 2014). Thus, by inoculating legumes with two synergistic microbial symbionts, the host can provide more resources that are essential to humans on Earth (or in space).

Under 1 g conditions, AMF in combination with rhizobia have beneficial effects on plant growth and health (Vazquez et al., 2001; Gao et al., 2012; Wang et al., 2012). Our findings show that nodule density is reduced in rhizobia + clinorotated treated plants compared to vertical gravitational control. In our studies, although the rhizobia + horizontal control is not significantly different from the clinorotated group, there is a substantial increase in nodules/cm within the horizontal group in comparison. Heath et al. (2010) describe nodule numbers as the trait most likely to be "the main determinant of rhizobium fitness during the symbiotic stage of the life cycle." The addition of AMF and clinorotation may reduce the ability of rhizobia to form nodules on the roots of a host within the time parameters of our experiment. Studies by other researchers show that under the growth conditions found on Earth, AMF can inhibit rhizobia species from nodulation when co-inoculated (Lisette et al., 2003; Mortimer et al., 2008; Wang and Dong, 2011).

Interestingly, it appears that rhizobia inoculation tends to have an overall negative effect on plant biomass accumulation, while AMF inoculation tends to have a more beneficial effect. In treatments where both microbes were present, the rhizobia effect of reduced biomass accumulation by the plant is still present. As part of the symbiotic relationship between rhizobia and *M. truncatula*, carbohydrates are provided to the bacteria in exchange for nitrogen. It is likely that the cost of this mutualistic relationship reduces the amount of carbohydrates available for overall biomass production of the plant.

In summary, our results clearly demonstrate that clinorotation reduces the growth and development of *M. truncatula* and significantly affects the tripartite symbiosis among *M. truncatula*, *S. meliloti*, and *R. irregularis*. Nodulation is affected in both groups of microbial and clinorotated treatments and the most notable effect of clinorotation is the reduction in nodule size and number. AMF colonization was not significantly altered by clinorotation in these studies, but further research into the relationship of the compatibility of the plant-microbe relationship under altered gravity is needed. In the long-term, these results on the effects of clinorotation on symbiosis need to be extended in spaceflight experiments. Furthermore, careful consideration of the symbiotic microbes of plants needs to be included in the design of self-sustaining life support systems needed for space travel.

AUTHOR CONTRIBUTIONS

AD designed and performed the experiments, analyzed the results, and wrote the manuscript. JV participated in the analysis of the results and the writing of the manuscript. JK supervised the project, and participated in the analysis and writing of the manuscript.

FUNDING

Financial support for this project was provided by NASA grant NNX15AK39A.

REFERENCES

Aarrouf, J., Darbelley, N., Demandre, C., Razafindramboa, N., and Perbal, G. (1999). Effect of horizontal clinorotation on the root system development and on lipid breakdown in rapeseed (*Brassica napus*) seedlings. *Plant Cell Physiol.* 40, 396–405. doi: 10.1093/oxfordjournals.pcp.a029555

Ashrafi, E., Zahedi, M., and Razmjoo, J. (2014). Co-inoculations of arbuscular mycorrhizal fungi and rhizobia under salinity in alfalfa. *Soil Sci. Plant Nutr.* 60, 619–629. doi: 10.1080/00380768.2014.936037

Bécard, G., and Fortin, J. A. (1988). Early events of vesicular–arbuscular mycorrhiza formation on Ri-T-DNA transformed roots. *New Phytol.* 108, 211–218. doi: 10.1111/j.1469-8137.1988.tb03698.x

Bishop, D. L., Levine, H. G., Kropp, B. R., and Anderson, A. J. (1997). Seedborne fungal contamination: consequences in space-grown wheat. *Phytopathology* 87, 1125–1133. doi: 10.1094/PHYTO.1997.87.11.1125

Colla, G., Battistelli, A., Proietti, S., Moscatello, S., Rouphael, Y., Cardarelli, M., et al. (2007). Rocket seedling production on the International Space Station: growth and nutritional properties. *Microgravity Sci. Tech.* 19, 118–121. doi: 10.1007/BF02919465

Connor, D. J., Loomis, R. S., and Cassman, K. G. (2011). *Crop Ecology: Productivity and Management in Agricultural Systems*. New York, NY: Cambridge University Press.

Ferl, R., Wheeler, R., Levine, H. G., and Paul, A. L. (2002). Plants in Space. *Curr. Opin. Plant Bio.* 5, 258–263. doi: 10.1016/S1369-5266(02)00254-6

Galkovskyi, T., Mileyko, Y., Bucksch, A., Moore, B., Symonova, O., Price, C. A., et al. (2012). GiA Roots: software for the high-throughput analysis of plant root system architecture. *BMC Plant Biol.* 12:116. doi: 10.1186/1471-2229-12-116

Gao, X., Lu, X., Wu, M., Zhang, H., Pan, R., Tian, J., et al. (2012). Co-inoculation with rhizobia and AMF inhibited soybean red crown rot: from field study to plant defense - related gene expression analysis. *PLoS ONE* 7:e33977. doi: 10.1371/journal.pone.0033977

Graham, P. H., and Vance, C. P. (2003). Legumes: importance and constraints to greater use. *Plant Physiol.* 131, 872–877. doi: 10.1104/pp.017004

Heath, K. D., Stock, A. J., and Stinchcombe, J. R. (2010). Mutualism variation in the nodulation response to nitrate. *J. Evol. Biol.* 23, 2494–2500. doi: 10.1111/j.1420-9101.2010.02092.x

Heath, K. D., and Tiffin, P. (2007). Context dependence in coevolution of plant and rhizobial mutualists. *Proc. Biol. Sci.* 274, 1905–1912. doi: 10.1098/rspb.2007.0495

Herranz, R., Anken, R., Boonstra, J., Braun, M., Christianen, P. C. M., de Geest, M., et al. (2013). Ground-based facilities for simulation of microgravity: organism-specific recommendations for their use, and recommended terminology. *Astrobiology* 13, 1–17. doi: 10.1089/ast.2012.0876

Hilaire, E. M., Peterson, B. V., Guikema, J. A., and Brown, C. S. (1996). Clinorotation affects morphology and ethylene production in soybean seedlings. *Plant Cell Physiol.* 37, 929–934. doi: 10.1093/oxfordjournals.pcp.a029041

Hogekamp, C., and Kuster, H. (2013). A roadmap of cell-type specific gene expression during sequential stages of arbuscular mycorrhiza symbiosis. *BMC Genomics* 14:306. doi: 10.1186/1471-2164-14-306

Johnson, C. M., Subramaniana, A., Edelmann, R. E., and Kiss, J. Z. (2015). Morphometric analyses of petioles of seedlings grown in a spaceflight experiment, *J. Plant Res.* 128, 1007–1016. doi: 10.1007/s10265-015-0749-0

Journet, E. P., de Carvalho-Niebel, F., Andriankaja, A., Huguet, T., and Barker, D. G. (2006). "Rhizobial inoculation and nodulation of *Medicago truncatula*," in *The Medicago truncatula Handbook*, eds U. Mathesius, E. P. Journet, and L. W. Sumner. Available online at: http://www.noble.org/MedicagoHandbook/

Kiss, J. Z. (2000). Mechanisms of the early phases of plant gravitropism. *Crit. Rev. Plant Sci.* 19, 551–573. doi: 10.1016/S0735-2689(01)80008-3

Kordyum, V. A., Man'ko, V. G., Popve, A. F., Mashinsky, A. L., Shcherbak, O. H., and Thoyk, N. G. (1983). Changes in symbiotic and associative interrelations in a higher plant-bacterial system during space flight. *Adv. Space Res.* 3, 265–268.

Kouchi, H., Imaizumi-Anraku, H., Hayashi, M., Hakoyama, T., Nakagawa, T., Umehara, Y., et al. (2010). How many peas in a pod? Legume genes responsible for mutualistic symbioses underground. *Plant Cell Physiol.* 51, 1381–1397. doi: 10.1093/pcp/pcq107

Kraft, T. F., van Loon, J. J., and Kiss, J. Z. (2000). Plastid position in *Arabidopsis* columella cells is similar in microgravity and on a random-positioning machine. *Planta* 211, 415–422. doi: 10.1007/s004250000302

Krikorian, A. D., and O'Connor, S. A. (1984). Karyological observations. *Ann. Bot.* 54, 49–63.

Larimer, A. L., and Bever, J. D. (2014). Synergism and context dependency of interactions between arbuscular mycorrhizal fungi and rhizobia with a prairie legume. *Ecology* 95, 1045–1054. doi: 10.1890/13-0025.1

Larimer, A. L., Bever, J. D., and Clay, K. (2010). The interactive effects of plant microbial symbionts a review and meta-analysis. *Symbiosis* 51, 139–148. doi: 10.1007/s13199-010-0083-1

Levine, H. G., and Piastuch, W. C. (2005). Growth patterns for etiolated soybeans germinated under spaceflight conditions. *Adv. Space Res.* 36, 1237–1243. doi: 10.1016/j.asr.2005.02.050

Lisette, J., Xavier, C., and Germida, J. J. (2003). Selective interactions between arbuscular mycorrhizal fungi and *Rhizobium leguminosarum* vc *Viceae* enhance pea yield and nutrition. *Biol. Fertil. Soil* 37, 261–267. doi: 10.1007/s00374-003-0605-6

Matia, I., Gonzalez-Camacho, F., Herranz, R., Kiss, J. Z., Gasset, G., van Loon, J. J., et al. (2010). Plant cell proliferation and growth are altered by microgravity conditions in spaceflight. *J. Plant Physiol.* 167, 184–193. doi: 10.1016/j.jplph.2009.08.012

McFall-Ngai, M. J., Nyholm, S. V., and Castillo, M. G. (2010). The role of the immune system in the initiation and persistence of the *Euprymna scolopes—Vibrio fischeri* symbiosis. *Semin. Immunol.* 22, 48–53. doi: 10.1016/j.smim.2009.11.003

Mhadhbi, H., Jebara, M., Limam, F., Huguet, T., and Aouani, M. E. (2005). Interaction between *Medicago truncatula* lines and *Sinorhizobium meliloti* strains for symbiotic efficiency and nodule antioxidant activites. *Physiol. Plant.* 124, 4–11. doi: 10.1111/j.1399-3054.2005.00489.x

Mortimer, P. E., Perez-Fernandez, M. A., and Valentine, A. J. (2008). The role of arbuscular mycorrhizal colonization in the carbon and nutrient economy of the tripartite symbiosis with nodulated *Phaseolus vularis*. *Soil Biol. Biochem.* 40, 1019–1027 doi: 10.1016/j.soilbio.2007.11.014

Mortley, D. G., Bonsi, C. K., Hill, W. A., and Morris, C. A. (2008). Influence of microgravity environment on root growth, soluble sugars, and starch concentration of sweetpotato stem cuttings. *J. Am. Soc. Hortic. Sci.* 133, 327–332.

Musgrave, M. E., Kuang, A., Xiao, Y., Stout, S. C., Bingham, G. E., Briarty, L. G., et al. (2000). Gravity independence in seed-to seed cycling in *Brassica rapa*. *Planta* 210, 400–406. doi: 10.1007/PL00008148

Nadeem, S. M., Ahmad, M., Zahir, Z. A., Javaid, A., and Ashraf, M. (2014). The role of mycorrhizae and plant growth promoting rhizobacteria (PGPR) in improving crop productivity under stressful environments. *Biotechnol. Adv.* 32, 429–448. doi: 10.1016/j.biotechadv.2013.12.005

National Research Council (2011). *Recapturing a Future for Space Exploration: Life and Physical Sciences Research for a New Era*. Washington, DC: The National Academies Press.

Oka, M., Ueda, J., Miyamoto, K., Yamamoto, R., Hoson, T., and Kamisaka, S. (1995). Effect of simulated microgravity on auxin polar transport in inflorescence axis in *Arabidopsis thaliana*. *Biol. Sci. Space* 9, 331–336. doi: 10.2187/bss.9.331

Paul, A. L., Amalfitano, C. E., and Ferl, R. J. (2012). Plant growth strategies are remodeled by spaceflight. *BMC Plant Bio.* 12:232. doi: 10.1186/1471-2229-12-232

Qian, A. R., Yin, D. C., Yang, P. F., Lv, Y., Tian, Z. C., and Shang, P. (2013). Application of diamagnetic levitation technology in biological science research. *IEEE Trans. Appl. Supercond.* 23, 1–5. doi: 10.1109/TASC.2012.2232919

Rasband, W. S. (2014). *ImageJ, U. S. National Institutes of Health*. Bethesda, MD. Available online at: http://imagej.nih.gov/ij/

Ryba-White, M., Nedukha, O., Hilaire, E., Guikema, J. A., Kordyum, E., and Leach, J. E. (2001). Growth in microgravity increases susceptibility of soybean to a fungal pathogen. *Plant Cell Physiol.* 42, 657–664. doi: 10.1093/pcp/pce082

Serraj, R., and Drevon, J. J. (1998). Effects of salinity and nitrogen source growth and nitrogen fixation on alfalfa. *J. Plant Nutr.* 21, 1805–1818. doi: 10.1080/01904169809365525

Serraj, R., Sinclair, T. R., and Purcell, L. C. (1999). Symbiotic N2 fixation response to drought. *J. Exp. Bot.* 50, 143–155. doi: 10.1093/jxb/50.331.143

Smith, S. E., and Read, D. J. (2008). *Mycorrhizal Symbiosis, 3rd Edn.* San Diego, CA: Academic Press.

Tripathy, B. C., Christopher, S. B., Levine, H. G., and Krikorian, A. D. (1996). Growth and photosynthetic response of wheat plants grown in space. *Plant Physiol.* 110, 801–806. doi: 10.1104/pp.110.3.801

Vandenbrink, J. P., and Kiss, J. Z. (2016). Space, the final frontier: A critical review of recent experiments performed in microgravity. *Plant Sci.* 243, 115–119. doi: 10.1016/j.plantsci.2015.11.004

Vazquez, M. M., Azcon, R., and Barea, J. M. (2001). Compatibility of a wild type and its genetically modified *Sinorhizobium* strain with two mycorrhizal fungi on *Medicago* species as affected by drought stress. *Plant Sci.* 161, 347–348. doi: 10.1016/S0168-9452(01)00416-2

Wang, D., and Dong, X. (2011). A highway for war and peace: the secretory pathway in plant-microbe interactions. *Mol. Plant* 4, 581–587. doi: 10.1093/mp/ssr053

Wang, D., Yang, S., Tang, F., and Zhu, H. (2012). Symbiosis specificity in the legume - rhizobial mutualism. *Cell. Microbiol.* 14, 334–342. doi: 10.1111/j.1462-5822.2011.01736.x

Wasson, A. P., Pellerone, F. I., and Mathesius, U. (2006). Silencing the flavonoid pathway in *Medicago truncatula* inhibits root nodule formation and prevents auxin transport regulation in Rhizobia. *Plant Cell* 18, 1617–1629. doi: 10.1105/tpc.105.038232

Wolverton, S. C., and Kiss, J. Z. (2009). An update on plant science biology. *Gravit. Space Biol. Bull.* 22, 13–20.

Conflict of Interest Statement: The authors declare that the research was conducted in the absence of any commercial or financial relationships that could be construed as a potential conflict of interest.

7

ROAM: A Radial-Basis-Function Optimization Approximation Method for Diagnosing the Three-Dimensional Coronal Magnetic Field

Kevin Dalmasse[1]*, Douglas W. Nychka[2], Sarah E. Gibson[3], Yuhong Fan[3] and Natasha Flyer[2]

[1] *CISL/HAO, National Center for Atmospheric Research, Boulder, CO, USA,* [2] *CISL, National Center for Atmospheric Research, Boulder, CO, USA,* [3] *HAO, National Center for Atmospheric Research, Boulder, CO, USA*

Edited by:
Xueshang Feng,
State Key Laboratory of Space
Weather, China

Reviewed by:
Gordon James Duncan Petrie,
National Solar Observatory, USA
Keiji Hayashi,
Nagoya University, Japan

*Correspondence:
Kevin Dalmasse
dalmasse@ucar.edu

Specialty section:
This article was submitted to
Stellar and Solar Physics,
a section of the journal
Frontiers in Astronomy and Space
Sciences

The Coronal Multichannel Polarimeter (CoMP) routinely performs coronal polarimetric measurements using the Fe XIII 10747 and 10798 lines, which are sensitive to the coronal magnetic field. However, inverting such polarimetric measurements into magnetic field data is a difficult task because the corona is optically thin at these wavelengths and the observed signal is therefore the integrated emission of all the plasma along the line of sight. To overcome this difficulty, we take on a new approach that combines a parameterized 3D magnetic field model with forward modeling of the polarization signal. For that purpose, we develop a new, fast and efficient, optimization method for model-data fitting: the Radial-basis-functions Optimization Approximation Method (ROAM). Model-data fitting is achieved by optimizing a user-specified log-likelihood function that quantifies the differences between the observed polarization signal and its synthetic/predicted analog. Speed and efficiency are obtained by combining sparse evaluation of the magnetic model with radial-basis-function (RBF) decomposition of the log-likelihood function. The RBF decomposition provides an analytical expression for the log-likelihood function that is used to inexpensively estimate the set of parameter values optimizing it. We test and validate ROAM on a synthetic test bed of a coronal magnetic flux rope and show that it performs well with a significantly sparse sample of the parameter space. We conclude that our optimization method is well-suited for fast and efficient model-data fitting and can be exploited for converting coronal polarimetric measurements, such as the ones provided by CoMP, into coronal magnetic field data.

Keywords: Sun: corona, Sun: magnetic fields, Sun: infrared, methods: statistical, methods: radial basis functions

1. INTRODUCTION

Modification to the polarization of light is one of the many signatures of a non-zero magnetic field in the solar corona, and more generally, in the solar atmosphere (e.g., Stenflo, 2015, and references therein). Several mechanisms producing or modifying the polarization of light have been observed and studied in the solar corona at different wavelengths including, but not limited to, the Zeeman and Hanle effects (see e.g., Hale, 1908; Hanle, 1924; Bird et al., 1985; White and Kundu, 1997; Casini and Judge, 1999; Lin et al., 2004; Gibson et al., 2016, and references therein). The former induces a frequency-modulated polarization while the latter induces a depolarization of scattered light (e.g., Sahal-Brechot et al., 1977; Bommier and Sahal-Brechot, 1982; Rachmeler et al., 2013; López Ariste, 2015). Both mechanisms allow us to probe the strength and direction of the coronal magnetic field. Coronal polarization associated with these two mechanisms is currently measured above the solar limb by the Coronal Multichannel Polarimeter from forbidden coronal lines such as the Fe XIII lines (10747 Å and 10798 Å; Tomczyk et al., 2008). For these two lines, the circular polarization signal is dominated by the Zeeman effect while the linear polarization signal is dominated by the Hanle effect (e.g., Judge et al., 2006).

Translating the polarization maps of CoMP into magnetic field maps is a challenging task. The main difficulty is that the solar corona is optically thin at these wavelengths (e.g., Rachmeler et al., 2012; Plowman, 2014). The observed signal is therefore the integrated emission of all the plasma along the line of sight (LOS). Hence, the polarization maps cannot, in general, be directly inverted into 2D maps of the plane-of-sky (POS) magnetic field. On the other hand, extracting individual magnetic information at specific positions along the LOS is extremely difficult without stereoscopic observations (e.g., Kramar et al., 2014). Another limitation is that the Hanle effect associated with the aforementioned forbidden infrared lines operates in the saturated regime (e.g., Casini and Judge, 1999; Tomczyk et al., 2008). Accordingly the linear polarization signal measured by CoMP is sensitive to the direction of the magnetic field but not its strength. Deriving the magnetic field associated with the polarization maps of CoMP therefore requires a different approach than the single point inversion that can be done with, e.g., photospheric polarimetric measurements.

The alternate approach we propose to follow is to combine a parameterized 3D magnetic field model with forward modeling of the polarization signal observed by CoMP. For that purpose, we take advantage of the Coronal Line Emission (CLE) polarimetry code developed by Casini and Judge (1999) and integrated into the FORWARD package. FORWARD[1] is a Solar Soft[2] IDL package designed to perform forward modeling of various observables including, e.g., visible/IR/UV polarimetry, EUV/X-ray/radio imaging, and white-light coronagraphic observations (Gibson et al., 2016). The goal is then to optimize a user-specified likelihood function comparing the polarization signal predicted by FORWARD to the real one and find the parameters of the magnetic field model such that the predicted signal fits the real data.

In the present paper, we develop and test a new method for performing fast and efficient optimization in a d-dimensional parameter space that may be used for converting the polarization observations of CoMP into magnetic field data. The optimization

method, called ROAM (Radial-basis-functions Optimization Approximation Method) is designed to be general enough so that it can be applied independently of the dimension and size of the parameter space, the 3D magnetic field model, the type of observables (provided that one can forward model them), and the form of the likelihood function used for comparing the predicted signal to the real one. ROAM is introduced in Section 2. Section 3 describes the results of multiple applications of ROAM to a synthetic test bed as validation of the optimization method. Our conclusions are then summarized in Section 4.

2. METHODS

The goal of this paper is to propose a model-data fitting method to be used for *near-real-time* 3D reconstruction of the solar coronal magnetic field. This requires developing a fast and efficient method for searching for the set of values of the model parameters that optimize a pre-defined function quantifying the differences between the predicted (or forward-modeled) and real data. Although similar approaches are standard in engineering (e.g., Jones et al., 1998), we propose a simplified version and tailored to the context of solar physics. The proposed method, ROAM, combines the computation of a log-likelihood function on a sparse sample of the parameter space with function approximation and is based on the five following steps:

1. Sparse sampling of the parameter space is performed using *Latin Hypercube Sampling* (LHS; McKay et al., 1979; Iman et al., 1981). LHS is a statistical method for generating a random sample of the parameter values in a d-dimensional space. For a d-dimensional space of n^d points (n is the number of points for each dimension), LHS creates a set, $\{x_i\}$, of n independent points or d-vectors of the parameter space (an example is given **Figure 1**) that will be referred to as the *design* in the following.

2. The model is computed for each point, x_i, of the design and used to generate the corresponding predicted observation, $y(x_i)$, to be compared with the ground truth, y_{gt} (which is either an actual observation or a synthetic one for test beds using analytical models or numerical simulations).

3. The set of predicted observations, $\{y(x_i)\}$, is then compared to the ground truth by means of a user-specified log-likelihood function

$$\ell(x_i) = \log \mathcal{L} = f\left(y(x_i) - y_{gt}\right), \qquad (1)$$

where $\mathcal{L}$ is the likelihood function, x_i is a d-vector of the design and f is a general, user-specified, well-behaved, scalar function. Typically, the likelihood function simplifies to depend on the difference between the observations and the predicted values and function f reflects that. An explicit expression of f is given in the section of each test considered in this paper (see Section 3).

4. This log-likelihood function is then approximated using radial-basis-function (RBF) decomposition (see e.g., Powell,

---

[1]http://www.hao.ucar.edu/FORWARD/
[2]http://www.lmsal.com/solarsoft/

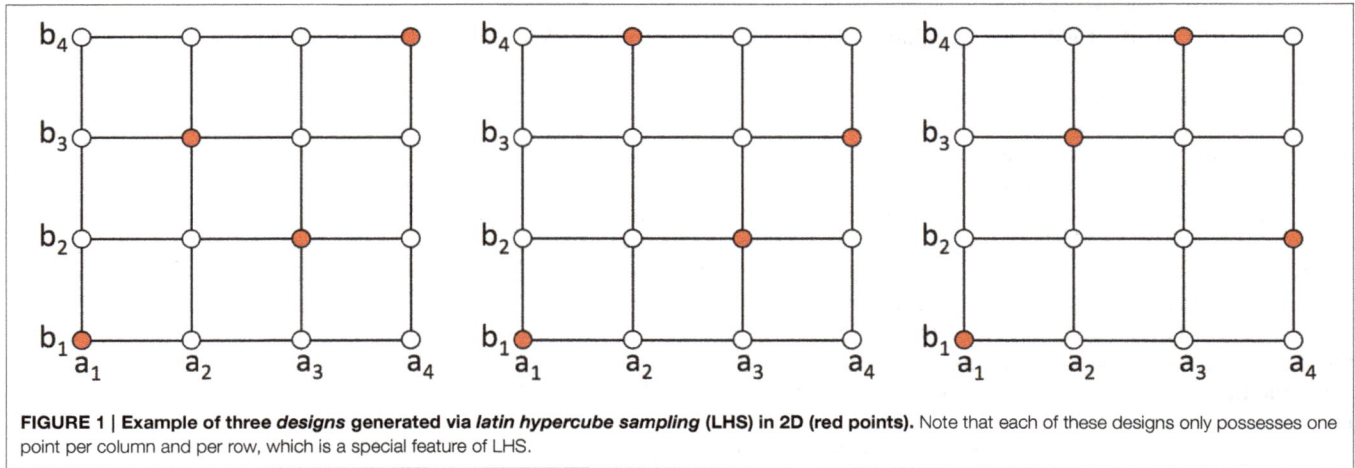

FIGURE 1 | Example of three *designs* generated via *latin hypercube sampling* (LHS) in 2D (red points). Note that each of these designs only possesses one point per column and per row, which is a special feature of LHS.

1977; Broomhead and Lowe, 1988; Buhmann, 2003; Nychka et al., 2015)

$$\ell(x) \approx \hat{\ell}(x) = \sum_{j=1}^{n} a_j \varphi_j \left(\|x - x_j\| \right) + \sum_{j=1}^{\binom{p+d}{p}} b_j \psi_j(x) , \quad (2)$$

$$\varphi_j(\|x - x_j\|) = \|x - x_j\|^{2m-d} \log \left(\|x - x_j\| \right) ,$$
$$\text{if } d \text{ is even}, \quad (3)$$
$$= \|x - x_j\|^{2m-d} , \text{ if } d \text{ is odd}, \quad (4)$$

where φ_j is the j-th RBF centered at point x_j of the design, $\| \cdot \|$ is the usual Euclidean norm, $m \in \mathbb{N}$ is such that $2m - d > 1$, and $\{\psi_j\}$ is a set of polynomials up to degree p in the dimension d of the problem with the constraint $p \leq m - 1$. In the following, we always use $p = m - 1$. When periodic components of the d-space exist, the value of d must be modified for the RBF decomposition to take the periodicities into account (an example and further details on handling periodic components are provided in Appendix 2 of Supplementary Material). Note that the particular choice of RBFs, φ_j, in Equations (3) and (4) is called a *Polyharmonic Spline* (see e.g., Duchon, 1977; Madych and Nelson, 1990) and that the polynomial term in Equation (2) is not a regularization term but an additional term that directly comes from the definition of Polyharmonic Splines as minimizers of the energy functional $\int_{\mathcal{V} \subset \mathbb{R}^d} |\nabla^m g|^2 dx$ (which is not modified by adding polynomials of order $p \leq m - 1$ to g). Although required from the definition of Polyharmonic Splines, this polynomial term is particularly beneficial for improving the fitting accuracy and extrapolation away from the RBF centers x_j, while also ensuring polynomial reproducibility. Note also that the a_j and b_j are coefficients determined from the set of n equations provided by the constraint (the detailed derivation of the coefficients is given in Appendix 1 of Supplementary Material)

$$\hat{\ell}(x_i) = \ell(x_i) . \quad (5)$$

5. Finally, we compute the set of values of the model parameters optimizing the approximated log-likelihood function using the DFPMIN IDL routine and take it as the maximum-likelihood estimator of the set of values optimizing the exact log-likelihood function. To ensure the reliability of the maximum likelihood estimator (MLE; see Section 3.2) obtained with DFPMIN, we apply the latter from (i) the point of the design that possesses the largest likelihood function value prior to step (4), (ii) N^d points spanning the entire parameter space and where N ($\neq n$) is a relatively low number of points (typically $N \lesssim 10$), and (iii) the likelihood-weighted average position of these N^d points (i.e., their center of mass). Starting from these $N^d + 2$ points ensures that at least one of them will lead DFPMIN to converge toward the global maximum when the approximated log-likelihood function contains multiple global and local maxima.

An RBF is a real-valued function that only depends on the Euclidean distance to a center whose location can be set arbitrarily. RBFs provide a class of functions that possess particularly interesting properties such as continuity, smoothness, and infinite differentiability. Their use is widely spread in various branches of applied mathematics and computer science including, e.g., function approximation (Powell, 1993; Buhmann, 2003), data mining and interpolation (Harder and Desmarais, 1972; Lam, 1983; Nychka et al., 2015), numerical analysis with meshfree methods for, e.g., solving partial differential equations in numerical simulations (Fasshauer, 2007; Flyer and Fornberg, 2011; Fornberg and Flyer, 2015; Flyer et al., 2016), computer graphics and machine learning (Broomhead and Lowe, 1988; Boser et al., 1992). Polyharmonic Splines (PHS) are a type of infinitely smooth RBFs that does not possess any free parameter requiring a manual tuning. PHS can therefore be easily implemented for automated calculations.

As previously stated, the goal behind combining sparse calculations of a log-likelihood with an RBF decomposition is to limit the number of model evaluations / forward calculations (n) to reduce the computational cost while maintaining a good accuracy on retrieving the exact maximum likelihood. Through low number of model evaluations, we mean to keep $n \lesssim 100 - 300$ *regardless of the dimension of the parameter*

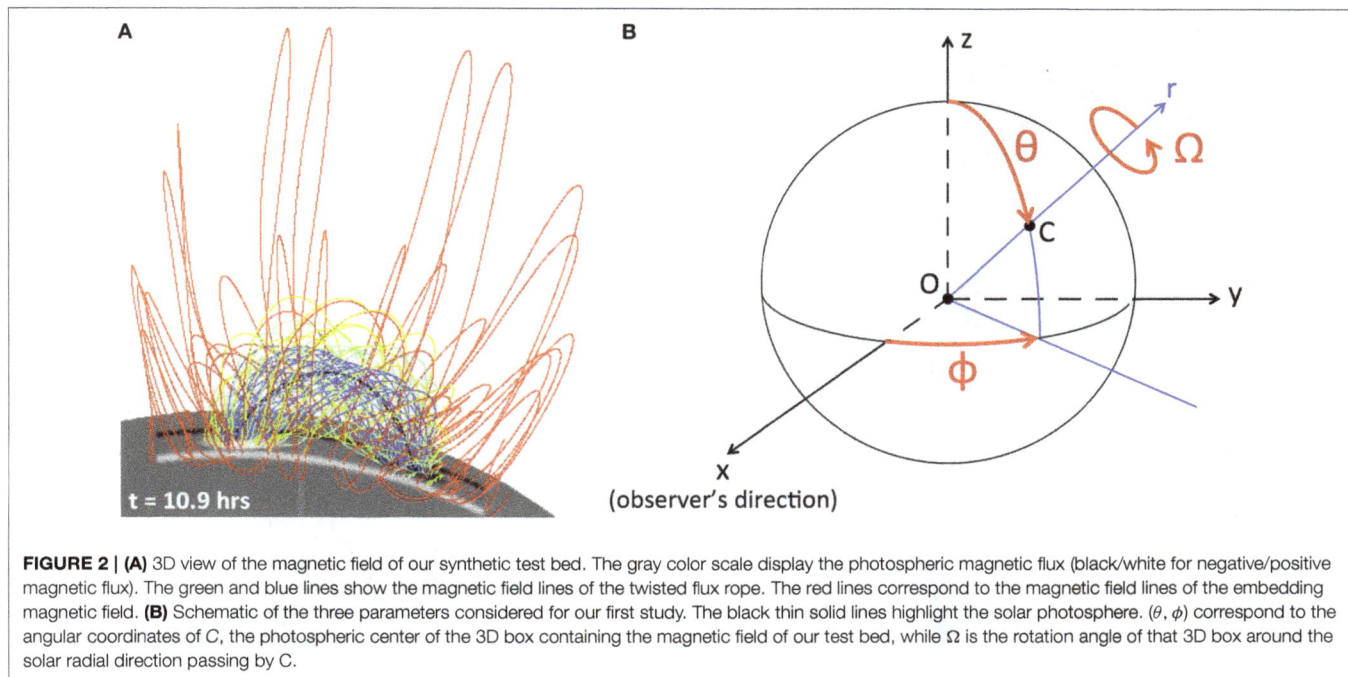

FIGURE 2 | (A) 3D view of the magnetic field of our synthetic test bed. The gray color scale display the photospheric magnetic flux (black/white for negative/positive magnetic flux). The green and blue lines show the magnetic field lines of the twisted flux rope. The red lines correspond to the magnetic field lines of the embedding magnetic field. **(B)** Schematic of the three parameters considered for our first study. The black thin solid lines highlight the solar photosphere. (θ, ϕ) correspond to the angular coordinates of C, the photospheric center of the 3D box containing the magnetic field of our test bed, while Ω is the rotation angle of that 3D box around the solar radial direction passing by C.

space, such that all model evaluations can easily be performed at once in parallel on a high-performance computing cluster. This provides us with a significant advantage as compared with more traditional sequential optimization methods since the effective computational time of our optimization method would only correspond to the computational time of *one* model evaluation (because steps 4 and 5 of the method only take up to $\lesssim$ 30 s as long as $n \lesssim$ 500). The optimization method we propose would, in general, also be more advantageous than a full grid search. Indeed, an accurate full grid search would typically require to sample each parameter of the *d*-space with about 50 − 100 points at the least. This rapidly leads to a number of model evaluations that is not practical even when using parallel computing. Finally, ROAM should be competitive with genetic algorithms. Genetic algorithms applied to small population samples, e.g., $\lesssim$ a few 100 points, typically require on the order of hundred generations to converge (e.g., Louis and Rawlins, 1993; Gibson and Charbonneau, 1998, and references therein), while faster convergence would require larger population sets. For ROAM, the equivalent of a population sample is a design of the parameter space and the equivalent of a generation would be an iteration of ROAM on a smaller parameter space region. For a population/design of *n*-points, ROAM should, in principle, be able to converge toward the solution without the need for iterations and, hence, we estimate would be at least 50–100 times faster than a genetic algorithm with the same population/design. In practice, preliminary tests of an iterative implementation of ROAM, which will be published in a subsequent paper, show robust and accurate convergence of ROAM within a few iterations(typically < 10).

3. RESULTS

In this section, we present a set of test cases performed on a synthetic test bed to validate ROAM (Section 2) prior to any observational application. The set of test cases aims at assessing the performance of our method in different circumstances and defining a framework of application that will make use of its strengths.

3.1. Numerical Setup for the Forward Calculations

Our goal is to use the proposed optimization method for data-constrained modeling of the solar coronal magnetic field using, in particular, coronal polarimetric observations (i.e., the four Stokes parameters, (I, Q, U, V), where Stokes I is the total line intensity, Stokes V is the circular polarization, and Stokes Q and U are the two components of the linear polarization). All our test cases are therefore applied to a 3D model of magnetic fields chosen to represent scenarios typically observed in the solar atmosphere. The considered magnetic model is that of a 3D coronal magnetic flux rope generated from a 3D MHD numerical simulation of the emergence of twisted magnetic fields in the solar corona (**Figure 2A**; Fan, 2012).

For the test cases, this magnetic field is assumed to depend on four parameters, i.e., height in the corona (*h*; monotonically depends on the time of the MHD simulation, though not linearly), co-latitude (θ), longitude (ϕ), and rotation angle[3]

---

[3]Note that $(\theta; \phi; \Omega)$ are the co-latitude, longitude, and rotation angle of the numerical box—containing the magnetic field of the MHD simulation—around the Sun, while h is the actual height of the flux rope in that numerical box (inside of which the solar photosphere is located at $h = 0$).

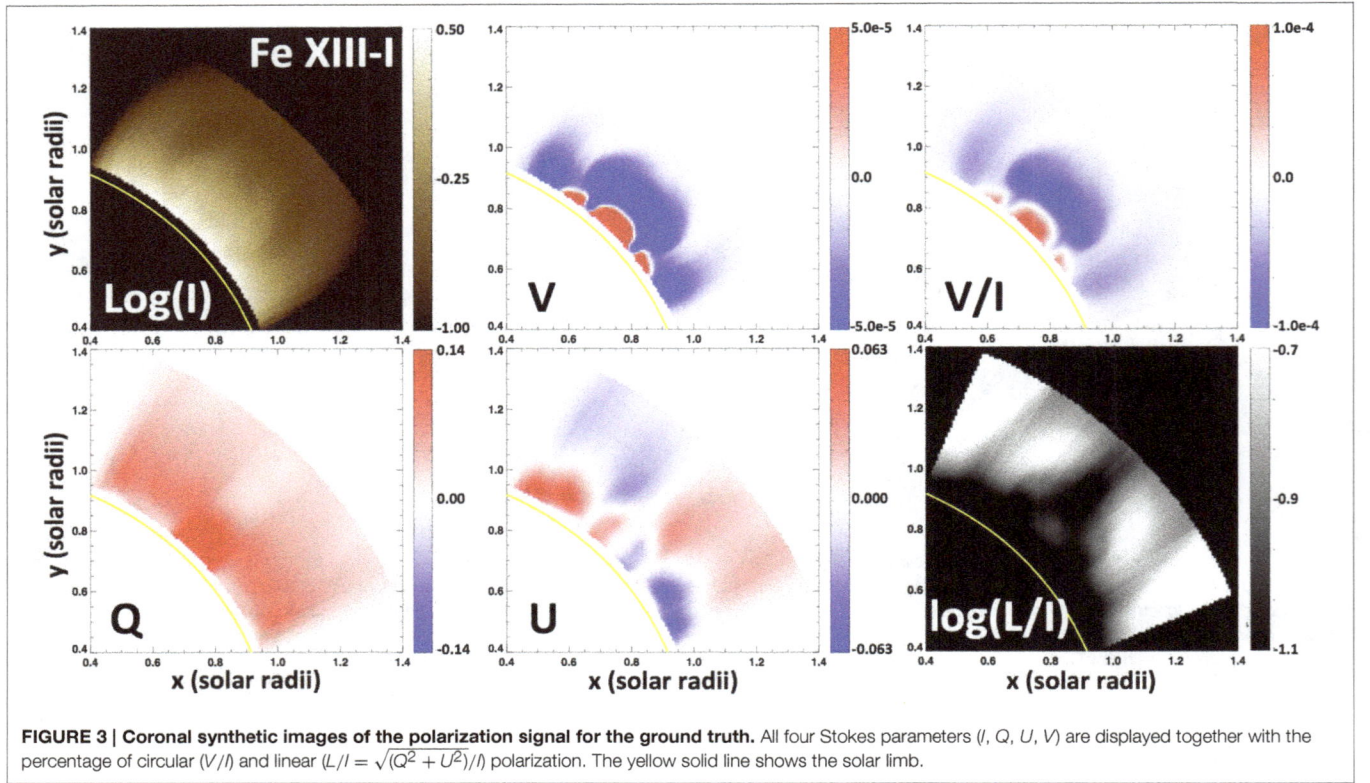

FIGURE 3 | Coronal synthetic images of the polarization signal for the ground truth. All four Stokes parameters (I, Q, U, V) are displayed together with the percentage of circular (V/I) and linear ($L/I = \sqrt{(Q^2 + U^2)}/I$) polarization. The yellow solid line shows the solar limb.

(Ω; of **Figure 2B**). A series of synthetic polarimetric data, referred to as the ground truth (GT) in the following, is generated for the flux rope associated with $(h; \theta; \phi; \Omega) = (0.16\ R_\odot; 45°; 90°; 30°)$ (see **Figure 3**; note that both Stokes Q and U are presented in a frame of reference relative to the local vertical, or radial coordinate). All synthetic data are computed using the FORWARD Solar Soft IDL package with a field-of-view (FOV) set to $y \times z = [0\ R_\odot; 1.5\ R_\odot]^2$ (where y and z are the POS coordinates) and $x = [-0.79; 0.79]\ R_\odot$ for the LOS. We use 192 points along both directions for the POS and 80 points for each LOS, leading to spatial resolutions of $7.6''$ and $19.3''$ respectively. We limit the forward calculations of the polarization signals to a radial range of $[1.03; 1.5]\ R_\odot$, i.e., the FOV of CoMP. Although the spatial resolution of CoMP is $4.5''$, we restrict ourselves to a spatial resolution of $7.6''$ to allow for relatively fast (about 4–5 min on a MacBook Pro with a 2.7 GHz Intel Core i7 processor) calculations of the polarization signals while maintaining a quasi-CoMP resolution. We impose this FOV and POS spatial resolution to show that CoMP data currently carry meaningful information that can be used to constrain 3D reconstructions of the solar coronal magnetic field.

Finally, it should be emphasized that the considered flux rope possesses a strong degree of symmetry, such that B $(\theta; \phi = 90; \Omega \pm 180°) = -B\ (\theta; \phi = 90; \Omega)$. We will exploit these symmetry properties to test ROAM when faced with a log-likelihood function containing multiple maxima.

3.2. Likelihood Function with a Single Maximum

We first apply ROAM in the context of a 3D likelihood function possessing a single maximum. The parameters considered for this study are the co-latitude, longitude, and rotation angle, i.e., $(\theta; \phi; \Omega)$. We then build a likelihood function that takes into account all four Stokes parameters, i.e., I, Q, U, and V. For the set $\{x_i\}$ of a design, we first define the log-likelihood function for a given Stokes parameter, $S = \{I, Q, U, V\}$, up to a constant, as

$$\ell_S(x_i) = f\left(S(x_i) - S_{gt}\right) = -\sum_k \left(S_k(x_i) - S_k^{gt}\right)^2, \quad (6)$$

where k is the k-th pixel of the Stokes, S, image. The final log-likelihood function is then constructed as

$$\ell(x_i) = w_I\ell_I(x_i) + w_Q\ell_Q(x_i) + w_U\ell_U(x_i) + w_V\ell_V(x_i), \quad (7)$$

where the weighting coefficients w_S were chosen to ensure that I, Q, U, and V similarly contribute to the log-likelihood function, which behavior would otherwise be dominated by the quantity possessing the largest values (here, Stokes I). We use $\left(w_I; w_Q; w_U; w_V\right) = \left(1.3 \times 10^{-4}; 1.9 \times 10^{-2}; 9.2 \times 10^{-2}; 1.2 \times 10^4\right)$.

With the log-likelihood function defined in Equation (7), we consider three test cases referred to as 3DN31, 3DN301, and 3DN31ZOOM (see **Table 1**). These three test cases each contain

TABLE 1 | Characteristics of the test with a likelihood function possessing a single maximum.

	n	$t_{elapsed}$ (hrs)	t_{full} (hrs)	h (R$_\odot$)	θ (°)	ϕ (°)	Ω (°)
3DN31	31	2.6	2.5×10^3	0.16	[24; 66]	[60; 120]	[0; 90]
3DN301	301	25	2.3×10^6	0.16	[24; 66]	[60; 120]	[0; 90]
3DN31ZOOM	31	2.6	2.5×10^3	0.16	[42; 48]	[75; 105]	[15; 45]

n is the number of points per design. $t_{elapsed}$ is the elapsed time for forwarding the Stokes images associated with the n points of a design in series, while t_{full} is the total elapsed time that would be required to compute Stokes images for the n^3 points of the 3D parameter space in series. Each test case contains 100 randomly-chosen different designs. The naming convention is such that "xD" indicates the dimension of the parameter space and "Nx" indicates the number of points per design (n). The polarimetric data for the ground-truth are associated with (h; θ; ϕ; Ω) = (0.16 R$_\odot$; 45°; 90°; 30°).

TABLE 2 | Optimization results for a likelihood function with a single maximum.

	θ_{rms}	ϕ_{rms}	Ω_{rms}
3DN31	47.1	92.8	51.0
3DN301	45.2	90.0	30.2
3DN31ZOOM	45.0	89.8	29.7

The ground-truth parameters are $(\theta; \phi; \Omega) = (45°; 90°; 30°)$. All angles are in degrees.

100 different designs and differ by the number of points in the designs (31 or 301) as well as by the size of the parameter space to allow us to investigate their role on the performances of ROAM. These test cases are designed to allow us to determine the criteria required for the method to ensure robustness and reliability of the results, i.e., such that the method provides a maximum likelihood estimator (MLE) that gives a good approximation of the parameters of the maximum of the exact likelihood function independently of the design and number of points used.

For each test case, the parameters of the RBF decomposition are $d = 3$, $m = 3$ and $p = m - 1 = 2$. We choose the minimum m satisfying the condition $2m - d > 1$ (see Section 2). Although θ, ϕ, and Ω all are periodic parameters, their corresponding range is smaller than half the associated period and, hence, no periodic effect is expected. As explained Appendix 2 in Supplementary Material, disregarding the periodicity and curvature of the d-space should not significantly affect the results in such circumstances. We therefore ignore the periodicity of θ, ϕ, and Ω in all 3D cases considered in this section, but return to the issue of periodicity in Section 3.3.

Figure 4 presents 2D dispersion plots of the MLEs obtained for each one of the 100 randomly-chosen designs of the 3DN31 (red), 3DN301 (blue), and 3DN31ZOOM (yellow) cases. For the 3DN31, the MLEs are fairly weakly dispersed for the θ parameter, spanning a range of roughly 10°. As summarized in **Table 2**, the root mean square (hereafter, rms) of the MLEs, θ_{rms}, is $\approx 47.1°$, which is only $\approx 2.1°$ different from $\theta_{GT} = 45°$. This suggests that θ_{MLEs} is not overly sensitive to the design used for the RBF decomposition. These conclusions contrast with both the ϕ and Ω parameters. Although $\phi_{rms} \approx 92.8°$ is very close to the ground-truth, $\phi_{GT} = 90°$, the ensemble of solutions, ϕ_{MLEs}, spans the entire ϕ-range considered for the 3DN31. Similarly poor results are obtained for the set of Ω_{MLEs}, whose rms is $\approx 21°$ off from the ground-truth, $\Omega_{GT} = 30°$ (see **Table 2**). **Figure 4** further shows that there is a strong coupling between ϕ and Ω. In particular, we find that Ω_{MLEs} provide a poor estimation of Ω_{GT} whenever ϕ_{MLEs} are themselves a poor estimation of ϕ_{GT} (and vice-versa). Such results trace very poor performances of our optimization method for the chosen setup of the 3DN31 case. The MLE strongly depends on the design used to perform the RBF decomposition. Hence, the MLE obtained from applying

our method to a single design is not reliable for the setup of the 3DN31 case.

When comparing 3DN31 to 3DN301 in **Figure 4**, we can see that increasing the number of points significantly improves the performances of ROAM (see dark blue crosses). For all three parameters, the rms is only $\approx 0.2°$ off the ground-truth for 3DN301. The range spanned by the ensemble of solutions is relatively smaller than for 3DN31, ≈ 5 *times smaller* for θ_{MLEs} and ≈ 3 *times smaller* for both ϕ_{MLEs} and Ω_{MLEs}. Again, the strong coupling between ϕ and Ω is still present but its effect on their uncertainties is strongly reduced as compared with the case 3DN31. Such results are not yet perfect since the set of solutions for both ϕ and Ω is spread over 20°, which is relatively significant considering the range of their values. However, they do show that increasing the number of points per design strongly helps in reducing (a) the dependence of the MLE on the design used for the RBF decomposition, and (b) the effect of (even strong) coupling of parameters on their uncertainty. Increasing the number of points per design therefore strongly helps in improving the reliability and robustness of the proposed optimization method.

Compared with 3DN31, the 3DN31ZOOM case is used to investigate the effect of focusing the parameter space around a region closer to the exact maximum while keeping the number of points per design constant. **Figure 4** shows that reducing the size of the parameter space is also beneficial for reducing the MLEs dispersion (see yellow crosses). The rms values for all three parameters are as close to the ground-truth as for 3DN301 (see **Table 2**) and the solutions are spanning a range that is ≈ 4 *times smaller* than for 3DN301 and ≈ 12 *times smaller* than for 3DN31. The MLEs are now independent of the design used for the RBF decomposition for θ and very weakly dependent on that design for both ϕ and Ω. The effect of the strong $\phi - \Omega$ coupling on their uncertainty is again strongly reduced and even smaller than for the 3DN301 case. These very good results prove that ROAM can perform very well and provide an accurate estimation of the ground-truth parameters when the setup is suitably defined.

Figure 5 displays 2D cuts of the exact log-likelihood function and the approximated ones associated with the designs of 3DN31 and 3DN301 giving the best MLEs (referred to as *best cases* in the following), as well as the approximated log-likelihood function associated with the design of 3DN301 giving the worst MLEs (referred to as *worst case*). Here, the best (worst) MLE is defined as the MLE minimizing (maximizing) the distance to the ground truth in the parameter space. The best MLE from 3DN31 is $(\theta; \phi; \Omega)_{best-MLE} = (44.6°; 92.9°; 30.1°)$ while the best MLE from 3DN301 is

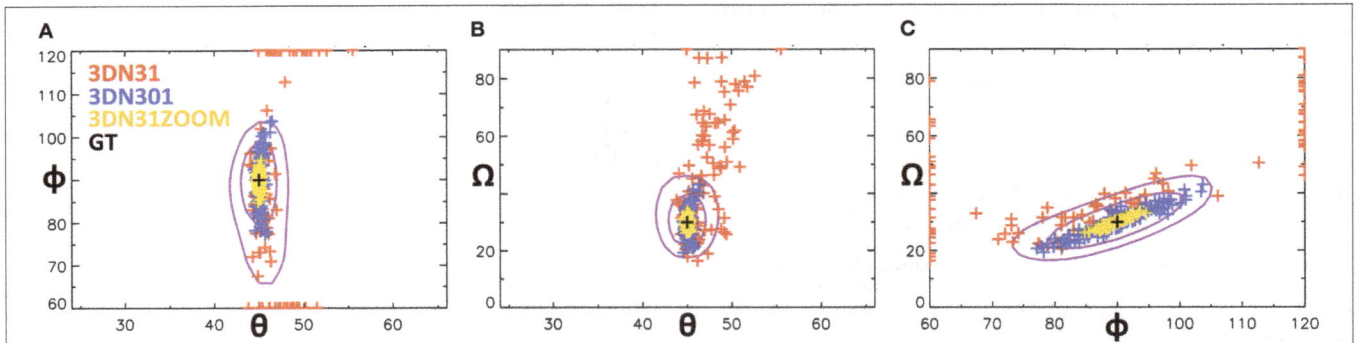

FIGURE 4 | 2D scatter plots of the maximum likelihood estimators (MLEs) found for each design of the 3DN31 (red crosses), 3DN301 (blue crosses), and 3DN31ZOOM (yellow crosses) cases. The black cross highlights the position of the exact maximum (i.e., ground-truth). The two purple solid lines show $[0.9; 0.95] \times \max(\ell)$ isocontours. Note that, in panel **(B)**, the red crosses (MLEs of 3DN31) outside the two log-likelihood function isocontours are the solutions associated with $\phi_{\text{MLEs}} = 60°$ and $\phi_{\text{MLEs}} = 120°$ from panels **(A,C)**.

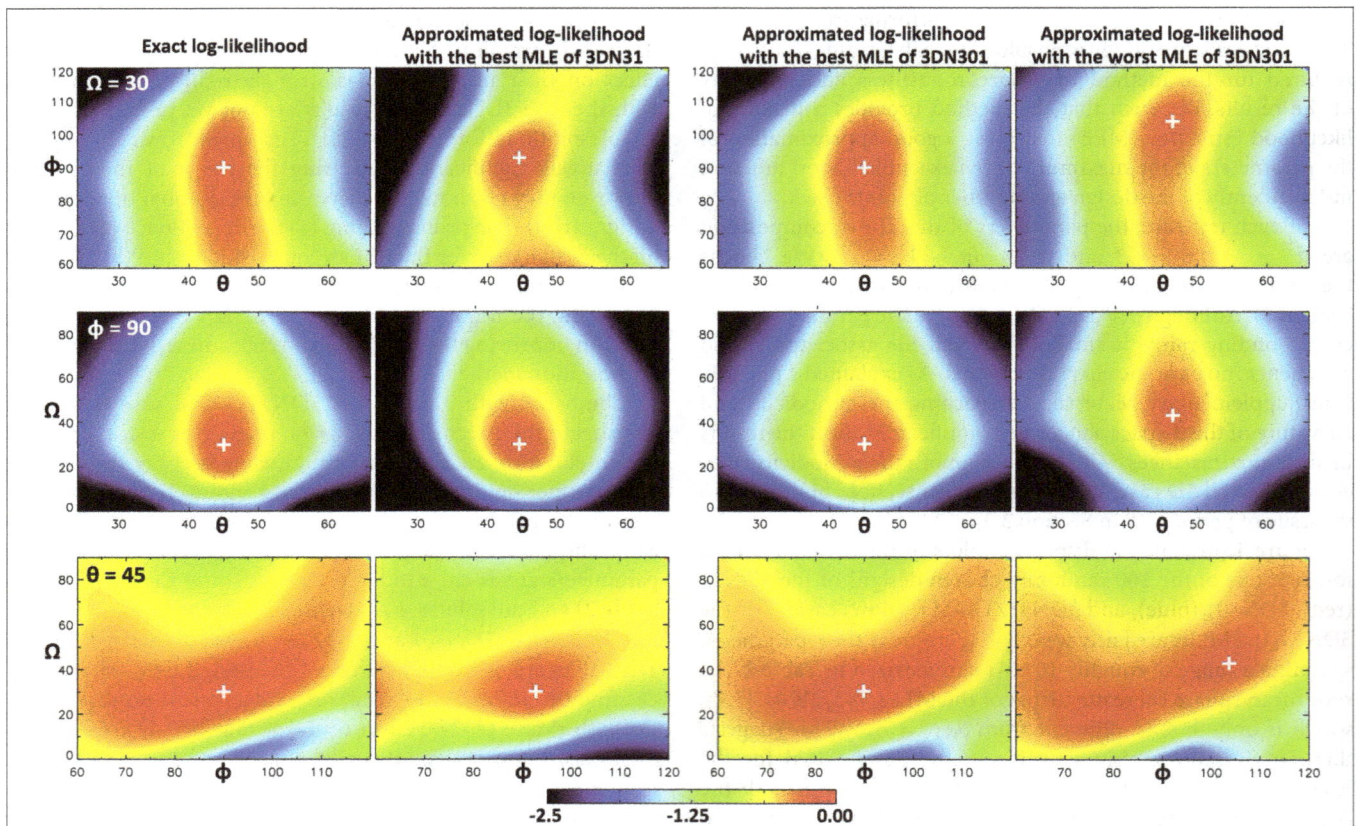

FIGURE 5 | Comparison between approximated and exact log-likelihood functions. The white "+" symbol indicates the position of the maximum log-likelihood.

$(\theta; \phi; \Omega)_{\text{best}-\text{MLE}} = (45.0°; 89.9°; 30.3°)$. The worst MLE from 3DN301 is $(\theta; \phi; \Omega)_{\text{best}-\text{MLE}} = (46.5°; 103.7°; 43.0°)$. The figure shows that the approximated log-likelihood function of the best case from 3DN31 gives an overall rough approximation of the exact one both in terms of values and shape. Note, though, that the rms error on the log-likelihood is 0.24, which is rather small given that $\max(|\ell(x)|) \approx 3.5$ for the considered parameter space. For the log-likelihood function of the best case from 3DN301, the results are very much better. The approximated log-likelihood function is able to accurately capture both the values and shape of the exact log-likelihood function; the rms error is 0.05, i.e., ≈ 5 times smaller than for the best case of 3DN31. For the worst case of 3DN301, the rms error on the log-likelihood is 0.25, which is very similar to that of the best case of 3DN31, and the MLE is far from the ground truth for both ϕ and Ω. However, we find that the worst case from 3DN301 provides a more accurate RBF decomposition of the exact log-likelihood function than the best case of 3DN31; the log-likelihood function surfaces display a similar pattern as for the best case of 3DN301 but shifted in the Ω direction. The difference with the 3DN31 lies in the density of points in the entire design, and in the vicinity of the exact maximum, with regard to the structuring, or gradients, of the exact log-likelihood function. This is because the goodness of the approximation is determined by that of the RBF decomposition, which depends on the number of constraints—and hence, points—brought by the design. In other words, the more structured the exact log-likelihood function, the stronger the effect of point density on the goodness of the RBF decomposition / log-likelihood function approximation, similar to what one would expect when discretizing a continuous functions that contains strong gradients. While not shown here, the combined effect of point density and log-likelihood function structuring on the quality of the RBF decomposition is further supported and illustrated by the best case of 3DN31ZOOM that provides the best approximation of the exact log-likelihood function in the vicinity of the exact maximum even though the corresponding design only includes 31 points.

The aforementioned results show that the RBF decomposition performed well from a sparse sampling of the parameter space, and hence ROAM, is able to capture both the values and variations of the exact log-likelihood function when suitable conditions are met, namely, when the design contains a high enough density of points in the surroundings of the exact maximum and in areas where the exact log-likelihood function is strongly structured. They further demonstrate that ROAM can perform well even with a very low number of points per design although not as robustly. The combined results from 3DN31 and 3DN31ZOOM indicate that an iterative application of ROAM with a smaller and smaller parameter space would be an interesting way to improve its robustness when used with a very sparse design. Such a robust approach has been successfully tested but is beyond the scope of this work and will be presented in a subsequent paper. In particular, the iterative implementation of ROAM strongly improves ϕ_{MLEs} and Ω_{MLEs}, leading to a better than 0.5° accuracy on both of these parameters in typically 4–5 iterations with designs of 31 points. The solution is quasi-independent of the design used for the RBF decomposition and the strong $\phi - \Omega$ coupling (previously mentioned and visible in **Figure 4C** and in the $\phi - \Omega$ cut of the exact log-likelihood function shown in **Figure 5**) is comfortably reduced and overcome [note that such coupling could also be overcome by separately optimizing one of the coupled parameters, e.g., Ω, and apply ROAM to the 2D parameter space $(\theta; \phi)$].

Note that, with the goal of increasing speed, and considering our previous comments on the acceptable degree of roughness in the log-likelihood function approximation and an iterative implementation of ROAM, a less sophisticated approach might be conceived. For instance, one could first go through steps 1 to 3 of the method (see Section 2). Then, step 4 (i.e., the

RBF decomposition) would be replaced by taking the point of the design associated with the highest log-likelihood function value as a temporary MLE and one would iterate the procedure by defining a smaller design centered around the temporary MLE until a convergence criterion is reached. There are several reasons for not making such a choice. The main reason is that such an initial guess can be far from the exact maximum likelihood, which would likely slow down the convergence by requiring unnecessary iterations and would make the final result more sensitive to local maxima. In addition, applying the RBF decomposition and the search for the maximum from the approximated log-likelihood function is computationally cheap when the number of RBFs is as small as for the cases considered in this study, i.e., typically takes <10 s for the designs of the 3DN301 case. The benefits of applying steps 4 and 5 of ROAM as proposed in Section 2 (i.e., the RBF decomposition and the search for the MLE from the RBFs approximated log-likelihood function) are illustrated in **Figure 6**. The figure displays Stokes images for the ground truth, the MLE obtained from fully applying our optimization method to one design of the 3DN301, and for the initial guess from that design. As one can see, the initial MLE guess from the design, $(\theta; \phi; \Omega)_{IG} = (46.0°; 101.4°; 45.9°)$, has a ϕ_{IG} and Ω_{IG} that are far off both the ground truth [$(\theta; \phi; \Omega)_{GT} = (45.0°; 90.0°; 30.0°)$] and the MLE obtained from the RBF decomposition [full application of ROAM; $(\theta; \phi; \Omega)_{MLEs} = (44.6°; 90.0°; 28.0°)$]. These strong differences in ϕ and Ω result in significantly different Stokes profiles. Iterations would then be needed for the results to be as close to the ground truth as the MLE from the full optimization, which (1) gives a very good estimation of the parameters of the exact maximum likelihood without any real need for iterations, and (2) only takes a few more seconds of calculations.

In practice, the current capabilities of the CoMP instrument and calibration software do not allow routine measurements of Stokes V since the signal-to-noise ratio is too small. We therefore perform an additional test to show that the current linear polarization signal from CoMP is sufficient to constrain the parameters of a magnetic model using ROAM. The log-likelihood function is defined as in Equation (7) keeping w_I, w_Q, w_U as before, but now setting $w_V = 0$. The results of that study are displayed in **Figure 7** for 3DN301. The figure presents scatter plots of the MLEs of a design obtained when using all four Stokes vs. obtained when using Stokes I, Q, and U only. In such plots, the points should form a line of equation $y = x$ whenever the solutions obtained one way or the other remain the same. As one can see from **Figure 7**, this is exactly the case for θ_{MLEs}. Most of the points are also forming a straight line, $y = x$, for both ϕ_{MLEs} and Ω_{MLEs}, with only about 7–8 points (out of 100) being off the line. Such results indicate that a log-likelihood function built from Stokes I, Q, and U contains sufficient information to constrain the three spatial location and orientation parameters considered here. We therefore conclude that the current linear polarization measurements from CoMP contain sufficient observational information to constrain *some* of the parameters of a given magnetic model.

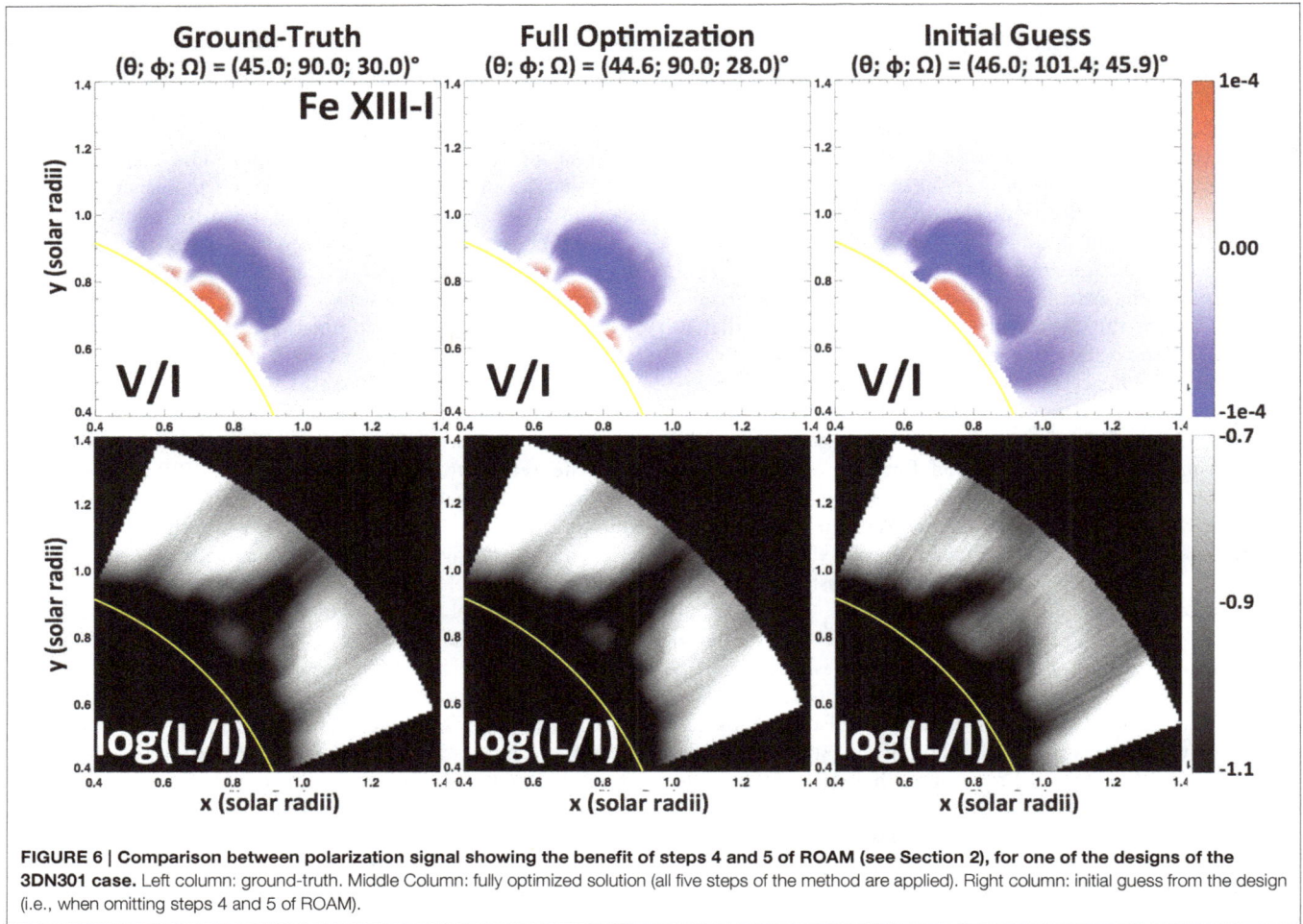

FIGURE 6 | Comparison between polarization signal showing the benefit of steps 4 and 5 of ROAM (see Section 2), for one of the designs of the 3DN301 case. Left column: ground-truth. Middle Column: fully optimized solution (all five steps of the method are applied). Right column: initial guess from the design (i.e., when omitting steps 4 and 5 of ROAM).

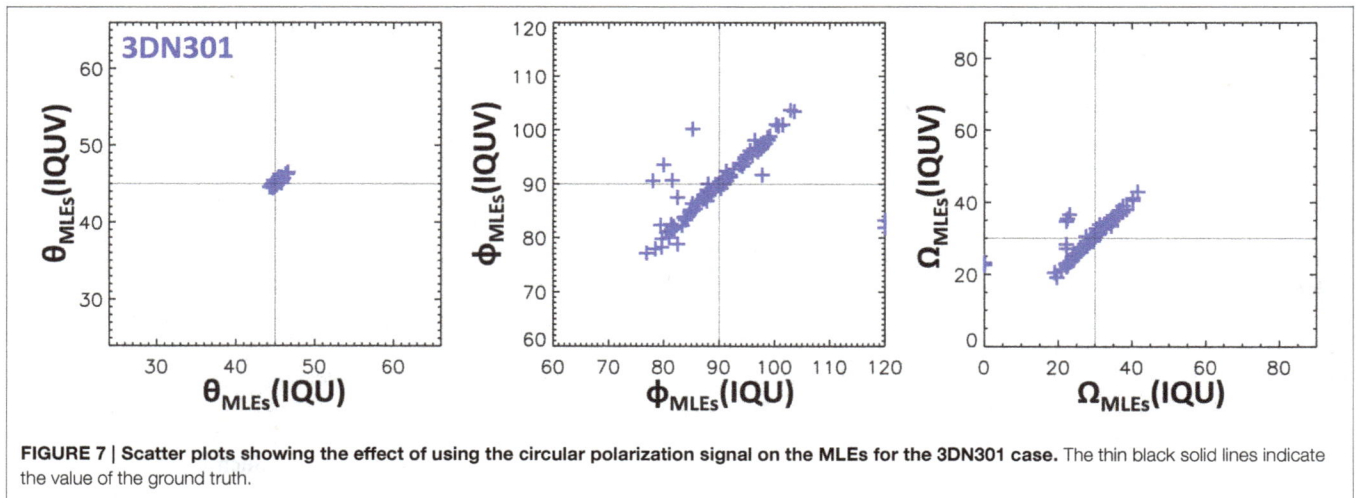

FIGURE 7 | Scatter plots showing the effect of using the circular polarization signal on the MLEs for the 3DN301 case. The thin black solid lines indicate the value of the ground truth.

3.3. Likelihood Function with Multiple Maxima

In this section, we test ROAM in the case of a log-likelihood function with multiple maxima having similar values. For that purpose, we only build the log-likelihood function with Stokes Q and U, setting the weight coefficients of Equation (7) to $(w_I; w_Q, w_U, w_V) = (0.; 2.0 \times 10^{-2}; 6.9 \times 10^{-2}; 0.)$. Only the height of the flux rope in the corona, h, and the tilt angle, Ω,

TABLE 3 | Characteristics of the test with a likelihood function possessing multiple maxima.

	n	$t_{elapsed}$ (hrs)	t_{full} (hrs)	h ($R_\odot$)	θ (°)	ϕ (°)	Ω (°)
2DN120	120	10	1.2×10^3	[0.04; 0.52]	45	90	[0; 357]

The test case contains 100 different designs. t_{full} is the total elapsed time that would be required to compute the Stokes images for the n^2 points of the 2D parameter space in series. The polarimetric data for the ground-truth are associated with $(h; \theta; \phi; \Omega) = (0.16\,R_\odot; 45°; 90°; 30°)$.

are considered for this test (see **Table 3** for the range of values considered for each parameter).

Stokes Q and U signals are associated with the transverse magnetic field, i.e., the component of a magnetic field perpendicular to the LOS. For a single point in the solar corona, the transverse magnetic field diagnosed from either the Hanle or Zeeman effect is subject to a 180° ambiguity (e.g., Casini and Judge, 1999; Judge, 2007). In terms of the parameters considered in our tests, it means that a single point magnetic field set with a rotation angle, Ω_{SP}, will give the same Stokes Q and U signals as when set with $\Omega_{SP} \pm 180°$. Considering that $\phi_{GT} = 90°$ (that is, the flux rope is centered at the solar limb) and the strong symmetry of our flux rope (see Section 3.1), we expect the LOS integrated Stokes Q and U to be the same for Ω and $\Omega \pm 180°$, resulting in a log-likelihood function with two maxima respectively located at Ω_{GT} and $\Omega_{GT} \pm 180°$; note that the symmetry of Stokes Q and U would be broken if the flux rope were not centered on the solar limb. This is indeed the case as shown **Figure 8B** where a maximum region can be observed at $\Omega = 30°$ and $\Omega = 210°$. **Figure 8B** further shows the presence of two additional maximum regions located at $\Omega = 150$ and 330°. These two solutions suggest a symmetry with regard to the plane $\Omega = 0°$ that is not expected. We find that the corresponding Stokes Q and U images are, as expected, different from those of the ground truth. However, the differences are small as compared with other values of Ω, resulting in a local maximum in those two regions. Note, though, that these four maximum regions are only possible because $\phi = 90°$, whereas any other value of ϕ would break the symmetry of the Stokes Q and U images.

In the present test case, the periodic parameter Ω varies on a range of values larger than half its period. In such circumstances, we must consider its periodicity for the RBF decomposition (see Appendix 2 in Supplementary Material). Accordingly, the parameters of the RBF decomposition are $d' = 3$, $m = 3$ and $p = m - 1 = 2$. As in Section 3.2, we run our optimization method on 100 different designs whose properties are given in **Table 3**. The results are summarized in a 2D dispersion plot in **Figure 8A**. The figure shows that the 100 MLEs are mainly, and almost equally, clustering around the two global Ω maximum regions, corresponding to the ground truth and its counterpart at 180°. We further find that, out of these 100 solutions, only four are associated with one of the two local maximum regions, here $\Omega \approx 150°$. As for the height of the MLEs, we find an average value of 1.6×10^{-1} $R_\odot$ with a 2σ dispersion level of 0.5×10^{-1} $R_\odot$, meaning that the height is well constrained even from using Stokes Q and U only. The dispersion plot from

Figure 8A therefore indicates that our optimization method is strongly sensitive to multiple global maxima and *can* be sensitive to local maxima. Note that the sensitivity to local maxima depends upon both the number of points used in the design and the value of these local maxima relatively to that of the global maxima.

Figure 8C displays a surface plot of the log-likelihood function from the best case of 2DN120. As one can see, the RBF decomposition is able to capture both the values and shapes of the exact log-likelihood function. We find an rms error of 0.04 on the log-likelihood. The RBF decomposition can therefore provide a good approximation of the exact log-likelihood function even with a periodic space and the presence of multiple maxima.

Finally, we show in **Figure 9** that using the Stokes V signal to build the log-likelihood function removes the Ω ambiguities that were observed in the log-likelihood function constructed from Stokes Q and U only. When using Stokes V, the optimization leads to $\Omega_{rms} \approx 28.8°$. This means that some additional observables might be worth considering to remove ambiguities in parameters when they exist. Another alternative to remove ambiguities is to reduce the parameter space to regions having a single maximum. Then, one can either study each region separately or use prior constraints to eliminate regions that are very unlikely. For instance, one can use the photospheric magnetograms, or Hα observations, prior to or after the passage of the flux rope at a limb to estimate the rotation angle (i.e., Ω) and put strong constraints on the values of rotation angle to consider for the parameter space.

3.4. Stability with Regard to Noise in the Data

In practice, any real data is subject to measurement errors. Such errors may prevent the retrieval of any meaningful information about the polarization, and hence, the magnetic field in regions of weak signals and/or when the signal-to-noise ratio is weak. The results of ROAM might be sensitive to such noise and we therefore need to investigate that sensitivity. For that reason, we now test our method when the synthetic observations associated with the ground truth contain some noise. In this regard, we build the log-likelihood function with Stokes U images only

$$\ell(x_i) = \frac{\ell_U(x_i)}{\sigma_U^2}, \tag{8}$$

where σ_U is the root mean square of the noise in the synthetic Stokes U signal of the ground truth.

For a given value of photon noise, σ_I, σ_Q, σ_U, σ_V are all different. As a consequence, if one uses more than one Stokes component, then varying the noise further changes the relative contribution of each Stokes parameter to the log-likelihood function due to the weighting by $1/\sigma_S$. We need to be free of the variation of relative contribution of the different Stokes in order to isolate the sole effect of noise on the robustness of our optimization method, which then implies using only one Stokes parameter to define the log-likelihood function. Considering CoMP capabilities and the current magnetic model and ground truth, we performed several tests with different levels of noise

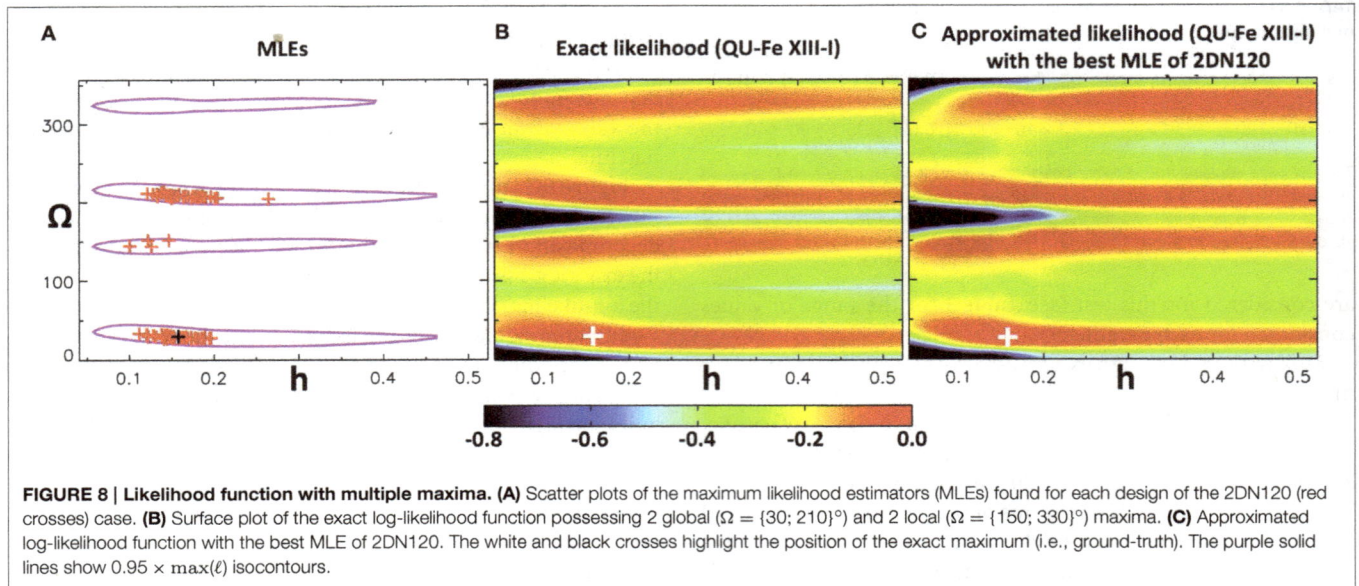

FIGURE 8 | Likelihood function with multiple maxima. (A) Scatter plots of the maximum likelihood estimators (MLEs) found for each design of the 2DN120 (red crosses) case. **(B)** Surface plot of the exact log-likelihood function possessing 2 global ($\Omega = \{30; 210\}°$) and 2 local ($\Omega = \{150; 330\}°$) maxima. **(C)** Approximated log-likelihood function with the best MLE of 2DN120. The white and black crosses highlight the position of the exact maximum (i.e., ground-truth). The purple solid lines show $0.95 \times \max(\ell)$ isocontours.

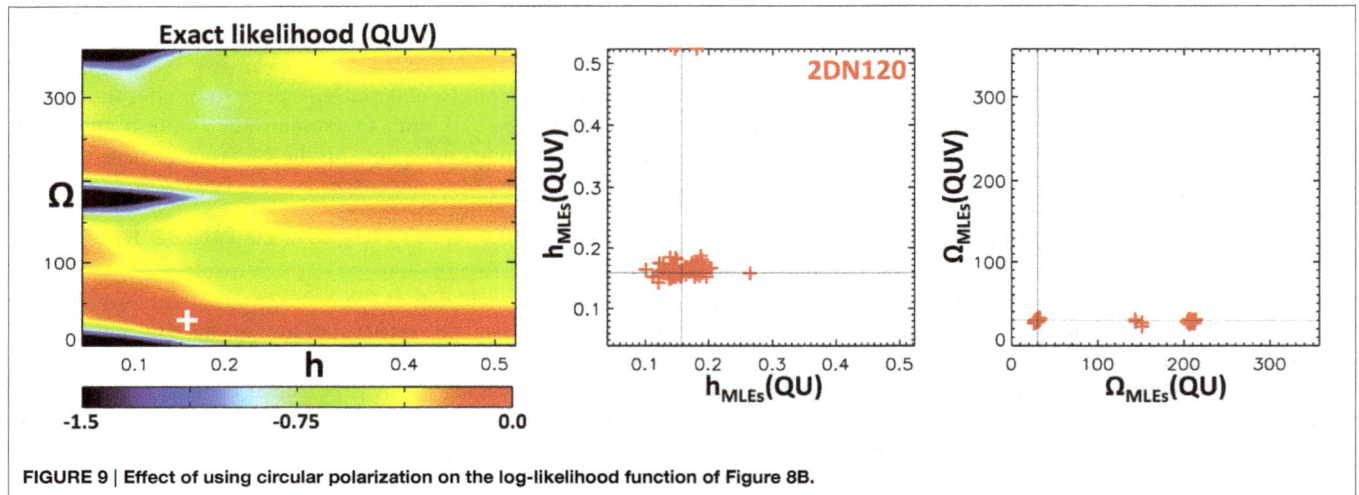

FIGURE 9 | Effect of using circular polarization on the log-likelihood function of Figure 8B.

(which can be added using FORWARD) and found that (1) Stokes V cannot be used for realistic exposure times because its values for our test bed are too weak and would require an unrealistic 4 days exposure time to reach a moderate level of noise for the particular choices of ground-truth parameters and pixel sizes (see corresponding values in Section 3.1), (2) Stokes I cannot be used because it is not sensitive enough to noise (even a 1 s exposure time leads to a very weak level of noise), and (3) Stokes Q and U are better suited for the noise test with exposure times of the order of 1–100 s. From this analysis, we chose Stokes U because it was slightly more sensitive to noise than Stokes Q for the setup considered in this paper (note that both Q and U are presented in a frame of reference relative to the local vertical, or radial coordinate).

FORWARD already implements a photon noise calculation for the infra-red lines under consideration (see e.g., Gibson et al.,

2016). The noise is calculated according to the specifications of the instrument considered (telescope aperture, detector efficiency), the background photon level, and the exposure time to obtain a forward calculation that includes the noise. For CoMP, the aperture is 20 cm, the efficiency is 0.05 throughput and the background is five parts per million of solar brightness. We perform three tests with different exposure time, $t_{\exp}$, hence noise level, i.e., $t_{\exp} = (1; 10; 300)$ seconds that respectively correspond to strong, moderate, and weak noise cases for the considered setup. The synthetic Stokes U images of the ground truth for these noise levels are displayed in **Figure 10**. These synthetic ground truth are used with all designs of the 3DN31, 3DN301, and 3DN31ZOOM cases.

Figure 11 presents scatter plots of the error on the MLEs obtained when noise is included in the ground truth Stokes U images as compared with the case when no noise is considered.

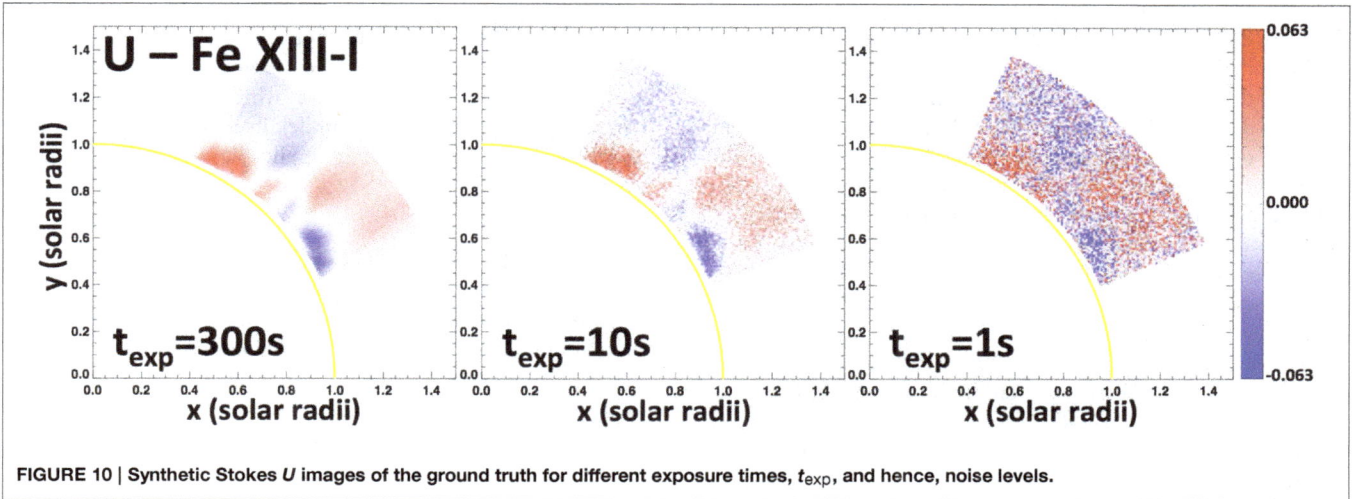

FIGURE 10 | Synthetic Stokes U images of the ground truth for different exposure times, t_{exp}, and hence, noise levels.

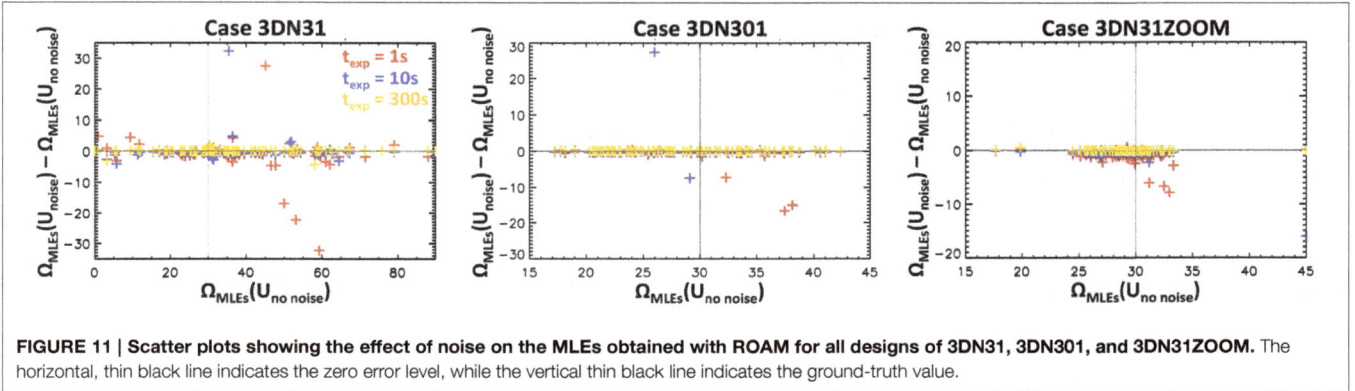

FIGURE 11 | Scatter plots showing the effect of noise on the MLEs obtained with ROAM for all designs of 3DN31, 3DN301, and 3DN31ZOOM. The horizontal, thin black line indicates the zero error level, while the vertical thin black line indicates the ground-truth value.

The plots are only shown for the Ω parameter because all three parameters θ, ϕ, and Ω display very similar results. In **Figure 11**, one can see a nearly perfect horizontal line at $y = 0$ for Ω_{MLEs} obtained with an exposure time of 300 s (yellow crosses) for all test cases (3DN31, 3DN301, and 3DN31ZOOM). This means that the $t_{exp} = 300$ s case is equivalent to the no-noise case. For cases $t_{exp} = 10$ and $t_{exp} = 1$ s, the plots show some departure from the $y = 0$ line, which increases with the level of noise. The figure also shows that the noise effect on the robustness of the MLEs depends on the density of points in the designs, i.e., 3DN31 is the most affected by the noise while 3DN31ZOOM is the least affected. That being said, we find that only less than $\approx 10 - 15$ points (out of 100) of 3DN31 exhibit a strong sensitivity to noise for the $t_{exp} = 1$ second case, i.e., with an error $>5°$. This number drops to ≈ 5 when $t_{exp} = 10$ s. For the $t_{exp} = 1$ s, **Figure 10** shows that the noise strongly masks the real Stokes U signal, although not entirely. We therefore deduce that our optimization method is very stable against the presence of noise in the data as long as the noise does not entirely mask the real signal. Considering that Stokes Q is similarly sensitive to noise as Stokes U and that Stokes I is much less sensitive to the noise, we conclude that our method can be robustly used with the Stokes I, Q, and U data provided by the CoMP.

4. CONCLUSIONS

In this paper, we introduced and validated a new optimization method for model-data fitting, ROAM (Radial-basis-functions Optimization Approximation Method). Our primary motivation for this work has been to develop a novel approach for diagnosing the solar coronal magnetic field by combining a parameterized 3D magnetic field model with forward modeling of coronal polarization. From various tests applied to the synthetic test bed of a coronal magnetic flux rope, we showed that ROAM allows for fast, efficient, and accurate model-data fitting in a d-dimensional parameter space. These test cases further enabled us to analyze and specify a framework for an optimal application of ROAM.

Applying our method with forward modeling of IR coronal polarimetry, we demonstrated that ROAM can be exploited for converting coronal polarimetric measurements into magnetic field data. The use of our model-data fitting method therefore opens new perspectives for the development and exploitation of coronal polarimetric measurements such as the ones routinely performed by CoMP (Tomczyk et al., 2008) and future telescopes such as the Daniel K. Inoue Solar Telescope[4] and the Coronal Solar Magnetism Observatory (Tomczyk et al., 2016), but also for

---

[4]http://www.ifa.hawaii.edu/~schad/dlnirsp/

a wider range of coronal observations including, e.g., UV (see e.g., Fineschi, 2001; Raouafi et al., 2009) and radio polarimetry (e.g., White and Kundu, 1997; Gelfreikh, 2004; see also Gibson et al., 2016, for discussion of multiwavelength magnetometry).

Beyond the analysis of coronal polarimetric measurements, ROAM offers interesting perspectives for magnetic field reconstruction models. Most of the current 3D diagnostics of the coronal magnetic field of solar active regions (ARs) are derived from the analysis of magnetic field reconstruction models including, e.g., force-free field extrapolations of the photospheric magnetic field (see e.g., Alissandrakis, 1981; Demoulin et al., 1989; Wheatland et al., 2000; Yan and Sakurai, 2000; Wiegelmann, 2004; Amari et al., 2006; Malanushenko et al., 2012, and references therein), and magneto-frictional methods (see e.g., van Ballegooijen, 2004; Valori et al., 2005, 2007; Jiang et al., 2011; Inoue et al., 2012; Titov et al., 2014, and references therein). ROAM could, in principle, be used to perform model-data fitting with such reconstruction models that either already are (i.e., through the poloidal and axial flux for the magneto-frictional methods with flux rope insertion) or could be (e.g., through the photospheric force-free parameter for both force-free field extrapolations and magneto-frictional methods without flux rope insertion) parameterized. The extensive work performed over the years in terms of forward modeling of various observables (see e.g., Gibson et al., 2016, and references therein) would then allow for using several types of different observations to constrain the parameters of the magnetic field reconstruction models. ROAM therefore opens new perspectives for including coronal polarimetric measurements into magnetic field reconstructions and, more generally, for data-optimized reconstruction of the solar coronal magnetic field. Such perspectives will be tackled in the framework of the Data Optimized Coronal Field Model[5] (DOCFM), a collaborative project that will make use of ROAM.

---

[5]http://www.hao.ucar.edu/DOCFM/

Finally, we wish to mention that ROAM is not limited to coronal magnetic field diagnostics and could be used for other optimization problems. The method will be of particular interest for model-data fitting for which a model evaluation (here, the evaluation of the model itself and/or the forward modeling of an observable if applicable) is computationally expensive.

AUTHOR CONTRIBUTIONS

All authors listed, have made substantial, direct and intellectual contribution to the work, and approved it for publication.

FUNDING

KD acknowledges funding from the Computational and Information Systems Laboratory and from the High Altitude Observatory, and along with SG and YF acknowledges support from the Air Force Office of Scientific Research under award FA9550-15-1-0030.

ACKNOWLEDGMENTS

We thank the two anonymous referees and Anna Malanushenko for a careful consideration of the manuscript and constructive comments. The National Center for Atmospheric Research is sponsored by the National Science Foundation.

REFERENCES

Alissandrakis, C. E. (1981). On the computation of constant alpha force-free magnetic field. *Astron. Astrophys.* 100, 197–200.

Amari, T., Boulmezaoud, T. Z., and Aly, J. J. (2006). Well posed reconstruction of the solar coronal magnetic field. *Astron. Astrophys.* 446, 691–705. doi: 10.1051/0004-6361:20054076

Bird, M. K., Volland, H., Howard, R. A., Koomen, M. J., Michels, D. J., Sheeley, N. R. Jr., et al. (1985). White-light and radio sounding observations of coronal transients. *Solar Phys.* 98, 341–368. doi: 10.1007/BF00152465

Bommier, V., and Sahal-Brechot, S. (1982). The Hanle effect of the coronal L-alpha line of hydrogen - Theoretical investigation. *Solar Phys.* 78, 157–178. doi: 10.1007/BF00151151

Boser, B. E., Guyon, I. M., and Vapnik, V. N. (1992). "A training algorithm for optimal margin classifiers," in *Proceedings of the Fifth Annual Workshop on Computational Learning Theory*, COLT '92, (New York, NY: ACM), 144–152. doi: 10.1145/130385.130401

Broomhead, D. S. and Lowe, D. (1988). Multivariable functional interpolation and adaptive networks. *Complex Syst.* 2, 321–355.

Buhmann, M. D. (2003). *Radial Basis Functions: Theory and Implementations.* Cambridge, UK: Cambridge Monographs on Applied and Computational Mathematics; Cambridge University Press.

Casini, R., and Judge, P. G. (1999). Spectral lines for polarization measurements of the coronal magnetic field. II. Consistent treatment of the stokes vector for magnetic-dipole transitions. *Astrophys. J.* 522, 524–539. doi: 10.1086/307629

Demoulin, P., Priest, E. R., and Anzer, U. (1989). A three-dimensional model for solar prominences. *Astron. Astrophys.* 221, 326–337.

Duchon, J. (1977). *Constructive Theory of Functions of Several Variables: Proceedings of a Conference Held at Oberwolfach April 25 - May 1, 1976.* Berlin; Heidelberg: Springer Berlin Heidelberg. doi: 10.1007/BFb0086566

Fan, Y. (2012). Thermal signatures of tether-cutting reconnections in pre-eruption coronal flux ropes: hot central voids in coronal cavities. *Astrophys. J.* 758:60. doi: 10.1088/0004-637X/758/1/60

Fasshauer, G. E. (2007). *Meshfree Approximation Methods with MATLAB*, Vol. 6 of *Interdisciplinary Mathematical Sciences*. Singapore: World Scientific Publishing Company.

Fineschi, S. (2001). "Space-based instrumentation for magnetic field studies of solar and stellar atmospheres," in *Magnetic Fields Across the Hertzsprung-Russell Diagram*, Vol. 248 of *Astronomical Society of the Pacific Conference Series*, eds G. Mathys, S. K. Solanki, and D. T. Wickramasinghe (San Francisco, CA: Astronomical Society of the Pacific), 597.

Flyer, N., Barnett, G. A., and Wicker, L. J. (2016). Enhancing finite differences with radial basis functions: experiments on the Navier-Stokes equations. *J. Comput. Phys.* 316, 39–62. doi: 10.1016/j.jcp.2016.02.078

Flyer, N., and Fornberg, B. (2011). Radial basis functions: developments and applications to planetary scale flows. *Comput. Fluids* 46, 32. doi: 10.1016/j.compfluid.2010.08.005

Fornberg, B. and Flyer, N. (2015). Solving PDEs with radial basis functions. *Acta Numerica* 24, 215–258. doi: 10.1017/S0962492914000130

Gelfreikh, G. B. (2004). "Coronal magnetic field measurements through Bremsstrahlung emission," in *Astrophysics and Space Science Library*, Vol. 314 of *Astrophysics and Space Science Library*, eds D. E. Gary and C. U. Keller (Dordrecht: Springer), 115. doi: 10.1007/1-4020-2814-8 6

Gibson, S., Kucera, T., White, S. M., Dove, J., Fan, Y., Forland, B., et al. (2016). FORWARD: a toolsel for multiwavelength coronal magnetometry. *Front. Astron. Space Sci.* 3:8. doi: 10.3389/fspas.2016.00008

Gibson, S. E., and Charbonneau, P. (1998). Empirical modeling of the solar corona using genetic algorithms. *J. Geophys. Res.* 103, 14511–14522. doi: 10.1029/98JA00676

Hale, G. E. (1908). On the probable existence of a magnetic field in sun-spots. *Astrophys. J.* 28, 315. doi: 10.1086/141602

Hanle, W. (1924). Über magnetische Beeinflussung der Polarisation der Resonanzfluoreszenz. *Z. Phys.* 30, 93–105. doi: 10.1007/BF01331827

Harder, R. L., and Desmarais, R. N. (1972). Interpolation using surface splines. *J. Aircraft* 9, 189–191. doi: 10.2514/3.44330

Iman, R. L., Helton, J. C., and Campbell, J. E. (1981). An approach to sensitivity analysis of computer models, part 1. introduction, input variable selection and preliminary variable assessment. *J. Qual. Technol.* 13, 174–183.

Inoue, S., Magara, T., Watari, S., and Choe, G. S. (2012). Nonlinear force-free modeling of a three-dimensional sigmoid observed on the sun. *Astrophys. J.* 747:65. doi: 10.1088/0004-637X/747/1/65

Jiang, C., Feng, X., Fan, Y., and Xiang, C. (2011). Reconstruction of the coronal magnetic field using the CESE-MHD Method. *Astrophys. J.* 727:101. doi: 10.1088/0004-637X/727/2/101

Jones, D. R., Schonlau, M., and Welch, W. J. (1998). Efficient global optimization of expensive black-box functions. *J. Glob. Optimization* 13, 455–492. doi: 10.1023/A:1008306431147

Judge, P. G. (2007). Spectral lines for polarization measurements of the coronal magnetic field. V. Information content of magnetic dipole lines. *Astrophys. J.* 662, 677–690. doi: 10.1086/515433

Judge, P. G., Low, B. C., and Casini, R. (2006). Spectral lines for polarization measurements of the coronal magnetic field. IV. Stokes signals in current-carrying fields. *Astrophys. J.* 651, 1229–1237. doi: 10.1086/507982

Kramar, M., Airapetian, V., Mikić, Z., and Davila, J. (2014). 3D coronal density reconstruction and retrieving the magnetic field structure during solar minimum. *Solar Phys.* 289, 2927–2944. doi: 10.1007/s11207-014-0525-7

Lam, N. S. N. (1983). Spatial interpolation methods: a review. *Am. Cartogr.* 10, 129–149.

Lin, H., Kuhn, J. R., and Coulter, R. (2004). Coronal magnetic field measurements. *Astrophys. J.* 613, L177–L180. doi: 10.1086/425217

López Ariste, A. (2015). "Magnetometry of prominences," in *Solar Prominences*, Vol. 415 of *Astrophysics and Space Science Library*, eds J.-C. Vial and O. Engvold (Cham: Springer International Publishing), 179.

Louis, S. J., and Rawlins, G. J. E. (1993). "Syntactic analysis of convergence in genetic algorithms," in *Foundations of Genetic Algorithms - 2*, ed L. D. Whitley (San Mateo, CA: Morgan Kauffman), 141–152.

Madych, W. R., and Nelson, S. A. (1990). Polyharmonic cardinal splines. *J. Approx. Theory* 60, 141–156. doi: 10.1016/0021-9045(90)90079-6

Malanushenko, A., Schrijver, C. J., DeRosa, M. L., Wheatland, M. S., and Gilchrist, S. A. (2012). Guiding nonlinear force-free modeling using coronal observations: first results using a Quasi-Grad-Rubin scheme. *Astrophys. J.* 756:153. doi: 10.1088/0004-637X/756/2/153

McKay, M. D., Beckman, R. J., and Conover, W. J. (1979). A comparison of three methods for selecting values of input variables in the analysis of output from a computer code. *Technometrics* 21, 239–245.

Nychka, D., Bandyopadhyay, S., Hammerling, D., Lindgren, F., and Sain, S. (2015). A multiresolution gaussian process model for the analysis of large spatial datasets. *J. Comput. Graphical Stat.* 24, 579–599. doi: 10.1080/10618600.2014.914946

Plowman, J. (2014). Single-point inversion of the coronal magnetic field. *Astrophys. J.* 792:23. doi: 10.1088/0004-637X/792/1/23

Powell, M. J. D. (1977). Restart procedures for the conjugate gradient method. *Math. Programming* 12, 241–254. doi: 10.1007/BF01593790

Powell, M. J. D. (1993). *Some Algorithms for Thin Plate Spline Interpolation to Functions of Two Variables.* Cambridge University Dept. of Applied Mathematics and Theoretical Physics technical report.

Rachmeler, L. A., Casini, R., and Gibson, S. E. (2012). "Interpreting Coronal Polarization Observations," in *Second ATST-EAST Meeting: Magnetic Fields from the Photosphere to the Corona.*, Vol. 463 of *Astronomical Society of the Pacific Conference Series*, eds T. R. Rimmele, A. Tritschler, F. Wöger, M. Collados Vera, H. Socas-Navarro, R. Schlichenmaier, M. Carlsson, T. Berger, A. Cadavid, P. R. Gilbert, P. R. Goode, and M. Knölker (San Francisco, CA: Astronomical Society of the Pacific), 227.

Rachmeler, L. A., Gibson, S. E., Dove, J. B., DeVore, C. R., and Fan, Y. (2013). Polarimetric properties of flux ropes and sheared arcades in coronal prominence cavities. *Solar Phys.* 288, 617–636. doi: 10.1007/s11207-013-0325-5

Raouafi, N.-E., Solanki, S. K., and Wiegelmann, T. (2009). "Hanle effect diagnostics of the coronal magnetic field: a test using realistic magnetic field configurations," in *Solar Polarization 5: In Honor of Jan Stenflo*, Vol. 405 of *Astronomical Society of the Pacific Conference Series*, eds S. V. Berdyugina, K. N. Nagendra, and R. Ramelli (San Francisco, CA: Astronomical Society of the Pacific), 429.

Sahal-Brechot, S., Bommier, V., and Leroy, J. L. (1977). The Hanle effect and the determination of magnetic fields in solar prominences. *Astron. Astrophys.* 59, 223–231.

Stenflo, J. O. (2015). History of solar magnetic fields since George Ellery Hale. *Space Sci. Rev.* 1–31. doi: 10.1007/s11214-015-0198-z

Titov, V. S., Török, T., Mikic, Z., and Linker, J. A. (2014). A method for embedding circular force-free flux ropes in potential magnetic fields. *Astrophys. J.* 790:163. doi: 10.1088/0004-637X/790/2/163

Tomczyk, S., Card, G. L., Darnell, T., Elmore, D. F., Lull, R., Nelson, P. G., et al. (2008). An instrument to measure coronal emission line polarization. *Solar Phys.* 247, 411–428.

Tomczyk, S., Landi, E., Burkepile, J. T., Casini, R., DeLuca, E. E., Fan, Y., et al. (2016). Scientific objectives and capabilities of the Coronal Solar Magnetism Observatory. *J. Geophys. Res. Space Phys.* doi: 10.1002/2016JA022871

Valori, G., Kliem, B., and Fuhrmann, M. (2007). Magnetofrictional extrapolations of low and lou's force-free equilibria. *Solar Phys.* 245, 263–285. doi: 10.1007/s11207-007-9046-y

Valori, G., Kliem, B., and Keppens, R. (2005). Extrapolation of a nonlinear force-free field containing a highly twisted magnetic loop. *Astron. Astrophys.* 433, 335–347. doi: 10.1051/0004-6361:20042008

van Ballegooijen, A. A. (2004). Observations and modeling of a filament on the Sun. *Astron. Astrophys.* 612, 519–529. doi: 10.1086/422512

Wheatland, M. S., Sturrock, P. A., and Roumeliotis, G. (2000). An optimization approach to reconstructing force-free fields. *Astron. Astrophys.* 540, 1150–1155. doi: 10.1086/309355

White, S. M., and Kundu, M. R. (1997). Radio observations of gyroresonance emission from coronal magnetic fields. *Solar Phys.* 174, 31–52. doi: 10.1023/A:1004975528106

Wiegelmann, T. (2004). Optimization code with weighting function for the reconstruction of coronal magnetic fields. *Solar Phys.* 219, 87–108.

Yan, Y. and Sakurai, T. (2000). New boundary integral equation representation for finite energy force-free magnetic fields in open space above the Sun. *Solar Phys.* 195, 89–109. doi: 10.1023/A:1005248128673

Conflict of Interest Statement: The authors declare that the research was conducted in the absence of any commercial or financial relationships that could be construed as a potential conflict of interest.

8

A Study of the Lunisolar Secular Resonance $2\dot{\omega} + \dot{\Omega} = 0$

Alessandra Celletti[1]* and Cătălin B. Galeş[2]

[1] Department of Mathematics, University of Rome Tor Vergata, Rome, Italy, [2] Department of Mathematics, University of Iasi, Iasi, Romania

The dynamics of small bodies around the Earth has gained a renewed interest, since the awareness of the problems that space debris can cause in the nearby future. A relevant role in space debris is played by lunisolar secular resonances, which might contribute to an increase of the orbital elements, typically of the eccentricity. We concentrate our attention on the lunisolar secular resonance described by the relation $2\dot{\omega} + \dot{\Omega} = 0$, where ω and Ω denote the argument of perigee and the longitude of the ascending node of the space debris. We introduce three different models with increasing complexity. We show that the growth in eccentricity, as observed in space debris located in the MEO region at the inclination about equal to 56°, can be explained as a natural effect of the secular resonance $2\dot{\omega} + \dot{\Omega} = 0$, while the chaotic variations of the orbital parameters are the result of interaction and overlapping of nearby resonances.

Keywords: space debris, lunisolar secular resonance, eccentricity growth

1. INTRODUCTION

Thousands of man-made objects, abandoned during space missions or remnants of operative satellites, orbit around the Earth at different altitudes. Their size varies from larger pieces, like old satellites or rocket stages, to dust-size particles given by fragmentation of satellites or even by collision events, like the impact between Kosmos 2251 and Iridium 33 in 2009, or the destruction of Fengyun-1C in 2007.

The dynamics of space debris strongly differs according to the altitude from the Earth. To this end, one distinguishes 4 main regions as follows:

(i) the LEO (Low Earth Orbit) region spans the altitude from 0 to 2000 km; here the objects feel, in order of importance, the gravitational attraction of our planet, the dissipation due to the atmospheric drag, the Earth's oblateness effect, the attraction of Moon and Sun, and the solar radiation pressure;

(ii) the MEO (Medium Earth Orbit) region goes from 2000 to 30,000 km of altitude; the forces felt by the debris are like in LEO, except that there is no atmospheric drag;

(iii) the GEO (Geostationary orbit) region is located around the value of 42,164.17 km from the Earth's center; geostationary objects move with an orbital period equal to the rotational period of the Earth;

(iv) HEO (High Earth orbit) region, refers to the space region with altitude above the geosynchronous orbit.

In this work we are interested in a particular type of motion, which corresponds to a so-called secular resonance. In particular, we consider the orbital elements which are solutions of the relation

$$2\dot{\omega} + \dot{\Omega} = 0 \,, \qquad (1)$$

where ω denotes the argument of perigee of the debris and Ω its longitude of the ascending node. A relation like Equation (1), involving quantities moving on long time-scales, is called a *secular resonance*. By considering the variations of ω and Ω as just due to the effect of the main spherical harmonics of the geopotential, one can show that Equation (1) can be written just in terms of the inclination. As shown in Hughes (1980), there can be several secular resonances which depend on the inclination only. Among such resonances, Equation (1) represents a very interesting case, since it has been shown that it affects the dynamics of objects in the MEO region (Rossi, 2008; Radtke et al., 2015; Sanchez et al., 2015). Chaotic motions arise from the interaction and overlapping of nearby resonances (Rosengren et al., 2015a,b; Daquin et al., 2016).

In this paper we introduce three different models with increasing complexity, apt to study the resonance Equation (1). The simplest model is described by a one degree-of-freedom autonomous Hamiltonian, which is obtained by averaging over the fast angles and by neglecting the rates of variation of the lunar longitude of the ascending node. This model provides the essential features, like the location of stable equilibria with large as well as with small libration amplitude. The growth of the eccentricity can be easily explained by this integrable model. In the second model one does not average over the fast angles, but still retains the assumption that the longitude of the ascending node of the Moon is constant. Circulation and libration regions can be located, as well as the chaotic separatrix, although the dynamics is very complicated: overlapping of resonances, bifurcations and, as a consequence, the existence of equilibria at large eccentricities as well as at small eccentricities, variation of the amplitude of the resonance. The last model includes the variation of the lunar longitude of the ascending node and shows that large chaotic regions can appear, contributing to an irregular variation of the orbital elements.

2. THE MODEL

We consider a space debris subject to the gravitational attraction of the Earth, including the oblateness potential, as well as the influence of Sun and Moon. This model is described by a Hamiltonian of the form

$$\mathcal{H} = \mathcal{H}_{Kep} + \mathcal{H}_{Geo} + \mathcal{H}_{Moon} + \mathcal{H}_{Sun} \,, \qquad (2)$$

which is the sum of different contributions that we are going to explain and express in terms of the Delaunay action–angle variables $(L, G, H, M, \omega, \Omega)$, where the actions are defined by

$$L = \sqrt{\mu_E a} \,, \quad G = L\sqrt{1 - e^2} \,, \quad H = G\cos I \,, \qquad (3)$$

with $\mu_E = \mathcal{G} m_E$ the product of the gravitational constant $\mathcal{G}$ and the Earth's mass m_E, a the semimajor axis, e the orbital eccentricity, I the inclination, while the angle variables are the mean anomaly M, the argument of perigee ω, the longitude of the ascending node Ω, which are expressed with respect to the equatorial plane.

The first term in Equation (2) represents the Keplerian part $\mathcal{H}_{Kep}$, which can be expressed as

$$\mathcal{H}_{Kep}(L) = -\frac{\mu_E^2}{2L^2} \,. \qquad (4)$$

The second term $\mathcal{H}_{Geo}$ describes the perturbation due to the Earth, when considering the shape of our planet. In particular, we will consider only the most important term of the expansion in spherical harmonics of the geopotential, the so-called J_2-term. Indeed, while studying the long-term dynamics of resonant orbits, the short-periodic terms that depend on the mean anomaly of the satellite (as well as the mean anomaly of the perturbing body, when dealing with Sun and Moon) can be averaged over from the disturbing function. Therefore, in the expression for $\mathcal{H}_{Geo}$ we take an average of the Hamiltonian over the mean anomaly of the space debris, which implies to consider only the most important contribution, corresponding to the J_2 gravity coefficient of the secular part (see, e.g., Celletti and Galeş, 2014, compare also with Celletti and Galeş, 2015). This leads to express $\mathcal{H}_{Geo}$ in the form:

$$\mathcal{H}_{Geo}(L, G, H) = \frac{R_E^2 J_2 \mu_E^4}{4} \frac{1}{L^3 G^3} \left(1 - 3\frac{H^2}{G^2}\right) , \qquad (5)$$

where R_E is the mean equatorial radius of the Earth and $J_2 = 1.08263 \times 10^{-3}$.

The contributions due to Moon and Sun are simplified by averaging over the fast angles, precisely the mean anomaly of the debris and the mean anomalies of the perturbers (Moon and Sun). Moreover, we truncate the potentials to second order in the ratio of semi-major axes (see Kaula, 1962; Lane, 1989; Celletti et al., 2016b for details), thus obtaining the expression for $\mathcal{H}_{Sun}$ and the (quite long) expression for $\mathcal{H}_{Moon}$, reported in Supplementary Section (see also Cook, 1962). Adding the contributions in Equations (4), (5) as well as $\mathcal{H}_{Sun}$ and $\mathcal{H}_{Moon}$, we obtain the Hamiltonian Equation (2).

Since the mean anomaly M is a cyclic variable, its conjugated action L (or equivalently the semi-major axis a) is constant. As a consequence, the Hamiltonian system described by Equation (2) is non-autonomous with two degrees of freedom. As it was remarked by Rosengren et al. (2015a), Daquin et al. (2016), and analytically shown in Celletti et al. (2016b), the Hamiltonian $\mathcal{H}$ depends on time just through the longitude of lunar ascending node Ω_M with a rate of variation equal to $\dot{\Omega}_M \simeq -0.053°/day$, which implies a periodicity of Ω_M over 18.6 years. More precisely, since $\dot{\Omega}_S = 0$, where Ω_S is the longitude of the solar ascending node, and the expansions of the lunar and solar potentials to second order in the ratio of semimajor axes are independent of the lunar and solar perigees, it follows that $\mathcal{H}$ depends on time only through Ω_M.

To a first approximation we assume that the Moon orbits on an elliptic trajectory with semimajor axis equal to $a_M = 384,748$ km, eccentricity $e_M = 0.0549006$ and inclination $I_M = 5°15'$; the mass m_M of the Moon, expressed in Earth's masses, is about equal to 0.0123. The orbital elements of the Moon are referred to the ecliptic plane.

As for the Sun, we can assume that its elements are constants and, precisely, $a_S = 149,597,871$ km, eccentricity $e_S = 0.01671123$ and inclination $I_S = 23°26'21.406''$; the mass of the Sun m_S, expressed in Earth's masses, is approximately equal to 333,060.4016. The orbital elements of the Sun are expressed with respect to the equatorial plane.

The model described by Equation (2) gives all the ingredients to capture the main dynamical features of the resonant structure within the MEO region (see Rosengren et al., 2015a for a comparison between various models).

3. THE SECULAR RESONANCE $2\dot{\omega} + \dot{\Omega} = 0$

In this Section we are interested to the so-called (lunar and solar) *secular resonances*, which occur whenever one has a commensurability between the arguments of perigee and the longitudes of the nodes of the debris and the perturbers, according to the following definition.

Definition 1. *A lunar gravity secular resonance occurs whenever there exists an integer vector $(k_1, k_2, k_3) \in \mathbb{Z}^3 \backslash \{0\}$, such that*

$$k_1\dot{\omega} + k_2\dot{\Omega} + k_3\dot{\Omega}_M = 0. \tag{6}$$

We have a solar gravity secular resonance whenever there exist $(k_1, k_2, k_3) \in \mathbb{Z}^3 \backslash \{0\}$, such that

$$k_1\dot{\omega} + k_2\dot{\Omega} + k_3\dot{\Omega}_S = 0. \tag{7}$$

We can assume that the rate of variation $\dot{\Omega}_S$ is zero, while for the Moon we will build different models according to which the rate $\dot{\Omega}_M$ is zero or it is rather equal to $\dot{\Omega}_M \simeq -0.053°/day$.

As for the debris, we can approximate $\dot{\omega}$, $\dot{\Omega}$ by considering only the effect of J_2 (Hughes, 1980):

$$\dot{\omega} \simeq 4.98 \left(\frac{R_E}{a}\right)^{\frac{7}{2}} (1 - e^2)^{-2} (5\cos^2 I - 1) \, °/day,$$

$$\dot{\Omega} \simeq -9.97 \left(\frac{R_E}{a}\right)^{\frac{7}{2}} (1 - e^2)^{-2} \cos I \, °/day. \tag{8}$$

Inserting Equation (8) in Equation (6) or Equation (7), we get an expression which involves the orbital elements a, e, I, thus providing the location of the secular resonance.

A remarkable fact (see Hughes, 1980) is that some resonances depend only on the inclination and are independent on a, e. Precisely, following Hughes (1980) we can identify the following classes of lunisolar secular resonances depending only on inclination (see **Figure 1**):

(i) $\dot{\omega} = 0$, which occurs at the critical inclinations $I = 63.4°$, $I = 116.6°$;
(ii) $\dot{\Omega} = 0$, which corresponds to polar orbits;
(iii) $\alpha\dot{\omega} + \beta\dot{\Omega} = 0$ for some nonzero $\alpha, \beta \in \mathbb{Z}$.

In this work we are interested to a specific resonance of type *(iii)* and precisely to the resonance

$$2\dot{\omega} + \dot{\Omega} = 0. \tag{9}$$

Using Equations (8) and (9), one can write this resonances as

$$2\dot{\omega} + \dot{\Omega} = \left(\frac{R_E}{a}\right)^{\frac{7}{2}} (1 - e^2)^{-2} \left[9.96(5\cos^2 I - 1) - 9.97\cos I\right] = 0,$$

whose solutions are $I = 56.1°$ and $I = 111.0°$, independently of the values of semimajor axis and eccentricity.

In writing Equation (9) we have implicitly assumed that $\dot{\Omega}_M = 0$ (as we mentioned, the other rates $\dot{\omega}_M$, $\dot{\omega}_S$, $\dot{\Omega}_S$ can be assumed to be equal to zero). However, Ω_M varies periodically and some arguments of $\mathcal{H}_{Moon}$ could depend also on Ω_M. Therefore, besides $2\dot{\omega} + \dot{\Omega} = 0$, one also has the commensurability relations

$$2\dot{\omega} + \dot{\Omega} + s\dot{\Omega}_M = 0, \qquad s = -2, -1, 1, 2. \tag{10}$$

This means that the secular resonance splits into a multiplet of resonances. This splitting phenomenon is responsible for the existence of a very complex web-like background of resonances in the phase space, which leads to a chaotic variation of the orbital elements. An analytical estimate of the location of the resonance corresponding to each component of the multiplet, as a function of eccentricity and inclination, can be obtained by using Equation (8) (see, for example, Figure 2 in Ely and Howell, 1997 or Rosengren et al., 2015a).

To describe properly the dynamics, it is convenient to use resonant variables, which are introduced through the symplectic transformation $(G, H, \omega, \Omega) \rightarrow (S, T, \sigma, \eta)$ defined by

$$\sigma = 2\omega + \Omega, \qquad S = \frac{G}{2},$$

$$\eta = \Omega, \qquad T = H - \frac{G}{2}. \tag{11}$$

Since we expressed the Hamiltonian in Delaunay variables, we represent in **Figure 1** the web structure of resonances in the space of the actions T–S introduced in Equation (11). To avoid confusions that might arise when we speak about a specific resonance, we will use the terminology *exact resonance* when we refer to the component of the multiplet characterized by $s = 0$ in Equation (10), while the expression *whole resonance* means that we refer to all components of the multiplet.

We underline that the units of length and time are normalized so that the geostationary distance is unity (it amounts to $42,164.17$ km) and that the period of the Earth's rotation is equal to 2π. As a consequence, from Kepler's third law it follows that $\mu_E = 1$. Therefore, unless the units are explicitly specified, the action variables L, S and T are expressed in the above units.

Figure 1 shows the structure of resonances for $a = 15,000$ km (top panels) and $a = 29,546$ km (bottom panels). The colored curves provide the location of the resonances, while the vertical black dashed line in the top-right panel is drawn to provide the value of T used in computing the FLI plot for $a = 15,000$ km (see **Figure 6**). In order to show graphical evidence of the splitting phenomenon, **Figure 1**, left panels, provide the resonant structure for $S \in [0, S_{max}]$, where $S_{max} = \frac{\sqrt{\mu_E a}}{2}$. These plots contain also the horizontal black line $S = S_{min}$, where S_{min} is computed from the condition that the distance of the perigee cannot be smaller than the radius of the Earth, that is

$$S_{min} = \frac{1}{2}\sqrt{\frac{(2a - R_E)\mu_E R_E}{a}}.$$

Therefore, the interval of interest is $[S_{min}, S_{max}]$. The right panels of **Figure 1** magnify the regions associated to the orbits that do not collide with the Earth (at least for a small interval of time). **Figure 1** shows the complicated interplay of the web of resonances, with multiple crossings of lines, which correspond to overlapping of resonances, possibly providing a mechanism for the onset of chaos (Chirikov, 1979; Daquin et al., 2016).

4. A COMPARISON OF DIFFERENT MODELS

In order to understand the complicated dynamics of the *whole resonance* $2\dot\omega + \dot\Omega = 0$, we shall simplify further the model described in the previous Section. In fact, we consider three different models, based on the Hamiltonian function introduced in Equation (2):

a) The one degree-of-freedom autonomous Hamiltonian, obtained by averaging $\mathcal{H}$ in Equation (2) over the *fast angle* η

and by neglecting the rates of variation of Ω_M. Indeed, we use the constant value $\Omega_M = 125.045°$, valid at epoch J2000.

b) The two degrees-of-freedom autonomous Hamiltonian, derived under the assumption that the rate of variation of Ω_M is negligible. Again, we use the constant value $\Omega_M = 125.045°$, valid at epoch J2000.

c) The non-autonomous Hamiltonian $\mathcal{H}$, defined by Equation (2).

The following sections describe in detail the results which are obtained using models a–c.

4.1. Results for model A

The results obtained integrating model a, the simplest model as possible, are shown in **Figure 2**, which provides the phase space portraits for $a = 15,000$ km and $a = 29,546$ km. In order to show more clearly the structure of the phase space, in all figures we represent the resonant angle $\sigma = 2\omega + \Omega$ on intervals longer than 360°.

Figure 2 shows that for sufficiently small values of the semimajor axis (left panel) the phase space has a pendulum-like structure, while for larger values of the semimajor axis (middle and right panels) the pendulum-like model is no longer valid. In fact, for $a = 29,546$ km, a bifurcation phenomenon appears, showing that there are some cases when a specific resonance

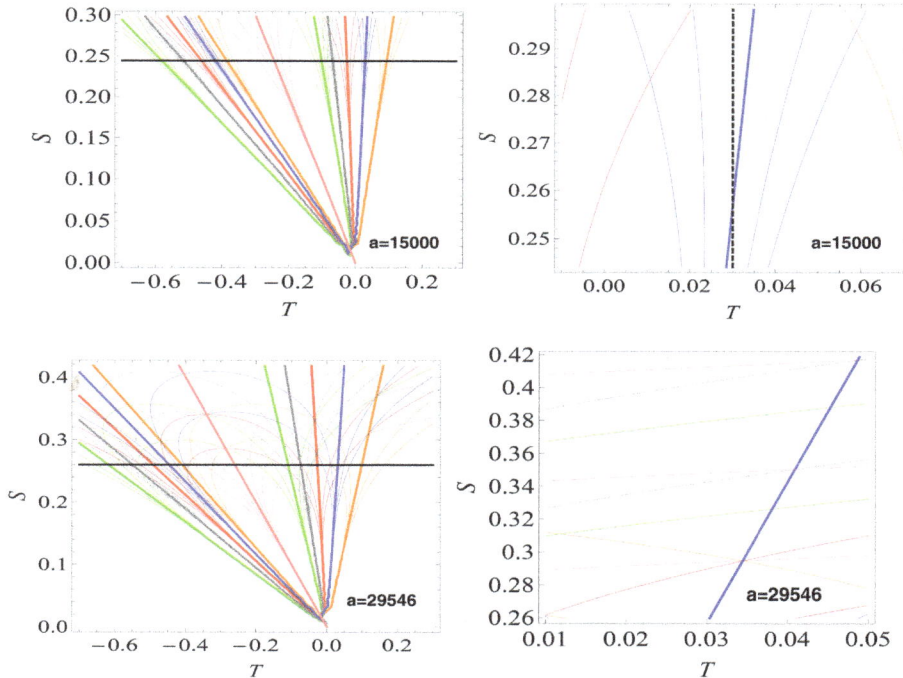

FIGURE 1 | The web structure of resonances in the space of the actions for $a = 15,000$ km (upper panels) and $a = 29,546$ km (bottom panels). The thick curves represent the location of the following *exact resonances* (the multiplet component having $s = 0$): $\dot\Omega = 0$ (pink color, $I = 90°$), $\dot\omega - \dot\Omega = 0$ (green color, $I = 73.2°$, $I = 133.6°$), $2\dot\omega - \dot\Omega = 0$ (gray color, $I = 69.0°$), $I = 123.9°$, $\dot\omega = 0$ (red color, $I = 63.4°$, $I = 116.6°$), $2\dot\omega + \dot\Omega = 0$ (blue color, $I = 56.1°$, $I = 111°$) and $\dot\omega + \dot\Omega = 0$ (orange color, $I = 46.4°$, $I = 106.9°$). The thin curves give the position of the resonances $(2 - 2p)\dot\omega + m\dot\Omega + s\dot\Omega_k = 0$ with $p, m = 0, 1, 2$ and $s = -2, -1, 1, 2$. The vertical black dashed line (top right panel) corresponds to the values of T used in computing the **Figure 6**. Left panels are obtained for $S \in [0, S_{max}]$, whereas in the right plots S varies from S_{min} to S_{max}, as explained in the text.

cannot be modeled by a pendulum type system, but one should use a more complex model, referred in the literature as the *extended fundamental model* (see Breiter, 2001; Celletti et al., 2016a for details).

Comparing the right panel of **Figure 2**, obtained for $T = 0.03$, with the middle panel of the same **Figure 2**, computed for $T = 0.05$, we notice the appearance of a new elliptic point, located at $\sigma = 180°$. Besides this phenomenon, it is important to note that the main stable point, which is located at $\sigma = 360°$ (or $0°$), changes its position in the action space as a function of T. For instance, for $T = 0.05$, this point is located at $S = 0.3407$ [or at $e = 0.581$, as it follows from Equations (3), (11)], while for $T = 0.03$, it is positioned at $S = 0.26$ (or $e = 0.784$). **Figure 2** middle plot reveals the fact that none of the orbits located inside the libration region of the elliptic point will collide with the Earth, while in **Figure 2** right plot, all orbits located inside the libration region associated with the main elliptic point are colliding orbits.

The integrable model a gives a clear explanation for the growth of the eccentricity of the satellites and space debris revolving around the Earth on orbits having an inclination about equal to $56°$. In fact, the growth of the eccentricity is mainly due to the dynamical feature of the resonance. Inside the libration region, the resonant angle $\sigma = 2\omega + \Omega$ and its conjugated action S vary periodically. Since, the eccentricity e is related to S through the relation $e = \sqrt{1 - \frac{4S^2}{T^2}}$, then it follows naturally that the eccentricity varies in time.

4.2. Results for Model B

To analyze model b we use the Fast Lyapunov Indicators (hereafter, FLI), which are defined as the largest Lyapunov characteristic exponents at a fixed time (compare with Celletti and Galeş, 2014). We provide the definition of FLI in Supplementary Materials. Their values provide a numerical indication of the stable (low values) and chaotic (high values) behavior of the dynamical system as the initial conditions or some internal parameters are varied.

We shall focus on $a = 29,546$ km, because for $a = 15,000$ km the phase plane σ–S, even in the case of the full model c, is similar to a pendulum, as it is shown in **Figure 6**.

The results for model b are given in **Figures 3–5**. Thus, given $a = 29,546$ km and a value for T, we compute a grid of 100×100 points of the σ–S plane, where the resonant angle ranges in the interval $[0°, 360°]$ (also here we use a larger interval just to show better the structure of the phase space), while S spans the interval $[S_{min}, S_{max}]$. However, instead of displaying S on the vertical axis, in each plot we show the eccentricity values (on the left) and the inclination values (on the right), computed by using the relations (3) and (11) for given values of T. In all plots that represent the FLI values, we use the ranges corresponding to those used in the right panels of **Figure 1**. The relation among S, T, e and I is trivial; for instance, the value $e = 0.784$ from the left panel of **Figure 3** corresponds to the value $S = 0.26$ from the top right panel of **Figure 1**, while the value $I = 52.02°$ from the same left panel of **Figure 3** corresponds to the values $S = 0.26$ and $T = 0.06$.

Although the initial conditions are set such that the initial orbits have the perigee larger than R_E, since we are interested in understanding the mean dynamical features of the $2\dot{\omega} + \dot{\Omega} = 0$ resonance, during the total time of integration, we neglect the Earth's dimensions. Namely, we propagate each orbit up to 465 years (equal to 25×18.6 years), even if at some intermediate time the perigee distance becomes smaller than the radius of the Earth.

As we mentioned in Section 4.1, for large values of the semimajor axis in model a, the phase space is much more complicated than the one associated to the pendulum model. The complexity increases when we consider the two degrees-of-freedom autonomous Hamiltonian of model b. In fact, the manifolds defined by $\mathcal{H}(S, T, \sigma, \eta) = const.$ have dimension three in the four dimensional phase space $\mathbb{R}^2 \times \mathbb{T}^2$. This makes difficult the visualization of phase portraits or even the interpretation of the FLI plots. However, we can draw some conclusions from **Figures 3, 5**, obtained by projecting the phase space on the plane (σ, S), for fixed values of T and η.

In fact, we underline three aspects concerning the global dynamics, which are revealed by the model b, namely: the amplitude of resonance depends on the values of both canonical variables T and η. For some values of the canonical variables, the resonances $2\dot{\omega} + \dot{\Omega} = 0$ and $\dot{\omega} = 0$ overlap;

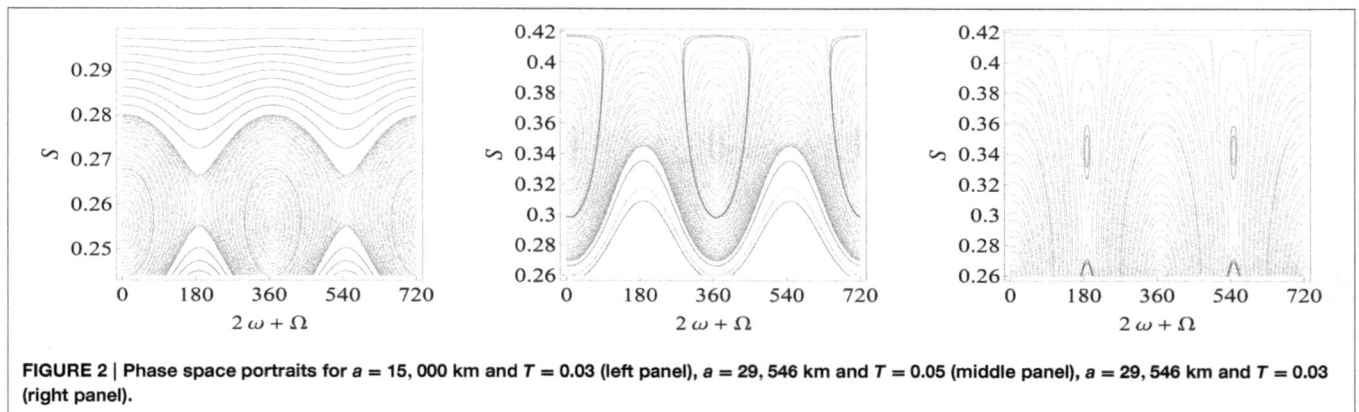

FIGURE 2 | Phase space portraits for $a = 15,000$ km and $T = 0.03$ (left panel), $a = 29,546$ km and $T = 0.05$ (middle panel), $a = 29,546$ km and $T = 0.03$ (right panel).

the bifurcation phenomenon, revealed by the model a, is observable both in this case but also in the case of the full model c.

The plots shown in **Figure 3** are obtained for $\eta = 180°$ and different values of its conjugated action T, while **Figure 5** shows some results obtained for the same value of T and various values

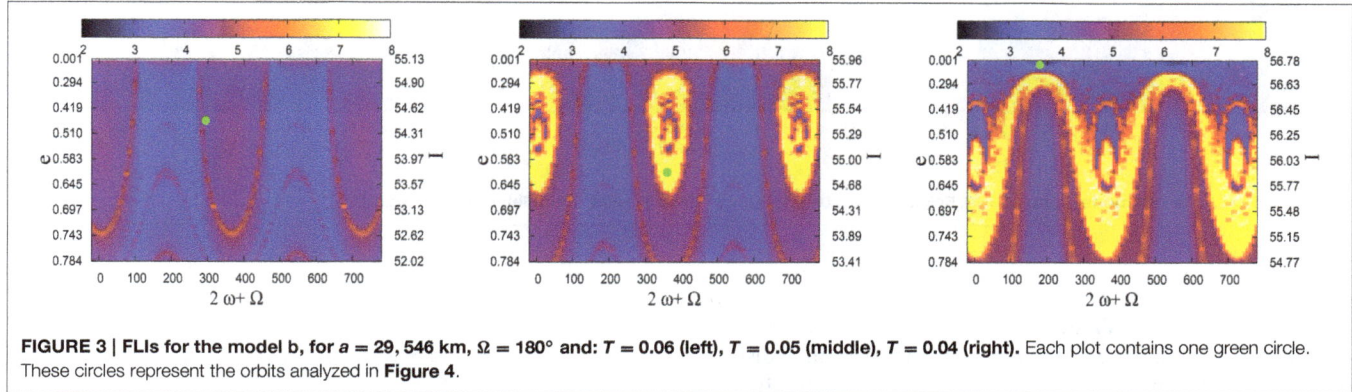

FIGURE 3 | FLIs for the model b, for $a = 29,546$ km, $\Omega = 180°$ and: $T = 0.06$ (left), $T = 0.05$ (middle), $T = 0.04$ (right). Each plot contains one green circle. These circles represent the orbits analyzed in **Figure 4**.

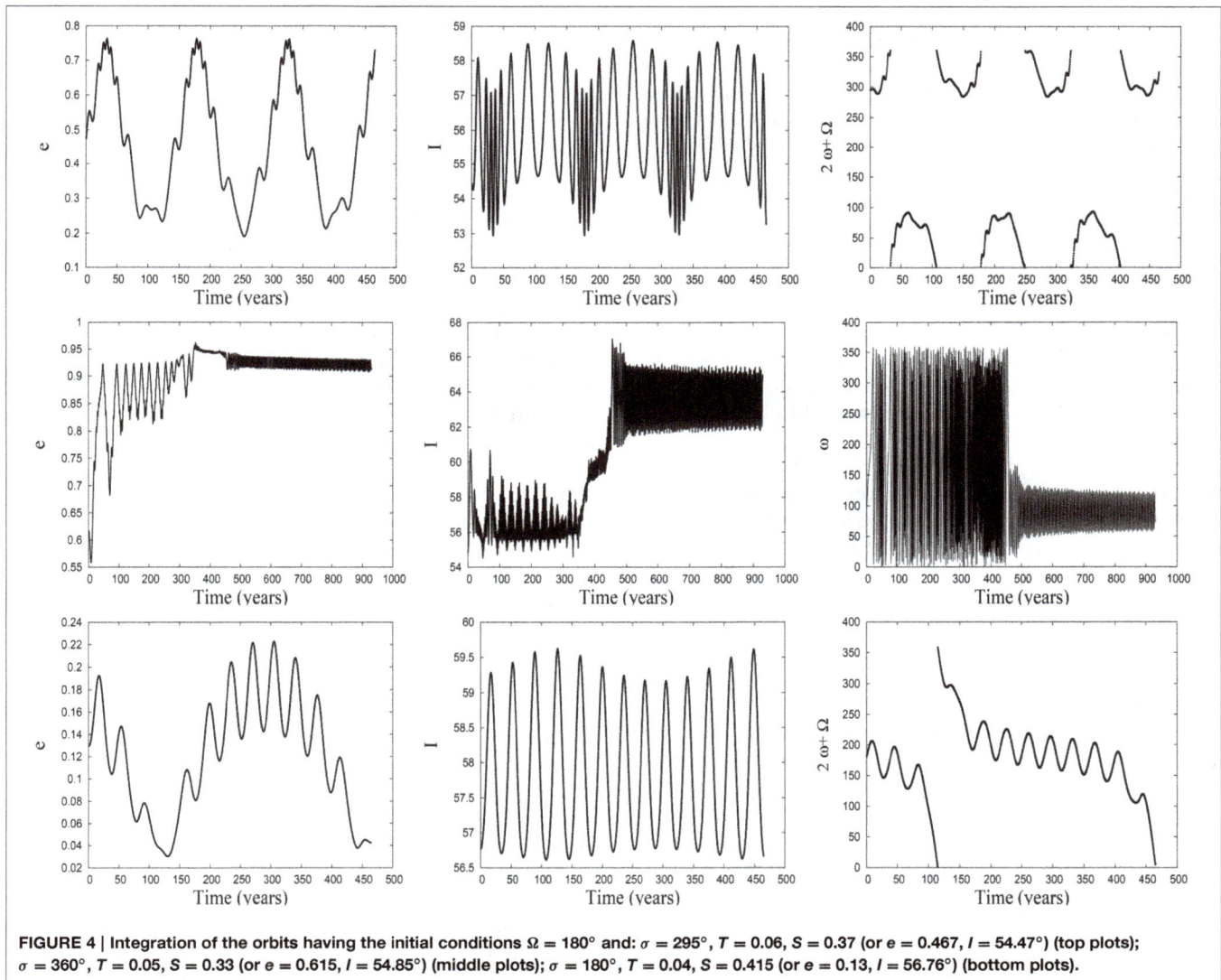

FIGURE 4 | Integration of the orbits having the initial conditions $\Omega = 180°$ and: $\sigma = 295°$, $T = 0.06$, $S = 0.37$ (or $e = 0.467$, $I = 54.47°$) (top plots); $\sigma = 360°$, $T = 0.05$, $S = 0.33$ (or $e = 0.615$, $I = 54.85°$) (middle plots); $\sigma = 180°$, $T = 0.04$, $S = 0.415$ (or $e = 0.13$, $I = 56.76°$) (bottom plots).

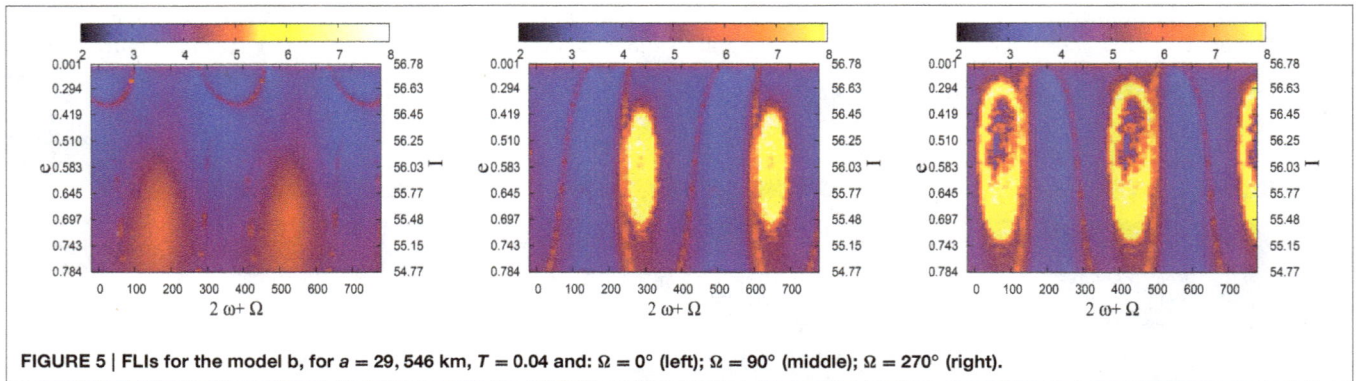

FIGURE 5 | FLIs for the model b, for $a = 29,546$ km, $T = 0.04$ and: $\Omega = 0°$ (left); $\Omega = 90°$ (middle); $\Omega = 270°$ (right).

of η. Moreover, in order to have a clear idea about the patterns shown in these plots, in **Figure 4** we represent the evolution of the eccentricity, inclination and the resonant angle for three distinct orbits. Thus, the orbit depicted by the top plots of **Figure 4** (the green circle in the left panel of **Figure 3**) is located inside the libration region; the eccentricity and resonant angle vary periodically. In the middle panels of **Figure 4** (see also the green circle of the middle panel of **Figure 3**) we consider an orbit located inside the region where the resonances $2\dot{\omega} + \dot{\Omega} = 0$ and $\dot{\omega} = 0$ are so close that there is a non negligible interaction; we integrate the orbit over a longer time (930 years), even if it is a colliding orbit just to show the strong interaction of the above mentioned resonances. Over a period of 350 years the orbit is located inside the libration region of the resonance $2\dot{\omega} + \dot{\Omega} = 0$, then, after an interval of time, it escapes from that resonance and it is rather captured into the critical inclination resonance. Finally, the bottom plots of **Figure 4** correspond also to a resonant orbit (the green circle of the right panel of **Figure 4**): they do not belong to the main resonant libration region, but rather to the resonant small region which appears as a result of the bifurcation phenomenon, already described by the model a.

In conclusion, the global dynamics revealed by the model b is very complex: overlapping of resonances (the yellow regions[1] in **Figures 3, 5**), bifurcations and, as a consequence, the existence of equilibria at large eccentricities as well as at small eccentricities, variation of the amplitude of the resonance as a function of T and η (compare, for instance, the small libration zone of the left plot of **Figure 5** with the large libration regions from the middle and right plots again of **Figure 5**).

4.3. Results for Model C

We finally consider the dynamics associated to the more complete model c, which is described by the non-autonomous Hamiltonian $\mathcal{H}$ introduced in Equation (2) The results are presented in **Figures 6–8**. As we already remarked above, for $a = 15,000$ km, the phase plane σ–S is very similar to the one described by model a, compare **Figure 6** with the left panel of **Figure 2**. However, for large a, the dynamics is much more complex. Roughly speaking, on the global dynamical

---

[1] For the critical inclination resonance, the stable equilibrium points are located at $\omega = 90°$ and $\omega = 270°$.

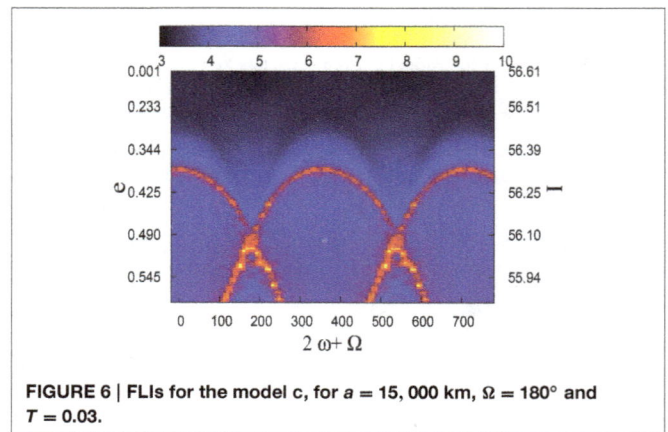

FIGURE 6 | FLIs for the model c, for $a = 15,000$ km, $\Omega = 180°$ and $T = 0.03$.

background described by model b, and which does not change significantly in a vicinity of several km from the nominal distance of $a = 29,546$ km, one should superimpose the exact resonances shown in different colors in the right bottom panel of **Figure 1**. These resonances are due to the variation of the lunar node, as noted by Ely and Howell (1997), Rosengren et al. (2015a), and their location depends on the value of the semimajor axis.

As a consequence, since the resonance $2\dot{\omega} + \dot{\Omega} = 0$ is crossed by multiple exact resonances, having different widths (see Daquin et al., 2016), the orbital elements vary chaotically. One gets large regions filled by chaotic motions, marked by larger yellow-red values of the FLI. In contrast with the model b, here the FLI values vary on a longer scale, from 2 to 14. **Figure 7** shows the results for $T = 0.04$ and for $\Omega = 0°$ (left), $\Omega = 90°$ (middle) and $\Omega = 180°$ (right). Comparing these plots with the corresponding ones obtained for model b, we remark that, besides the large yellow-red regions obtained as effect of the overlapping of resonances (either the superposition of the exact resonances shown in the right bottom panel of **Figure 1** with the exact resonance $2\dot{\omega} + \dot{\Omega} = 0$, or with the critical inclination resonance $\dot{\omega} = 0$), some blue regions are noticeable, which account for the libration regions associated to the equilibrium points. For instance, in the left plot of **Figure 7**, we have a stable equilibrium point at about $\sigma = 360°$ and $e = 0.294$ with a libration island (blue color) small in width (compare also with

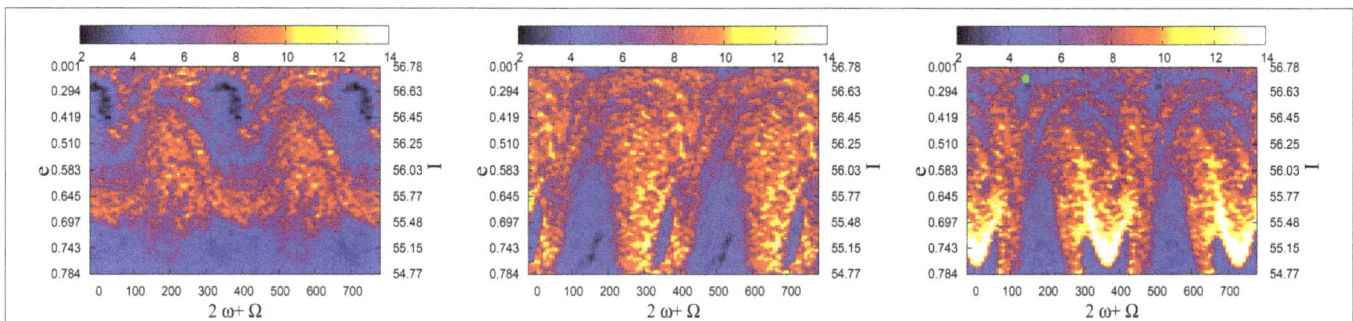

FIGURE 7 | FLIs for the model c, for $a = 29,546$ km $T = 0.04$ and: $\Omega = 0°$ (left); $\Omega = 90°$ (middle); $\Omega = 180°$ (right). The green circle in the right plot represents an orbit analyzed in **Figure 8**.

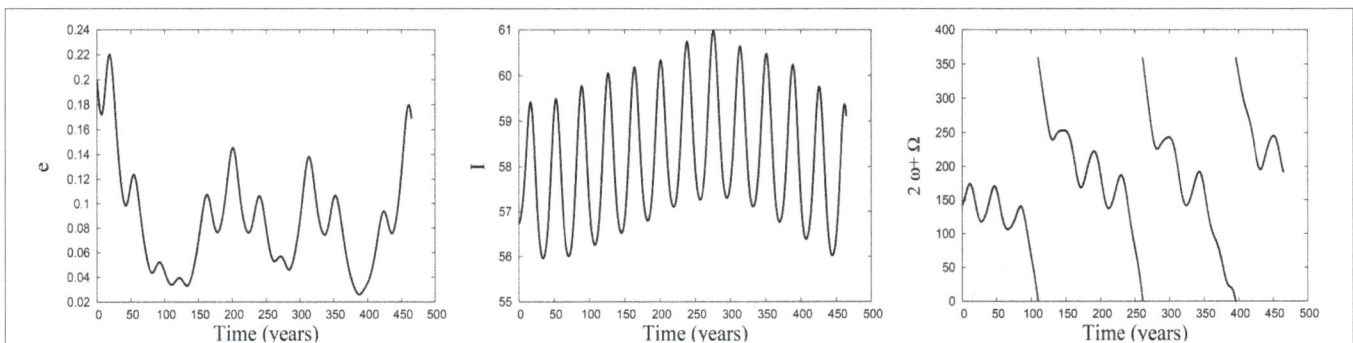

FIGURE 8 | Integration of the orbit having the initial conditions $T = 0.04$, $\Omega = 180°$, $\sigma = 142°$ and $S = 0.41$ (or $e = 0.201$, $I = 56.71°$).

the left plot of **Figure 5**). Numerical tests show that an initial condition inside this region remains there, even if the variations of e and σ are not regular. The red-yellow regions visible for eccentricities larger than 0.5 are due to the interaction of the exact resonances depicted in **Figure 1**, bottom right plot, with the critical inclination resonance.

In both the middle and right panels of **Figure 7**, we notice two important blue (libration) regions: one at small eccentricities (the orbit marked with a green circle in the right panel of **Figure 7** and analyzed in **Figure 8** is within this region) and one at large eccentricities (at about $\sigma = 360°$ and $e = 0.784$ in the right panel of **Figure 7**). These regions show that the bifurcation phenomenon described by the model a is still valid for the more complete model c.

As a final remark, one should clarify what is happening inside the yellow-red region, for example in the middle panel of **Figure 7**. The answer is the following: usually one obtains an irregular growth in eccentricity. The growth is due, in essence, to the resonance $2\dot\omega + \dot\Omega = 0$ (as the models a and b infer) and the irregular (chaotic) behavior is obtained as an effect of the overlapping of the resonance $2\dot\omega + \dot\Omega = 0$ with the resonances shown in **Figure 1**. We made several other experiments and found that colliding orbits can occur as a byproduct of the eccentricity growth due to the interaction with the resonance $2\dot\omega + \dot\Omega = 0$: the increase of the eccentricity leads to have a distance at perigee less than the Earth's radius. On the other hand, initial data in a chaotic region can undergo the effect of the interaction between different resonance, but without leading to collisions.

5. CONCLUSIONS

Lunisolar resonances might contribute to shape the dynamics of small bodies around the Earth (Breiter, 2001; Rosengren et al., 2015a; Daquin et al., 2016). Among such resonances, that corresponding to $2\dot\omega + \dot\Omega = 0$ is responsible for the growth in eccentricity. To explain this phenomenon, we compare three different models with increasing complexity, obtained averaging over fast angles (model a), or just by neglecting the rate of variation of Ω_M (model b), or rather including the variation of Ω_M (model c). A comparison among these models provide us with the ingredients which lead to chaos and which provide an increase of the eccentricity.

By comparing the results of models a–c, we infer that the dynamics around the stable equilibria at large values of the eccentricity is well represented by all models. On the contrary, for small values of the eccentricity the effect of the variation of the lunar longitude of the node plays a relevant role and, even if it occurs on long time scales, cannot be neglected for an accurate description of the dynamics.

Finally, it is worth noticing that the growth in eccentricity provoked by the resonance $2\dot\omega + \dot\Omega = 0$ can be used as an effective strategy to move space debris into non-operative or graveyard orbits.

AUTHOR CONTRIBUTIONS

The paper has been written by both authors in equal parts.

ACKNOWLEDGMENTS

AC was partially supported by the European Grant MC-ITN Stardust, PRIN-MIUR 2010JJ4KPA_009 and GNFM/INdAM; CG was supported by a grant of the Romanian National Authority for Scientific Research and Innovation, CNCS - UEFISCDI, project number PN-II-RU-TE-2014-4-0320 and by GNFM/INdAM.

REFERENCES

Breiter, S. (2001). Lunisolar resonances revisited. *Celest. Mech. Dyn. Astr.* 81, 81–91. doi: 10.1007/978-94-017-1327-6_10

Celletti, A., and Galeş, C. (2014). On the dynamics of space debris: 1:1 and 2:1 resonances. *J. Nonlin. Sci.* 24, 1231–1262. doi: 10.1023/A:1013363221377

Celletti, A., and Galeş, C. (2015). A study of the main resonances outside the geostationary ring. *Adv. Space Res.* 56, 388–405. doi: 10.1016/j.asr.2015.02.012

Celletti, A., Galeş, C., and Pucacco, G. (2016a). Bifurcation of lunisolar secular resonances for space debris orbits. arXiv: 1512.02178.

Celletti, A., Galeş, C., Pucacco, G., and Rosengren, A. (2016b). Analytical development of the lunisolar disturbing function and the critical inclination secular resonance. arXiv: 1511.03567.

Chirikov, B. V. (1979). A universal instability of many-dimensional oscillator systems. *Phys. Rep.* 52, 263–379. doi: 10.1016/0370-1573(79)90023-1

Cook, G. E. (1962). Luni-solar perturbations of the orbit of an Earth satellite. *Geophys. J.* 6, 271–291. doi: 10.1111/j.1365-246X.1962.tb00351.x

Daquin, J., Rosengren, A. J., Alessi, E. M., Deleflie, F., Valsecchi, G. B., Rossi, A., et al. (2016). The dynamical structure of the MEO region: long-term stability, chaos, and transport. *Celest. Mech. Dyn. Astr.* 124, 335–366. doi: 10.1007/s10569-015-9665-9

Ely, T. A., and Howell, K. C. (1997). Dynamics of artificial satellite orbits with tesseral resonances including the effects of luni-solar perturbations. *Dyn. Stabil. Syst.* 12, 243–269. doi: 10.1080/02681119708806247

Froeschlé, C., Lega, E., and Gonczi, R. (1997). Fast Lyapunov indicators. Application to asteroidal motion. *Celest. Mech. Dyn. Astr.* 67, 41–62. doi: 10.1023/A:1008276418601

Hughes, S. (1980). Earth satellite orbits with resonant lunisolar perturbations. I. Resonances dependent only on inclination. *Proc. R. Soc. Lond. A* 372, 243–264. doi: 10.1098/rspa.1980.0111

Kaula, W. M. (1962). Development of the lunar and solar disturbing functions for a close satellite. *Astron. J.* 67, 300–303. doi: 10.1086/108729

Lane, M. T. (1989). On analytic modeling of lunar perturbations of artificial satellites of the Earth. *Celest. Mech. Dyn. Astr.* 46, 287–305. doi: 10.1007/BF00051484

Radtke, J., Dominguez-Gonzalez, R., Flegel, S. K., Sanchez-Ortiz, N., and Merz, K. (2015). Impact of eccentricity build-up and graveyard disposal Strategies on MEO navigation constellations. *Adv. Space Res.* 56, 2626–2644. doi: 10.1016/j.asr.2015.10.015

Rosengren, A. J., Alessi, E. M., Rossi, A., and Valsecchi, G. B. (2015a). Chaos in navigation satellite orbits caused by the perturbed motion of the Moon. *Mon. Not. R. Astron. Soc.* 449, 3522–3526. doi: 10.1093/mnras/stv534

Rosengren, A. J., Daquin, J., Alessi, E. M., Deleflie, F., Rossi, A., Valsecchi, G. B., et al. (2015b). Galileo disposal strategy: stability, chaos and predictability. arXiv:1512:05822v1

Rossi, A. (2008). Resonant dynamics of Medium Earth Orbits: space debris issues. *Celest. Mech. Dyn. Astr.* 100, 267–286. doi: 10.1007/s10569-008-9121-1

Sanchez, D. M., Yokoyama, T., and de Almeida Prado, A. F. B. (2015). Study of some strategies for disposal of the GNSS satellites. *Math. Probl. Eng.* 2015:382340. doi: 10.1155/2015/382340

9

Connecting VLBI and Gaia Celestial Reference Frames

Zinovy Malkin [1,2,3]*

[1] Department of Radio Astronomy Research, The Pulkovo Astronomical Observatory, St. Petersburg, Russia, [2] Institute of Earth Sciences, St. Petersburg State University, St. Petersburg, Russia, [3] Astronomy and Cosmic Geodesy Department, Kazan Federal University, Kazan, Russia

The current state of the link problem between radio and optical celestial reference frames is considered. The main objectives of the investigations in this direction during the next few years are the preparation of a comparison and the mutual orientation and rotation between the optical *Gaia* Celestial Reference Frame (GCRF) and the 3rd generation radio International Celestial Reference Frame (ICRF3), obtained from VLBI observations. Both systems, ideally, should be a realization of the ICRS (International Celestial Reference System) at micro-arcsecond level accuracy. Therefore, the link accuracy between the ICRF and GCRF should be obtained with similar error level, which is not a trivial task due to relatively large systematic and random errors in source positions at different frequency bands. In this paper, a brief overview of recent work on the GCRF–ICRF link is presented. Additional possibilities to improve the GCRF–ICRF link accuracy are discussed. The suggestion is made to use astrometric radio sources with optical magnitude to 20^m rather than to 18^m as currently planned for the GCRF–ICRF link. In addition, the use of radio stars is also a prospective method to obtain independent and accurate orientation between the Gaia frame and the ICRF.

Keywords: astrometry, reference systems, International Celestial Reference Frame (ICRF), Gaia Celestial Refernce Frame (GCRF), link between radio and optical frames

Edited by:
Ludwig Combrinck,
Hartebeesthoek Radio Astronomy
Observatory, South Africa

Reviewed by:
Alberto Vecchiato,
INAF - Astrophysical Observatory of
Torino, Italy
Robert Heinkelmann,
Helmholtz-Zentrum Potsdam
Deutsches GeoForschungsZentrum
GFZ, Germany

*Correspondence:
Zinovy Malkin
malkin@gao.spb.ru

Specialty section:
This article was submitted to
Fundamental Astronomy,
a section of the journal
Frontiers in Astronomy and Space
Sciences

1. INTRODUCTION

The ESA's *Gaia* space astrometry mission (Perryman et al., 2001; Lindegren et al., 2008) commenced successfully in December 2013 and its main scientific program in July 2014. One of the most important results of the *Gaia* mission will be a new highly-accurate optical celestial reference frame; *Gaia* Celestial Reference Frame (GCRF). Although the final GCRF version is expected to be available in the early 2020s, intermediate releases are planned, the first of them (DR1) is expected to be released in 2016. A new release of the VLBI-based celestial reference frame of similar accuracy, the 3rd realization of the International Celestial Reference Frame (ICRF3) is planned for 2018 (Jacobs et al., 2014).

Both radio (ICRF) and optical (GCRF) frames must be realizations of the same concept of the International Celestial Reference System (ICRS), Arias et al. (1995) with an expected accuracy at the level of a few tens of μas. The link between the ICRF and GCRF should be realized at a similar level of accuracy, which is not a trivial task. This problem is similar to that of the link between the *Hipparcos* Celestial Reference Frame (HCRF) and the ICRF (Kovalevsky et al., 1997). Generally speaking, both GCRF and ICRF object positions are time-dependent. Therefore, analogously to HCRF, both the orientation and rotation of the GCRF with respect to ICRF are to be defined. In this paper, the general term "orientation" is used to avoid a non-principal discussion related to spin. Interested readers can find more theoretical and practical details in Lindegren and Kovalevsky (1995) and Kovalevsky et al. (1997).

On the one hand, the link task is more straightforward for the GCRF than for the HCRF, as most of the ICRF objects will be directly observed by *Gaia*. On the other hand, the task is much more complicated due to the requirement that a much higher level of accuracy for the GCRF–ICRF link is needed so as to not compromise the high level of precision of the two frames.

The basic method to tie the *Gaia* catalog to the ICRF, and hence to the ICRS, is using *Gaia* observations of compact extragalactic ICRF objects that have accurate radio astrometric positions. With the help of these common objects, the orientation angles between the ICRF and GCRF will be determined. Finally, the GCRF catalog will be aligned to the ICRS by applying these orientation angles. The accuracy of this link depends on many factors, such as random and systematic errors of both radio and optical catalogs.

The objective of this paper is to briefly overview recent work on the ICRF–GCRF link and to discuss new possibilities to improve the link accuracy. It should be noted that the link between the GCRF and ICRF is not a task currently planned for completion before the end of this decade (Jacobs et al., 2014). Based on the *Hipparcos* experience, it can be envisioned that the work on improving such a link will be continued for a prolonged period after completion of the *Gaia* mission. The improvements will be primarily based on a new VLBI-based celestial reference frame (CRF) realization of which the accuracy can improve over time. New after-mission *Gaia* data reductions are also possible. Therefore, research and development of the methods for the linking of radio and optical reference frames will remain one of the primary tasks of fundamental astrometry throughout the next decade.

The paper is structured as follows. In Section 2, basic equations used to link two CRF realizations are described. Section 3 contains a brief overview of recent investigations regarding aspects concerning the ICRF–GCRF link. The following three sections are devoted to a discussion of new possibilities and possible improvements in both theoretical analysis and the final ICRF–GCRF link accuracy, such as the choice of the ICRF realization used for modeling and simulation (Section 4), using more link sources (Section 5), using radio stars (Section 6), and proper accounting for the galactic aberration in proper motions (Section 7).

2. BASIC EQUATIONS

Each catalog of the positions of celestial objects (CRF realization), be it ICRF or GCRF, represents its own coordinate frame linked to the ICRS at some degree of accuracy. Mutual orientation between these frames is defined by the three orientation angles A_1, A_2, and A_3 around the three ICRS Cartesian axes. Since the catalogs under consideration are close to each other at sub-arcsecond level, the orientation of a vector (X, Y, Z), can be written in the following simple form:

$$\begin{pmatrix} X_1 \\ Y_1 \\ Z_1 \end{pmatrix} = \begin{pmatrix} 1 & A_3 & -A_2 \\ -A_3 & 1 & A_1 \\ A_2 & -A_1 & 1 \end{pmatrix} \begin{pmatrix} X_2 \\ Y_2 \\ Z_2 \end{pmatrix}. \qquad (1)$$

Taking into account that on the celestial sphere

$$\begin{pmatrix} X \\ Y \\ Z \end{pmatrix} = \begin{pmatrix} \cos\alpha \cos\delta \\ \sin\alpha \cos\delta \\ \sin\delta \end{pmatrix}, \qquad (2)$$

and turning to the differences between the object positions in two catalogs $\Delta\alpha = \alpha_1 - \alpha_2$ and $\Delta\delta = \delta_1 - \delta_2$, the final expression can be derived:

$$\begin{aligned} \Delta\alpha &= A_1 \cos\alpha \tan\delta + A_2 \sin\alpha \tan\delta - A_3, \\ \Delta\delta &= -A_1 \sin\alpha + A_2 \cos\alpha. \end{aligned} \qquad (3)$$

The system of Equation (3) for all or selected common objects in two catalogs is solved by the least squares method (LSM) to determine the orientation angles A_1, A_2, and A_3 between two CRF realizations and their errors (uncertainties).

Generally speaking, the differences between the two catalogs include not only the rotational part but also other, mostly coordinate-dependent terms that describe the systematic errors in the compared catalogs, including distortion of the celestial frames realized by the catalogs. Determination of the systematic errors of the celestial object positions in catalogs is a traditional and well developed astrometric task, see, e.g., Sokolova and Malkin (2007) and papers cited therein. Since these systematic errors might influence the orientation angles, they should be estimated during the GCRF–ICRF alignment procedure.

3. OVERVIEW OF RECENT ACTIVITY

The GCRF astrometric catalog will join both galactic stars and extragalactic objects in a single highly-accurate system. As a next stage, this catalog will be aligned to the ICRS using common GCRF and ICRF objects. Two tasks should be solved to provide such an alignment, analogously to what was done for the *Hipparcos* catalog (Lindegren and Kovalevsky, 1995; Kovalevsky et al., 1997):

1. Determination of mutual orientation between the two frames at an initial epoch, most naturally at the *Gaia* mean observation epoch, which is expected to be ∼2017.0 or somewhat later if the mission will be prolonged after its 5-year initially planned duration.
2. Determination of the mutual rotation between two systems.

For the *Hipparcos* catalog, the achieved accuracy was 0.6 mas at the epoch 1991.25 for orientation and 0.25 mas/yr for rotation (Kovalevsky et al., 1997). In the case of the GCRF, the desired accuracy is about an order better, which is quite a challenge.

On the outset, the accuracy of the link between GCRF and ICRF will depend on the number of common objects and their astrometric quality, i.e., accuracy of their coordinates in two catalogs. If Gaia just observes all the objects it can detect with nearly uniform accuracy over the sky depending mainly on the object's optical brightness, the situation with the ICRF sources is more complicated. The ICRF2 catalog is very inhomogeneous in respect to the radio source position errors due to the large difference in number of observations from 3 to

337,322, see Section 4 for more detail. The situation has improved substantially with realization of the project on re-observation of the VCS (VLBA Calibrator Survey) sources (Gordon et al., 2016), which allowed a significant improvement in the accuracy of about 2000 ICRF source positions (see Section 4), but it is still far from ideal.

Current activities in preparation toward aligning the GCRF with the ICRF are developing in several directions:

1. Selection of prospective link radio sources and their intensive observations.
2. Photometric optical observations of radio sources.
3. Optical astrometric observations of radio sources with ground-based telescopes.
4. Creation of data banks of optical images of ICRF objects.

Several years ago, Bourda et al. (2008) started work on selecting optically bright radio sources of good radio astrometric quality, i.e., having sufficient flux density and compact structure. Finally, 195 prospective link sources were selected. The various stages of this work were described by Bourda et al. (2008) (selection of optically bright radio sources), Bourda et al. (2010) (source detection on the long VLBI baselines), Bourda et al. (2011) (source imaging to estimate their radioastrometric quality), Le Bail et al. (2016) (improving radio positions of selected sources). The program is being continued.

Zacharias and Zacharias (2014) obtained accurate optical positions of 413 AGN, and investigated their errors and their impact on the accuracy of the link between radio and optical frames. Comparison of the optical positions with radio positions showed that the differences statistically exceed the known errors in the observations. The physical offset between the optical and radio emission centers was identified as a likely cause. This effect, called by the authors detrimental, astrophysical, random noise (DARN), was found to be at ~10 mas level. The authors came to the conclusion that the GCRF–ICRF orientation angles can hardly be determined with an error better than 0.5 mas, without a substantial increase in the number of the link objects. This estimate was based on ground-based results and can be somewhat improved with Gaia observations, but the DARN can prevail in this case too.

The new catalog URAT1 contains positions of over 228 million objects in the magnitude range of about $R = 3$–18.5 with a typical error of 10–30 mas and proper motions of over 188 million objects with a typical error of 5–7 mas/yr (Zacharias et al., 2015). There are two shortcomings of this catalog limiting its usefulness for studies on the GCRF–ICRF link. Firstly, the catalog covers only just over a half of the sky, namely the region with $\delta \geq 15°$. Secondly, most of the astrometric radio sources are absent in the URAT1 catalog as they are fainter than 18.5^m (see **Figure 4**). However, this work is of great importance as it provides very valuable information on the shift between radio and optical positions for hundreds of radio sources.

A dedicated program of photometric observations of the ICRF sources have been conducted under coordination of the Paris Observatory Taris et al. (2013, 2015, 2016). These observations provide much new information about the optical brightness of the astrometric radio sources as well as about their optical variability. The results reported by this group showed that the peak-to-peak change in optical magnitude is typically at a level of several tenths of magnitude, and exceed 1^m for many objects reaching sometimes 3^m, e.g., for the source B0716+714.

Andrei et al. (2012, 2014) have been working on the compilation of the *Gaia* QSO catalog. Data from different surveys and catalogs are assembled to complete QSO characteristics including positions, photometry, morphology, and imaging. This work, in particular, provides useful data for investigation of the radio minus optics position shift. The most complete bank of optical images of the ICRF objects was created by Andrei et al. (2015).

Souchay et al. (2015) presented the 3rd release of the Large Quasar Astrometric Catalog (LQAC-3). This catalog contains accurate positions, magnitudes in 9 bands *UBVGRIZJK*, radio flux in 5 bands from 750 MHz to 30 GHz, morphology indexes in *BRI* bands, and absolute magnitude in *B* and *R* band for 321,957 objects, primarily quasars, including ~5% of other AGN types.

All these works are directed, in particular, toward better link source selection and the improvement in the accuracy of their radio astrometric positions, which provides better accuracy of the GCRF–ICRF link. In the next sections, new possibilities to improve the quality of the link will be discussed.

4. RADIO FRAME

All the recent studies discussed in Section 3 are based on using the ICRF2 as a radio CRF. However, the ICRF2 catalog created in 2009 (Fey et al., 2015) is already outdated. It is expected that the first link between the GCRF and ICRF will be performed during 2018–2019 using the next VLBI-based ICRF realization, ICRF3 (Jacobs et al., 2014). Currently, the radio source position catalog gsf2015b[1] derived at the NASA Goddard Space Flight Center (GSFC) VLBI Group appears to be the closest to the future ICRF3. It is computed using about the same data analysis strategy as was used for computation of ICRF2 (also at GSFC) but involves many more observations (VLBI delays). **Table 1** contains a statistical comparison between the gsf2015b and ICRF2 catalogs.

Figure 1 illustrates the position uncertainty distribution in two catalogs. One can see that the overall level of the position uncertainty in the gsf2015b catalog is much smaller than that in the ICRF2 catalog.

It can be seen from **Table 1** and **Figure 1** that the GSFC catalog is more advanced when compared with the ICRF2. It should be noted that the gsf2015b catalog provides the original position errors obtained from the LSM solution, while ICRF2 position errors were inflated following the formula $\sigma_{inflated}^2 = (1.5\,\sigma_{computed})^2 + (0.04 \text{ mas})^2$ (Fey et al., 2015), see **Figure 2**. The error floor of 0.04 mas is mostly effective for small original errors less than 0.1–0.2 mas; for larger original errors inflated errors may be taken as original errors multiplied by factor 1.5. Nevertheless, even taking this factor into account, gsf2015b source positions are

---

[1]http://gemini.gsfc.nasa.gov/solutions/astro/.

TABLE 1 | Basic statistics of the ICRF2 and gsf2015b catalogs.

Catalog	Sources	Sessions	Delays	Period	Median error, μas	
					$\alpha \cos \delta$	δ
ICRF2	3414	4540	6,495,553	1979.08.03–2009.03.16	397	739
gsf2015b	4089	5836	10,453,527	1979.08.03–2015.11.09	123	210

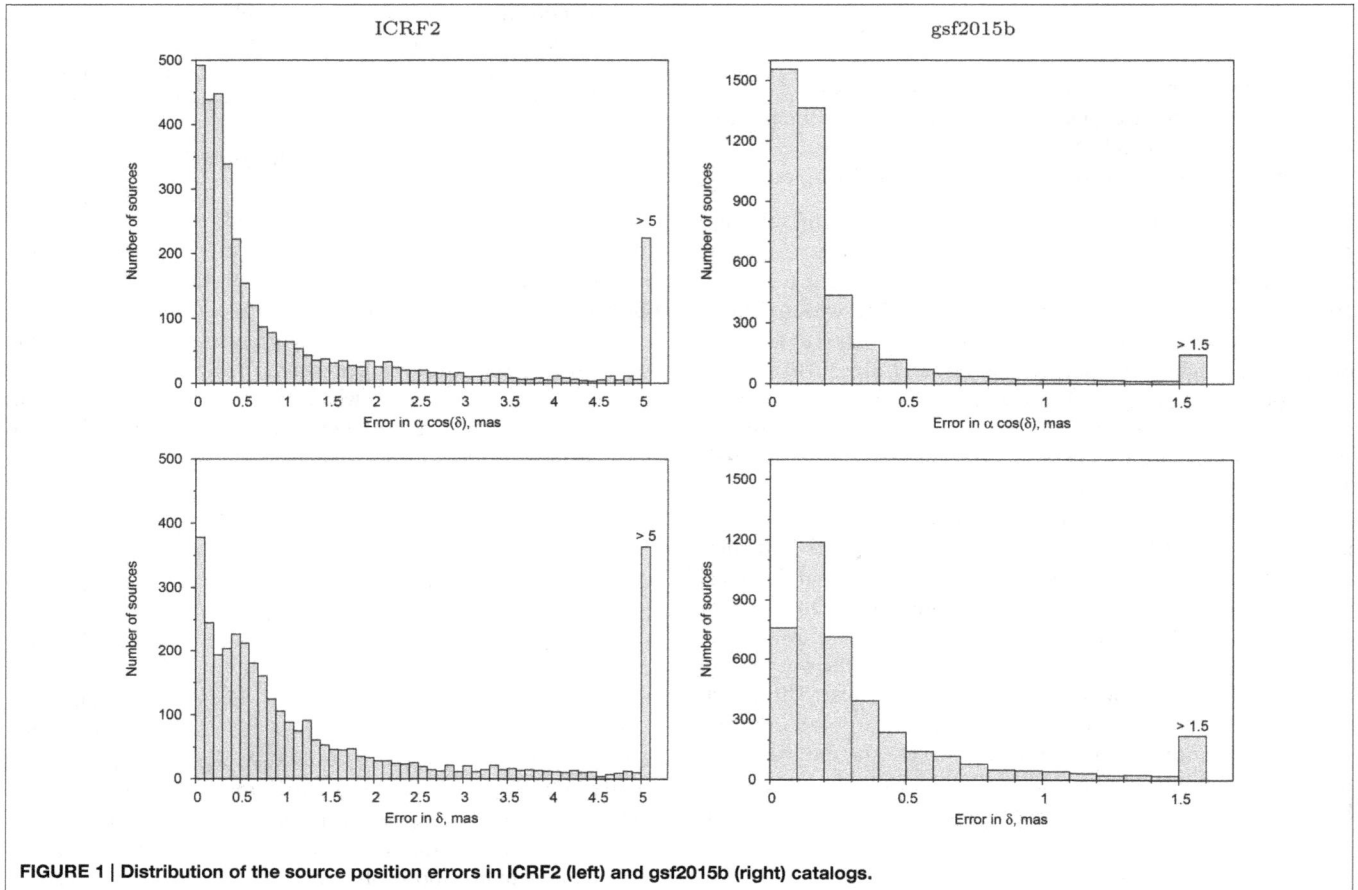

FIGURE 1 | Distribution of the source position errors in ICRF2 (left) and gsf2015b (right) catalogs.

substantially more precise than those of the ICRF2. The primary reason for the significant improvement in source position errors is re-observation of about 2000 VCS (VLBA Calibrator Survey) sources (Gordon et al., 2016), which is ~2/3 of the total number of the ICRF2 sources.

Based on these considerations, the gsf2015b catalog was used in this study as a prototype of ICRF3.

One of the permanent tasks of the VLBI community is improving accuracy of the radio source positions. Due to limited VLBI resources, proper planning of the observations is important. **Figure 3** illustrates the dependence of the source position uncertainty on the number of observations (delays and sessions) for the ICRF2 and gsf2015b catalogs. The error floor of 0.04 mas introduced in the final catalog (Fey et al., 2015) can clearly be seen in the plot for ICRF2. This analysis shows that the source position uncertainties depend primarily on the number of delays, and to lesser extent on the number of sessions. It also

suggests that to reliably achieve sub-mas position error, about 100 observations (VLBI delays) should be obtained.

5. INCREASING THE NUMBER OF LINK RADIO SOURCES

A criterion for the initial source selection in Bourda et al. (2008), which was the base for the consequent works Bourda et al. (2010, 2011); Le Bail et al. (2016), was using ICRF2 sources with optical magnitude $\leq 18^m$. The latter limit was defined to use the objects with a small *Gaia* position error $<$~70 μas as was estimated during *Gaia* pre-launch analysis (Lindegren et al., 2008). To select the optically bright radio sources, the catalog of Véron-Cetty and Véron (2006) was used as the source of the photometric data.

It seems that the approach applied by Bourda et al. (2008) for the link source selection can be substantially improved in view of new data that became available during recent years. First, as was discussed in Section 4, the ICRF2 does not provide the best choice of astrometric radio sources with reliable highly-accurate

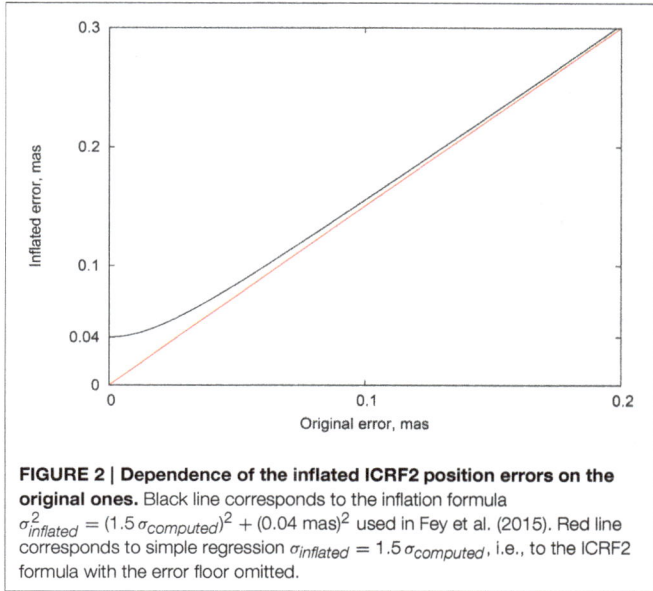

FIGURE 2 | Dependence of the inflated ICRF2 position errors on the original ones. Black line corresponds to the inflation formula $\sigma^2_{inflated} = (1.5\,\sigma_{computed})^2 + (0.04\ mas)^2$ used in Fey et al. (2015). Red line corresponds to simple regression $\sigma_{inflated} = 1.5\,\sigma_{computed}$, i.e., to the ICRF2 formula with the error floor omitted.

coordinates. Second, the catalog Véron-Cetty and Véron (2006) does not contain sufficiently complete photometric data as compared with the latest catalogs. So, an improved strategy for preliminary link source selection can be suggested. A new approach can include the use of the latest VLBI-based CRF solutions containing more radio sources with accurate positions, especially in the Southern Hemisphere, and the catalog OCARS (Optical Characteristics of Astrometric Radio Sources, Malkin, 2016b) that contains the most complete photometric data for astrometric radio sources and thus provides the maximum choice for selection of optically bright radio sources.

The second option that is worth investigating is using radio sources with a magnitude $18 < G \leq 20^m$, where $G = 20^m$ is the threshold for *Gaia* observations (Perryman et al., 2001; Lindegren et al., 2008). Although fainter sources will have much larger *Gaia* positional error, the number of these sources may compensate for such a loss of precision. An analysis based on the rotation covariance matrix analysis was performed by Mignard (2014). It was found that moving from 18^m to 20^m threshold leads to about a twofold increase of ICRF2 link sources from ~500 to ~1000, and reduction of the errors of the orientation angles from ~7 μas to ~6 μas.

Another approach based on Monte Carlo simulation was used in the current study. A list of the link sources for these computations was composed of all AGN with the optical

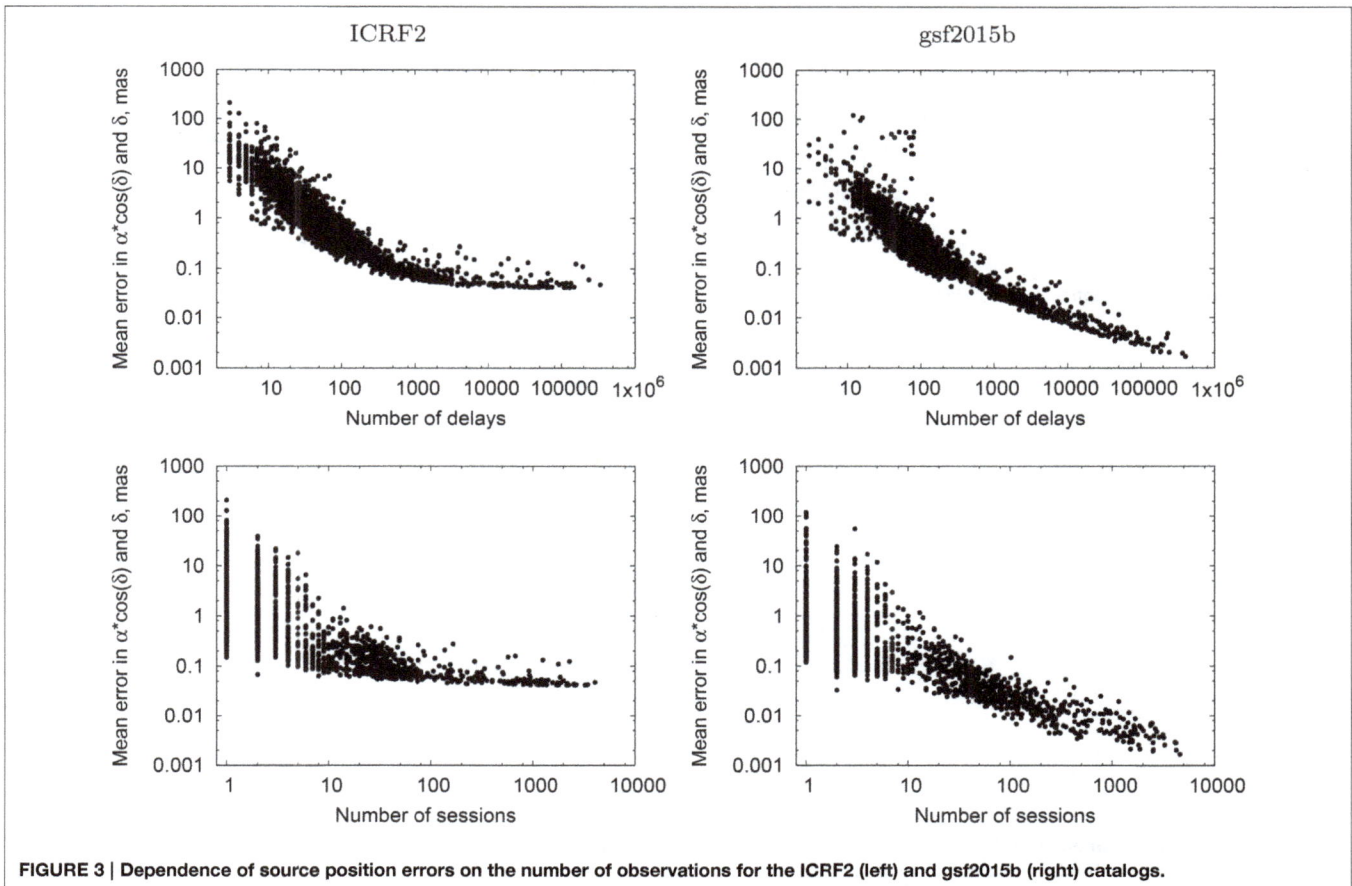

FIGURE 3 | Dependence of source position errors on the number of observations for the ICRF2 (left) and gsf2015b (right) catalogs.

magnitude $\leq 20^m$ and galaxies with the optical magnitude $\geq 16^m$ selected from the gsf2015b catalog. It is supposed that faint galaxies can be expected to be unresolved objects for *Gaia* (de Souza et al., 2014). Optical magnitudes for these sources were taken from the OCARS catalog (Malkin, 2016b). **Figure 4** illustrates the distribution of the optical magnitudes in the source set used in analysis.

The median uncertainty of the gsf2015b source positions for different sets of sources selected with different upper limits for the optical magnitude is depicted in **Figure 5**. One can see that moving to the optically fainter sources improves the overall position precision.

For the VLBI position errors, the uncertainty estimates given in the gsf2015b catalog were used. The *Gaia* position errors used during Monte Carlo simulation were estimated in the following way. Expected *Gaia* parallax standard error σ_π depending on the optical brightness of the object G is given by de Bruijne et al. (2014) (points in **Figure 6**). The authors also provide a rather complicated formula representing σ_π depending on the optical

brightness. However, this formula appears to be unnecessarily complicated for the simulation studies, it depends not only on the G magnitude but also on the $V - I$ color, which is not known for most of the astrometric radio sources. So, a simpler approximation of the σ_π was derived (in μas):

$$\sigma_\pi = \begin{cases} 6.7, & G < 12.1, \\ 6.7 + 4.86\,(G - 12.1), & 12.1 \leq G < 13, \\ 10^{(1.044 + 0.1528\,(G-13) + 0.01373\,(G-13)^2)}, & G \geq 13. \end{cases} \tag{4}$$

The *Gaia* position error is computed as $\sigma_0 = 0.743\,\sigma_\pi$ (de Bruijne et al., 2014). This approximation function is depicted in **Figure 6**.

The result obtained with Monte Carlo simulation (10,000 iterations) is shown in **Figure 7**. This result is generally similar to that obtained by Mignard (2014) using a different method,

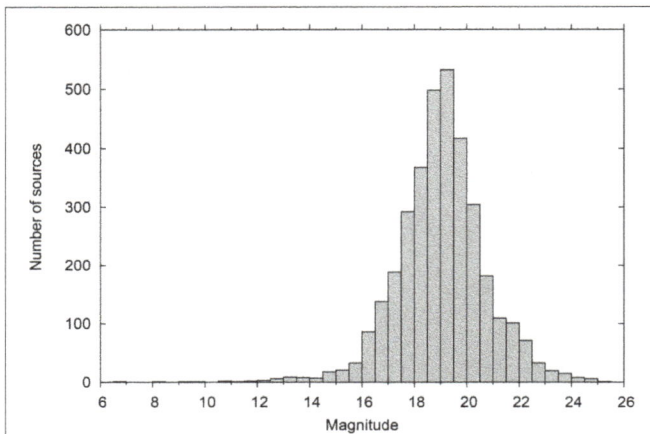

FIGURE 4 | Distribution of optical magnitudes in the gsf2015b catalog.

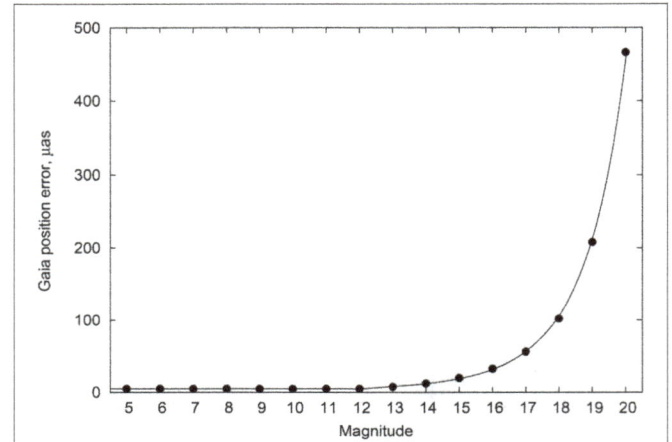

FIGURE 6 | Expected *Gaia* position error according to de Bruijne et al. (2014) (points). Solid line corresponds to the approximation function Equation (4) multiplied by 0.743.

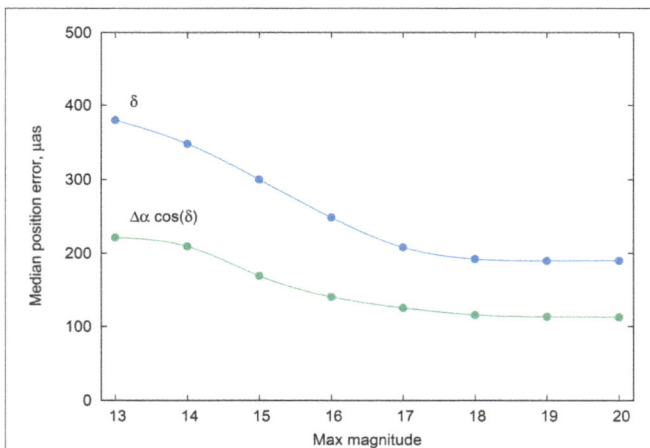

FIGURE 5 | Median errors of the gsf2015b positions depending on source selection based on the faint-end optical magnitude threshold.

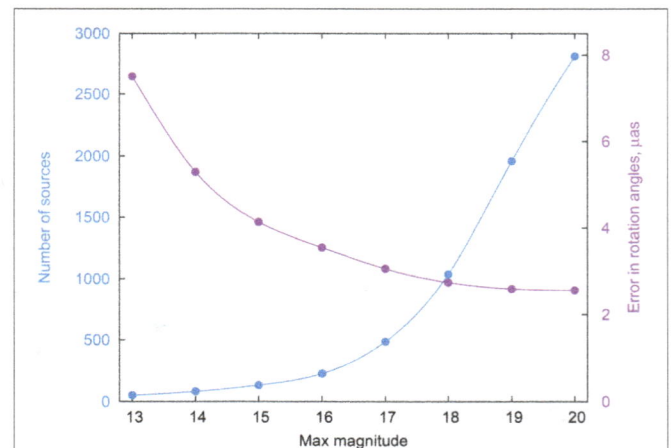

FIGURE 7 | Simulated error in orientation angles depending on source selection based on the faint-end optical magnitude threshold.

however the errors in the orientation angles are about two times smaller in the current work, in particular, due to larger number of sources used.

Surely, both results are too optimistic because they were obtained without taking into account the radio–optics position shift, such as the DARN mentioned in Section 3. To estimate the impact of this effect, an additive error σ_{added} was added in quadrature to the simulated ICRF–GCRF coordinate differences. The result of this test obtained by Monte Carlo simulation in the same way as the previous one is presented in **Figure 8**. The obtained error in the orientation angles agrees with the value of 0.5 mas predicted by Zacharias and Zacharias (2014) for $\sigma_{added} = 10$ mas.

Finally, all the results presented in this section showed that it is advisable to use all compact radio sources up to 20^m for the ICRF–GCRF link. The distribution of 2815 gsf2015b sources with the optical magnitude $\leq 20^m$ over the sky is shown in **Figure 9**. Indeed, this is the first stage of selection of the best ICRF–GCRF link sources that should follow by estimation of their radioastrometric quality as discussed in Bourda et al. (2008). However, it allows one to have several times more prospective link sources at this stage than was considered earlier: 2815 sources vs. 535 sources selected by Bourda et al. (2008).

6. USING RADIO STARS

The problem of aligning the GCRF to ICRF is similar to the problem of aligning the HCRF to the ICRF. One of the methods applied for this purpose during the *Hipparcos* mission was using radio stars. It proved to be the most accurate method among others considered for the orientation of the *Hipparcos* catalog (Kovalevsky et al., 1997). The error in the orientation angles between the two frames obtained from VLBI observations of 12 radio stars was estimated to be about 0.5 mas.

Later, Boboltz et al. (2007) reported on the results of observations of 46 radio stars with flux density of 1–10 mJy obtained with the VLA (Very Large Array) plus the Pie Town

antenna of the VLBA (Very Long Baseline Array) narrow regional network. The position of radio stars were determined by means of phase referencing to close ICRF sources with an error of about 10 mas on average. The HCRF–ICRF orientation angles were estimated with an error of $\sim$2.7 mas.

The current accuracy of both VLBI and optical (*Gaia*) observations is much better. Consequently, properly scheduled radio star observations using large regional or global VLBI networks can provide much smaller position errors and hence better accuracy of the link between optical and radio frames. A Monte Carlo simulation was performed by Malkin (2016a) to estimate an error in the orientation angle between GCRF and ICRF obtained from radio stars observation. Results of this study showed that VLBI observations of radio stars can provide an independent and accurate method to link the GCRF to the ICRF. A properly organized VLBI program for radio star observations can lead to the realization of the GCRF–ICRF link with an error of about 0.1 mas with a reasonable load on the VLBI network. Thus, this method can provide a valuable contribution to the improvement of the GCRF–ICRF link.

Details of this work are given in Malkin (2016a).

7. GALACTIC ABERRATION

Comparison of the ICRF and GCRF catalogs should be made at a certain epoch t_0. Two natural choices are the current standard epoch of the astronomical equations and quantities $t_0 = $ J2000.0, and the mean epoch of the *Gaia* catalog $t_0 = \sim$2017.0, supposing a 5-year period of *Gaia* operations starting from July 2014. It can be reasonably supposed that the *Gaia* positions will be brought to t_0 using its own proper motions.

The situation with the ICRF positions is not so simple. The ICRF concept is based on the absence of detectable motions of ICRF sources. It is definitely not the case at the micro-arcsecond level of accuracy. Several works showed that astrometric VLBI observations are capable of revealing statistically meaningful apparent motions of the ICRF objects, see, e.g., MacMillan (2005)

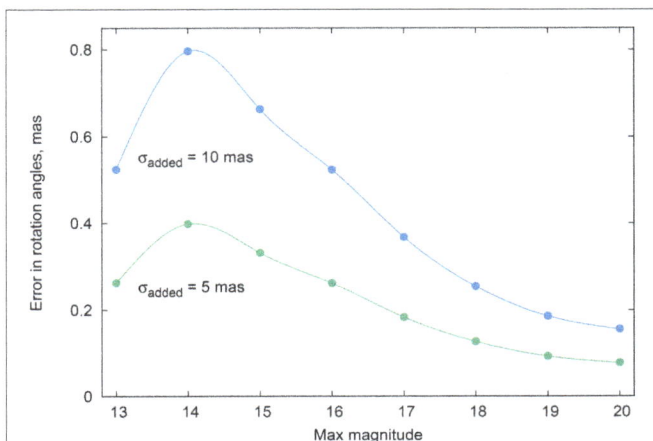

FIGURE 8 | Simulated error in orientation angles depending on the additive error due to radio–optics position shift.

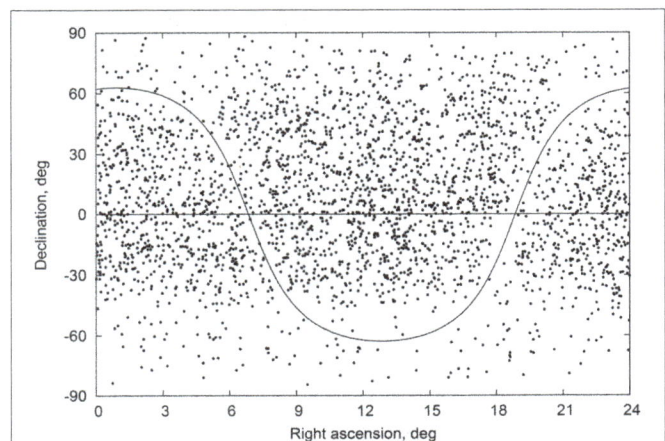

FIGURE 9 | Distribution of the gsf2015b sources with optical magnitude $\leq 20^m$ over the sky.

and Titov and Lambert (2013). The question of the nature of these motions is very complicated, and the consistency between various estimates is not satisfactory yet. The overall problem of source motions is beyond the scope of this paper. Here, only one systematic component of source motion pattern, namely galactic aberration in proper motions (GA) is discussed. The theory of this effect is considered in Kovalevsky (2003), Kopeikin and Makarov (2006), Liu et al. (2012), and Liu et al. (2013). The proper motion caused by the GA is given by Malkin (2011):

$$\mu_l \cos b = -A \sin l,$$
$$\mu_b = -A \cos l \sin b, \tag{5}$$

where l and b are the galactic longitude and latitude of the object, respectively, and A is the GA constant depending on the galactic rotation parameters. The most probable value of the GA constant is $A = 5.0 \pm 0.3$ μas/yr (Malkin, 2014).

To estimate the impact of the GA on the orientation angles between ICRF and GCRF, a special test was performed. The idea of this test was proposed by Malkin (2015b) and is extended here. Two catalogs of radio source positions were used for the simulation. The first catalog comprises 688 ICRF2 sources of AGN type and with visual magnitude 18^m or brighter following the principles of the link source selection proposed by Bourda et al. (2008). The second catalog consists of 2815 gsf2015b sources discussed in Section 4.

All the existing catalogs of radio source positions are derived without accounting for the GA during data processing. Therefore, to bring the selected link source positions to the epoch t_0 the following equations should be used:

$$\alpha(t_0) = \alpha(t) - \mu_\alpha(t - t_0),$$
$$\delta(t_0) = \delta(t) - \mu_\delta(t - t_0), \tag{6}$$

where t is the mean epoch of observations of the source in the ICRF2 or GSFC catalogs, $\alpha(t)$ and $\delta(t)$ are source coordinates (right ascension and declination, respectively) in the catalog, μ_α and μ_δ are GA-induced motions in right ascension and declination, respectively, computed by Equation (5) and transformed to the equatorial system as described by Malkin (2014) and Malkin (2015a). All the computations were made for $t_0 = $ J2000.0 and $t_0 = 2017.0$.

Then, we have two catalogs for each of four variants (two initial catalogs and two t_0 epochs). The first catalog in each pair is merely the initial catalog. Such a catalog would be used for the ICRF–GCRF link if GA is not taken into account, which is currently the case. The second catalog contains positions of the same sources transferred to the epoch t_0 for the GA-induced proper motions. This catalog would correspond to the radio source positions computed with taking into account the GA effect during VLBI data processing. Consequently the orientation angles between the two catalogs were computed. Results are presented in **Table 2**.

The first line of **Table 2** corresponds to computations made in Malkin (2015b). The three other lines are the test extension that allowed us to correct the earlier preliminary conclusion. After the preliminary test (first line) a conclusion was drawn

TABLE 2 | Impact of galactic aberration on the orientation angles between the ICRF and GCRF.

Catalog	t_0	A_1	A_2	A_3
ICRF2	J2000.0	1.3 ± 0.5	-0.1 ± 0.5	0.3 ± 0.4
	2017.0	26.8 ± 2.4	-4.9 ± 2.3	4.0 ± 2.0
gsf2015b	J2000.0	-19.4 ± 0.5	1.9 ± 0.5	0.6 ± 0.4
	2017.0	36.1 ± 0.6	-4.2 ± 0.6	2.7 ± 0.6

Unit: μas.

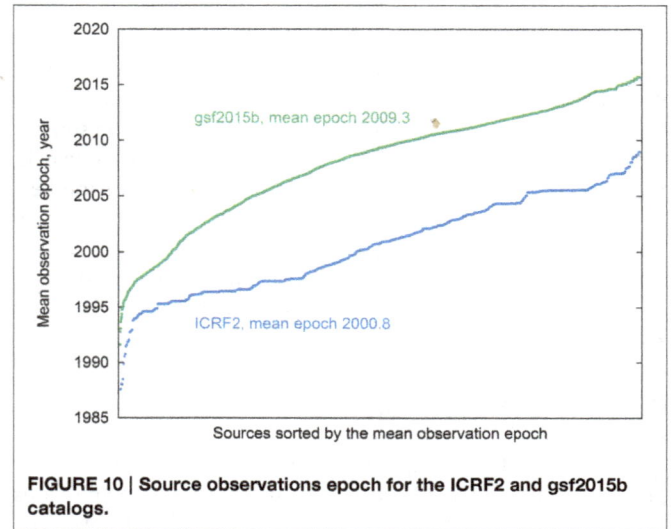

FIGURE 10 | Source observations epoch for the ICRF2 and gsf2015b catalogs.

that the impact of the GA on the ICRF–GCRF link is practically negligible. More detailed test performed in this study has shown that the GA effect can be substantial. This can be seen from both the significant value of the orientation angles and their uncertainties. The errors in the orientation angles depend on both the number of common sources in the catalogs and the time interval between the mean source observation epoch in the catalog (let us call it catalog epoch) and t_0. Mean epochs for two catalogs are shown in **Figure 10**. The ICRF2 catalog epoch is closer to J2000.0 than the gsf2015b epoch, but the latter contains four times more sources, which resulted in about the same errors in the orientation angles for $t0 = $ J2000.0. In contrast, the gsf2015b catalog epoch is closer to 2017.0, which together with larger number of sources gives much smaller errors in the orientation angles as compared with the ICRF2.

8. DISCUSSION AND CONCLUSION

A great milestone in the construction of the celestial reference frame is the *Gaia* mission, which will result in the GCRF with expected accuracy of a few tens of micro-arcsecond for final release in the early 2020s. A new ICRF release based on VLBI observations of extragalactic radio sources of similar accuracy can be also expected by that time. Constructing of a single multi-frequency celestial reference frame based on the ICRF and GCRF is the primary task of fundamental astronomy for the next decade (Gaume, 2015).

The first step toward this goal is an accurate alignment of the *Gaia* catalog to the ICRF. The desired goal is to achieve mutual orientation between the two frames with an accuracy of 0.1 mas or better, which is a challenging task.

Currently, the International VLBI Service for Geodesy and Astrometry (IVS) is conducting a special program on observation of prospective link sources not having a sufficient number of observations in the framework of regular observing programs (Le Bail et al., 2016). This forms part of the plan to prepare to achieve the GCRF–ICRF link using 195 selected sources selected on basis of several criteria, such as, inclusion in the ICRF2, optical magnitude $\leq 18^m$, symmetric compact structure, sufficient radio flux density. The two first criteria seem to be outdated. Firstly, the ICRF2 is currently not the most appropriate radio source position catalog that can serve as an ICRF3 prototype, which is planned to be used for initial alignment of the *Gaia* catalog to the ICRS as discussed in Section 4. Secondly, as shown in Section 5, using optically fainter sources up to 20^m provides more precise determination of the orientation angles. It is even more important that the use of more sources with reliable VLBI positions is necessary to mitigate the impact of most of the negative factors mentioned below, see **Figure 8** and related text. As discussed, there are possibilities to multiply the number of link sources taking into account the substantial increase of the number of radio sources with reliably determined positions and the number of sources having photometry measurements.

The following problems were identified in the literature that dilute the accuracy of the link between radio and optical frames:

- Structure effects (discussed below).
- Systematic errors of radio position catalogs.
- Multiple radio sources related to a single object in optics, e.g., binary black holes, and vice versa.
- Gravitational lenses.
- Errors in ICRF–GCRF cross-identification.

Most probably, the main factor that will deteriorate the accuracy of the link between the radio and optical frames, is the source structure. It can manifest at both radio and optical wavelengths as complex, asymmetric distribution of the brightness over the source map, spatial bias between the optical and radio brightness centroids, core-shift effects, spatial bias between the core/AGN brightness centroid and with respect to the optical centroid of the host galaxy. Moreover, the structure effects are often variable. Although many studies are devoted to this problem, see, Fey and Charlot (1997); da Silva Neto et al. (2002); Moór et al. (2011); Bouffet et al. (2013); Zacharias and Zacharias (2014); Berghea et al. (2016) and papers cited therein, there are insufficient data to quantify the impact of source structure on the accuracy of the orientation angles between optical and radio frames. Evidently, the most complete study is provided by Zacharias and Zacharias (2014) used in the current work. It should be noted that though the structure effect can reach several mas for an individual source, it can hardly be systematic and thus will be averaged over the sky during the computation of the orientation angles between the

GCRF and ICRF. However, supplementary observations and theoretical considerations are needed to quantify this effect more accurately.

Two main methods to obtain the link between the GCRF and ICRF were considered in this paper. The first method is direct *Gaia* observations of the sufficiently optically bright ICRF sources. This method allows for a straightforward solution of the task. However, there are serious constraints on the accuracy of this method caused by the previously mentioned factors. These factors can limit the real accuracy of the GCRF–ICRF link to 0.1 mas or worse.

Observations of radio stars can serve as an alternative equally accurate method. It was successfully used to link the HCRF to the ICRF. However, this method is also affected by some accuracy-limiting factors (Malkin, 2016a). Many radio stars comprise double or multiple systems, and thus their orbital motions must be accounted for. The accurate *Gaia*-derived orbits can be used for this purpose. Moreover, radio stars may have complex and variable structures, which might cause a time-dependent bias between the radio and optical positions. Lestrade et al. (1995) estimated the impact of radio star structure and possible variations in the radio star emitting center to be within the error budget of ∼0.5 mas. Lestrade et al. (1999) found that the structure-induced systematic errors in the VLBI positions of 12 stars ranged from 0.07 mas to 0.54 mas, with a median value of 0.18 mas. Provided that several tens of radio stars have been observed, this factor should not significantly impact on the errors in the orientation angles.

Combination of two methods of linking the GCRF to the ICRF should facilitate improvement of the systematic accuracy of the link between radio and optical frames.

Finally, it should be noted that the preparation for the aligning of the *Gaia* catalog to ICRF3 planned for 2018–2019 is only an intermediate stage in construction of the multi-frequency celestial reference frame. Certainly, the main work on the link between radio and optical frames is to be done in the early 2020s, after the final *Gaia* catalog is prepared. It is desirable to plan preparation of a new ICRF realization, will it be called ICRF4 or ICRF3 extension, at the same time, i.e., immediately before the final GCRF link to ICRF. Comparison of gsf2015b with ICRF2 has clearly shown that the latest radio catalog should be used for this work due to the accuracy of the VLBI-derived CRF solutions which rapidly improves with time. In the framework of this activity, it appears to be very important to fast-track the plans on the ICRF improvements in the Southern Hemisphere (Jacobs et al., 2014; Plank et al., 2015), and to start a program of observations of radio stars (Malkin, 2016a).

As the *Hipparcos* experience has shown, it can be expected that the work on improving the link between radio and optical frames will be continued during a long period after completing the *Gaia* mission. Continuous quality improvement of a VLBI-based ICRF promises corresponding improvement of this link with time.

AUTHOR CONTRIBUTIONS

The author confirms being the sole contributor of this work and approved it for publication.

ACKNOWLEDGMENTS

The author would like to thank the two reviewers and the handling editor for their valuable help in improving the manuscript. This work was partially supported by the Russian Government Program of Competitive Growth of Kazan Federal University. This research has made use of NASA's Astrophysics Data System.

REFERENCES

Andrei, A. H., Anton, S., Barache, C., Bouquillon, S., Bourda, G., Le Campion, J.-F., et al. (2012). Gaia initial QSO catalogue: the variability and compactness indexes. *Mem. Soc. Astron.* 83:930.

Andrei, A. H., Antón, S., Taris, F., Bourda, G., Souchay, J., Bouquillon, J., et al. (2014). "The Gaia initial quasar catalog," in *Journées 2013 "Systèmes de référence spatio-temporels"*, ed N. Capitaine (Paris: Paris Observatory), 84–87.

Andrei, A. H., Taris, F., Anton, S., Bourda, G., Damljanovic, G., Souchay, J., et al. (2015). A complete bank of optical images of the ICRF QSOs. *IAU Gen. Assem.* 22:2243437.

Arias, E. F., Charlot, P., Feissel, M., and Lestrade, J.-F. (1995). The extragalactic reference system of the International Earth Rotation Service, ICRS. *Astron. Astrophys.* 303, 604–608.

Berghea, C. T., Makarov, V. V., Frouard, J., Hennessy, G. S., Dorland, B. N., Veillette, D. R., et al. (2016). A Global Astrometric Solution for Pan-STARRS referenced to ICRF2. arXiv:1606.03446.

Boboltz, D. A., Fey, A. L., Puatua, W. K., Zacharias, N., Claussen, M. J., Johnston, K. J., et al. (2007). Very large array plus pie town astrometry of 46 radio stars. *Astron. J.* 133, 906–916. doi: 10.1086/510154

Bouffet, R., Charlot, P., and Lambert, S. (2013). "Radio source structure and VLBI position instabilities," in *SF2A-2013: Proceedings of the Annual Meeting of the French Society of Astronomy and Astrophysics*, eds L. Cambresy, F. Martins, E. Nuss, and A. Palacios (Montpelier, VT), 161–164.

Bourda, G., Charlot, P., and Le Campion, J.-F. (2008). Astrometric suitability of optically-bright ICRF sources for the alignment with the future Gaia celestial reference frame. *Astron. Astrophys.* 490, 403–408. doi: 10.1051/0004-6361:200810667

Bourda, G., Charlot, P., Porcas, R. W., and Garrington, S. T. (2010). VLBI observations of optically-bright extragalactic radio sources for the alignment of the radio frame with the future Gaia frame. I. Source detection. *Astron. Astrophys.* 520:A113. doi: 10.1051/0004-6361/201014248

Bourda, G., Collioud, A., Charlot, P., Porcas, R., and Garrington, S. (2011). VLBI observations of optically-bright extragalactic radio sources for the alignment of the radio frame with the future Gaia frame. II. Imaging candidate sources. *Astron. Astrophys.* 526:A102. doi: 10.1051/0004-6361/201014249

da Silva Neto, D. N., Andrei, A. H., Vieira Martins, R., and Assafin, M. (2002). A pattern of noncoincidence between radio and optical positions of international celestial reference frame sources. *Astron. J.* 124, 612–618. doi: 10.1086/341163

de Bruijne, J. H. J., Rygl, K. L. J., and Antoja, T. (2014). Gaia astrometric science performance - post-launch predictions. *EAS Publ. Ser.* 67, 23–29. doi: 10.1051/eas/1567004

de Souza, R. E., Krone-Martins, A., dos Anjos, S., Ducourant, C., and Teixeira, R. (2014). Detection of galaxies with Gaia. *Astron. Astrophys.* 568:A124. doi: 10.1051/0004-6361/201423514

Fey, A. L., and Charlot, P. (1997). VLBA observations of radio reference frame sources. II. Astrometric suitability based on observed structure. *Astrophys. J. Suppl. Ser.* 111, 95–142.

Fey, A. L., Gordon, D., Jacobs, C. S., Ma, C., Gaume, R. A., Arias, E. F., et al. (2015). The Second Realization of the International Celestial Reference Frame by Very Long Baseline Interferometry. *Astron. J.* 150:58. doi: 10.1088/0004-6256/150/2/58

Gaume, R. (2015). The IAU Division A Working Group on the Third Realization of the ICRF: Background, Goals, Plans. *IAU Gen. Assem.* 22:2256777.

Gordon, D., Jacobs, C., Beasley, A., Peck, A., Gaume, R., Charlot, P., et al. (2016). Second epoch VLBA calibrator survey observations: VCS-II. *Astron. J.* 151:154. doi: 10.3847/0004-6256/151/6/154

Jacobs, C. S., Arias, F., Boboltz, D., Boehm, J., Bolotin, S., Bourda, G., et al. (2014). "ICRF-3: roadmap to the next generation ICRF," in *Proceeding Journées 2013 Systèmes de Référence Spatio-temporels, Observatoire de Paris*, ed N. Capitaine (Paris), 51–56.

Kopeikin, S., and Makarov, V. (2006). Astrometric effects of secular aberration. *Astron. J.* 131, 1471–1478. doi: 10.1086/500170

Kovalevsky, J. (2003). Aberration in proper motions. *Astron. Astrophys.* 404, 743–747. doi: 10.1051/0004-6361:20030560

Kovalevsky, J., Lindegren, L., Perryman, M. A. C., Hemenway, P. D., Johnston, K. J., Kislyuk, V. S., et al. (1997). The HIPPARCOS catalogue as a realisation of the extragalactic reference system. *Astron. Astrophys.* 323, 620–633.

Le Bail, K., Gipson, J. M., Gordon, D., MacMillan, D. S., Behrend, D., Thomas, C. C., et al. (2016). IVS observation of ICRF2-Gaia transfer sources. *Astron. J.* 151:79. doi: 10.3847/0004-6256/151/3/79

Lestrade, J.-F., Jones, D. L., Preston, R. A., Phillips, R. B., Titus, M. A., Kovalevsky, J., et al. (1995). Preliminary link of the HIPPARCOS and VLBI reference frames. *Astron. Astrophys.* 304:182.

Lestrade, J.-F., Preston, R. A., Jones, D. L., Phillips, R. B., Rogers, A. E. E., Titus, M. A., et al. (1999). High-precision VLBI astrometry of radio-emitting stars. *Astron. Astrophys.* 344, 1014–1026.

Lindegren, L., Babusiaux, C., Bailer-Jones, C., Bastian, U., Brown, A. G. A., Cropper, M., et al. (2008). "The Gaia mission: science, organization and present status," in *IAU Symposium*, Vol. 248, eds W. J. Jin, I. Platais, and M. A. C. Perryman (Shanghai), 217–223.

Lindegren, L., and Kovalevsky, J. (1995). Linking the HIPPARCOS Catalogue to the extragalactic reference system. *Astron. Astrophys.* 304:189.

Liu, J.-C., Capitaine, N., Lambert, S. B., Malkin, Z., and Zhu, Z. (2012). Systematic effect of the Galactic aberration on the ICRS realization and the Earth orientation parameters. *Astron. Astrophys.* 548:A50. doi: 10.1051/0004-6361/201219421

Liu, J.-C., Xie, Y., and Zhu, Z. (2013). Aberration in proper motions for stars in our Galaxy. *Mon. Not. Roy. Astron. Soc.* 433, 3597–3604. doi: 10.1093/mnras/stt1006

MacMillan, D. S. (2005). "Quasar apparent proper motion observed by geodetic VLBI networks," in *Future Directions in High Resolution Astronomy, volume 340 of Astronomical Society of the Pacific Conference Series*, eds J. Romney and M. Reid (San Francisco, CA: Astronomical Society of the Pacific), 477–481.

Malkin, Z. (2014). On the implications of the Galactic aberration in proper motions for the Celestial Reference Frame. *Mon. Not. Roy. Astron. Soc.* 445, 845–849. doi: 10.1093/mnras/stu1796

Malkin, Z. (2015a). Erratum: on the implications of the Galactic aberration in proper motions for celestial reference frame. *Mon. Not. R. Astron. Soc.* 447, 4028. doi: 10.1093/mnras/stv032

Malkin, Z. (2015b). How much can Galactic aberration impact the link between radio (ICRF) and optical (GCRF) reference frames. arXiv:1509.07245.

Malkin, Z. (2016a). Using radio stars to link the Gaia and VLBI reference frames. *Mon. Not. R. Astron. Soc.* 461, 1937–1942. doi: 10.1093/mnras/stw1488

Malkin, Z. (2016b). Second version of the catalog of optical characteristics of astrometric radio sources OCARS. *Astron. Rep.* 60. doi: 10.1134/S1063772916110032

Malkin, Z. M. (2011). The influence of Galactic aberration on precession parameters determined from VLBI observations. *Astron. Rep.* 55, 810–815. doi: 10.1134/S1063772911090058

Mignard, F. (2014). "Building a CRF with gaia," in *Dynamical Astronomy in Latin-America - 2014*. Available online at: http://adela2014.das.uchile.cl/talks/Workshop_Classes/FMignard_Lecture_2_ADeLA2014@SCL.pdf

Moór, A., Frey, S., Lambert, S. B., Titov, O. A., and Bakos, J. (2011). On the connection of the apparent proper motion and the VLBI structure of compact radio sources. *Astron. J.* 141, 178. doi: 10.1088/0004-6256/141/6/178

Perryman, M. A. C., de Boer, K. S., Gilmore, G., Høg, E., Lattanzi, M. G., Lindegren, L., et al. (2001). GAIA: composition, formation and evolution of the Galaxy. *Astron. Astrophys.* 369, 339–363. doi: 10.1051/0004-6361:20010085

Plank, L., Lovell, J. E. J., Shabala, S. S., Böhm, J., and Titov, O. (2015). Challenges for geodetic VLBI in the southern hemisphere. *Adv. Space Res.* 56, 304–313. doi: 10.1016/j.asr.2015.04.022

Sokolova, J., and Malkin, Z. (2007). On comparison and combination of catalogues of radio source positions. *Astron. Astrophys.* 474, 665–670. doi: 10.1051/0004-6361:20077450

Souchay, J., Andrei, A. H., Barache, C., Kalewicz, T., Gattano, C., Coelho, B., et al. (2015). The third release of the Large Quasar Astrometric Catalog (LQAC-3): a compilation of 321 957 objects. *Astron. Astrophys.* 583:A75. doi: 10.1051/0004-6361/201526092

Taris, F., Andrei, A., Klotz, A., Vachier, F., Côte, R., Bouquillon, S., et al. (2013). Optical monitoring of extragalactic sources for linking the ICRF and the future Gaia celestial reference frame. I. Variability of ICRF sources. *Astron. Astrophys.* 552:A98. doi: 10.1051/0004- 6361/201219686

Taris, F., Andrei, A., Roland, J., Klotz, A., Vachier, F., and Souchay, J. (2016). Long-term R and V-band monitoring of some suitable targets for the link between ICRF and the future Gaia celestial reference frame. *Astron. Astrophys.* 587:A112. doi: 10.1051/0004-6361/201526676

Taris, F., Damljanovic, G., Andrei, A., Klotz, A., and Vachier, F. (2015). "Optical monitoring of QSO in the framework of the Gaia space mission," in *Journées 2014 "Systèmes de référence spatio-temporels"*, eds Z. Malkin and N. Capitaine, 42–43.

Titov, O., and Lambert, S. (2013). Improved VLBI measurement of the solar system acceleration. *Astron. Astrophys.* 559:A95. doi: 10.1051/0004-6361/201321806

Véron-Cetty, M.-P., and Véron, P. (2006). A catalogue of quasars and active nuclei: 12th edition. *Astron. Astrophys.* 455, 773–777. doi: 10.1051/0004-6361:20065177

Zacharias, N., Finch, C., Subasavage, J., Bredthauer, G., Crockett, C., Divittorio, M., et al. (2015). The first U.S. naval observatory robotic astrometric telescope catalog. *Astron. J.* 150:101. doi: 10.1088/0004-6256/150/4/101

Zacharias, N., and Zacharias, M. I. (2014). Radio-optical reference frame link using the U.S. naval observatory astrograph and deep CCD imaging. *Astron. J.* 147:95. doi: 10.1088/0004-6256/147/5/95

Conflict of Interest Statement: The author declares that the research was conducted in the absence of any commercial or financial relationships that could be construed as a potential conflict of interest.

10

FORWARD: A Toolset for Multiwavelength Coronal Magnetometry

Sarah E. Gibson[1]*, Therese A. Kucera[2], Stephen M. White[3], James B. Dove[4], Yuhong Fan[1], Blake C. Forland[5], Laurel A. Rachmeler[6,7], Cooper Downs[8] and Katharine K. Reeves[9]

[1] High Altitude Observatory, National Center for Atmospheric Research, Boulder, CO, USA, [2] Goddard Space Flight Center, National Aeronautics and Space Administration (NASA), Greenbelt, MD, USA, [3] Air Force Research Labs, Kirtland Air Force Base, Albuquerque, NM, USA, [4] Department of Physics, Metro State University Denver, Denver, CO, USA, [5] Department of Physics, Indiana University, Bloomington, IN, USA, [6] Royal Observatory of Belgium, Brussels, Belgium, [7] Marshall Space Flight Center, National Aeronautics and Space Administration (NASA), Huntsville, AL, USA, [8] Predictive Science Inc., San Diego, CA, USA, [9] Department of High Energy Astrophysics, Harvard-Smithsonian Center for Astrophysics, Cambridge, MA, USA

Edited by:
Xueshang Feng,
National Space Science Center, China

Reviewed by:
Satoshi Inoue,
Max-Planck Institute for Solar System
Research, Germany
Jiansen He,
Peking University, China

*Correspondence:
Sarah E. Gibson
sgibson@ucar.edu

Specialty section:
This article was submitted to
Stellar and Solar Physics,
a section of the journal
Frontiers in Astronomy and Space
Sciences

Determining the 3D coronal magnetic field is a critical, but extremely difficult problem to solve. Since different types of multiwavelength coronal data probe different aspects of the coronal magnetic field, ideally these data should be used together to validate and constrain specifications of that field. Such a task requires the ability to create observable quantities at a range of wavelengths from a distribution of magnetic field and associated plasma—i.e., to perform forward calculations. In this paper we describe the capabilities of the FORWARD SolarSoft IDL package, a uniquely comprehensive toolset for coronal magnetometry. FORWARD is a community resource that may be used both to synthesize a broad range of coronal observables, and to access and compare synthetic observables to existing data. It enables forward fitting of specific observations, and helps to build intuition into how the physical properties of coronal magnetic structures translate to observable properties. FORWARD can also be used to generate synthetic test beds from MHD simulations in order to facilitate the development of coronal magnetometric inversion methods, and to prepare for the analysis of future large solar telescope data.

Keywords: sun: corona, sun: magnetic fields, sun: x-rays, sun: radio, sun: infrared, sun: EUV

1. INTRODUCTION

In essence, the goal of coronal magnetometry is to solve an inverse problem. Given magnetically-sensitive coronal observations (including, but not limited to polarimetry), the challenge is to determine the magnetic field distribution that generates them. Solving such an inverse problem requires three things: a means of specifying the physical state (e.g., the distribution of density, temperature, velocity, and magnetic field), a well-defined forward calculation (i.e., the physical process relating the physical state and the observations), and the observations themselves.

FORWARD is a set of more than 200 IDL procedures and functions that form a SolarSoft (Freeland and Handy, 1998) package for synthesizing observables and comparing them to coronal data from EUV/Xray imagers, UV/EUV spectrometers, visible/IR/UV polarimeters, white-light coronagraphs, and radio telescopes. It may be called from the command line (i.e., `for_drive`), or via a widget interface (i.e., `for_widget`; Forland et al., 2014). The standard output product is a 2D plane-of-sky map, 2D latitude-longitude (Carrington) map, or user-specified spatial sampling

(**Figure 1**). Image field of view and resolution is user-controlled, as is "viewer" position and line-of-sight (LOS) integration spacing and limits. Details on how to run and install FORWARD are available at http://www.hao.ucar.edu/FORWARD/.

This paper describes how FORWARD addresses all three of the requirements for coronal magnetometric inverson and gives examples of how it may be used. Section 2 demonstrates how the physical state may be defined through analytic or numerical models, either user-inputted or generated by FORWARD through included codes or via its interface with online coronal simulations. Section 3 describes the multiwavelength forward calculations that predict observational manifestations of physical processes such as *Thomson scattering, collisional excitation, continuum absorption, resonance scattering, Zeeman and Hanle effects, Doppler shift, thermal bremsstrahllung, gyroresonance,* and *Faraday rotation,* and discusses the magnetic diagnostic potential of each. Section 4 describes how FORWARD enables the access and manipulation of observations and converts them to a format directly comparable to the predictions of forward calculations. Section 5 shows how FORWARD may be applied to validate models, build intuition regarding coronal magnetic signatures, tune models to match data, and generally guide the development of multiwavelength magnetometric inversion techniques. Finally, in Section 6 we present our conclusions.

2. THE PHYSICAL STATE

When discussing solar-coronal forward analysis, it is important to differentiate between the model of the *physical state* of the corona, which addresses the distribution of magnetic fields and plasma throughout 3D space, and the model of how these fields and plasma operate in the presence of a *physical process,* which enables the synthesis of an observed quantity. We will treat the latter in Section 3 as the heart of the forward calculation.

Models of the physical state essentially create synthetic Suns—generally through solutions of the MHD equations. FORWARD includes several analytic models in its distribution (i.e., Low and Hundhausen, 1995; Lites and Low, 1997; Gibson et al., 2010; **Figure 1**; Gibson and Low, 1998; **Figure 2**). It is straightforward to expand it to incorporate other analytic models. Alternatively, a user may input a numerical data cube describing the 3D distribution of plasma and fields. If the data cube is not global, options are provided regarding what to do outside the cube (e.g., zero, constant, or dipolar field, and hydrostatic atmospheres—either isothermal-exponential or power-law). If the data cube only provides a magnetic field, hydrostatic atmospheres can be applied throughout space. (See http://www.hao.ucar.edu/FORWARD/FOR_SSW/idl/MODELS/NUMCUBE/make_my_cube.pro for instructions on how to convert a numerical data cube to FORWARD format.) In addition to the forward-calculated observables discussed in Section 3, FORWARD allows easy display of the parameters of the physical state, e.g., density, temperature, magnetic field, velocity (see e.g., **Figure 2**).

Given a calendar date, FORWARD can also automatically interface with the SolarSoft Potential Field Source Surface (PFSS) package (http://www.lmsal.com/~derosa/pfsspack/) and

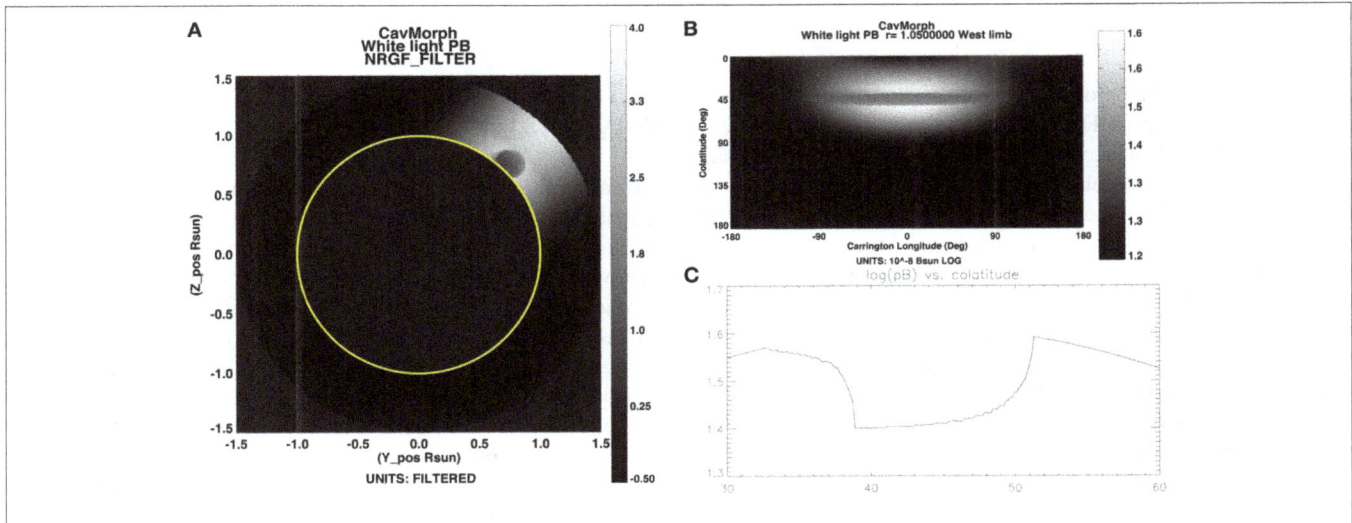

FIGURE 1 | Examples of FORWARD output of LOS-integrated white-light polarized Brightness (pB) for a morphological model of a cavity embedded in a coronal streamer (Gibson et al., 2010). (A) Cavity in plane of sky at limb (plotted with non-radial gradient filter Morgan et al., 2006). Plot obtained by FORWARD line command: `for_drive,'cavmorph',inst='wl',line='pb',thcs=45,cavlength=150,rfilter='NRGF_FILTER'`. **(B)** Cavity in latitude-longitude Carrington map. Plot obtained as for **(A)**, but without `rfilter` keyword and with `,gridtype='Carrmap',cmer=0,charsize=.85` added. **(C)** Cavity in constant radius latitudinal cut. Plot obtained as for **(A)**, but with removal of `rfilter` keyword and addition of keywords: `,gridtype='user',ruser=dblarr(201)+1.05,thuser=dindgen(201)*.15+30,phuser=dblarr(201)-30.,quantmap=quantmap` and followed by command: `plot,dindgen(201)*.15+30,alog10(quantmap.data),yrange=[1.3,1.7],xrange=[30.,60.],title='log(pB) vs. colatitude'`. Note that this and other IDL commands provided in figure captions below can be accessed via `$FORWARD_DOCS/EXAMPLES/examples_forwardpaper.html`.

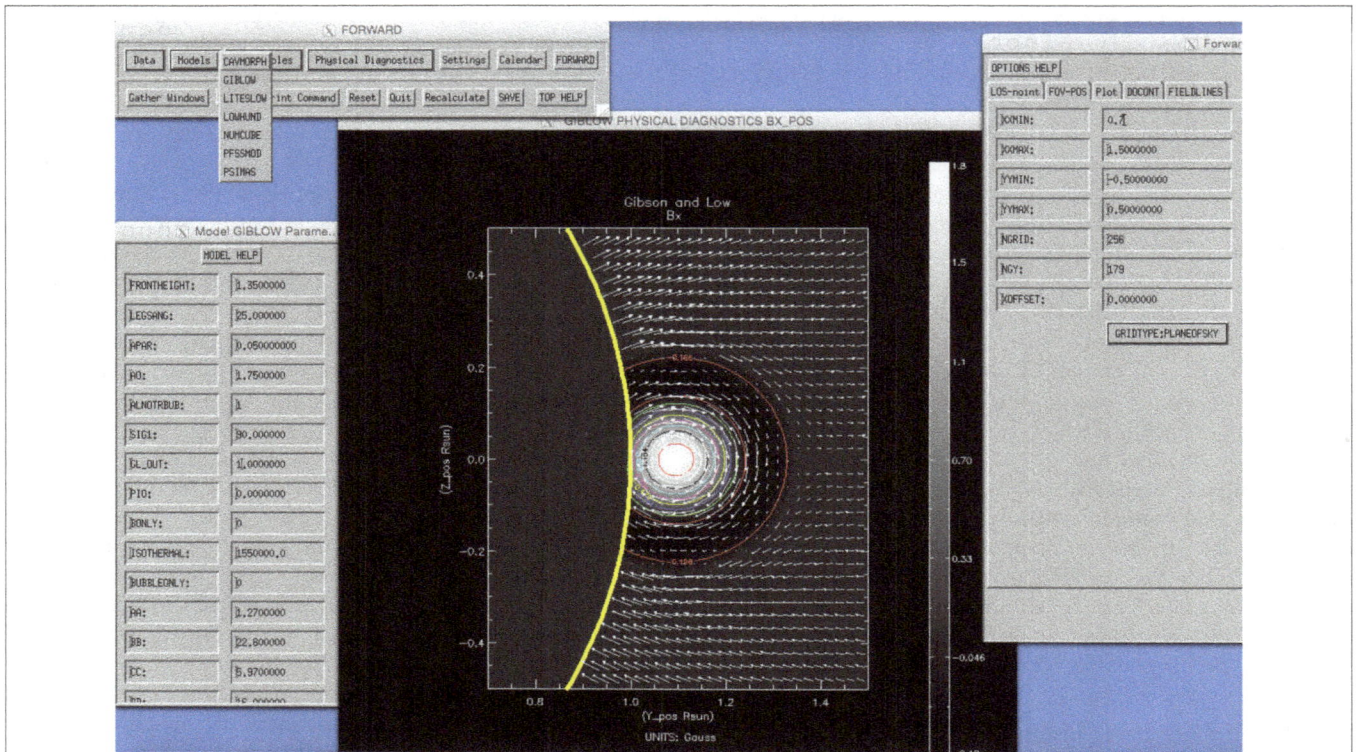

FIGURE 2 | Example of analytic model of a spheromak flux rope embedded in an otherwise open bipolar global magnetic field (Gibson and Low, 1998), provided within the FORWARD distribution and demonstrated here using the for_widget interface. The user chooses the model via the drop-down menu in the top-left widget as shown, and then may choose model parameters (bottom-left widget), and display (as in line-of-sight magnetic field example shown here) model diagnostics (top-left widget, drop-down menu for Physical Diagnostics) with various plotting choices such as plane-of-sky field lines (white vectors; set in right widget). Doing an actual forward calculation of a coronal observable (not shown) is done by choosing one of the Observables (top-left widget, drop-down menu). All calculations are intitiated by clicking on the FORWARD button (top-left widget).

the web-served Magnetohydrodynamic Algorithm outside a Sphere (MAS)-corona MHD simulation data cubes (http://www.predsci.com/hmi/data_access.php; **Figure 3**; Lionello et al., 2009). This enables global descriptions of the 3D coronal magnetic field, and for the MAS model also the plasma in MHD force balance, specific to a given time/day and viewer position.

FORWARD also allows the user to specify the physical state of two populations of plasma that may need to be treated independently in the forward calculation. For example, a user may specify a population of plasma at a coronal temperature, and another population of cooler, chromospheric plasma subject to continuum absorption in EUV images (see Section 3.3). This provides capability, for example, for depicting models where cool solar prominences exist in the context of surrounding coronal temperature material, such as those produced by Luna et al. (2012) or Xia et al. (2014) (see also the cavity-prominence test-bed simulation shown in Section 5). Another application would be to allow models where two different coronal populations lie along the line of sight, each with different abundance properties. It is also possible to set a filling factor for one or both populations. This capability allows exploitation of the diagnostic potential of comparing emissions which may have different dependencies on density (as we discuss in Section 3).

3. THE FORWARD CALCULATION: PHYSICAL PROCESSES

Given a modeled physical state, i.e., a specification of the distribution of density, temperature, magnetic field and velocity in the corona, FORWARD is able to produce many different synthetic observables. These observables arise from various physical processes manifesting at different wavelengths of light in the corona. They depend upon the viewer's line of sight, along which (for example) optically-thin emission must be integrated. FORWARD establishes these lines of sight either through keyword definition of an observer's heliographic latitude and longitude, or through keyword setting of a calendar date from which the position of the Earth (or STEREO spacecraft) can be determined. In this section, we will discuss a range of physical processes relevant to the corona, describe how they translate to observables that FORWARD synthesizes, and consider their potential for coronal magnetometry. **Table 1** provides a summary.

3.1. Thomson Scattering

Thomson scattering is the main physical process responsible for illuminating the continuum, or "K" corona. Photospheric

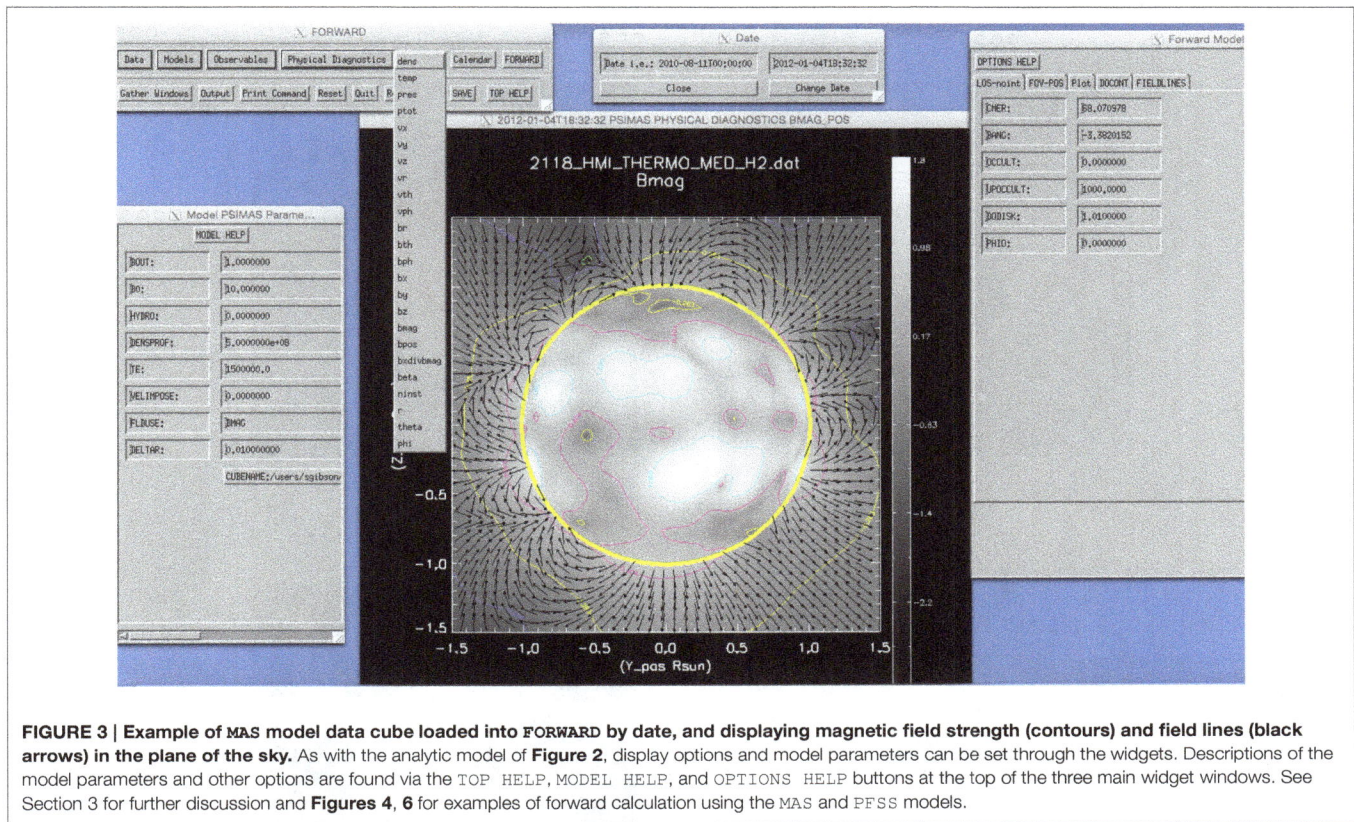

FIGURE 3 | Example of MAS model data cube loaded into FORWARD by date, and displaying magnetic field strength (contours) and field lines (black arrows) in the plane of the sky. As with the analytic model of **Figure 2**, display options and model parameters can be set through the widgets. Descriptions of the model parameters and other options are found via the TOP HELP, MODEL HELP, and OPTIONS HELP buttons at the top of the three main widget windows. See Section 3 for further discussion and **Figures 4, 6** for examples of forward calculation using the MAS and PFSS models.

TABLE 1 | Physical processes as defined in Section 3, highlighting dependency on attributes of the physical state, which observations are sensitive to them, and diagnostic sensitivity to the 3D coronal magnetic field.

Process	Physical-state dependency	Observation	Magnetic quantity probed
Thomson scattering	Electron density	White-light pB, TB	Plasma structured by field (e.g., closed vs. open field boundaries, flux surfaces)
Collisional excitation	Electron density, temperature	IR/Visible/EUV/SXR emission	Plasma structured by field (incl. loops, closed/open boundaries, flux surfaces)
Continuum absorption	Chromospheric population density, electron density, temperature	EUV absorption features	Can indicate magnetic geometry suitable for prominence formation
Resonance scattering; polarization	Electron density, temperature, vector magnetic field	Visible/IR spectra	B_{los} from Stokes V; Magnetic field direction from Stokes Q, U
Doppler shift	Electron density, temperature, velocity	Visible/IR spectra	B_{pos} and field line direction from waves; flux surfaces from bulk flows
Thermal bremsstrahllung	Electron density, temperature, vector magnetic field	Radio emission (intensity and circular polarization) as a function of frequency	B_{los} from Stokes V
Gyroresonance	Electron density, temperature, vector magnetic field	Radio emission (intensity and circular polarization) as a function of frequency	Surfaces of constant magnetic field strength at each frequency
Faraday rotation	Electron density, temperature, vector magnetic field	Rotation of plane of polarization	B_{los} from rotation measure

light scatters off of free coronal electrons and results in both unpolarized and linearly polarized emission. Both the total brightness (*TB*) and polarized brightness (*pB*) of white light are proportional to n_e and to a scattering function that depends upon radial distance from the photosphere (Billings, 1966). They are also integrated along the line of sight in the optically-thin corona.

Given a distribution of electron density, FORWARD can synthesize images of *TB*, *pB*, and degree of polarization *p* (**Figures 4A**, **5A**), comparable to observations from white light coronagraphs such as SOHO/LASCO, STEREO/SECCHI, and MLSO/KCOR. If keyword *fcor* is set, FORWARD will call upon SolarSoft function *fcorpol_KL.pro* in order to add a model

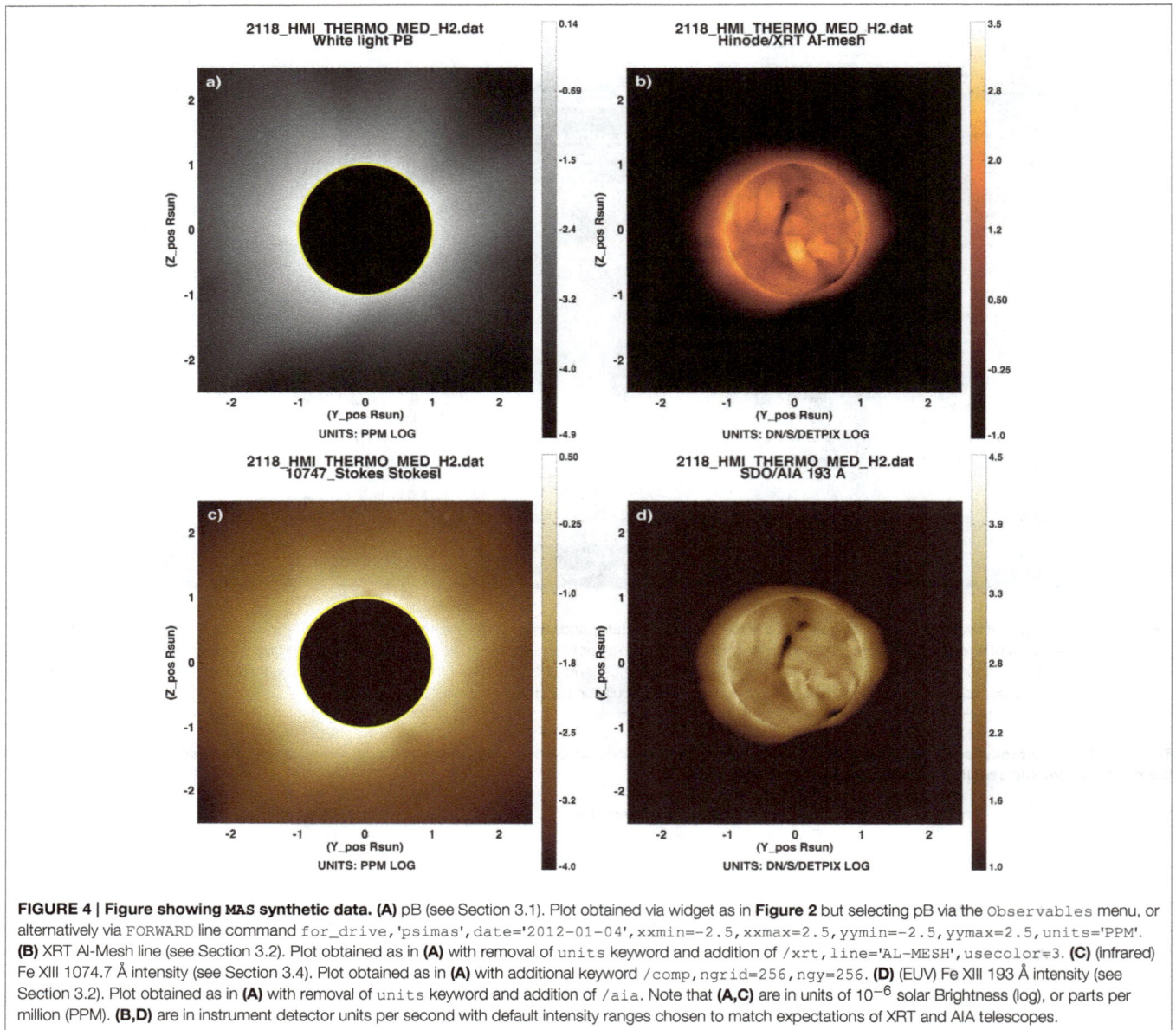

FIGURE 4 | Figure showing MAS synthetic data. (A) pB (see Section 3.1). Plot obtained via widget as in **Figure 2** but selecting pB via the `Observables` menu, or alternatively via `FORWARD` line command `for_drive,'psimas',date='2012-01-04',xxmin=-2.5,xxmax=2.5,yymin=-2.5,yymax=2.5,units='PPM'`. **(B)** XRT Al-Mesh line (see Section 3.2). Plot obtained as in **(A)** with removal of `units` keyword and addition of `/xrt,line='AL-MESH',usecolor=3`. **(C)** (infrared) Fe XIII 1074.7 Å intensity (see Section 3.4). Plot obtained as in **(A)** with additional keyword `/comp,ngrid=256,ngy=256`. **(D)** (EUV) Fe XIII 193 Å intensity (see Section 3.2). Plot obtained as in **(A)** with removal of `units` keyword and addition of `/aia`. Note that **(A,C)** are in units of 10^{-6} solar Brightness (log), or parts per million (PPM). **(B,D)** are in instrument detector units per second with default intensity ranges chosen to match expectations of XRT and AIA telescopes.

distribution of F-coronal brightness (Koutchmy and Lamy, 1985). This arises from light diffracting through interplanetary particles in the plane of the ecliptic, and is also known as the zodiacal light. It is essentially unpolarized in the first few solar radii (Mann, 1992).

Thomson scattering has no direct dependency on magnetic field, but there is sensitivity to magnetic topology through its dependence on density. For example, bright (dense) coronal streamers generally correspond to closed magnetic fields, and dark (sparse) coronal holes generally correspond to open magnetic fields. For this reason, white light coronagraph data have been used to qualitatively validate features of coronal magnetic models (Newkirk and Altschuler, 1970), and more quantitatively, to define the average nonradial expansion of

magnetic fields in coronal holes (Kopp and Holzer, 1976; Munro and Jackson, 1977). Magnetic flux surfaces also may be delineated by white light structures, such as three-part CME features (Low and Hundhausen, 1995), and prominence cavities (e.g., **Figure 5A**; see also Gibson and Fan, 2006).

3.2. Collisional Excitation

Solar coronal radiation in the extreme ultraviolet (EUV) and soft X-ray (SXR) is produced by collisionally-excited atoms in thermal and ionization equilibrium. The intensity of this emission is proportional to n_e^2 and the temperature response of the line(s). For spectrographs, FORWARD calculates the integrated intensities of lines in physical units to be compared with processed spectral data. For waveband imagers, FORWARD

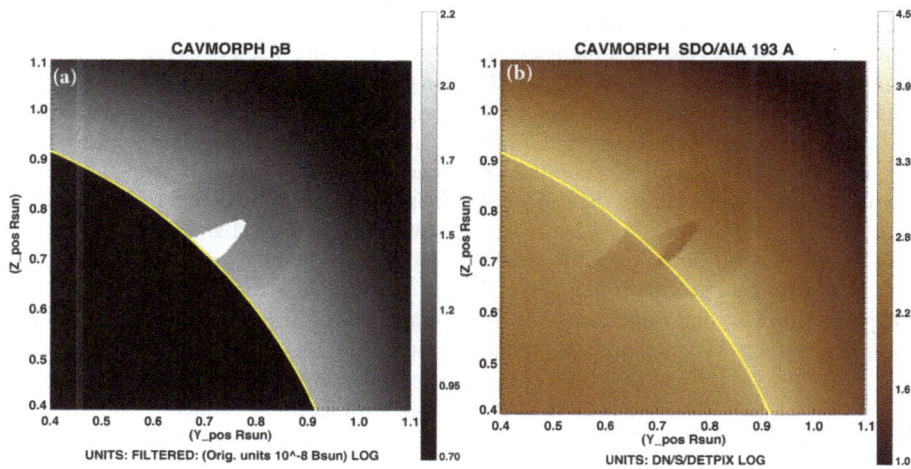

FIGURE 5 | A simplified cavity and prominence produced using the Gibson et al. (2010) model. (A) synthesized white light polarized brightness (see Section 3.1), with high-density prominence appearing as enhanced intensity. Plot obtained by FORWARD line command: `for_drive,'cavmorph',` `/nougat,thcs=45,cdens=1e10,cff_noug=[.2,8.,0,0,0,0],nougwidth=.025,nougtop_r=.9,pop2T=2,xxmin=0.4,yymin=0.4,yymax=1.1,` `xxmax=1.1.` **(B)** SDO/AIA 193 Å emission, with dark prominence because of continuum absorption (see Sections 3.2–3.3). Plot obtained by above command with the addition of keyword `/aia`.

incorporates the wavelength-response function of the instrument into its calculated intensities. The emission is integrated along the line of sight in the optically-thin corona.

FORWARD synthesizes images comparable to those produced by numerous EUV and Soft X-ray imagers and also many spectral line intensities in these wave bands. Currently simulated imagers include SOHO/EIT, STEREO/EUVI, Hinode/XRT (e.g., **Figure 4B**), ProbA-2/SWAP, and SDO/AIA (e.g., **Figures 4D, 5B**). Adding new imagers is straightforward if the wavelength response function is available. Count rates for imagers are calculated by convolving the wavelength response function of the imager with pre-calculated spectra at various temperatures and densities, produced using the Chianti atomic data base and related software (Dere et al., 1997; Del Zanna et al., 2015). For the imagers, users may select from pre-calculated abundance options; current selections include coronal abundances determined by Feldman et al. (1992) or Schmelz et al. (2012) or photospheric abundances of Caffau et al. (2011). The code uses the Chianti ionization equilibrium calculations (Dere et al., 2009).

Spectral line intensities can be calculated for any line between 1 and 1410 Å, again under the assumption that the coronal plasma is collisionally excited and in thermal and ionization equilibrium. Based on a user-specified instrumental line-width, the code includes any blended lines included in the Chianti spectral line calculations. For spectral lines, users may specify any abundance or ionization table in the Chianti database format. Default line widths are provided for particular instruments like Hinode/EIS, IRIS, and SoHO/CDS, but the code is not limited to lines observed by these instrument. The wavelength range covers the IRIS far ultraviolet (FUV) range, but not the near ultraviolet (NUV) range, which does not include lines that can be modeled by FORWARD.

As with Thomson scattering, radiation from collisional excitation does not have a direct dependence on magnetic fields. However, the suppression of conductivity across magnetic field lines means that magnetic field lines are essentially traced out in coronal emission. EUV and SXR structures thus often provide a diagnostic of local magnetic field geometry—for example when coronal loops are lit up in active regions (see e.g., Savcheva et al., 2013; Malanushenko et al., 2014; also Savcheva and Malanushenko, in preparation). Magnetic boundaries or flux surfaces are also often delineated due to sharp density/temperature gradients, as in the case of open vs. closed fields (e.g., **Figures 4B,D**), and prominence cavities (e.g., **Figure 5B**).

3.3. Continuum Absorption

Relatively cool, chromospheric temperature material suspended in the corona (e.g., a prominence) results in Lyman continuum absorption by neutral hydrogen and by neutral and once-ionized helium (see Kucera, 2015 for further details).

The observed intensity is

$$I = I_f + I_b e^{-\tau} \tag{1}$$

where I_f is the foreground radiation, I_b is the background radiation, and τ is the continuum absorption summed over the three absorbing species:

$$\tau = \sum_i \sigma_i \int n_i dh \tag{2}$$

where h is the distance along the line of sight, n_i is the number density of each species, and σ_i is the absorbing cross section as a function of wavelength (values calculated with the formulation of Keady and Kilcrease, 2000).

Through the definition of a second population of low-temperature plasma with a specified density distinct from the primary coronal population, FORWARD calculates the effect of continuum absorption on the total intensity in EUV. **Figure 5B** shows a model of a simple prominence inside a cavity in the 193 Å band of SDO/AIA. The cavity is darker than its surroundings because it has a lower density, but the central prominence is darker because of continuum absorption of background emission.

Again, continuum absorption has no direct dependence on magnetic fields. However, magnetic field geometry (e.g., dipped or flat field lines) is expected to play an important role in establishing where prominences form (see Karpen, 2014 and references therein).

3.4. Resonance Scattering and Polarization

Emission from the coronal forbidden lines arises both from collisional excitation as described above in Section 3.2, and also from resonance scattering. In resonance scattering, anisotropic radiation from the underlying photosphere excites coronal ions and leads to reemitted light with a characteristic polarization signature (Casini and Judge, 1999). This emission depends linearly upon ion density (and thus electron density), as opposed to quadratically as in collisional excitation. **Figure 4C** vs. **Figure 4D** illustrates the difference between the collisionally-excited/resonantly-scattered infrared Fe XIII line and the predominantly collisionally-excited EUV Fe XIII line. Intensity of the latter drops off quickly, while the former shows similarity in the outer field of view to the Thomson-scattered white light (**Figure 4A**), which is also linearly-dependent upon electron density at these heights (see Habbal et al., 2011 for further discussion).

FORWARD employs the Coronal Line Emission (CLE) Fortran-77 polarimetry code developed by Judge and Casini (2001) to synthesize Stokes (I, Q, U, V) line profiles for the visible and infrared forbidden lines including Fe XIII 1074.7 and 1079.8 nm (currently observed by MLSO/CoMP as discussed in Section 4), Fe XIV 530.3 nm, Si IX 393.4 nm, and Si X 1430.5 nm. Stokes I indicates the total intensity of the line, Q and U together constitute its linearly polarized intensity, and V is the circularly polarized intensity. The CLE code models the lines under the combined influence of resonance scattering and particle collisions in the presence of coronal magnetic fields.

Because of its sensitivity to magnetic fields, the Stokes I, Q, U, V polarization vector can be used as a direct diagnostic of coronal magnetism (subject to intensity-weighted line-of-sight integration). For example, the *Zeeman effect* generates circularly polarized light (Stokes V) proportional to line-of-sight-oriented magnetic field B_{los}. Since the coronal visible/infrared forbidden lines treated by CLE have a Larmor frequency $\nu_L \approx \mu_B B/h$ that is much larger than the inverse lifetime of the atomic transitions being modeled, they lie in the strong field (or saturation) limit of the *Hanle effect* (see Raouafi et al., submitted; Dima et al., submitted) for discussion of magnetometry in the UV "unsaturated" Hanle regime). In the saturated regime, linear polarization provides a probe of the direction of the magnetic

field in the plane-of-sky (POS), but not its strength. In particular, the direction of the linear polarization vector [or azimuth, $Az = -0.5 * atan(U/Q)$] is parallel to the POS component of the magnetic field, as long as the local magnetic vector field has an angle relative to the solar radial direction (ϑ_B) less than the critical *"van Vleck"* angle, at which point the azimuth becomes perpendicular to the POS field. This occurs because the atomic alignment upon which the linear polarization depends goes through zero (and changes sign) when $3cos^2(\vartheta_B) = 1$, i.e., when $\vartheta_B = 54.74°$ (van Vleck, 1925). The location of van Vleck nulls in linear polarization $L = \sqrt{(Q^2 + U^2)}$ thus also acts as a diagnostic of magnetic field direction (see Section 5.1 for further discussion).

Figures 6A,C,E,G shows I, V/I, Az and L/I calculated from the MAS coronal model. From this it is clear that, despite the line-of-sight superposition of optically-thin coronal plasma, Stokes polarimetry can provide a quantitative measure of coronal magnetic field strength and direction. The Stokes V/I (**Figure 6C**) represents a line-of-sight intensity-weighted average of B_{los}. The dark linear-polarization features shown in **Figure 6E**) are generally signatures of magnetic fields oriented at the van Vleck angle (although note that the presence of strong B_{los} field can also result in linear polarization nulls; see Section 5.1 for further discussion). Even with LOS integration, the linear polarization vectors (blue) are largely aligned with the POS magnetic field vectors (red) (see **Figures 6A,E**), except when at the van Vleck angle they flip 90°. **Figure 6G**) also illustrates this sensitivity to POS magnetic field direction, showing magnitude of departure from radial-orientation in Az (red = counterclockwise, blue = clockwise). The coronal hole in the south/southwest is evident as a broad blue/red interface in Az, indicative of diverging magnetic fields, while closed field structures exhibit a red-black-blue interface (e.g., south/southeast) which indicates converging fields.

Strongly nonradial azimuths (represented as green in the local-vertical reference frame of **Figure 6G**) are rare. They can occur if the local magnetic vector > 54.74° as measured from the solar radial direction but the POS projection is close to radial, as in the case of magnetic fields that are oriented largely along the LOS. In order for such a nearly-perpendicular azimuth to survive LOS integration, either the plasma must be localized to a magnetic structure oriented in this manner, or a larger-scale magnetic structure must possess a symmetry along the LOS. Such symmetries are fairly common in large-scale POS-oriented fields extended along the LOS, e.g., arcade fields or coronal holes, and because such structures are POS-oriented, they possess a strong linear-polarization signal and so the azimuth survives LOS integration (see further discussion in Section 5.1). Even if LOS-oriented fields are localized or exist with orientation extended along the LOS, however, because they do not possess a strong linear polarization signal they are likely to be obscured by any POS-oriented fields lying along their integration path.

Figures 6B,D,F,H shows the polarization for a potential field model extrapolation for the same day and using similar (although not identical) photospheric magnetic boundary data as the MAS model of **Figures 6A,C,E,G**. The differences between MAS and

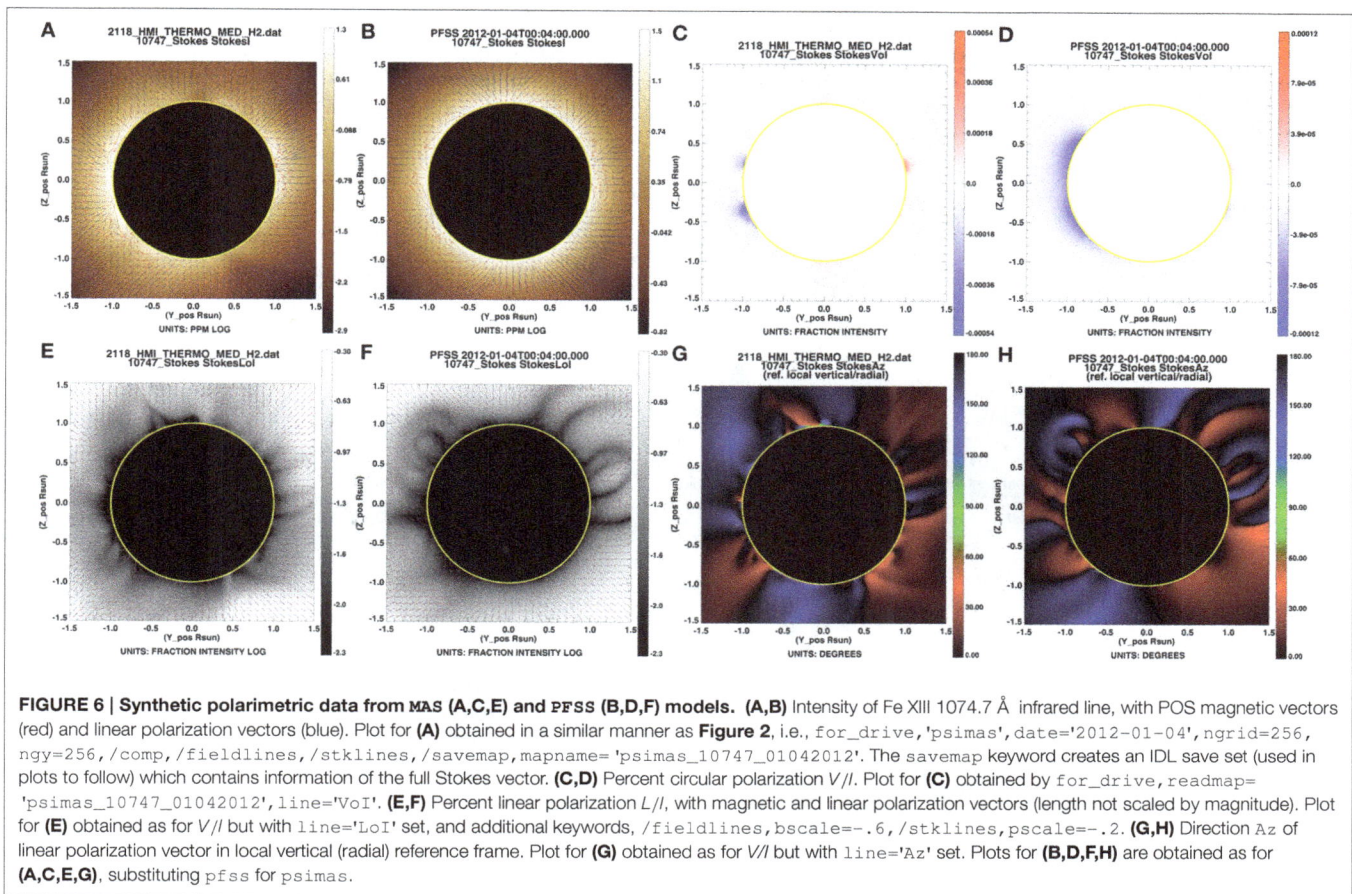

FIGURE 6 | Synthetic polarimetric data from MAS (A,C,E) and PFSS (B,D,F) models. (A,B) Intensity of Fe XIII 1074.7 Å infrared line, with POS magnetic vectors (red) and linear polarization vectors (blue). Plot for **(A)** obtained in a similar manner as **Figure 2**, i.e., `for_drive,'psimas',date='2012-01-04',ngrid=256,` `ngy=256,/comp,/fieldlines,/stklines,/savemap,mapname= 'psimas_10747_01042012'`. The `savemap` keyword creates an IDL save set (used in plots to follow) which contains information of the full Stokes vector. **(C,D)** Percent circular polarization V/I. Plot for **(C)** obtained by `for_drive,readmap= 'psimas_10747_01042012',line='VoI'`. **(E,F)** Percent linear polarization L/I, with magnetic and linear polarization vectors (length not scaled by magnitude). Plot for **(E)** obtained as for V/I but with `line='LoI'` set, and additional keywords, `/fieldlines,bscale=-.6,/stklines,pscale=-.2`. **(G,H)** Direction Az of linear polarization vector in local vertical (radial) reference frame. Plot for **(G)** obtained as for V/I but with `line='Az'` set. Plots for **(B,D,F,H)** are obtained as for **(A,C,E,G)**, substituting `pfss` for `psimas`.

PFSS predictions for circular and linear polarization result from differences at the lower boundary, from the non-potentiality of the MAS model magnetic field, and also to some degree from differences in intensity-weighting along the line of sight (the PFSS solution requires a density/temperature distribution that is spherically-symmetric). The significance of intensity weighting is also evident in **Figure 7**, where the LOS-integrated Stokes V differs depending on the wavelength used to observe it (visible, IR, radio). Since the same (MAS) model is used for all four forward calculations, variation must be due to the different sensitivities to temperature and density for the four wavelength regimes, which in turn means that different distributions of plasma are contributing to the integrals along the line of sight. In Section 5 we will discuss the importance of making full use of such multiwavelength magnetic dependencies in choosing between models.

3.5. Doppler Shift

If light-emitting plasma is moving, spectral lines are subject to a Doppler shift proportional to the line-of-sight component of the plasma velocity (v_{los}). For optically-thin plasma, this v_{los} is further weighted by the distribution of intensity along the line of sight. From the line profiles in the visible and IR generated by CLE (see Section 3.4), FORWARD determines Doppler shift and integrates

along the line of sight to get a synthetic observable comparable to observations.

Doppler velocity observations in the IR by the MLSO/CoMP telescope have proved to be a good resource for measuring ubiquitous waves in the corona (Tomczyk et al., 2007). The phase speeds of these waves are expected to be proportional to the plane-of-sky component of the magnetic field strength, and the direction of propagation of the waves will be aligned with the magnetic field direction. In general, the flux-freezing condition forces plasma flows to follow the direction of the magnetic field, so bulk velocity flows also can act as a probe of magnetic structure (**Figure 10C**; Bąk-Stęślicka et al., 2013, also Bąk-Stęślicka et al., submitted).

3.6. Radio Emission: Thermal Bremstrahllung and Gyroresonance

The two thermal emission mechanisms that dominate non–flaring solar radio emission are bremsstrahlung (also known as "free–free emission") and gyroresonance emission. Bremsstrahlung is produced by all plasma in the solar atmosphere and is strongest in dense regions, while gyroresonance emission requires strong magnetic fields in the corona and is usually confined to locations above sunspots. The Jansky Very Large Array, the Expanded Owens Valley Solar Array, the

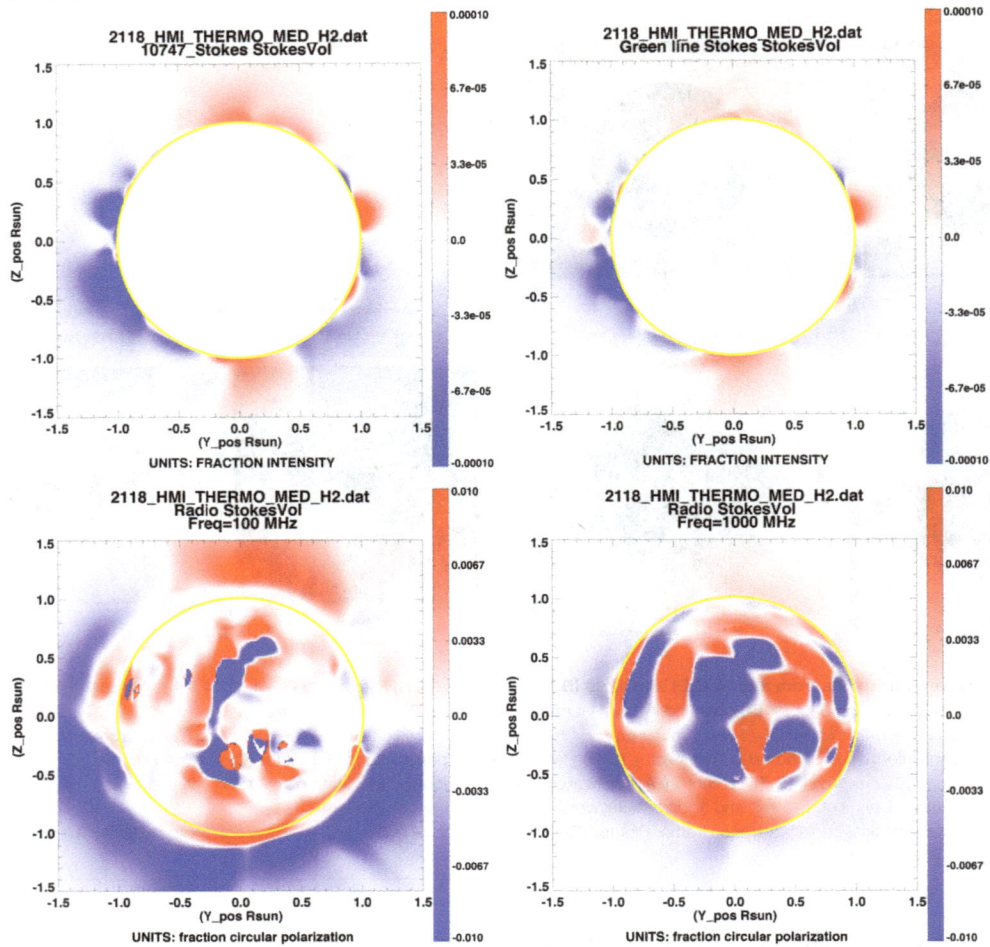

FIGURE 7 | Because of different dependencies on plasma along the line of sight, the representation of integrated circular polarization (dependent upon line-of-sight magnetic field strength) appears differently at different wavelengths. (Top left) MAS model *V/I* for Fe XIII 1074.7. Plot obtained as in **Figure 6**, with additional keywords `imin=-0.00001,imax=0.00001`. (Top right) Same for Fe XIV green line. Plot obtained by following the process outlined in **Figure 6**, but substituting `/greencomp` for `/comp`. (Bottom left) Same for radio bremsstrahllung, at a frequency of 100 MHz. Plot obtained as in **Figure 6**, but with `/radio` instead of `/comp` and `frequency_MHz=100,imin=-0.001,imax=0.001`. (Bottom right) Same, but substituting `frequency_MHz=1000`. Unlike the visible and IR lines, radio frequencies can be observed above the solar disk as well as at the limb.

Nobeyama Radioheliograph and the Mingantu Ultrawide Spectral Radioheliograph are examples of radio telescopes capable of high-resolution, high-dynamic-range imaging, including circular polarization imaging, in the frequency range (1–20 GHz) where these two mechanisms are important diagnostics of the magnetic field in the solar atmosphere.

Optical depths are generally significant in the solar atmosphere at radio wavelengths and in order to calculate the radio emission arising from either of these physical processes one must carry out a radiative transfer calculation (as for continuum absorption in Section 3.3). It is convenient to do the calculation in terms of brightness temperature, T_B, because radio emission takes place in the Rayleigh–Jeans limit where the effective radiative temperature of an optically thick source is the physical temperature of that source. Brightness temperature may be converted to flux density S via the relation

$$S = k_B \frac{f^2}{c^2} \int T_B \, d\Omega \qquad (3)$$

where k_B is Boltzmann's constant, c is the speed of light and the integral is over the solid angle Ω of interest. Note that brightness temperature is a local quantity whereas flux density is integrated over a source area.

The radiative transfer calculation for radio emission in FORWARD follows standard methods: the brightness temperature transfer is governed by the differential equation (e.g., Dulk, 1985):

$$\frac{dT_B}{ds} = \kappa \, (T_e - T_B) \qquad (4)$$

where κ is the opacity per unit distance s along the line of sight and T_e is the local electron temperature. We solve radiative transfer by determining κ and T_e in each pixel along the line of sight and integrate Equation 4 across each pixel as follows:

$$T_B{}' = T_B\, e^{-d\tau} + T_e\,(1 - e^{-d\tau}) \qquad (5)$$

where T_B is the incident brightness temperature, and $T_B{}'$ is the emergent brightness temperature—integrated across the line-of-sight pixel and serving as the incident brightness temperature to the next pixel. $d\tau = \kappa\, ds$ is the opacity change across the pixel.

Radio emission from the solar atmosphere is strongly influenced by the magnetic field in the emitting regions and provides valuable diagnostics of solar magnetic fields that complement other techniques. The magnetic field plays a role in the absorption coefficients κ: electrons interact more strongly with the sense of circular polarization that matches the sense of rotation of an electron as it spirals along magnetic field lines under Larmor motion. The polarization that interacts more strongly with electrons is the *extraordinary* or *x* mode, with the other polarization being labeled the *ordinary* (*o*) mode. Under most conditions in the solar corona, and following propagation to terrestrial observers, the *x* and *o* modes are 100% circularly polarized with opposite sense of polarization. FORWARD solves the radiative transfer equations as described above for each of the circular polarizations separately. The difference between the *x* and *o* modes is then Stokes V (modulo a sign), while the sum is the total intensity, Stokes I (For radio emission from the solar atmosphere, we may ignore any weak linear polarization present due to the fact that the large Faraday rotation in the solar atmosphere wipes out linear polarization over a finite observing bandwidth, see below).

For thermal bremsstrahlung, which is always included in a FORWARD radio emission calculation, opacity results from collisions between electrons and ions. We use the simple expression (Dulk, 1985; Gelfreikh, 2004)

$$\kappa = 0.2\,\frac{n_e^2}{T_e^{1.5}\,(f \pm f_B|\cos\theta|)^2} \qquad (6)$$

which is appropriate for coronal temperatures, where $f_B = 2.8 \times 10^6 B_{\text{gauss}}$ Hz is the electron gyrofrequency and the factor in parentheses deals with polarization (with the assumption that $f \gg f_B$): θ is the angle between the magnetic field direction and the line of sight, and the minus sign refers to the *x* mode while the plus sign refers to the *o* mode. Thus, the magnetic field information present in bremsstrahlung emission resides in the circular polarization and represents the line-of-sight component of *B*.

The dependencies in Equation 6 mean that bremsstrahlung is strongly favored in dense regions of the atmosphere and weighted toward cooler material (since in Equation 4, $\kappa T_e \propto T_e^{-0.5}$, e.g., White, 2000). The f^{-2} dependence of bremsstrahlung opacity also means that optical depth decreases rapidly as frequency increases, and at low frequencies one is likely to be optically thick such that the lower the frequency, the higher in the atmosphere one sees. This is evident in **Figure 8**, where polarization extends much higher above the photosphere at 100 MHz (lower left panel) than at 1000 MHz (lower right panel). When optically thick,

the circular polarization produced by bremsstrahlung emission actually depends on the presence of a temperature gradient. If one has well–calibrated brightness temperature measurements across a continuous frequency range, one can in fact determine both the temperature gradient and the magnetic field from the data (Grebinskij et al., 2000, disproving a comment in White, 2000). In **Figure 7** the lower degrees of polarization over much of the disk at 100 MHz reflect the fact that the temperature gradient is weaker higher in the corona.

The gyroresonance calculation is more complex. Gyroresonance opacity results from the acceleration of electrons in a magnetic field under the Lorentz force, and is only significant in narrow layers where the observing frequency f is a low integer multiple s of the electron gyrofrequency f_B (e.g., White and Kundu, 1997). The optical depth τ of a thermal gyroresonance layer (the absorption coefficient integrated through the layer) is

$$\tau_{x,o}(s,f,\theta) \propto \frac{n_e\, L_B(\theta)}{f}\,\frac{s^2}{s!}\left(\frac{s^2\sin^2\theta}{2\,\mu}\right)^{s-1} F_{x,o}(\theta) \qquad (7)$$

where $L_B(\theta)$ is the scale length of the magnetic field $(B/\frac{\partial B}{\partial l})$ evaluated along the line of sight and $\mu = m_e c^2/k_B T_e$. For coronal conditions $\mu \approx 2000$, and the μ^{-s} dependence in Equation 7 produces a dramatic change in opacity as harmonic number s changes. $F_{x,o}(\theta)$ is a function of angle which is of order unity for the *x* mode near $\theta = 90°$, but decreases sharply at smaller θ, and is smaller in the *o* mode than in the *x* mode. FORWARD uses a more exact approximation for $\tau_{x,o}$ due to Robinson and Melrose (1984) which requires a careful calculation of the cold plasma properties of the electromagnetic modes under the conditions that apply in the gyroresonance layer. FORWARD incorporates gyroresonance emission by testing for harmonic layer crossings along the line of sight in the range $s = 1$ to 5, and calculating the resulting opacity as shown above: significant gyroresonance opacity at higher harmonics generally requires mildly relativistic electrons which puts emission in the gyrosynchrotron limit in which harmonics are much broader and Equation 7 is no longer valid [Note the simulation package *GX_Simulator* can handle gyrosynchrotron emission (Nita et al., 2015)].

Gyroresonance emission is most commonly seen in the strong magnetic fields above solar active regions. At frequencies above a few GHz, bremsstrahlung does not produce enough opacity to make the corona optically thick, while the large change in gyroresonance opacity as s decreases (typically a factor of order 1000) means that a given harmonic layer is usually either very optically thick or very optically thin. When optically thick, gyroresonance produces million-K coronal brightness temperature features in radio images. In practice, we see down to the highest optically thick layer (usually $s = 3$ in *x* mode and $s = 2$ in *o* mode), and the brightness temperature variations across the surface (of constant field strength for a given frequency) represent actual temperature variations across that surface. Thus, for this mechanism the magnetic field information contained in the emission morphology is somewhat complex and does not simply reside in the polarization (White and Kundu, 1997).

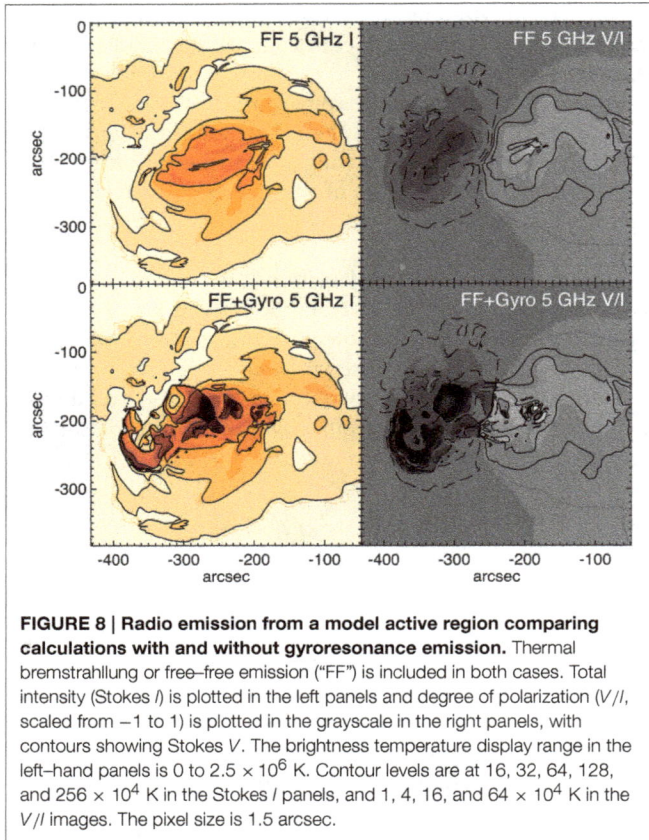

FIGURE 8 | Radio emission from a model active region comparing calculations with and without gyroresonance emission. Thermal bremsstrahllung or free–free emission ("FF") is included in both cases. Total intensity (Stokes *I*) is plotted in the left panels and degree of polarization (*V/I*, scaled from −1 to 1) is plotted in the grayscale in the right panels, with contours showing Stokes *V*. The brightness temperature display range in the left-hand panels is 0 to 2.5×10^6 K. Contour levels are at 16, 32, 64, 128, and 256×10^4 K in the Stokes *I* panels, and 1, 4, 16, and 64×10^4 K in the *V/I* images. The pixel size is 1.5 arcsec.

Internally, FORWARD carries out radiative transfer for radio emission in the *x* and *o* modes by summing the requested absorption coefficients in each pixel: bremsstrahlung is the default opacity and gyroresonance opacity may be turned on or off. The brightness temperatures in the two modes are summed and differenced to report Stokes I and V, or one can display V/I, the degree of circular polarization, as in **Figures 7, 8**. Mode coupling between the *x* and *o* modes, which can result in reversal in the sense of circular polarization at points where the magnetic field direction along the line of sight reverses (e.g., White et al., 1992), is not yet included in the FORWARD calculation but will be in future releases. **Figure 8** shows an example of a FORWARD radio emission calculation using a three-dimensional hydrodynamic active region model (density, vector magnetic field and temperature) with thermal conduction and radiative cooling (Lionello et al., 2013). The upper panels show the model radio emission obtained with just bremsstrahlung opacity included, while the lower panels include gyroresonance opacity. The brightness temperatures are much higher when gyroresonance opacity is included, and the polarization structure becomes more complex.

FORWARD does not currently include plasma emission: this is the dominant emission in low–frequency solar radio bursts (e.g., Kundu, 1965), but it is a coherent emission mechanism and there is no simple way to calculate it (e.g., see Schmidt and Cairns, 2012a,b, for a detailed calculation). In addition, fundamental plasma emission occurs at frequencies where the

refractive index may be significantly different from unity and refraction can play a major role in determining ray paths. FORWARD assumes linear ray paths along lines of sight and does not currently handle refraction at low radio frequencies which will produce curved ray paths in a realistic solar atmosphere.

3.7. Faraday Rotation

In general radio emission can be elliptically polarized. Electromagnetic radiation in a magnetized plasma can be decomposed into two natural modes with orthogonal polarizations, and as radiation propagates the two intrinsic polarizations have different refractive indices and slightly different phase speeds. This effect causes the plane of linear polarization to rotate, with the amount of rotation being a function of frequency. In the lower regions of the solar atmosphere the rotation is so large that, as mentioned in the previous subsection, when averaged across a finite observing bandwidth the linear polarization is washed out. However, further out in the solar wind where the magnetic field is lower, Faraday rotation can be measured as a function of frequency, and such measurements are one of the few techniques that can be used as a remote probe of the magnetic field in the solar wind. This technique has been applied to both communication transmissions from satellites (e.g., Bird et al., 1985; Jensen et al., 2005, 2013) as well as polarized background cosmic sources (e.g., Bird et al., 1980; Mancuso and Spangler, 2000; Ord et al., 2007; You et al., 2012; Kooi et al., 2014). In FORWARD it can be used to simulate the contributions of CME and solar wind plasma to an observable diagnostic of the magnetic field.

The expression for Faraday rotation is relatively straightforward: the angle of rotation *FR* (in radians) at wavelength λ is

$$FR = RM \lambda^2 \qquad (8)$$

where the rotation measure *RM* (measured in radians per square meter) is the wavelength-independent measure of Faraday rotation, calculated from the integral of the product of electron density and line–of–sight magnetic field along the ray path:

$$RM = 2.6 \times 10^{-13} \int n_e B \cos\theta \, ds \; (\text{rad m}^{-2}) \qquad (9)$$

with *B* measured in Gauss and electron density in cm^{-3}. FORWARD carries out this integral and will report either *FR* at a specific frequency or *RM*.

A related and useful quantity that is obtained in conjunction with Faraday rotation measurements of pulsars is dispersion measure,

$$DM = \int n_e \, ds \qquad (10)$$

which in radio astronomy is usually measured in units of cm^{-3} parsecs. The dispersion measure is used to remove the frequency-dependent delay of pulsar pulses introduced by the variation of refractive index in the interstellar and interplanetary media with frequency, so that pulses can be aligned across the full observing bandwidth, a necessary step in measuring the rotation of the plane of polarization vs. frequency. FORWARD provides the same quantity but referred to as a column density (accessed via keyword /colden or in the Physical Diagnostics drop-down menu of the widget, in units of cm^{-2}). Variations in DM on timescales of tens of minutes to hours are dominated by density variations in the solar wind, and knowledge of DM is valuable when trying to assess the relative roles of density and magnetic field in observed RM variability.

4. OBSERVATIONS

As described above in Section 3 and summarized in **Table 1**, multiple physical processes operating in the corona have sensitivities to the coronal magnetic field and manifest observable signatures from radio to soft-Xray emission. There is thus clear value in obtaining observations at a broad range of wavelengths for intercomparison and use in constraining and defining models.

To facilitate model-data comparisons, FORWARD enables the access and manipulation of observations in a form designed to match the output of forward calculations. To this end, FORWARD extracts SolarSoft IDL maps from FITS-format observational data files, and preserves these along with associated structures in a standard format. Observational data are accessed either "by date" or "by file"—via keywords if using the command-line version of FORWARD, or a calendar/directory search if

using the widget interface (see **Figure 9**). Note that FORWARD looks first in a user-defined working directory for existing FORWARD-formatted maps or fits files before downloading or processing new files.

Since the focus of this paper is magnetometry, we first describe how FORWARD enables access and manipulation of Stokes polarimetric data. The Mauna Loa Solar Observatory (MLSO) Coronal Multi-channel Polarimeter (CoMP) on Hawaii (Tomczyk et al., 2008) is a 20-cm aperture coronagraph with a full field of view of the corona from 1.05 to 1.38 solar radii. It utilizes a narrow-band imaging polarimeter to observe the Fe XIII coronal line at 1074.7 and 1079.8 nm and the chromospheric HeI line at 1083 nm. Data products currently served include intensity, Doppler velocity, line width, and Stokes linear polarization [Q, U as well as $L = (Q^2 + U^2)^{\frac{1}{2}}$ and $Az = 0.5 * atan(\frac{U}{Q})$]. CoMP linear polarization is currently the only direct magnetic diagnostic of the corona publicly available on a near-daily basis (subject to weather, etc.). These data are available online, beginning from May 2011, and can be downloaded in FITS or image format via the MLSO web pages (*http://www2.hao.ucar.edu/mlso*).

FORWARD offers another means of downloading CoMP data, and moreover acts as a tool for its display and analysis (see Gibson, 2015b for further details). **Figures 10B,D** illustrate FORWARD linear polarization output given a specified calendar date and field of view. In this case, CoMP standard "Quick Invert" data file for that date is automatically accessed, which represents an averaged image and may not include all CoMP data products.

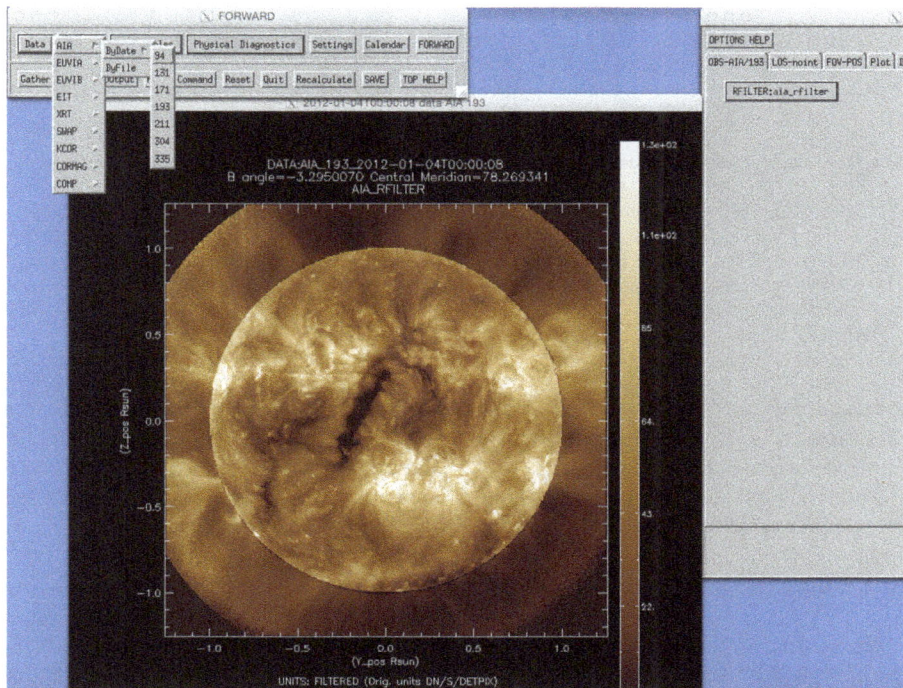

FIGURE 9 | Example of automatically uploading SDO/AIA data via FORWARD widget by date access. Data are accessed via either the Virtual Solar Observatory SolarSoft interface, or an observatory's own web interface. Data may also be uploaded by file from locally stored FITS files, and in some cases (e.g., the CorMag telescope—Fineschi et al., in preparation), this is currently the only available choice.

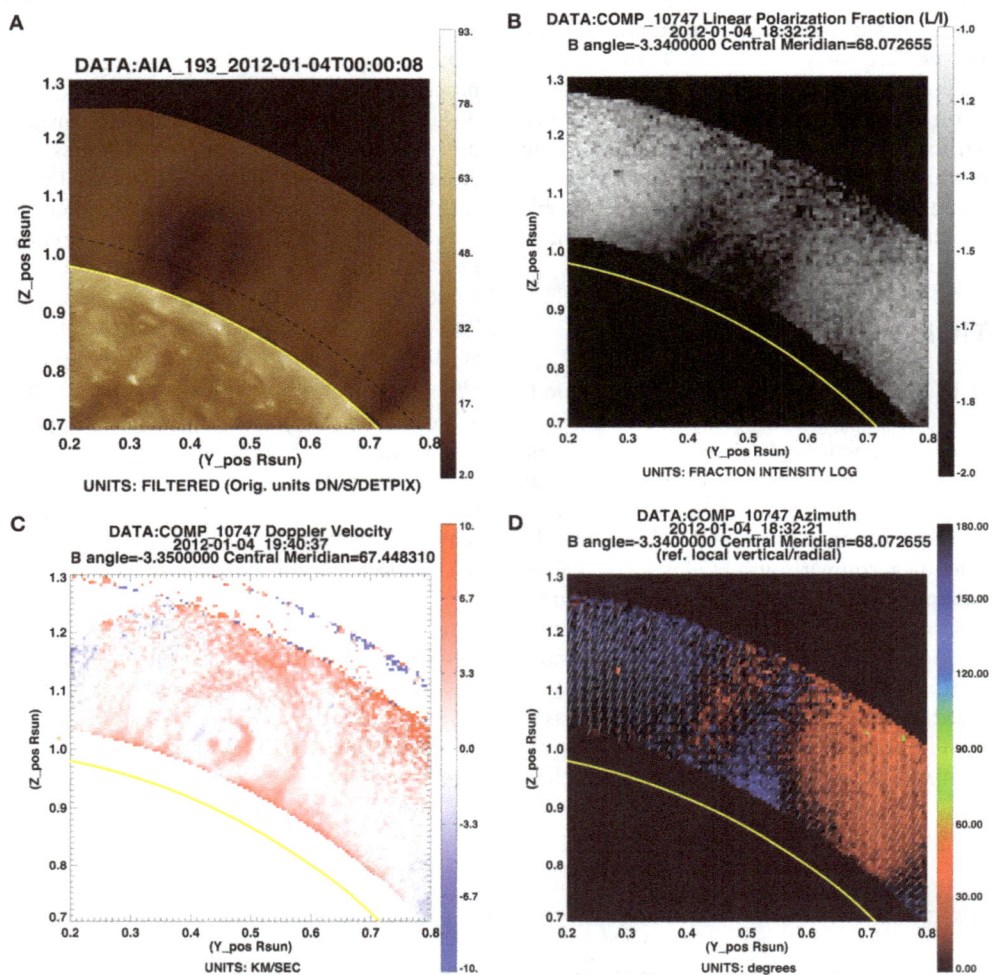

FIGURE 10 | Example of FORWARD-displayed data products for a coronal cavity. (A) SDO/AIA 193 Angstrom. Plot obtained via widget as described in the text, or through IDL line command `for_plotfits,date='2012-01-04',/aia,xxmin=.2,xxmax=.8,yymin=.7,yymax=1.3,occult=-1.05, upoccult=1.29`. **(B)** CoMP fraction of linearly-polarized light *L*/*I*. Plot obtained via widget and utilizing `moreplots` option as described in the text, or command as in **(A)** substituting `/comp` for `/aia`, removing `occult` and `upoccult` keywords, and adding `line='LoI',imin=-2.,imax=-1.`. **(C)** CoMP Doppler velocity (partially corrected for solar rotation Tian et al., 2013). Plot displays data from CoMP Dynamics fits file downloaded from MLSO web page (http://www2.hao.ucar.edu/mlso), through widget "By File" option or command as in **(A)** substituting `/comp`, removing `date` and `occult`, `upoccult` keywords, and adding `filename='20120104.194037.comp.1074.dynamics.3.fts',line='DOPPLERVLOS'`. **(D)** CoMP linear polarization azimuth. Plot obtained by widget or command line as in **(A)**, but substituting `/comp`, and adding `line='Az'`.

Comprehensive, non-averaged data are available in the "Daily Dynamics" and "Daily Polarization" FITS archives on the MLSO web pages, and once these are downloaded to a local directory they may be displayed "by file" using FORWARD. This has been done to show the CoMP Doppler velocity image of **Figure 10C**. White light data from the MLSO K-coronagraph (KCOR) can similarly be downloaded and displayed through the FORWARD widget tools.

A range of extreme ultraviolet (EUV) and soft Xray (SXR) imager data is available through the Virtual Solar Observatory (VSO; Hill et al., 2009) and accessed by FORWARD. These include data from the currently operating Solar Dynamics Observatory Atmospheric Imaging Assembly (SDO/AIA) and Hinode X-ray Telescope (XRT), along with prior data from the

Solar and Heliospheric Observatory Extreme ultraviolet Imaging Telescope (SOHO/EIT) and Transition Region and Coronal Explorer (TRACE), which provide observations at wavelengths spanning coronal and transition region temperatures. Data from the ProbA-2 "Sun Watcher using APS and Image Processing" (SWAP) EUV imager (Halain et al., 2013; Seaton et al., 2013) provide an extended (54 arcminute) field of view (FOV), and the Solar-Terrestrial Relations Observatory Extreme Ultraviolet Imagers (STEREO/EUVIA and EUVIB) provide additional viewing options for EUV coronal structures.

It is particularly simple to intercompare observations using the widget interface. For example, AIA data may be loaded by date as in **Figure 9**, and then a particular structure such as the cavity shown in **Figure 10A** may be zoomed in on using the field of view (FOV-POS) tab of the right-hand widget. By switching on the keyword moreplots (located in the Output tab of the top-left widget), the zoomed-in FOV is retained and closest date/time sought in subsequent loading of CoMP data (b-d). This capability extends to forward-modeled synthetic data: if the moreplots option is turned on and followed by choice of a model and a click on the FORWARD button (top left widget), the field of view, viewer's position, and observable (instrument and line) are all preserved in subsequent forward calculations unless explicitly changed. In this manner, as demonstrated by **Figures 11, 12** and discussed in the next section, model predictions may be directly compared to observations.

5. MULTIWAVELENGTH MAGNETOMETRY

Having described how FORWARD incorporates the three essential components of physical state, physical process, and observation, we now discuss how it may be used to further

multiwavelength magnetometry. FORWARD can be applied to validating models, to building intuition into how magnetic fields manifest in observations, to forward-fitting models to data, and ultimately, to developing coronal magnetic inversion methods that take full advantage of multiwavelength observations.

5.1. Comparing Models and Data

Figure 11 shows the CoMP observations for the day profiled in most of the Figures so far, allowing validation of the predictions of the MAS model. Inspection shows that while some regions match very well, others do not. For example, **Figure 12** illustrates that for the southeast quadrant, the open region of diverging field (just below the equator) is well-captured by the model. Some of the details south of this are not captured, but the red-black-blue interface characteristic of a large-scale closed structure is reproduced. In contrast, the northwest quadrant shows considerably more structure in linear polarization observations (**Figure 11A**) than in the model (**Figure 11C**). **Figure 10** further

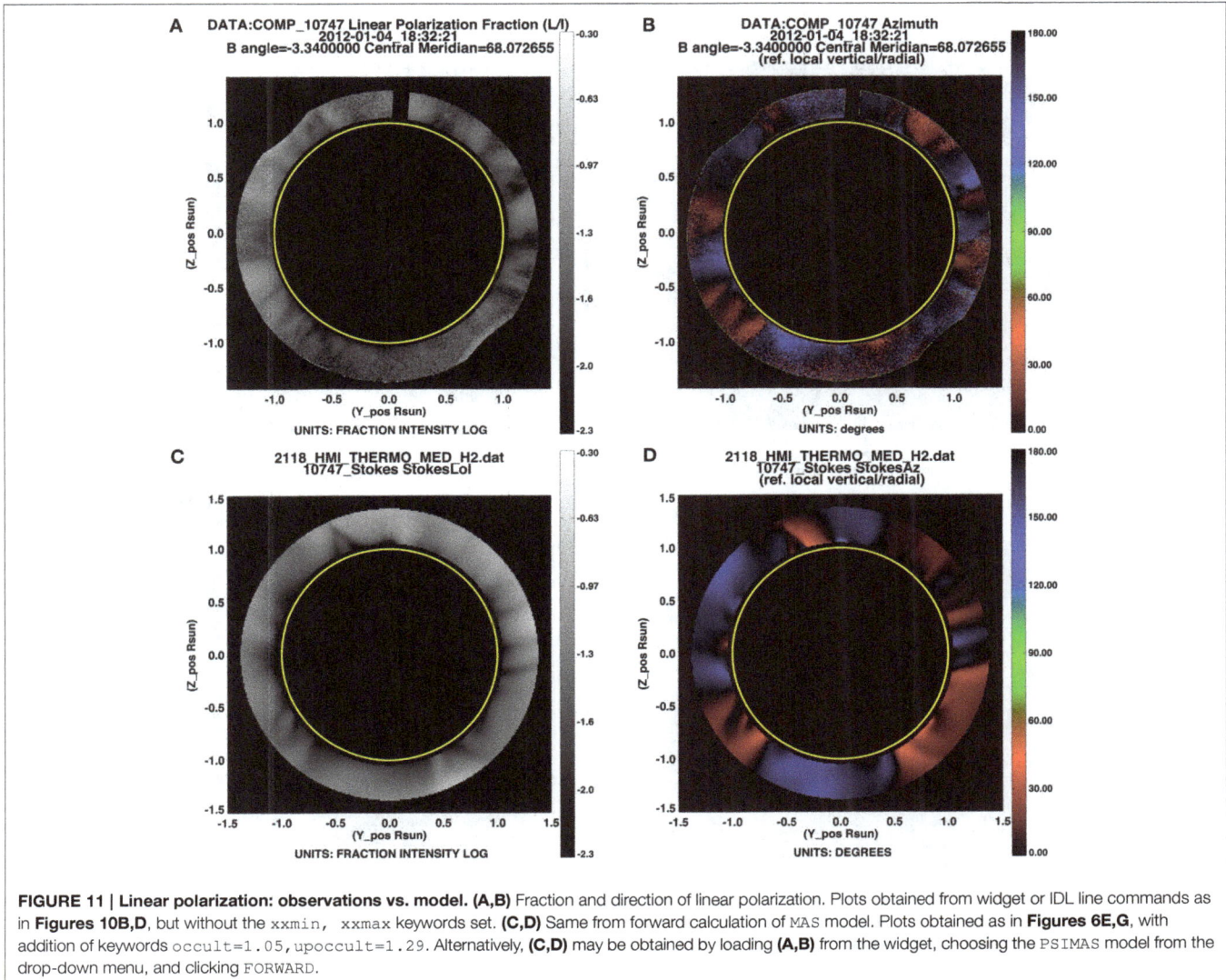

FIGURE 11 | Linear polarization: observations vs. model. (A,B) Fraction and direction of linear polarization. Plots obtained from widget or IDL line commands as in **Figures 10B,D**, but without the xxmin, xxmax keywords set. **(C,D)** Same from forward calculation of MAS model. Plots obtained as in **Figures 6E,G**, with addition of keywords occult=1.05, upoccult=1.29. Alternatively, **(C,D)** may be obtained by loading **(A,B)** from the widget, choosing the PSIMAS model from the drop-down menu, and clicking FORWARD.

magnifies the data in this region, and demonstrates that the northern-most of these linear polarization structures is associated with a coronal cavity. The difference between model and data in this region likely arises because, although the MAS simulation we have shown is non-potential, it does not capture all coronal currents. In particular, currents that slowly build up over time are not reproduced. Such a buildup of currents is expected in polar crown regions (Yeates and Mackay, 2012), and is thus likely in the region of the cavity of **Figure 10**.

The linear polarization observations of the CoMP telescope represent a unique observational resource, and one that has benefited greatly from the intuition built via forward modeling. In advance of CoMP's synoptic operation at MLSO, the CLE code was used to demonstrate that the presence of currents in the corona should be observable in Fe XIII linear polarization (Judge et al., 2006). Indeed, this has proved the case, and comparisons of CoMP data to FORWARD-generated images have shown linear polarization to be a useful diagnostic of magnetic topologies, including spheromaks (Dove et al., 2011),

pseudostreamers (Rachmeler et al., 2014), and cylindrical flux ropes (Bąk-Stęślicka et al., 2013; Rachmeler et al., 2013). The CoMP linear polarization structure shown in **Figure 10B** is an example of a "lagomorph" (named for its rabbit's-head shape). Bąk-Stęślicka et al. (2013, 2014) studied dozens of examples of CoMP lagomorphs, and showed clear correlation with the size and location of associated EUV prominence cavities (e.g., **Figure 10A** vs. **Figure 10B**). The authors also used forward modeling to demonstrate that a magnetic flux rope model results in a lagomorph: the van Vleck angles within the outer portions of the flux rope and the overlying arcade creates a dark structure framing the rabbit's ears and the sides of its head, and sheared or twisted fields at the flux rope's axis, being oriented perpendicular to the plane of sky, are also relatively dark in linear polarization and form the center of the head.

The presence of a wide V shape is expected even for a potential field arcade. Using an analytic model of a flux rope included in FORWARD (Low and Hundhausen, 1995), **Figure 13** shows how the addition of coronal currents above the magnetic neutral

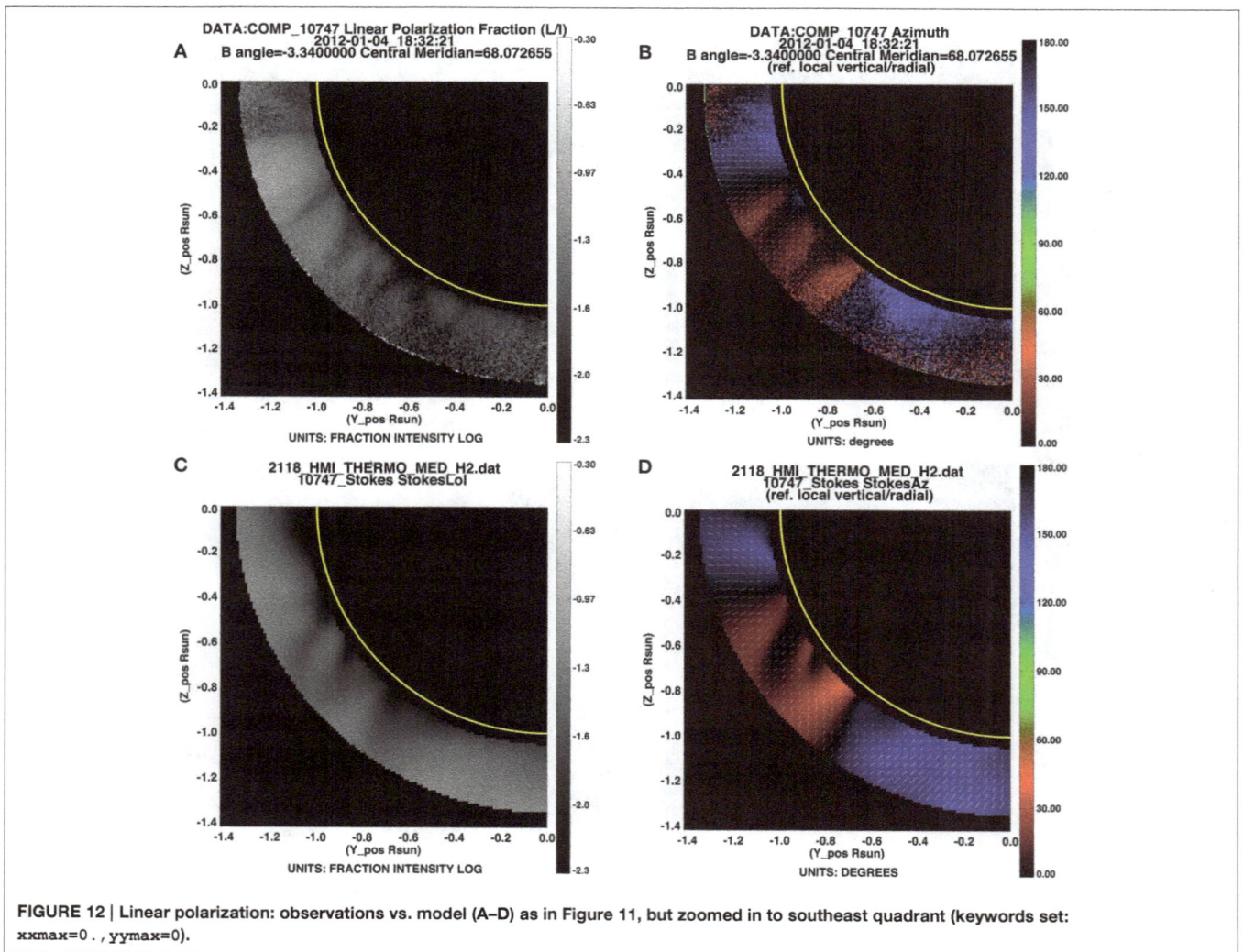

FIGURE 12 | Linear polarization: observations vs. model (A–D) as in Figure 11, but zoomed in to southeast quadrant (keywords set: xxmax=0., yymax=0).

line narrows this V and introduces a dark central structure. The bottom row of **Figure 13** may be compared to **Figure 10**, noting that the top of the cavity/flux rope is near the top of the CoMP field of view, so that the ears are not captured in this case.

As we have discussed above, there are a range of multiwavelength observations that can be used to constrain coronal magnetic fields. Indeed, coronal-cavity white light and emission observations have been interpreted as largely independent indicators of a flux-rope magnetic structure (see

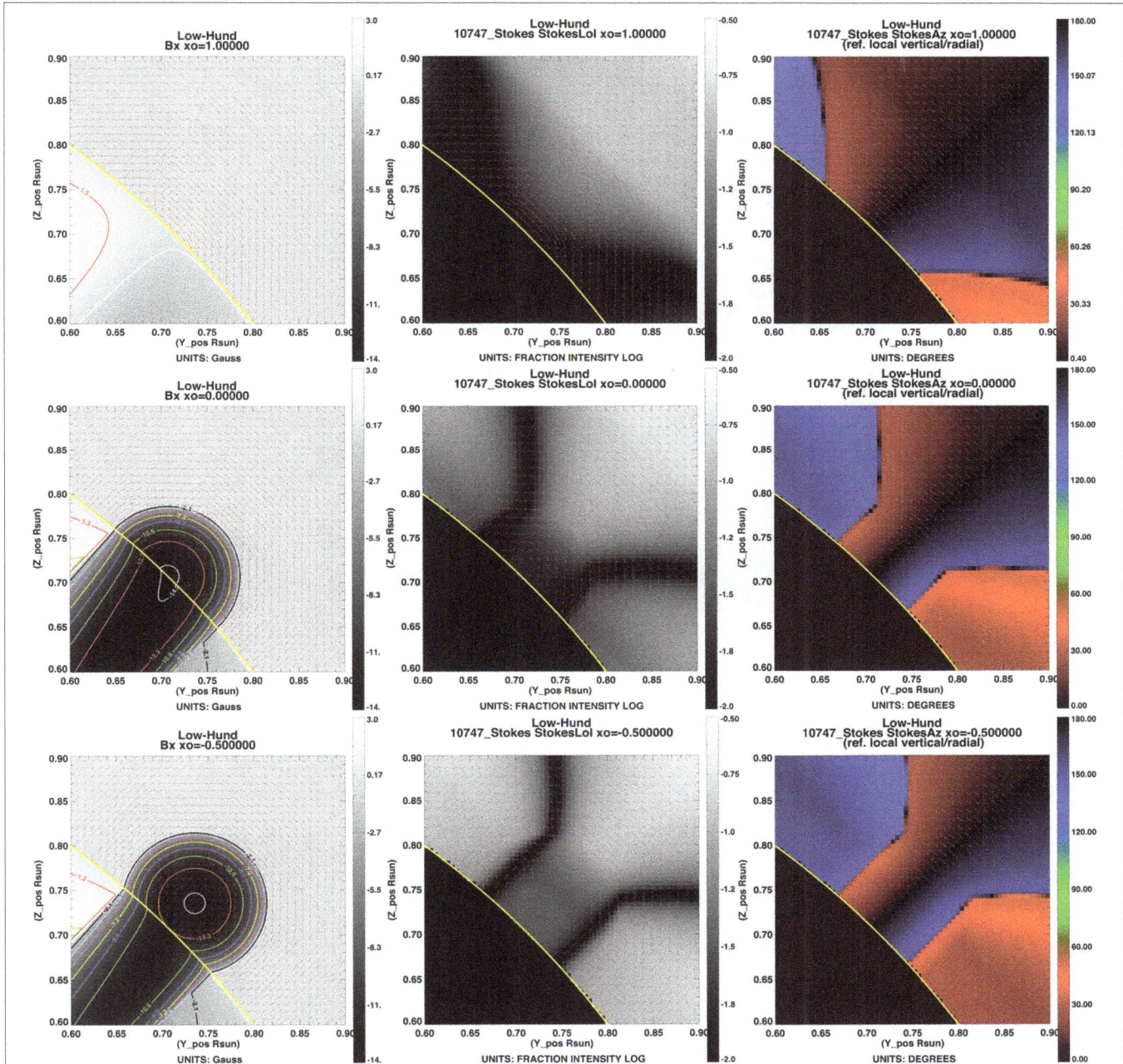

FIGURE 13 | Using an analytic flux-rope model (Low and Hundhausen, 1995), we see that the presence of currents above the underlying neutral line narrows the ears and introduces the dark central head structure to a linear polarization lagomorph. Left column: LOS-oriented magnetic field strength. Plots obtained through commands `for_drive,'lowhund',line='bx',thetao=45.,x_oinput=xo,xxmin=0.6,xxmax=0.9,yymin=0.6,yymax=0.9,/fieldlines, imin=-14,imax=3` for values of `xo = 1.,0.,-.5`. Middle column: LOS-integrated linear polarization fraction. Plots obtained from same commands, substituting `line='LoI',imax=-.5,imin=-2.` and adding keyword `/comp`. Right column: LOS-integrated linear polarization direction (azimuth). Plots obtained as for L/I, but substituting `line='Az'`.

discussion in Gibson, 2014, 2015a; see also Bąk-Stęślicka et al., in preparation for discussion of LOS flows as indications of magnetic-flux-rope topology). Quantification of the three-dimensional morphology, substructure, and plasma properties of cavities have served to justify such interpretations. These quantifications were obtained by fitting the "CAVMORPH" analytic model included within the FORWARD distribution to observations of cavities in white light, EUV, and SXR (Gibson et al., 2010; Schmit and Gibson, 2011; Kucera et al., 2012; Reeves et al., 2012).

Such "forward fitting" goes beyond intuition building, and in fact is a means of inverting observations to quantify properties of the physical state. It does require specification of a parameterized model, such that through iteration best-fit parameters are determined. Dalmasse et al. (in preparation) provides an example

of a statistical method applied to forward fitting a flux-rope model to visible/IR polarimetric data, including both linear and circular polarization (see also Jibben et al., submitted). Other inversion methods applicable to these data are also under development (Kramar et al., 2006, 2013, 2014; Plowman, 2014; also Kramar et al., in preparation).

5.2. Synthetic Testbeds and Beyond

The method described in Dalmasse et al. (in preparation) employs synthetic Fe XIII linear polarization data generated using FORWARD. This work represents an area of active development for FORWARD, i.e., the creation of multiwavelength synthetic data for coronal magnetic structures ranging from active regions (e.g., M. Rempel, private communication), to polar crown prominence/cavity systems

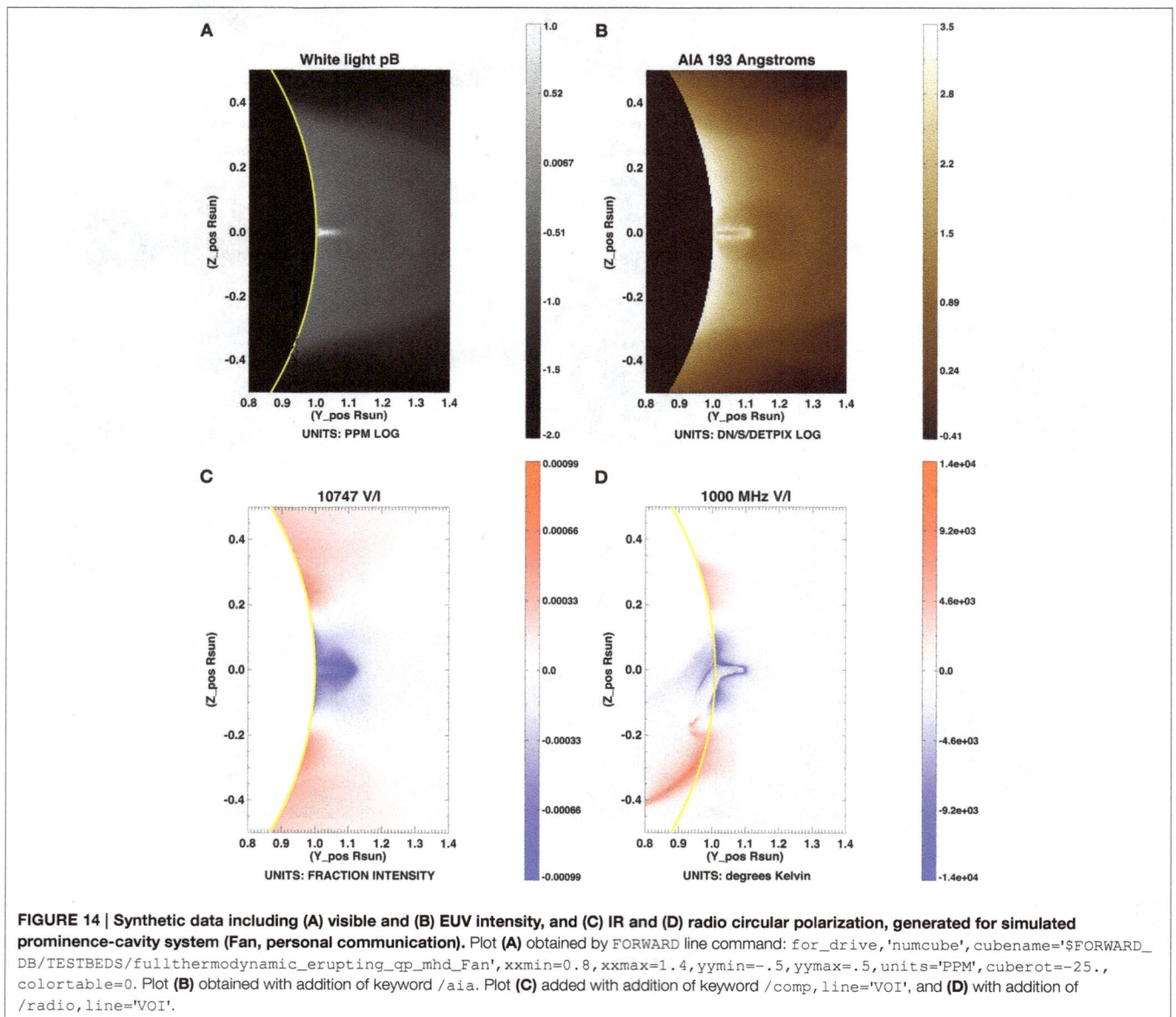

FIGURE 14 | Synthetic data including (A) visible and (B) EUV intensity, and (C) IR and (D) radio circular polarization, generated for simulated prominence-cavity system (Fan, personal communication). Plot (A) obtained by FORWARD line command: for_drive,'numcube',cubename='$FORWARD_DB/TESTBEDS/fullthermodynamic_erupting_qp_mhd_Fan',xxmin=0.8,xxmax=1.4,yymin=-.5,yymax=.5,units='PPM',cuberot=-25.,colortable=0. Plot (B) obtained with addition of keyword /aia. Plot (C) added with addition of keyword /comp,line='VOI', and (D) with addition of /radio,line='VOI'.

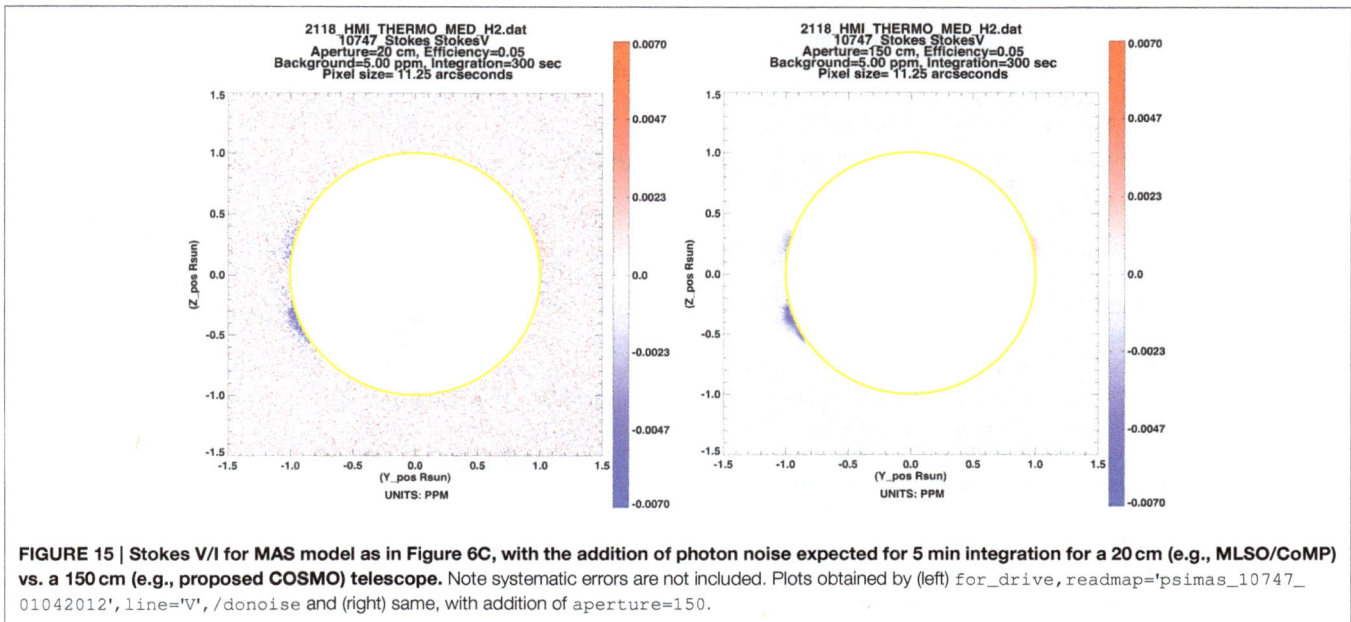

FIGURE 15 | Stokes V/I for MAS model as in Figure 6C, with the addition of photon noise expected for 5 min integration for a 20 cm (e.g., MLSO/CoMP) vs. a 150 cm (e.g., proposed COSMO) telescope. Note systematic errors are not included. Plots obtained by (left) `for_drive,readmap='psimas_10747_01042012',line='V',/donoise` and (right) same, with addition of `aperture=150`.

(e.g., Fan, personal communication; see **Figure 14**, to a global corona containing a variety of currents, e.g., D. Mackay, private communication). These numerical simulations will be included in future FORWARD SolarSoft distributions, and from them synthetic data ranging from radio to SXR wavelengths can be generated as community testbeds to aid in the development of inversion methods and in analyses of the sensitivity of different types of observations to physical parameters.

The images shown in **Figures 1–14** are idealized. Real data has noise, and inversions must take this into account. Sensitivity, field of view and spatial resolution of the telescope used to obtain the data may all contribute to noise. For polarimetry, sensitivity is usually a constraining factor because (at least at optical, IR and UV wavelengths) the polarized signals are much weaker than the total intensities and subject to cross-talk that requires careful time-consuming calibration to ensure robust measurements. With modern telescopes one is also often trading large fields of view for high spatial resolution, and for the observation of spatially extended features this can be a problem.

At radio wavelengths, one has to deal with the fact that spatial resolution is always frequency dependent: for a fixed effective aperture dimension, the size of a resolution element is inversely proportional to frequency, and for typical modern radio observations taken over a wide frequency range, the spatial resolution can vary by factors of several from high to low frequencies. For the purpose of measuring coronal magnetic fields with gyroresonance emission where field strength is proportional to frequency, this means that one generally has poorer spatial resolution for the study of weak fields than for strong fields. Fortunately this is in the right direction since strong-field regions are usually smaller, but it does limit our ability to study three-dimensional fields with uniform resolution.

At the moment, FORWARD only implements noise for visible/IR spectropolarimetry (Section 3.4), and then only photon

noise (see **Figure 15**). Efforts are underway to allow "instrument personality profiles" in which a loss of resolution appropriate to a particular observation could be overlaid on the forward calculation, given details of a telescope and its observing configuration. This capability would enable the design and use of future large telescopes such as the Daniel K. Inouye Solar Telescope (DKIST), the Frequency Agile Solar Telescope (FASR), and the Coronal Solar Magnetism Observatory (COSMO; e.g., **Figure 15**; see also Lin, 2016).

6. CONCLUSIONS

Our primary motivation in developing FORWARD has been to enable multiwavelength coronal magnetometry. The coronal magnetic field lies at the heart of many of the mysteries of solar physics, including coronal heating, solar wind acceleration, and flare and coronal mass ejection onset and evolution. It holds the key to progress in predictive capability for space-weather events: in particular, the direction of the magnetic field at 1 AU depends crucially on the magnetic field at its coronal source, and on the context of this source in both time and space. In this paper we have demonstrated how different physical processes effectively highlight different aspects of the coronal magnetic field, and how these manifest in observations at different wavelengths. Because the photospheric magnetic field is not force-free, our ability to find a meaningful solution to coronal magnetic field through extrapolations from this boundary is limited (De Rosa et al., 2009). We therefore *must* make use of multiwavelength observations of the solar atmosphere to further constrain the global coronal magnetic field.

FORWARD represents a community effort to design and gather a library of codes for the synthesis of multiwavelength coronal data from physical models. Our philosophy has been to incorporate as many existing resources as possible, and to make use of the comprehensive and ever-growing resources available via SolarSoft IDL. We note complementary capabilities available for forward modeling in radio wavelengths, i.e., the *GX_Simulator* package referred to above in Section 3.6 (Nita et al., 2015), and for forward modeling coronal waves, i.e., the *FoMo* codes described in van Doorsselaere et al. (2016).

FORWARD continues to be developed. New subroutines for ultraviolet spectropolarimetry in the unsaturated Hanle regime are being tested (Fineschi, 2001; see also Raouafi et al., submitted; Dima et al., submitted). We are also expanding our numerical interface to allow varied-grid models (currently numerical datacubes must be on a regular grid). A future goal will be to add capability for synthesizing heliospheric images, which would complement current capability for Faraday rotation, and enable connections between imaging and *in situ* observations during the era of Solar Probe and Solar Orbiter. The wide variety of multiwavelength data currently and soon to be available, in combination with ongoing efforts to develop comprehensive and efficient inversion methods, makes us confident that ultimately the goal of quantifying the coronal magnetic field will be achieved.

AUTHOR CONTRIBUTIONS

SG was primary author of this paper, wrote or contributed to most of the subroutines in FORWARD, and is responsible for the oversight of its ongoing development. TK wrote Sections 3.2, 3.3, and contributed to Sections 5, 6. She wrote or otherwise coordinated all FORWARD codes and databases related to EUV/SXR imaging and spectroscopy, and contributed to several other FORWARD subroutines. SW wrote Sections 3.6, 3.7, and contributed to Section 5, 6. He wrote or otherwise coordinated all FORWARD codes related to radio spectropolarimetry. JD was the main author of the integration of CLE into FORWARD, and the initial developer of several of the backbone subroutines of FORWARD as well as the analytic model interface. LR developed the numerical interface and incorporated the PFSS model into FORWARD, and contributed to the data-download interface codes and other FORWARD subroutines. YF contributed to the numerical interface effort, provided an axisymmetric flux rope model to the FORWARD package for demonstration purposes, and is providing numerical simulations for synthetic test-beds as discussed in Section 5. BF was the initial developer of the `for_widget` interface. CD contributed a subroutine for interfacing with the PSI MAS model and assisted with its implementation. KR contributed the XRT response subroutine. All authors read and critically revised the paper, approved the final version, and agreed to be accountable for all aspects of the work.

FUNDING

NCAR is supported by the National Science Foundation. SG acknowledges support from the Air Force Office of Space Research, FA9550-15-1-0030. TK acknowledges support from NASA. LR acknowledges support from the Belgian Federal Science Policy Office (BELSPO) through the ESA-PRODEX program, grant No. 4000103240. KR acknowledges support from contract NNM07AB07C from NASA to SAO.

ACKNOWLEDGMENTS

We thank Phil Judge and Roberto Casini for their assistance with the utilization and interpretation of the FORTRAN CLE code incorporated within FORWARD, and Roberto Casini for his internal review of this paper. We thank the International Space Science Institute (ISSI) teams on coronal cavities (2008–2010) and coronal magnetism (2013–2014) who were key to initiating and guiding FORWARD development efforts. In particular, we thank Tim Bastian for contributions to the radio codes within FORWARD, and Jeff Kuhn for interesting discussions pertaining to instrument personality profiles (currently under development). We acknowledge Matthias Rempel and Duncan Mackay for contributions of synthetic test-beds for FORWARD. We also acknowledge the Data-Optimized Coronal Field Model (DOC-FM) team and especially Kevin Dalmasse, Silvano Fineschi, Natasha Flyer, Nathaniel Mathews, and Doug Nychka for current engagement in FORWARD development, and Haosheng Lin, Don Schmit, and Chris Bethge for past code contributions. In addition, we acknowledge Enrico Landi for assistance with CHIANTI interfacing, Dominic Zarro for assistance with VSO interfacing, and Sam Freeland for assistance with SolarSoft interfacing. FORWARD makes use of a range of SolarSoft codes developed by others—we are grateful to Marc de Rosa for his PFSS SolarSoft codes, and Andrew Hayes for his routines for the F corona. We acknowledge Leonard Sitongia for initial help with MLSO data access, and Joan Burkepile, Giuliana de Toma, and Steve Tomczyk for ongoing assistance. MLSO is operated by the High Altitude Observatory, as part of the National Center for Atmospheric Research (NCAR). AIA data are courtesy of NASA/SDO and the AIA, EVE, and HMI science teams.

REFERENCES

Bąk-Stęślicka, U., Gibson, S. E., Fan, Y., Bethge, C., Forland, B., and Rachmeler, L. A. (2013). The magnetic structure of solar prominence cavities: new observational signature revealed by coronal magnetometry. *Astrophys. J.* 770:28. doi: 10.1088/2041-8205/770/2/L28

Bąk-Stęślicka, U., Gibson, S. E., Fan, Y., Bethge, C., Forland, B., and Rachmeler, L. A. (2014). The spatial relation between EUV cavities and linear polarization

signatures. *Proc. Int. Astronom. Union* 300, 395–396. doi: 10.1017/S1743921 313011253

Billings, D. E. (1966). *A Guide to the Solar Corona.* New York, NY: Academic Press.

Bird, M. K., Schruefer, E., Volland, H., and Sieber, W. (1980). Coronal faraday rotation during solar occultation of PSR0525 + 21. *Nature* 283, 459. doi: 10.1038/283459a0

Bird, M. K., Volland, H., Howard, R. A., Koomen, M. J., Michels, D. J., Sheeley, Jr., N. R., et al. (1985). White-light and radio sounding observations of coronal transients. *Solar Phys.* 98, 341–368. doi: 10.1007/BF001 52465

Caffau, E., Ludwig, H.-G., Steffen, M., Freytag, B., and Bonifacio, P. (2011). Solar chemical abundances determined with a CO5BOLD 3D model atmosphere. *Solar Phys.* 268, 255–269. doi: 10.1007/s11207-010-9541-4

Casini, R., and Judge, P. G. (1999). Spectral Lines for Polarization Measurements of the Coronal Magnetic Field. II. Consistent Treatment of the Stokes Vector forMagnetic-Dipole Transitions. *Astrophys. J.* 522, 524. doi: 10.1086/307629

De Rosa, M. L., Schrijver, C. J., Barnes, G., Leka, K. D., Lites, B. W., Aschwanden, M. J., et al. (2009). A critical assessment of nonlinear force-free field modeling of the solar corona for active region 10953. *Astrophys. J.* 696, 1780–1791. doi: 10.1088/0004-637X/696/2/1780

Del Zanna, G., Dere, K. P., Young, P. R., Landi, E., and Mason, H. E. (2015). CHIANTI - An atomic database for emission lines. Version 8. *Astron. Astrophys.* 582:A56. doi: 10.1051/0004-6361/201526827

Dere, K. P., Landi, E., Mason, H. E., Monsignori Fossi, B. C., and Young, P. R. (1997). CHIANTI - an atomic database for emission lines. *Astron. Astrophys. Suppl.* 125, 149–173. doi: 10.1051/aas:1997368

Dere, K. P., Landi, E., Young, P. R., Del Zanna, G., Landini, M., and Mason, H. E. (2009). CHIANTI - an atomic database for emission lines. IX. Ionization rates, recombination rates, ionization equilibria for the elements hydrogen through zinc and updated atomic data. *Astron. Astrophys.* 498, 915–929. doi: 10.1051/0004-6361/200911712

Dove, J., Gibson, S., Rachmeler, L. A., Tomczyk, S., and Judge, P. (2011). A ring of polarized light: evidence for twisted coronal magnetism in cavities. *Astrophys. J. Lett.* 731:1. doi: 10.1088/2041-8205/731/1/L1

Dulk, G. A. (1985). Radio emission from the sun and stars. *Ann. Rev. Astron. Astrophys.* 23:169. doi: 10.1146/annurev.aa.23.090185.001125

Feldman, U., Mandelbaum, P., Seely, J. F., Doschek, G. A., and Gursky, H. (1992). The potential for plasma diagnostics from stellar extreme-ultraviolet observations. *Astrophys. J. Supp.* 81, 387–408. doi: 10.1086/191698

Fineschi, S. (2001). "Space-based instrumentation for magnetic field studies of solar and stellar atmospheres," in *Magnetic Fields Across the Hertzsprung-Russell Diagram, ASP Conference Proceedings*, Vol. 248, eds G. Mathys, S. K. Solanki, and D. T. Wickramasinghe (San Francisco, CA: Astronomical Society of the Pacific), 597.

Forland, B. F., Gibson, S. E., Dove, J. B., and Kucera, T. A. (2014). The solar physics forward codes: now with widgets. *Proc. Int. Astronom. Union* 300, 414–415. doi: 10.1017/S1743921313011332

Freeland, S. L., and Handy, B. N. (1998). Data analysis with the solarSoft system. *Solar Phys.* 182, 497–500. doi: 10.1023/A:1005038224881

Gelfreikh, G. B. (2004). "Coronal magnetic field measurements through bremsstrahlung emission," in *Solar and Space Weather Radiophysics*, eds D. E. Gary and C. U. Keller (Dordrecht. Kluwer Academic Publishers), 115–133. doi: 10.1007/1-4020-2814-8_6

Gibson, S. E. (2014). "Magnetism and the invisible man: the mysteries of coronal cavities," in *IAU Symposium, Volume 300 of IAU Symposium*, eds B. Schmieder, J.-M. Malherbe, and S. T. Wu, 139–146.

Gibson, S. (2015a). "Coronal cavities: observations and implications for the magnetic environment of prominences," in *Astrophysics and Space Science Library, Vol. 415 of Astrophysics and Space Science Library*, eds J.-C. Vial and O. Engvold (Springer), 323–353. doi: 10.1007/978-3-319-10416-4_13

Gibson, S. (2015b). Data-model comparison using FORWARD and CoMP. *Proc. Int. Astronom. Union* 305, 245–250. doi: 10.1017/S1743921315004846

Gibson, S. E., and Fan, Y. (2006). Coronal prominence structure and dynamics: a magnetic flux rope interpretation. *J. Geophys. Res.* 111:A12103. doi: 10.1029/2006ja011871

Gibson, S. E., Kucera, T. A., Rastawicki, D., Dove, J., de Toma, G., Hao, J., et al. (2010). Three-dimensional morphology of a coronal prominence cavity. *Astrophys. J.* 723:1133. doi: 10.1088/0004-637X/724/2/1133

Gibson, S. E., and Low, B. C. (1998). A time-dependent three-dimensional magnetohydrodynamic model of the coronal mass ejection. *Astrophys. J.* 493, 460. doi: 10.1086/305107

Grebinskij, A., Bogod, V., Gelfreikh, G., Urpo, S., Pohjolainen, S., and Shibasaki, K. (2000). Microwave tomography of solar magnetic fields. *Astron. Astrophys. Supp* 144:169. doi: 10.1051/aas:2000202

Habbal, S. R., Morgan, H., and Druckmüller, M. (2011). "A new view of coronal structures: implications for the source and acceleration of the solar wind," in

Astronomical Society of India Conference Series, Vol. 2 of Astronomical Society of India Conference Series, eds A. R., Choudhuri and D. Banerjee, 259–269.

Halain, J.-P., Berghmans, D., Seaton, D. B., Nicula, B., De Groof, A., Mierla, M., et al. (2013). The SWAP EUV imaging telescope. Part II: in-flight performance and calibration. *solphys* 286, 67–91. doi: 10.1007/s11207-012-0183-6

Hill, F., Martens, P., Yoshimura, K., Gurman, J., Hourclé, J., Dimitoglou, G., et al. (2009). The virtual solar observatory - a resource for international heliophysics research. *Earth Moon Plan.* 104, 315–330. doi: 10.1007/s11038-008-9274-7

Jensen, E. A., Bird, M. K., Asmar, S. W., Iess, L., Anderson, J. D., and Russell, C. T. (2005). The cassini solar faraday rotation experiment. *Adv. Space Res.* 36, 1587–1594. doi: 10.1016/j.asr.2005.09.039

Jensen, E. A., Bisi, M. M., Breen, A. R., Heiles, C., Minter, T., and Vilas, F. (2013). Measurements of faraday rotation through the solar corona during the 2009 solar minimum with the MESSENGER spacecraft. *Solar Phys.* 285, 83–95. doi: 10.1007/s11207-012-0213-4

Judge, P. G., and Casini, R. (2001). "A synthesis code for forbidden coronal lines," in *Advanced Solar Polarimetry—Theory, Observation, and Instrumentation, 20TH NSO/Sac Summer Workshop, ASP Conference Proceedings, Vol. 236, of Astronomical Society of the Pacific Conference Series* (San Francisco, CA: Astronomical Society of the Pacific), 503.

Judge, P. G., Low, B. C., and Casini, R. (2006). Spectral lines for polarization measurements of the coronal magnetic field. iv. stokes signals in current-carrying fields. *Astrophys. J.* 651:1229. doi: 10.1086/507982

Karpen, J. (2014). "Plasma structure and dynamics," in *Solar Prominences, Vol. 415, of Astrophysics and Space Science Library*, eds O. Engvold and J. C. Vial (Springer), 237–257. doi: 10.1007/978-3-319-10416-4_10

Keady, J. J., and Kilcrease, D. P. (2000). "Radiation," in *Allens Astrophysical Quantities*, ed A. N. Cox (New York, NY: AIP), 95–120.

Kooi, J. E., Fischer, P. D., Buffo, J. J., and Spangler, S. R. (2014). Measurements of coronal faraday rotation at 4.6 $R_\odot$. *Astrophys. J.* 784:68. doi: 10.1088/0004-637X/784/1/68

Kopp, R. A., and Holzer, T. E. (1976). Dynamics of coronal hole regions. I - Steady polytropic flows with multiple critical points. *Solar Phys.* 49, 43–56.

Koutchmy, S., and Lamy, P. L. (1985). "The F-corona and the circum-solar dust evidences and properties," in *IAU Colloq. 85: Properties and Interactions of Interplanetary Dust, Vol. 119 of Astrophysics and Space Science Library*, eds R. H. Giese and P. Lamy (Springer), 63–74.

Kramar, M., Airapetian, V., Mikić, Z., and Davila, J. (2014). 3D coronal density reconstruction and retrieving the magnetic field structure during solar minimum. *Solar Phys.* 289, 2927–2944. doi: 10.1007/s11207-014-0525-7

Kramar, M., Inhester, B., Lin, H., and Davila, J. (2013). Vector tomography for the coronal magnetic field. II. Hanle effect measurements. *Astron. Astrophys.* 775, 25. doi: 10.1088/0004-637x/775/1/25

Kramar, M., Inhester, B., and Solanki, S. K. (2006). Vector tomography for the coronal magnetic field. I. Longitudinal Zeeman effect measurements. *Astron. Astrophys.* 456, 665–673. doi: 10.1051/0004-6361:20064865

Kucera, T. (2015). *Solar Prominences, Astrophysics and Space Science Library, Vol. 415*, eds O. Engvold and J. C. Vial (Switzerland: Springer International Publishing Switzerland), 79. doi: 10.1007/978-3-319-10416-4_4

Kucera, T. A., Gibson, S. E., Schmit, D. J., Landi, E., and Tripathi, D. (2012). Temperature and euv intensity in a coronal prominence cavity. *Astrophys. J.* 757:73. doi: 10.1088/0004-637X/757/1/73

Kundu, M. R. (1965). *Solar Radio Astronomy*. New York, NY: Interscience Publishers.

Lin, H. (2016). mxCSM: A 100-slit, 6-wavelength wide-field coronal spectropolarimeter for the study of the dynamics and the magnetic fields of the solar corona. *Front. Astron. Space Sci.* 3:9. doi: 10.3389/fspas.2016.00009

Lionello, R., Linker, J. A., and Mikic, Z. (2009). Multispectral emission of the sun during the first whole sun month. *Astrophys. J.* 690:902. doi: 10.1088/0004-637X/690/1/902

Lionello, R., Winebarger, A. R., Mok, Y., Linker, J. A., and Mikić, Z. (2013). Thermal non-equilibrium revisited: a heating model for coronal loops. *Astrophys. J.* 773:134. doi: 10.1088/0004-637X/773/2/134

Lites, B. W., and Low, B. C. (1997). Flux emergence and prominences: a new scenario for 3-dimensional field geometry based on observations with the advanced stokes polarimeter. *Solar Phys.* 174, 91. doi: 10.1023/A:1004936204808

Low, B. C., and Hundhausen, J. R. (1995). Magnetostatic structures of the solar corona. ii. the magnetic topology of quiescent prominences. *Astrophys. J.* 443, 818. doi: 10.1086/175572

Luna, M., Karpen, J. T., and DeVore, C. R. (2012). Formation and evolution of a multi-threaded solar prominence. *Astrophys. J.* 746, 30. doi: 10.1088/0004-637X/746/1/30

Malanushenko, A., Schrijver, C. J., DeRosa, M. L., and Wheatland, M. S. (2014). Using coronal loops to reconstruct the magnetic field of an active region before and after a major flare. *Solar Stell. Astrophys.* 783, 102. doi: 10.1088/0004-637x/783/2/102

Mancuso, S., and Spangler, S. R. (2000). Faraday rotation and models for the plasma structure of the solar corona. *Astrophys. J.* 539, 480–491. doi: 10.1086/309205

Mann, I. (1992). The solar F-corona - calculations of the optical and infrared brightness of circumsolar dust. *Solar Phys.* 261, 329–335.

Morgan, H., Habbal, S. R., and Woo, R. (2006). The depiction of coronal structure in white-light images. *Astrophysics* 236, 263–272. doi: 10.1007/s11207-006-0113-6

Munro, R. H., and Jackson, B. V. (1977). Physical properties of a polar coronal hole from 2 to 5 solar radii. *Astrophys. J.* 213, 874. doi: 10.1086/155220

Newkirk, G., and Altschuler, M. D. (1970). Magnetic fields and the solar corona. III: the observed connection between magnetic fields and the density structure of the corona. *Solar Phys.* 13, 131–152. doi: 10.1007/BF00963948

Nita, G. M., Fleishman, G. D., Kuznetsov, A. A., Kontar, E. P., and Gary, D. E. (2015). Three-dimensional radio and X-ray modeling and data analysis software: revealing flare complexity. *Astrophys. J.* 799, 236. doi: 10.1088/0004-637X/799/2/236

Ord, S. M., Johnston, S., and Sarkissian, J. (2007). The magnetic field of the solar corona from pulsar observations. *Solar Phys.* 245, 109–120. doi: 10.1007/s11207-007-9030-6

Plowman, J. (2014). Single-point inversion of the coronal magnetic field. *Astrophys. J.* 792, 23. doi: 10.1088/0004-637X/792/1/23

Rachmeler, L. A., Gibson, S. E., Dove, J. B., DeVore, C. R., and Fan, Y. (2013). Polarimetric properties of flux ropes and sheared arcades in coronal prominence cavities. *Solar. Phys.* 288, 617–636. doi: 10.1007/s11207-013-0325-5

Rachmeler, L. A., Platten, S. J., Bethge, C., Seaton, D. B., and Yeates, A. R. (2014). Observations of a hybrid double-streamer/pseudostreamer in the solar corona. *Solar Stell. Astrophys.* 787, L3. doi: 10.1088/2041-8205/787/1/l3

Reeves, K. K., Gibson, S. E., Kucera, T. A., and Hudson, H. S. (2012). Thermal properties of coronal cavities observed with the X-ray telescope on Hinode. *Astrophys. J.* 746:146. doi: 10.1088/0004-637X/746/2/146

Robinson, P. A., and Melrose, D. B. (1984). Gyromagnetic emission and absorption: approximate formulas of wide validity. *Aust. J. Phys.* 37, 675. doi: 10.1071/PH840675

Savcheva, A. A., McKillop, S. C., McCauley, P. I., Hanson, E. M., and DeLuca, E. E. (2013). A new sigmoid catalog from hinode and the solar dynamics observatory: statistical properties and evolutionary histories. *Solar Phys.* 289, 3297–3311. doi: 10.1007/s11207-013-0469-3

Schmelz, J. T., Reames, D. V., von Steiger, R., and Basu, S. (2012). Composition of the solar corona, solar wind, and solar energetic particles. *Astrophys. J.* 755, 33. doi: 10.1088/0004-637X/755/1/33

Schmidt, J. M., and Cairns, I. H. (2012a). Type II radio bursts: 1. New entirely analytic formalism for the electron beams, Langmuir waves, and radio emission. *J. Geophys. Res.* 117, 4106. doi: 10.1029/2011ja017318

Schmidt, J. M., and Cairns, I. H. (2012b). Type II radio bursts: 2. Application of the new analytic formalism. *J. Geophys. Res.* 117, 11104. doi: 10.1029/2012JA017932

Schmit, D. J., and Gibson, S. E. (2011). Forward modeling cavity density: a multi-instrument diagnostic. *Astrophys. J.* 733:1. doi: 10.1088/0004-637X/733/1/1

Seaton, D. B., Berghmans, D., Nicula, B., Halain, J.-P., De Groof, A., Thibert, T., et al. (2013). The SWAP EUV imaging telescope part I: instrument overview and pre-flight testing. *Solar.* 286, 43–65. doi: 10.1007/s11207-012-0114-6

Tian, H., Tomczyk, S., McIntosh, S. W., Bethge, C., de Toma, G., and Gibson, S. (2013). Observations of coronal mass ejections with the coronal multichannel polarimeter. *Solar Phys.* 288, 637–650. doi: 10.1007/s11207-013-0317-5

Tomczyk, S., Card, G. L., Darnell, T., Elmore, D. F., Lull, R., Nelson, P. G., et al. (2008). An instrument to measure coronal emission line polarization. *Solar Phys.* 247:411. doi: 10.1007/s11207-007-9103-6

Tomczyk, S., McIntosh, S. W., Keil, S. L., and Judge, P. G. (2007). Alfven waves in the solar corona. *Science* 317:1192. doi: 10.1126/science.1143304

van Doorsselaere, T., Antolin, P., Yuan, D., Reznikova, V., and Magyar, N. (2016). Forward modeling of EUV and gyrosynchrotron emission from coronal plasmas with FoMo. *Front. Astron. Space Sci.* 3:4. doi: 10.3389/fspas.2016.00004

van Vleck, J. H. (1925). On the quantum theory of the polarization of resonance radiation in magnetic fields. *Proc. Natl. Acad. Sci. U.S.A.* 11:612. doi: 10.1073/pnas.11.10.612

White, S. M. (2000). Radio versus euv/x-ray observations of the solar atmosphere. *Solar Phys.* 190, 309. doi: 10.1023/A:1005253501584

White, S. M., and Kundu, M. R. (1997). Radio observations of gyroresonance emission from coronal magnetic fields. *Solar Phys.* 174, 31. doi: 10.1023/A:1004975528106

White, S. M., Thejappa, G., and Kundu, M. R. (1992). Mode coupling in the solar corona and bipolar noise storms. *Solar Phys.* 138, 163. doi: 10.1007/BF00146202

Xia, C., Keppens, R., Antolin, P., and Porth, O. (2014). Simulating the *in situ* condensation process of solar prominences. *Solar Stell. Astrophys.* 792, L38. doi: 10.1088/2041-8205/792/2/l38

Yeates, A. R., and Mackay, D. H. (2012). Chirality of high-latitude filaments over solar cycle 23. *Solar Stell. Astrophys.* 753, L34. doi: 10.1088/2041-8205/753/2/l34

You, X. P., Coles, W. A., Hobbs, G. B., and Manchester, R. N. (2012). Measurement of the electron density and magnetic field of the solar wind using millisecond pulsars. *Monthly Notices R. Astron. Soc.* 422, 1160–1165. doi: 10.1111/j.1365-2966.2012.20688.x

Conflict of Interest Statement: The authors declare that the research was conducted in the absence of any commercial or financial relationships that could be construed as a potential conflict of interest.

11

Waves and Magnetism in the Solar Atmosphere (WAMIS)

Yuan-Kuen Ko[1]*, John D. Moses[2], John M. Laming[1], Leonard Strachan[1], Samuel Tun Beltran[1], Steven Tomczyk[3], Sarah E. Gibson[3], Frédéric Auchère[4], Roberto Casini[3], Silvano Fineschi[5], Michael Knoelker[3], Clarence Korendyke[1], Scott W. McIntosh[3], Marco Romoli[6], Jan Rybak[7], Dennis G. Socker[1], Angelos Vourlidas[8] and Qian Wu[3]

[1] Space Science Division, Naval Research Laboratory, Washington, DC, USA, [2] Heliophysics Division, Science Mission Directorate, NASA, Washington, DC, USA, [3] High Altitude Observatory, Boulder, CO, USA, [4] Institut d'Astrophysique Spatiale, CNRS Université Paris-Sud, Orsay, France, [5] INAF - National Institute for Astrophysics, Astrophysical Observatory of Torino, Pino Torinese, Italy, [6] Department of Physics and Astronomy, University of Florence, Florence, Italy, [7] Astronomical Institute, Slovak Academy of Sciences, Tatranska Lomnica, Slovakia, [8] Applied Physics Laboratory, Johns Hopkins University, Laurel, MD, USA

Edited by:
*Mario J. P. F. G. Monteiro,
Institute of Astrophysics and Space
Sciences, Portugal*

Reviewed by:
*Gordon James Duncan Petrie,
National Solar Observatory, USA
Robertus Erdelyi,
University of Sheffield, UK
João José Graça Lima,
Institute of Astrophysics and Space
Sciences, Portugal*

*Correspondence:
*Yuan-Kuen Ko
yuan-kuen.ko@nrl.navy.mil*

Specialty section:
*This article was submitted to
Stellar and Solar Physics,
a section of the journal
Frontiers in Astronomy and Space
Sciences*

Comprehensive measurements of magnetic fields in the solar corona have a long history as an important scientific goal. Besides being crucial to understanding coronal structures and the Sun's generation of space weather, direct measurements of their strength and direction are also crucial steps in understanding observed wave motions. In this regard, the remote sensing instrumentation used to make coronal magnetic field measurements is well suited to measuring the Doppler signature of waves in the solar structures. In this paper, we describe the design and scientific values of the Waves and Magnetism in the Solar Atmosphere (WAMIS) investigation. WAMIS, taking advantage of greatly improved infrared filters and detectors, forward models, advanced diagnostic tools and inversion codes, is a long-duration high-altitude balloon payload designed to obtain a breakthrough in the measurement of coronal magnetic fields and in advancing the understanding of the interaction of these fields with space plasmas. It consists of a 20 cm aperture coronagraph with a visible-IR spectro-polarimeter focal plane assembly. The balloon altitude would provide minimum sky background and atmospheric scattering at the wavelengths in which these observations are made. It would also enable continuous measurements of the strength and direction of coronal magnetic fields without interruptions from the day–night cycle and weather. These measurements will be made over a large field-of-view allowing one to distinguish the magnetic signatures of different coronal structures, and at the spatial and temporal resolutions required to address outstanding problems in coronal physics. Additionally, WAMIS could obtain near simultaneous observations of the electron scattered K-corona for context and to obtain the electron density. These comprehensive observations are not provided by any current single ground-based or space observatory. The fundamental advancements achieved by the near-space observations of WAMIS on coronal field would point the way for future ground based and orbital instrumentation.

Keywords: corona, magnetic field, MHD waves, spectroscopy, polarization

INTRODUCTION

Since the advent of the Solar and Heliospheric Observatory (SOHO; Domingo et al., 1995) followed by the Transition Region and Coronal Explorer (TRACE; Handy et al., 1999; Schrijver et al., 1999) and the Solar Dynamics Observatory (SDO; Pesnell et al., 2012), the solar atmosphere has come to be increasingly appreciated as a dynamic and complex environment. The concept of a static quiescent atmosphere and corona has given way to an environment where waves play a much larger role in shaping the plasma properties than hitherto assumed, and can have non-negligible energy densities compared to the thermal gas in the low β corona. Periodic oscillations in the solar atmosphere have long been observed (e.g., Chapman et al., 1972; Roberts et al., 1983; Antonucci et al., 1984; Aschwanden, 1987; Harrison, 1987), and various oscillation modes of coronal loops have been identified (e.g., Aschwanden et al., 1999; Nakariakov et al., 1999; Wang et al., 2009). Compressive waves connected to slow mode or fast mode waves, or their analogs in inhomogeneous media, have been readily detected, but the non-compressive Alfvén wave has proven more elusive. Early claims of Alfvén wave detections (Cirtain et al., 2007; De Pontieu et al., 2007; Tomczyk et al., 2007) have been discussed (Erdélyi and Fedun, 2007; Van Doorsselaere et al., 2008), and as this last reference emphasizes, the realization that Alfvén or fast mode waves (loosely collectively referred to as "Alfvénic" when close to parallel propagation where magnetic tension is the dominant restoring force) are ubiquitous in the solar upper atmosphere (McIntosh et al., 2011) signifies an important new development with profound consequences for our understanding of the corona and solar wind. More recently Jess et al. (2009) have detected Alfvén waves lower in the solar atmosphere.

Comprehensive measurements of magnetic fields in the solar corona have a longer history as an important scientific goal (e.g., Dulk and McLean, 1978; House et al., 1982; Arnaud and Newkirk, 1987; Lin et al., 2004; Tomczyk et al., 2007). As well as being crucial to understanding coronal structures and the Sun's generation of space weather which can affect communications, GPS systems, space flight, and power transmission (Hanslmeier, 2003; Lambour et al., 2003; Iucci et al., 2006), the measurement of its strength and direction is also a crucial step in understanding observed wave motions. Most forms of solar activity, including high energy electromagnetic radiation, solar energetic particles, flares, and coronal mass ejections (CMEs), derive their energy from magnetic fields. The corona is also the most plausible source of the solar wind with its embedded magnetic field that engulfs the Earth. The ability to measure coronal magnetic fields will lead to improved predictions of hazardous space weather effects on Earth because of further understanding of the underlying physical processes.

Magnetic fields in the corona have been extremely difficult to measure for three important reasons: (1) the magnetic fields in the corona are intrinsically weak compared to the rest of the sun; (2) coronal spectroscopic lines are dimmer than their photospheric counterparts; and (3) the optically thin corona requires interpretation of magnetic signatures integrated along extended path lengths. Most knowledge to date has been derived from extrapolations from photospheric magnetograms (see e.g., the review by Wiegelmann and Sakurai, 2012; Régnier, 2013). Recently, the HAO-NCAR Coronal Multi-channel Polarimeter (CoMP) instrument (Tomczyk et al., 2008) made breakthrough measurements of the coronal magnetic field that lead to discoveries of coronal Alfvén waves (Tomczyk et al., 2007), as well as advancement in the magnetic structure in prominences and coronal cavities (Dove et al., 2011; Bak-Steślicka et al., 2013). However, such ground observations are still limited by the sky background, atmospheric seeing effect and the day–night and weather related interruptions. With such observations from space still lacking and the prospect of such instrumentation on a space mission still uncertain, the most sensible way is to take the measurements from above the atmosphere with long-duration balloon flights.

In this paper, we describe the design and scientific values of the Waves and Magnetism in the Solar Atmosphere (WAMIS) investigation. In The Importance of Magnetic Field and Waves Measurements in the Corona we describe the importance of coronal magnetic field and waves measurements in answering current outstanding questions in solar physics. WAMIS Instrument Concept describes the observational requirements, methodology and the WAMIS Instrument Design for making breakthroughs in the coronal field measurements. Concluding remarks gives some concluding remarks.

THE IMPORTANCE OF MAGNETIC FIELD AND WAVES MEASUREMENTS IN THE CORONA

In this section, we describe major outstanding questions in solar physics research that illustrate the importance of direct measurements of the coronal magnetic field in its strength, structure and dynamics.

What Determines the Magnetic Structure of the Corona?

The large-scale coronal structure is a consequence of surface field advection, differential rotation, and photospheric flux emergence. Information on the evolution and interactions between magnetically closed and open regions could shed light into understanding the changing structure of the heliospheric magnetic field and how the slow solar wind is formed. The fast solar wind has been known for some time to originate in open field regions, i.e., coronal holes (e.g., Krieger et al., 1973). The origins of the slow wind are more obscure, but are thought to be at the interface between open and closed field where reconnection opens up previously closed regions (e.g., Fisk and Schwadron, 2001). This idea has been refined recently by Antiochos et al. (2011) in terms of the S-web ("S" stands for separatrix), where extensions from the polar coronal holes reach down to lower latitudes, allowing open field and closed field regions to interact.

Coronal magnetic field measurements would allow a reconstruction of magnetic field in the extended corona from which the topology of the S-web could be estimated. Also, turbulence in the slow wind is known to be more "balanced" than in the fast wind (e.g., Bruno and Carbone, 2005), meaning that the amplitudes of waves propagating in opposite directions along the magnetic field are more nearly equal than in the fast wind. The existing claims for coronal Alfvén wave detections (Cirtain et al., 2007; De Pontieu et al., 2007; Tomczyk et al., 2007; Tomczyk and McIntosh, 2009; Okamoto and De Pontieu, 2011) often see a preponderance of waves propagating in one direction (i.e., upwards), more consistent with fast wind. More recently De Moortel et al. (2014) and Liu et al. (2014) have seen upward propagating waves at both loop footpoints meeting at the apex and generating higher frequency (presumably balanced) turbulence. In the likely case that this difference in turbulence has its origin in the solar wind source regions, measurement of waves in the corona has unique potential to distinguish between slow and fast wind in this way, and thus investigate their interface. Such investigations are ideally suited to solar minimum conditions when polar coronal holes are better defined and magnetic topology is less complex than at solar maximum.

Another distinction between fast and slow solar wind lies in their elemental compositions. The fast wind is relatively unfractionated, while the slow wind exhibits an enhancement in abundance of elements with first ionization potential (FIP) less than about 10 eV (the so-called "FIP Effect"; e.g., see von Steiger et al., 1995; Feldman and Laming, 2000). This effect is most convincingly explained in terms of the ponderomotive force in the chromosphere, resulting from the propagation through or reflection from the chromosphere of Alfvén waves (Laming, 2004, 2009, 2012, 2015; Rakowski and Laming, 2012) with peak amplitudes in the corona of $25-100\,\mathrm{km\,s^{-1}}$, depending on the chromospheric model and coronal density. This amplitude is larger than that typically associated with nonthermal mass motions inferred from spectral line broadening by a factor of up to 4, but evidence for such motions has more recently been documented (e.g., Peter, 2001, 2010). CoMP sees much lower Doppler velocity amplitudes than these (Tomczyk et al., 2007), but McIntosh and De Pontieu (2012) argue that this is due to line of sight (LoS) superposition effects "hiding" the true coronal wave flux in enhanced non-thermal broadening. FIP fractionated closed loops should show more balanced waves than less fractionated open field regions, due to repeated Alfvén wave reflection from the chromosphere, consistent with presumed origin of the slow wind in a fractionated closed loop and the fast wind in a relatively unfractionated open field region. De Moortel et al. (2014) and Liu et al. (2014) see something like this in CoMP observations of coronal loops, though the balanced turbulence is restricted to the apex region, and is observed at higher frequencies possibly indicating an onset of turbulent cascade where upcoming waves from each footpoint meet. The interpretation of decreasing spectral line widths with height above a coronal hole in terms of Alfvén wave damping (Hahn et al., 2012; Hahn and Savin, 2013) would lead to the prediction of a similar phenomenon in open fields, if the Alfvén wave damping proceeds by turbulent cascade. Counter propagating Alfvénic waves have recently been detected in coronal holes (Morton et al., 2015), supporting this inference. Further, Alfvén and fast mode waves also behave differently around coronal null points, also represented by separatrices or quasi-separatrix layers. Fast

mode waves refract across field lines and accumulate at the null point, leading to increased wave heating, while Alfvén waves are confined to magnetic field lines (Thurgood and McLaughlin, 2013). Detecting these waves directly would provide valuable information for distinguishing the solar wind formation and acceleration mechanisms in both the fast and slow solar wind.

How are Flux Ropes Formed, How Do they Evolve, and How are they Related to CMEs?

Opposing views exist regarding the nature of flux tubes in active regions. Some authors suggest that coronal loop must have twisted field, in order to give it a distinct identity, separate from other coronal magnetic field (e.g., Hood et al., 2009; Vasheghani Farahani et al., 2010), and to be sufficiently buoyant to emerge from the convection zone (e.g., Archontis, 2008). Others argue that newly emerged flux is untwisted, and the flux rope signatures seen in *in situ* observations of ICMEs arise due to reconnection of a sheared arcade during the CME eruption (e.g., Lynch et al., 2004) or a pre-CME flare (Patsourakos et al., 2013). Sakai et al. (2001) found that torsional waves in twisted high β (≈1) loops propagate preferentially in a direction that unwinds the twist. Observations of constant cross sectional loops have been interpreted as being due to circular loops necessarily exhibiting significant twist (Klimchuk, 2000). More recent studies suggest that this is an observational selection effect, and that coronal magnetic field is asymmetric and untwisted (Malanushenko and Schrijver, 2013). The distinction is important, relating to the mechanisms by which flux emerges, CMEs erupt, and the nature of waves on such structures. Pure torsional Alfvén waves may only propagate on untwisted flux tubes (Vasheghani Farahani et al., 2010). The twist necessarily introduces mixing between Alfvén and kink modes, with consequences for the wave damping and coronal heating. The absence of intensity oscillations seen by CoMP (Tomczyk et al., 2007) suggests that any torsional waves in the solar corona must have been propagating on untwisted magnetic field.

Observing the magnetic structure of various loops, including prominences, and prominence flows, e.g., through He I 1083.0 nm, together with observations of the chromospheric magnetic fields under the same structure, once inverted (e.g., Orozco Suárez et al., 2014), could place constraints on prominence densities and determine how prominence and coronal magnetic fields interact, how and where magnetic energy is stored (e.g., flux and helicity transport) and how it is released (e.g., instabilities, reconnection, dissipative heating). In particular, measurements of prominence cavities obtained by the CoMP instrument indicate a characteristic "lagomorphic" signal (i.e., morphologically shaped like a rabbit's ear) in linear polarization consistent with twisted magnetic flux tubes, or ropes. Bak-Steślicka et al. (2013) showed that coronal prominence cavities, for example as observed by SDO/AIA193Å channel, are observed by CoMP to have such linear polarization signature in the Fe XIII 1074.7 nm line. This signature can be explained as arising from an arched magnetic flux rope with axis oriented along the LoS (e.g., Fan, 2010; Gibson et al., 2010). When integrated along the LoS, a combination of linear polarization nulls occurring where the flux rope magnetic field is oriented at the van Vleck angle θ, ($\cos^2\theta = 1/3$), or where the axial magnetic field is oriented completely out of the plane of sky (PoS), leads to a forward-modeled signal of the same characteristic shape as observed. Linear polarization is sensitive to PoS magnetic field, so a "polarization ring" may occur when magnetic field winds around a central LoS-oriented axis. Indeed, early CoMP observations of a prominence cavity showed just such a structure (Dove et al., 2011). LoS effects play an important role so that such a pure ring may be rare, and may also indicate a different magnetic topology to the flux ropes in Bak-Steślicka et al. (2013). However, analysis of CoMP observations indicate a truly ubiquitous lagomorphic structure in linear polarization observed in Fe XIII associated with prominence cavities that matches expectations for the linear polarization signal of a forward-modeled, arched magnetic flux rope fully integrated along the LoS (e.g., Rachmeler et al., 2013). Analysis of the circular polarization of such structures would confirm the presence of a magnetic axis and quantify its field strength. Inside prominences, similar studies may be undertaken with the He I 1083.0 nm line, while outside a coronal line, such as Fe XIII 1074.7 nm, can be used.

Where Do CME-Associated Shocks Form?

Particles accelerated by CME-driven shocks have the highest particle energies of all suprathermal species and pose the greatest space weather hazards to spaceborne instrumentation and humans. The very highest SEP energies arise when acceleration begins very close to the sun. Gopalswamy et al. (2001) and Mann et al. (2003) use model magnetic fields and density profiles for different solar regions, with particular attention to active regions, to estimate the heliocentric radius where a CME driven disturbance becomes a shock (i.e., Alfvén Mach number, $M_A > 1$), and where it becomes supercritical ($M_A >$ 2–3, depending on plasma β and shock obliquity). This last transition is crucial, because it determines where the shock begins to become turbulent and may begin to reflect and accelerate particles (Edmiston and Kennel, 1984), in the absence of pre-existing seed particles.

Measurements of the electron density and constraints on the magnetic field within about 1.25 $R_\odot$ (where the Fe XIII 1074.7 nm/1079.8 nm line pair is sensitive to electron density) would remove the ambiguities introduced in the model for these quantities entering in the calculation of the Alfvén speed. Further the magnetic field PoS direction allows inference of the CME shock obliquity. Such shocks are currently believed to propagate close to the Sun as quasi-perpendicular, evolving to quasi-parallel further out (Tylka and Lee, 2006; Rouillard et al., 2011). This has consequences for particle acceleration (Laming et al., 2013). More sophisticated current shock acceleration theories generally treat only the parallel case (e.g., Ng and Reames, 2008).

Coronal magnetic field measurements during the passage of a CME can detect compressions and distortions in the magnetic field as well as associated waves due to the formation and passage of a CME shock. Density compressions resulting from shock formation can be large enough to produce intensity enhancements in white light coronal images as well

(Vourlidas et al., 2003) that are complementary to coronal field measurement for understanding the CME shock and its role in producing SEPs.

How is Energy Stored and Released by Reconnection in Coronal Heating, Flares and CMEs?

Most observations designed to detect signatures of magnetic reconnection in the solar corona to date have focused on observing high temperature plasma, specifically high electron temperatures. While observations of CME current sheets have been successful in this respect (e.g., Ciaravella et al., 2002; Ko et al., 2003; Ciaravella and Raymond, 2008; Savage et al., 2010), searches of the solar corona for evidence of nanoflare heating have been less clear (e.g., Brooks et al., 2009). Observations by Hi-C (Cirtain et al., 2013) suggest that both nanoflares (e.g., Testa et al., 2013) and steady heating (e.g., Warren et al., 2010; Peter et al., 2013), presumably associated with waves, are present. Such waves are often suggested to propagate up from the convection zone (e.g., Asgari-Targhi et al., 2013), although reconnection is also a potential source of waves (e.g., Sturrock, 1999; Longcope et al., 2009; Kigure et al., 2010; Liu et al., 2011), and offers another interpretation of the results of De Moortel et al. (2014). Hence it appears that the detection and identification of different modes of magneto-hydrodynamic (MHD) waves, combined with observations of magnetic topology, would be highly constraining on the nature and existence of magnetic reconnection at the heart of an active region. The magnetic free energy could be calculated from the non-potential field (e.g., extrapolated from photospheric or chromospheric magnetograms), and compared with energetics of the active region.

Direct experimental knowledge of LoS magnetic field strengths before and after a CME eruption would allow estimation of the magnetic energy released, for comparison with measurements of CME kinetic, thermal and gravitational energy. This would be complementary to the estimates derived from field extrapolations. It would also allow an assessment of the likely mechanism of CME eruption (see e.g., Ugarte-Urra et al., 2007) from the change in magnetic topology. The LoS field strength and PoS direction together provide diagnostics for magnetic topologies, including magnetic nulls and current sheets that can be compared to the location of high temperature emission.

WAMIS INSTRUMENT CONCEPT

The Waves and Magnetism in the Solar Atmosphere (WAMIS) investigation is a long duration balloon (LDB) based 20 cm aperture coronagraph designed to meet challenges of answering these outstanding questions. WAMIS builds on the heritage of CoMP (Tomczyk et al., 2008), and could obtain continuous measurements over at least 2 weeks of the strength and direction of coronal magnetic fields within a large field-of-view (FOV) at the spatial and temporal resolutions required to address outstanding problems in coronal physics. The key WAMIS characteristics are summarized in **Table 1**. Additionally, the WAMIS investigation would make near simultaneous

Telescope type	Internally occulted Lyot coronagraph
Objective lens	f/10 singlet, aperture 20 cm, focal length 203.3 cm
Objective Stray Light	<0.2 $\mu B_\odot$ goal, 1.2–2.8 $R_\odot$ ($B_\odot = 9.34 \times 10^6$ erg/cm^2/s/sr/nm)
Overall Throughput	≈5%
Plate Scale	4.5"/pixel low magnification mode; 1.5"/pixel high magnification mode
Fe XIII (1074.7 nm) Count Rate @ 1.1 $R_\odot$	1×10^5 photons/pixel/s @1.5"/pixel magnification
Detector	Goodrich Visible+SWIR camera, 15 micron pixels, 1280 × 1024 format
Inner FOV Limit	1.02 $R_\odot$
Outer FOV	±2.8 $R_\odot$ @4.5"/pixel Sun Centered; 1.8 $R_\odot$ @1.5"/pixel Limb Centered
Primary Lines of Interest	Fe XIII (1074.7, 1079.8 nm); AR observations; Fe X (637.5 nm), Fe XI (789.2 nm); CH&CHB observations; He I (1083.0 nm); prominence/flux rope observations
Filter	Tunable Lyot filter, 3.8 cm aperture, 530–1100 nm range
Duration of Continuous Observational Sequence	2 weeks minimum; ≥4 weeks optimum

observations of the electron scattered K-corona for context and to establish electron density out to greater radial distances than those accessible with the Fe XIII 1074.7 nm/1079.8 nm intensity ratio (~1.25 $R_\odot$; wavelengths given here and elsewhere are values in air). These comprehensive observations are not provided by any current single observatory. The visible-IR spectral range covers emission lines for understanding the magnetic field strength and structure in the active region (AR) and coronal hole (CH), the coronal hole boundary (CHB) region relevant for solar wind studies, as well as the prominence/flux rope structures. Observations of MHD waves would address fundamental issues in coronal heating and the sources and acceleration of the solar wind. In particular, the cross helicity of observed waves (from Fe XIII together with Fe X 637.5 or Fe XI 789.2 nm) should allow an empirical distinction between "balanced" turbulence in the slow wind, compared to more directed turbulence in the fast wind (e.g., Bruno and Carbone, 2005). Since coronal waves detected so far are ubiquitously "Alfvénic," their direction of propagation indicates the magnetic field vector [either because of their natural properties (Alfvén mode), or because of refraction to high density that follows the magnetic field (fast mode)], and such observations obviously complement direct magnetic field measurements.

Observational Requirements

The science questions in Section The Importance of Magnetic Field and Waves Measurements in the Corona could be addressed through observation of coronal magnetic fields and waves over a 1.02–1.8 $R_\odot$ FOV with high spatial resolution of 1.5" or an alternative 1.02–2.8 $R_\odot$ large-FOV mode with a spatial resolution of 4.5", at a temporal cadence of 5 min (see Section Advantages of Measurements Outside of the Atmosphere for more detail). At this cadence, WAMIS would be sensitive to magnetic field

strengths of <10 Gauss for the faintest detectable coronal structures and one Gauss for the brightest. A spatial resolution of 4.5" is sufficient to address many of the scientific questions pertaining to the global magnetic structure of the corona, while the 1.5" resolution allows a more detailed view of individual coronal features. The capability for the higher spatial resolution observations (1.5") over a more limited FOV would be achieved with interchangeable magnification lenses. This new capability of WAMIS is potentially important for observations of active region loops, and represents a significant improvement on the capabilities of CoMP. According to McIntosh and De Pontieu (2012), WAMIS should see correspondingly higher Alfvén wave amplitudes than CoMP, due to reduced confusion caused by spatial-averaging. The large FOV of WAMIS (up to 2.8 $R_\odot$ from Sun center) would be needed to observe the global properties of the corona and is required to address the science questions, concerned with tracking the outward motion and other basic properties of the solar wind, CMEs and prominences as they are ejected from the corona. It is also necessary to observe as low in the corona as possible to understand how the corona continually reacts to changes occurring in the photosphere and chromosphere. High temporal cadence is needed to capture the relentless dynamic evolution of the coronal plasma structure and explosive disturbances (e.g., MHD waves, shocks, prominence eruptions, CMEs). The physical properties of CMEs and eruptive prominences are best determined from the polarization signal of broadband filtered white-light observations because the scattered light from the corona is partially polarized. Furthermore, the absence of all atmospheric seeing effects (not just sky brightness) could prove to be a critical advantage to balloon-borne as opposed to ground-based instrumentation.

Note that a 5 min cadence for magnetic field measurement does not limit the minimum wave frequency WAMIS can measure, because wave observations (unpolarized) would not need the same accumulated exposure time for a given signal-to-noise ratio (S/N, see discussion of magnetic field S/N in Section Improved Sensitivity for Measurement of the Magnetic Field Strength). Depending upon the intensity of the target structure for wave observations, the image cadence could be as short as 2 s, thus detecting waves of period as low as 1 min. The images required for a wave investigation at this cadence would not necessarily require a different observational program from the magnetic field observing program. Thus the wave and magnetic field observing programs could run simultaneously, even though the cadence of the two measurements generated by post-flight analysis would be very different.

Table 2 shows the science traceability matrix. There is a specific need to directly measure the magnetic field both in the corona and in the chromosphere. Recent observations suggest that MHD waves in the upper chromosphere have sufficient energy to accelerate the solar wind outside of active regions (Aschwanden et al., 2007; De Pontieu et al., 2007; McIntosh et al., 2011), if they can escape into the corona. WAMIS would provide routine magnetic field and wave measurements in this key region and would complement observations of activity lower in the solar atmosphere such as by the Interface Region Imaging Spectrograph (IRIS, De Pontieu et al., 2014) and Chromosphere and Prominence Magnetometer (ChroMag, de Wijn et al., 2012), already under development, and planned for deployment at Mauna Loa Solar Observatory (MLSO) by 2016. These measurements altogether would provide critical information on the magnetic and plasma conditions to couple the coronal magnetic fields with those measured at photospheric heights.

Advantages of Measurements Outside of the Atmosphere

Improved sensitivity for measurement of the magnetic field strength

In order to achieve such observational requirements, WAMIS would need to observe features down to a few millionths of the brightness of the solar disk (2×10^{-6} $B_\odot$ or $2 \mu B_\odot$), which requires an effective sky background an order of magnitude lower (e.g., $0.2 \mu B_\odot$) For example, bright loops above active regions are typically of the order of $20-25 \mu B_\odot$ while in coronal holes the brightness is typically $1-2 \mu B_\odot$. One objective of the WAMIS instrument would be to perform coronal magnetometry using the forbidden emission lines of Fe XIII at 1074.7 and 1079.8 nm and the He I emission line at 1083.0 nm. While coronal magnetometry would not be the only objective of WAMIS, it is instructive to consider the details of this measurement from outside the atmosphere.

Because the amplitude of the Zeeman-induced circular polarization (Stokes V) signal is $\approx 10^{-3}$ for a 10 G field for the Fe XIII 1074.7 nm line [the linear polarization signal is typically 2 orders of magnitude higher (Lin et al., 2000; Tomczyk et al., 2008)], the S/N requirement for the circular polarization measurements drives the requirement for coronal magnetometry (such as by enlarging the aperture of the coronagraph or longer integration time to obtain better counting statistics). For ground-based observations, stray light in the form of sky brightness is usually the dominant noise source. The expected noise in the LoS component of the coronal field due to the combination of photon counting statistics in the signal and in the stray light background can be derived from consideration of the propagation of errors in the circular polarization measurements (Penn et al., 2004), and is given by

$$\sigma_B[G] = \frac{8500}{\sqrt{I_{\text{line}}}} \sqrt{1 + 2\frac{I_{\text{sky}}}{I_{\text{line}}}} \qquad (1)$$

where I_{line} and I_{sky} are the number of photons in the emission line and background, respectively. This equation assumes photon noise limited observations in the Fe XIII 1074.7 nm (Landé g factor = 1.5) emission line. Note this equation ignores all other atmospheric seeing effects generating noise in the polarization measurement except for atmospheric stray light (sky brightness). Equation (1) shows that the presence of sky background (including instrument scattered light) reduces the effective aperture of the telescope. That is the reason why long integration times are needed from the ground to achieve the desired polarimetric sensitivity. From MLSO, where CoMP is located, the sky brightness is nominally $5.0 \mu B_\odot$. However, for balloon-borne observations, internally generated stray light of the coronagraph dominates the sky background. For the internally occulted design of WAMIS a conservative estimate of the stray light-generated sky background is equivalent to $I_{\text{sky}} = 0.2 \mu B_\odot$. We can combine Equation (1) with a flux budget for the corona and compute the expected noise level as a function of coronagraph aperture size and coronal brightness. This is illustrated in **Figure 1**, assuming a system throughput of 5%, a pixel of 5″, and an integration time of 5 min. On the left are balloon-borne observations where the stray light is strictly generated by the coronagraph (e.g., $0.2\mu B_\odot$). On the right are

TABLE 2 | Science traceability matrix; see text for details.

Science Objective	FoV/Spatial Resolution	Physical Observable
1. Fast/Slow Wind, Coronal B structure	1.02–1.8 $R_\odot$/1.5″ pix.; 1.02–2.8 $R_\odot$/4.5″ pix.	Waves: Doppler velocity, plasma density, B-field direction
2. Prominences, flux ropes	1.02–1.8 $R_\odot$/1.5″ pix	B-field magnitude & direction from He I and Fe XIII
3. CME Shocks	1.02–2.8 $R_\odot$/4.5″ pix.	B-field magnitude & direction, Waves: Doppler velocity, plasma density
4. Reconnection	1.02–1.8 $R_\odot$/1.5″ pix.	B-field magnitude & direction, Waves: Doppler velocity, plasma density

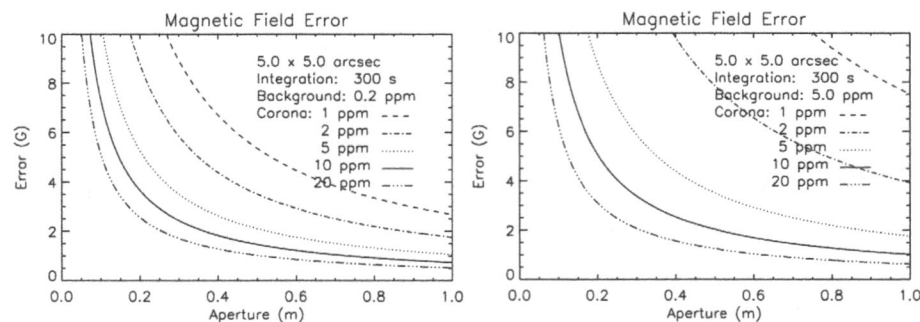

FIGURE 1 | These plots illustrate the impact of stray light on the relationship between LoS magnetic field strength sensitivity in Fe XIII 1074.7 nm and telescope aperture (see text for detail).

observations at MLSO where a typically good sky background is $5.0\,\mu B_\odot$. The lines plotted correspond to various intensities of the corona in $\mu B_\odot$ units. With balloon-borne observations, it is possible to achieve magnetic field sensitivities of 3.5 G in 5 min for coronal structures with brightness of $10\,\mu B_\odot$ using a 20 cm aperture telescope. From a similar instrument from the ground (i.e., CoMP at MLSO), with 5 min integration one achieves a sensitivity of 5 G. The needed integration time to reach comparable σ_B on the ground would be two times longer.

Figure 2 compares the modeled Fe XIII 1074.7 nm Stokes V signal integrated for 1 h for the sky background expected for WAMIS vs. that for CoMP. The larger circular polarization signal of a balloon-borne WAMIS is obvious. For example, Rachmeler et al. (2013) modeled the linear and circular polarizations from a flux rope and a sheared arcade. They showed that the true disambiguator between the two magnetic models is the circular polarization. A flux rope will have a clearly defined magnetic axis, where the circular polarization (proportional to the LoS magnetic field) will peak. This demonstrates the importance and advantage of performing IR coronal magnetometry from a balloon-borne or space platform.

Eliminating seeing effect through the Earth's atmosphere

Near-space observations on a balloon platform would eliminate all polarization noise (variability) introduced by the Earth's atmosphere which is difficult to quantify for ground-based observations. This would provide a fundamental advantage in the interpretation of all coronal magnetometry observations, including observations with future larger aperture ground-based instruments, e.g., the Daniel K. Inouye Solar Telescope (DKIST), and the Coronal Solar Magnetism Observatory (COSMO). A second advantage would be to allow WAMIS

to forego simultaneous continuum observations required for ground-based instruments (such as the Wollaston prism on CoMP, Tomczyk et al., 2008). Therefore, the full imager plane can be used for the line image resulting in an increase in routine spatial resolution for a given detector format and FOV. While continuum images do not need to be obtained simultaneously with the line images, they could be obtained at appropriate intervals to provide reference background and K-corona imaging.

Enabling continuous observations

The continuity of balloon-borne observations (independent of atmospheric variation, and no day/night cycle and weather-related interruptions) could be used to integrate signals over extended periods of time and beat down the photon noise. Coronal cavities associated with polar crown prominences would be good candidates for such a study, since they tend to be dynamically stable and are extended along the LoS, so that they can be essentially unchanged for up to several days of limb observations. Since most coronal lines observed by WAMIS are optically thin plasma projected against the PoS, this advantage is critical in separating the 3-D structure of the corona from short-term evolution of the corona. Uninterrupted observations also increase the probability of detecting and following solar transient events. In addition, uninterrupted observations of over 2 weeks with the expected magnetic field sensitivity of WAMIS could in principle enable tomographic inversions for 3D magnetic field vector (Kramar et al., 2006, 2013; Judge et al., 2013).

Advantages of the Large Field-of-View

The FOV of WAMIS (1.02–2.8 $R_\odot$) would enable analysis of the global magnetic topology of the corona. For example, Rachmeler

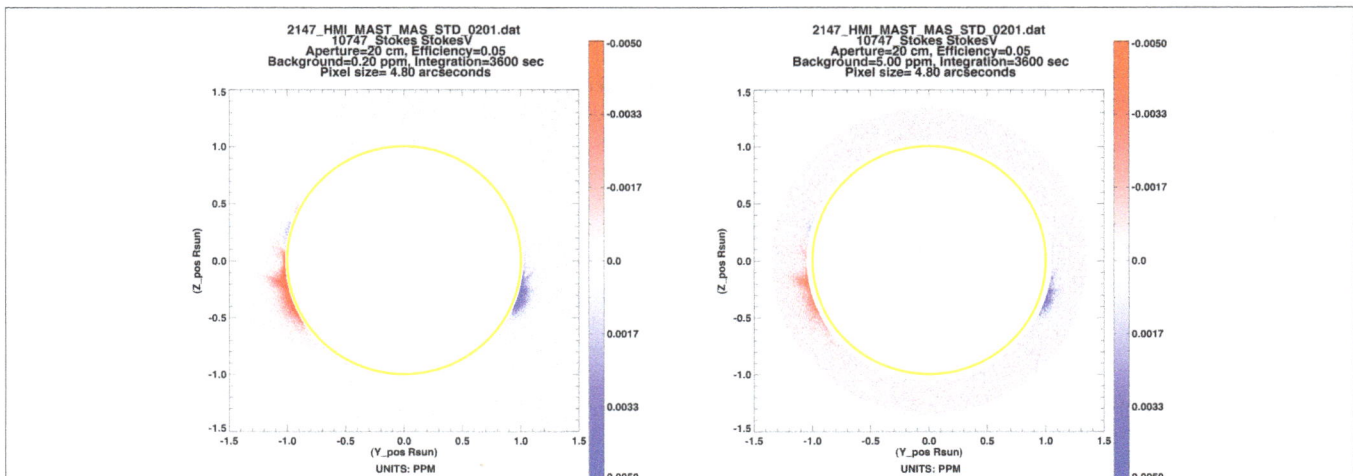

FIGURE 2 | Comparison of forward-modeled circular polarization for a global coronal MHD model (Predictive Science Inc. MAS model for Carrington Rotation 2147, from http://www.predsci.com/hmi/data_access.php), and applying the FORWARD SolarSoft codes (http://www.hao.ucar.edu/ FORWARD/). Photon noise is added based on telescope aperture, efficiency, background, pixel size, and integration time. Left: Background = 0.2 PPM (i.e., $0.2\,\mu B_\odot$, appropriate for WAMIS), Right: Background = 5 PPM (appropriate for CoMP), all integrated for 1 h. WAMIS vs. CoMP FOVs are explicitly applied to these images as internal/external occulters (at 1.02/1.05, 2.8/1.35 $R_\odot$ respectively). Note that sources of systematic errors are not considered, but are expected to be significant since the signal shown here is on the order of 0.1% of intensity.

et al. (2014) forward modeled Fe XIII signals for pseudostreamers vs. double streamers and found clear distinctions between these magnetic topologies arising from the null point lying above the pseudostreamer (at $\sim$1.4 $R_\odot$). Analysis of PROBA2 Sun Watcher using Active Pixel System Detector and Image Processing (SWAP) data yielded structures in EUV that aligned with the expected distinctions in morphology, but the Fe XIII 1074.7 nm linear polarization topological measurement was limited by the CoMP FOV. WAMIS will be able to unambiguously reveal the pseudostreamer topology. The large FOV would also allow better tracing of the evolution of the CME dynamics from the line intensity, Doppler shift and width (Tian et al., 2013). As a comparison, the up to 5 arcmin FOV of DKIST is not designed for studies of the global coronal structure and CMEs, and the 1.05–1.35 $R_\odot$ FOV of CoMP would miss a significant fraction of the null points in coronal structures and post-CME current sheet, as well as the likely formation of SEP-produced CME shocks above 1.5 $R_\odot$.

Coronal Magnetometry via Zeeman and Hanlé Effect

The Zeeman effect in forbidden coronal lines can provide information on coronal magnetic fields with strengths as low as a fraction of a Gauss and as high as several thousand Gauss. This is very important as it provides information on both the large-scale "quiet" coronal fields as well as the active region fields. The observations are restricted to off-limb observations obtained with coronagraphs (or at total solar eclipses) and in which LoS integration issues arise, because of the small optical depths in the corona. However, this is not an overwhelming issue as argued by Judge et al. (2013). To address the LoS confusion of coronagraph observations one can use coincident white-light and EUV observations such as those from the Solar Terrestrial Relations Observatory (STEREO), the Large Angle and Spectrometric Coronagraph Experiment (LASCO, Brueckner et al., 1995) on SOHO, and SDO observations to determine the distribution and the emission measure of material along the LoS. Also, persistent observations over at least 2 weeks under identical conditions can be used as rotational "tomography" on long-lived structures.

The Zeeman effect in circular polarization only gives information on the magnetic field projection along the LoS. Thus it provides a lower bound for the true magnetic strength. These measurements are challenging because the circular polarization signal is typically very small in the quiet corona (about 0.1% of the intensity for field strengths of 10 G). Thus a rigorous calibration of all possible sources of polarization noise is fundamental. The extended sequence of observations without atmospheric noise will allow an examination of the ultimate return that can be achieved by these techniques and guide the design and use of future large aperture ground based coronal magnetographs.

Using the saturated Hanlé effect (scattering polarization) applicable in coronal conditions, the PoS direction of the magnetic field can be determined from the linear polarization signal of the scattered radiation, subject in general to a 90° ambiguity (e.g., Lin and Casini, 2000). Where the Stokes parameters $U = 0$ and $Q \neq 0$, the field coincides with one of axes defined by Q. The ambiguity can be resolved due to the "Van Vleck" effect which causes nulls of linear polarization to occur at a specific angle θ, where $\cos^2\theta = 1/3$, between the magnetic field and the solar vertical. On either side of the null, the polarization direction changes by 90°. Identification of nulls can then be used to tightly constrain the morphology of the magnetic structure. These measurements are much easier, because the linear polarization of the forbidden coronal lines is typically of a few percent. The Fe XIII 1074.7 nm line is chosen because its emission is dominated by scattered disk radiation, and depolarizing effects due to collisions and radiative cascades are insignificant (Judge, 1998). The Fe XIII 1079.8 nm line conversely is dominated by collisions, yielding the density sensitivity of the ratio to the 1074.7 nm line. WAMIS magnetic field observations would additionally be supplemented by the ChroMag full-disk measurements of the chromospheric vector magnetic field, and by measurements of wave propagation in the PoS. Thus the PoS Alfvén speed will be inferred, and with the density from the 1074.7 /1079.8 nm intensity ratio, or from white light polarization brightness, the PoS magnetic field can also be calculated.

Measurements using both Zeeman effect and Hanlé effect can yield vector field information when the polarized light originates from a defined volume (Casini and Judge, 1999). In those cases constraints can be placed on the inclination of the magnetic field and therefore the total field strength and direction. In cases when a structure possesses substantial uniformity along the LoS (such as polar-crown-filaments and their cavities), magnetic field strength and structure may likewise be determined. **Table 3** lists key observables by WAMIS and the means to obtain them. See **Table 2** for connecting these observables to the science objectives.

WAMIS Instrument Design
CoMP as Heritage Instrumentation
The techniques described in Section Coronal Magnetometry via Zeeman and Hanlé Effect have been fully demonstrated with CoMP on the 20 cm aperture OneShot coronagraph originally at National Solar Observatory's Sacramento Peak Observatory in New Mexico (Tomczyk et al., 2008), and later has been operating on a daily basis at MLSO since 2011. The CoMP instrument was designed to observe the coronal magnetic field with a FOV in the low corona ($\sim$1.03 to 1.4 $R_\odot$), as well as to obtain information about the plasma density and motion.

TABLE 3 | Key WAMIS observables.

Line-of-sight B, field strength	Circular polarization	Longitudinal Zeeman effect
Plane-of-sky B, field direction	Linear polarization	Resonance scattering effect (Hanlé effect)
Line of sight velocity	Intensity vs. wavelength	Doppler effect
Plasma density	Fe XIII 1074.7 nm/1079.8 nm intensity ratio, IR continuum	Atomic physics, radiation transfer

The CoMP instrument is a combination polarimeter and narrowband tunable filter that can measure the Doppler shift and complete polarization state of the Fe XIII infrared coronal emission lines at 1074.7 and 1079.8 nm and the chromospheric 1083 nm He I line. The polarimeter function is achieved by a pair of Liquid Crystal Variable Retarders (LCVRs) followed by a linear polarizer that allows the selection of a polarization state characterized by Stokes parameters (I, Q, U, V). The filter is a four-stage, wide-field calcite birefringent filter with a bandwidth of 0.14 nm at 1074.7 nm. It is tuned in wavelength by four additional LCVRs. Both the polarization and filter bandpass selections are accomplished electro-optically. The CoMP filter has a transmission to unpolarized light of about 30%. The camera for CoMP is a liquid nitrogen cooled Rockwell Scientific (now Teledyne) 1024 × 1024 HgCdTe Infrared detector array. A filter wheel holding three order-blocking filters selects the emission line to be observed. See Tomczyk et al. (2008) for detailed descriptions.

The CoMP observations have a spatial sampling of 4.5″ per pixel and required 30 min of integration time to acquire a measure of the LoS magnetic field strength. As an example shown in Tomczyk et al. (2008), data were obtained in groups of 60 images in quick succession. For linear polarization, five images were taken at each of the four polarization states I+Q, I−Q, I+U, I−U, and at the three wavelengths, 1074.52, 1074.65, and 1074.78 nm, across the line. For circular polarization, 10 images were taken in each of the two polarization states of I+V and I−V at the same three sample wavelengths. The exposure time for the individual images was 250 ms and the two image groups were each obtained at a cadence of ~15 s with a duty cycle of 52%. The driver of the number of image groups (thus integration time) required for a single observation set is the level of sky brightness at MLSO (typically 5 $\mu B_\odot$). The WAMIS coronal magnetometer will profit much more from the unique observation conditions obtained at long duration balloon altitudes, because of the absence of sky brightness background as well as seeing-induced polarization cross-talk and atmospheric-induced source intensity fluctuations (Section Advantages of Measurements Outside of the Atmosphere).

WAMIS Filter/Polarimeter

In the time since the completion of the CoMP instrument, technological advances in broad-band polarizers and super-achromatic waveplates now present the possibility to construct a compact coronal polarimeter capable of observing coronal and prominence emission lines over a much wider wavelength range than the CoMP instrument. These advances have been incorporated into a filter/polarimeter called "CoMP-S" built by HAO for the Astronomical Institute of the Slovak Academy of Sciences (AISAS) coronagraph on Lomnicky Peak. The CoMP-S filter has been operational on the Lomnicky Peak Observatory 20 cm Zeiss coronagraph since April 2013. WAMIS filter will use the same CoMP-S filter/polarimeter design (Kucera et al., 2010; Koza et al., 2013; Rybak et al., 2013).

The CoMP-S design was primarily chosen to enable the WAMIS instrument to observe over the range between the Fe XIV coronal green line at 530.3 nm and the He I line at 1083 nm. While the target lines for WAMIS are the IR coronal emission lines 1074.7 nm Fe XIII and 1079.8 nm Fe XIII, and the IR prominence emission line 1083.0 nm He I, this filter will have the additional capability to observe corresponding visible lines - including 530.3 nm Fe XIV, 637.5 nm Fe X and 789.2 nm Fe XI in the corona and 587.6 nm He I and 656.3 nm Hα in prominences—for context and additional diagnostics. One particularly important diagnostic achieved with this extension of the wavelength range is the capability to observe coronal waves at temperatures other than Fe XIII, e.g., wave observations in coronal hole and coronal hole boundary with Fe X and Fe XI to study the origin of the slow solar wind. Each line will require a pre-filter be inserted in the optical system with a bandpass that depends on the free spectral range of the birefringent filter at that wavelength.

A secondary motivation for using the CoMP-S design is the use of Ferroelectric Liquid Crystals (FLC) in the polarization modulation instead of the usual LCVR. The objective of using FLC is their faster response time, which will significantly increase the duty cycle of the WAMIS instrument. The polarization modulator will consist of two FLC retarders and a fixed retarder followed by a linear polarizer acting as an analyzer. As for CoMP, the polarizer will also act as the entrance polarizer to the birefringent filter. Note that neither the FLCs nor the fixed retarder are achromatic and the value of retardation and orientation must be selected to optimize the Stokes modulation efficiency over a very broad wavelength range.

For ground observations, simultaneous images in the emission line and continuum bandpasses is important in that it allows the instrument to be insensitive to variations in the intensity of the background caused by image motion and the passage of atmospheric aerosols through the FOV. For a balloon flight above the atmosphere, no such simultaneous observations are required. However, the K-corona can be sequentially imaged in the continuum at a variety of wavelengths to obtain the electron density information.

WAMIS Coronagraph

The WAMIS coronagraph follows the classic internally occulted Lyot coronagraph design principles (**Figure 3**). The solar radiation is incident on the entrance aperture. A high quality solar image is produced at the occulter. Coronal radiation from heights of 1.02 to 2.8 $R_\odot$ passes through the coronagraph front end and is collimated before entering into the polarizer/filter assembly. A reimaging lens produces a high quality coronal image at the detector. A compensator plate located at the Lyot stop removes spherical aberration in the coronal image at the occulter plane. A polarization calibration optic will be incorporated in front of the instrument aperture to assure the instrumental polarization effects are understood to a level well under 10^{-4}. **Table 4** lists the optical design. The resulting plate scale is 4.5″ for low magnification mode, and 1.5″ for high magnification mode. The WAMIS coronagraph uses a 20 cm entrance aperture, f/10 objective which is sufficient to achieve the scientific requirements in **Table 2**. The diffraction limit corresponding to this aperture is 1.3″ at a wavelength of 1 μm, so diffraction effects are below the plate scale.

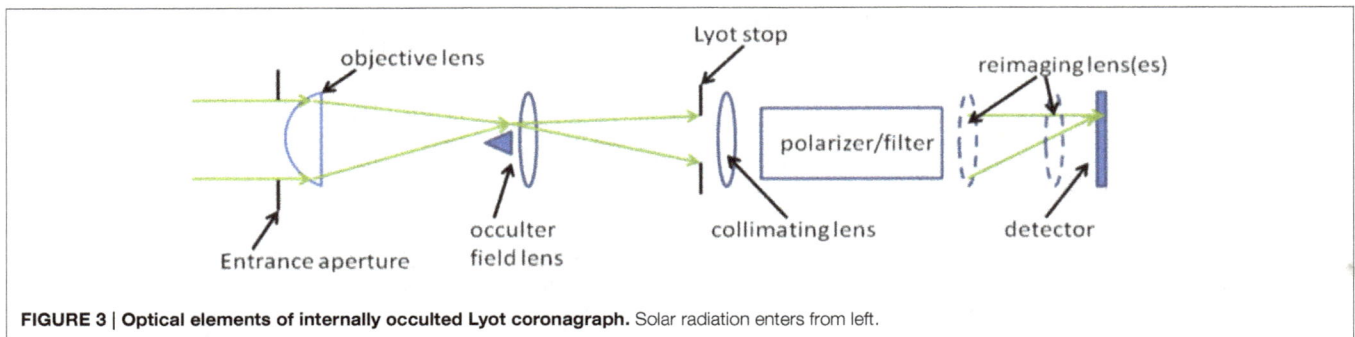

FIGURE 3 | Optical elements of internally occulted Lyot coronagraph. Solar radiation enters from left.

TABLE 4 | Optical design.

Objective	203.3 cm fl.
Field lens	31.0 cm fl.
Collimating lens	38.0 cm fl.
Re-imaging lens	High Mag. 38.0 cm fl.
	Low Mag. 12.9 cm fl.

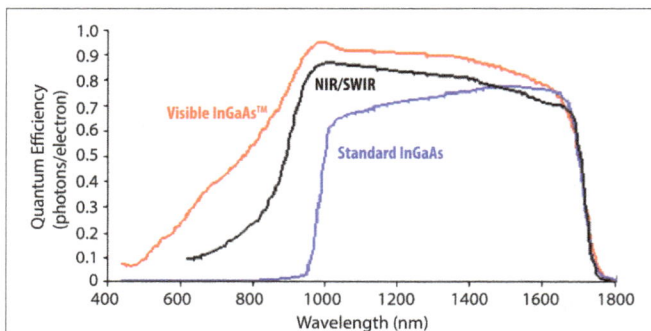

FIGURE 4 | Detector quantum efficiency (red curve) of the Goodrich Corp. InGaAs High Resolution Visible + SWIR Camera for WAMIS, compared to other camera models (NIR/SWIR and standard InGaAs).

WAMIS Detector System

WAMIS will use the Goodrich Corp. InGaAs High Resolution Visible+SWIR (short wavelength infrared) Camera as the detector instead of the Teledyne HgCdTe hybrid camera used on CoMP. A single detector for the wide range of wavelength coverage is also beneficial over separate detectors for the visible and IR. The 1280 × 1024 pixel array focal plane hybrid detector has a 15 micron pitch, achieves 80% quantum efficiency in the target wavelength regime (**Figure 4**).

Gondola and Pointing Control

There are at least a couple of options to carry instrument under the balloon. An HAO-developed gondola system was successfully used for two Sunrise solar observations in 2009 and 2012 (Barthol et al., 2011). The NASA Wallops Flight Facility (WFF) Balloon

Program Office (BPO) recently has available a gondola system under the Columbia Science Balloon Facility (CSBF), successfully pioneered by the University of Colorado HyperSpectral Imager for Climate Science (HySICS) investigation in 2013. WAMIS high resolution imaging of 1.5″ requires a matching performance in the pointing control during the balloon flight. The NASA Wallops Arc Second Pointer (WASP) system has achieved a pointing performance of better than 0.25″ RMS error in both pitch and yaw. This will easily satisfy the WAMIS requirements.

CONCLUDING REMARKS

The fundamental advance of the balloon-borne WAMIS beyond any ground-based coronal magnetograph will come from an effectively complete absence of variability in the polarization background and the extension of the duration of uninterrupted observing by over an order of magnitude. A good analogy for this advancement is the way the SOHO Michelson Doppler Imager (MDI, Scherrer et al., 1995) made fundamental discoveries on photospheric field even though it was preceded by decades of ground based observations by larger instruments. The freedom to explore different temporal regimes without seeing variability or day–night cycle interruptions was the key to these discoveries. Similarly, the fundamental advancements achieved by the near-space observations of WAMIS on coronal magnetic field and waves will point the way for future ground based and orbital instrumentation.

AUTHOR CONTRIBUTIONS

All authors contributed to the science and design of the WAMIS instrument. The first seven authors contributed significantly to the write-up of this manuscript.

ACKNOWLEDGMENTS

This research is supported by the Chief of Naval Research. NCAR is supported by the National Science Foundation.

REFERENCES

Antiochos, S. K., Mikić, Z., Titov, V. S., Lionello, R., and Linker, J. A. (2011). A model for the sources of the slow solar wind. *Astrophys. J.* 731, 112–122. doi: 10.1088/0004-637X/731/2/112

Antonucci, E., Gabriel, A. H., and Patchett, B. E. (1984). Oscillations in EUV emission lines during a loop brightening. *Solar Phys.* 93, 85–94.

Archontis, V. (2008). Magnetic Flux Emergence in the Sun. *J.Geophys. Res.* 113, A03S04. doi: 10.1029/2007ja012422

Arnaud, J., and Newkirk, G. Jr. (1987). Mean properties of the polarization of the Fe XIII 10747 Å coronal emission line. *Astron. Astrophys.* 178, 263–268.

Aschwanden, M. J. (1987). Theory of radio pulsations in coronal loops. *Solar Phys.* 111, 113–136. doi: 10.1007/BF00145445

Aschwanden, M. J., Fletcher, L., Schrijver, C. J., and Alexander, D. (1999). Coronal loop oscillations observed with the transition region and coronal explorer. *Astrophys. J.* 520, 880–894. doi: 10.1086/307502

Aschwanden, M. J., Winebarger, A., Tsiklauri, D., and Peter, H. (2007). The coronal heating paradox. *Astrophys. J.* 659, 1673–1681. doi: 10.1086/513070

Asgari-Targhi, M., van Ballegooijen, A. A., Cranmer, S. R., and DeLuca, E. E. (2013). The spatial and temporal dependence of coronal heating by alfvén wave turbulence. *Astrophys. J.* 773, 111–122. doi: 10.1088/0004-637X/773/2/111

Bak-Steślicka, U., Gibson, S. E., Fan, Y., Bethge, C., Forland, B., and Rachmeler, L. A. (2013). The magnetic structure of solar prominence cavities: new observational signature revealed by coronal magnetometry. *Astrophys. J.* 770, L28–L32. doi: 10.1088/2041-8205/770/2/L28

Barthol, P., Gandorfer, A., Solanki, S. K., Schüssler, M., Chares, B., Curdt, W., et al. (2011). The Sunrise mission. *Solar Phys.* 268, 1–34. doi: 10.1007/s11207-010-9662-9

Brooks, D. H., Warren, H. P., Williams, D. R., and Watanabe, T. (2009). Hinode/Extreme ultraviolet imaging spectrometer observations of the temperature structure of the quiet corona. *Astrophys. J.* 705, 1522–1532. doi: 10.1088/0004-637X/705/2/1522

Brueckner, G. E., Howard, R. A., Koomen, M. J., Korendyke, C. M., Michels, D. J., Moses, J. D., et al. (1995). The Large Angle Spectroscopic Coronagraph (LASCO): visible light coronal imaging and spectroscopy. *Solar Phys.* 162, 357–402. doi: 10.1007/BF00733434

Bruno, R., and Carbone, V. (2005). The solar wind as a turbulence laboratory. *Living Rev. Solar Phys.* 2:4. doi: 10.12942/lrsp-2005-4

Casini, R., and Judge, P. G. (1999). Spectral lines for polarization measurements of the coronal magnetic field. II. consistent treatment of the stokes vector for magnetic-dipole transitions. *Astrophys. J.* 522, 524–539. doi: 10.1086/307629

Chapman, R. D., Jordan, S. D., Neupert, W. M., and Thomas, R. J. (1972). Evidence for the 300-SECOND Oscillation from OSO-7 Extreme-Ultraviolet Observations. *Astrophys. J.* 174, L97–L99. doi: 10.1086/180957

Ciaravella, A., and Raymond, J. C. (2008). The current sheet associated with the 2003 November 4 coronal mass ejection: density, temperature, thickness, and line width. *Astrophys. J.* 686, 1372–1382. doi: 10.1086/590655

Ciaravella, A., Raymond, J. C., Li, J., Reiser, P., Gardner, L. D., Ko, Y.-K., et al. (2002). Elemental abundances and post-coronal mass ejection current sheet in a very hot active region. *Astrophys. J.* 575, 1116–1130. doi: 10.1086/341473

Cirtain, J. W., Golub, L., Lundquist, L., van Ballegooijen, A., Savcheva, A., Shimojo, M., et al. (2007). Evidence for Alfvén Waves in Solar X-ray jets. *Science* 318, 1580–1582. doi: 10.1126/science.1147050

Cirtain, J. W., Golub, L., Winebarger, A. R., de Pontieu, B., Kobayashi, K., Moore, R. L., et al. (2013). Energy Release in the solar corona from spatially resolved magnetic braids. *Nature* 493, 501–503. doi: 10.1038/nature11772

De Moortel, I., McIntosh, S. W., Threlfall, J., Bethge, C., and Liu, J. (2014). Potential evidence for the onset of alfvénic turbulence in trans-equatorial coronal loops. *Astrophys. J.* 782, L34–L39. doi: 10.1088/2041-8205/782/2/l34

De Pontieu, B., Title, A. M., Lemen, J. R., Kushner, G. D., Akin, D. J., Allard, B., et al. (2014). The Interface Region Imaging Spectrograph (IRIS). *Solar Phys.* 289, 2733–2779. doi: 10.1007/s11207-014-0485-y

De Pontieu, B., McIntosh, S. W., Carlsson, M., Hansteen, V. H., Tarbell, T. D., Schrijver, C. J., et al. (2007). Chromospheric alfvénic waves strong enough to power the solar wind. *Science* 318, 1574–1577. doi: 10.1126/science.1151747

de Wijn, A. G., Bethge, C., Tomczyk, S., and McIntosh, S. (2012). "The chromosphere and prominence magnetometer," in *Proceedings of SPIE 8446,*

844678, eds I. S. McLean, S. K. Ramsey, and H. Takami (Amsterdam). doi: 10.1117/12.926395

Domingo, V., Fleck, B., and Poland, A. I. (1995). The SOHO mission: an overview. *Solar Phys.* 162, 1–37. doi: 10.1007/BF00733425

Dove, J. B., Gibson, S. E., Rachmeler, L. A., Tomczyk, S., and Judge, P. (2011). A ring of polarized light: evidence for twisted coronal magnetism in cavities. *Astrophys. J.* 731, L1–L5. doi: 10.1088/2041-8205/731/1/L1

Dulk, G. A., and McLean, D. J. (1978). Coronal magnetic fields. *Solar Phys.* 57, 279–295. doi: 10.1007/BF00160102

Edmiston, J. P., and Kennel, C. F. (1984). A parametric survey of the first critical mach number for a fast MHD shock. *J. Plasma Phys.* 32, 429–441. doi: 10.1017/S002237780000218X

Erdélyi, R., and Fedun, V. (2007). Are there alfvén waves in the solar atmosphere? *Science* 318, 1572–1574. doi: 10.1126/science.1153006

Fan, Y. (2010). On the eruption of coronal flux ropes. *Astrophys. J.* 719, 728–736. doi: 10.1088/0004-637X/719/1/728

Feldman, U., and Laming, J. M. (2000). Element abundances in the upper atmosphere of the sun and stars: update of observational results. *Phys. Scripta* 61, 222–252. doi: 10.1238/Physica.Regular.061a00222

Fisk, L. A., and Schwadron, N. A. (2001). Origin of the solar wind: theory. *Space Sci. Rev.* 97, 21–33. doi: 10.1023/A:1011805606787

Gibson, S. E., Kucera, T. A., Rastawicki, D., Dove, J., de Toma, G., Hao, J., et al. (2010). Three-dimensional morphology of a coronal prominence cavity. *Astrophys. J.* 724, 1133–1146. doi: 10.1088/0004-637X/724/2/1133

Gopalswamy, N., Lara, A., Kaiser, M. L., and Bougeret, J.-L. (2001). Near-sun and near-earth manifestations of solar eruptions. *J.Geophys. Res.* 106, 25261–25278. doi: 10.1029/2000JA004025

Hahn, M., Landi, E., and Savin, D. W. (2012). Evidence of wave damping at low heights in a polar coronal hole. *Astrophys. J.* 753, 36–44. doi: 10.1088/0004-637X/753/1/36

Hahn, M., and Savin, D. W. (2013). Observational quantification of the energy dissipated by alfvén waves in a polar coronal hole: evidence that waves drive the fast solar wind. *Astrophys. J.* 776, 78–87. doi: 10.1088/0004-637X/776/2/78

Handy, B. N., Acton, L. W., Kankelborg, C. C., Wolfson, C. J., Akin, D. J., Bruner, M. E., et al. (1999). The transition region and coronal explorer. *Solar Phys.* 187, 229–260. doi: 10.1023/A:1005166902804

Hanslmeier, A. (2003). Space weather - effects of radiation on manned space missions. *Hvar Observatory Bull.* 27, 159–170.

Harrison, R. A. (1987). Solar soft X-ray pulsations. *Astron. Astrophys.* 182, 337–347.

Hood, A. W., Archontis, V., Galsgaard, K., and Moreno-Insertis, F. (2009). The emergence of toroidal flux tubes from beneath the solar photosphere. *Astron. Astrophys.* 503, 999–1011. doi: 10.1051/0004-6361/200912189

House, L. L., Querfeld, C. W., and Rees, D. E. (1982). Coronal emission-line polarization from the statistical equilibrium of magnetic sublevels. II. Fe XIV 5303 Å. *Astrophys. J.* 255, 753–763. doi: 10.1086/159874

Iucci, N., Dorman, L. I., Levitin, A. E., Belov, A. V., Eroshenko, E. A., Ptitsyna, N. G., et al. (2006). Spacecraft operational anomalies and space weather impact hazards. *Adv. Space Res.* 37, 184–190. doi: 10.1016/j.asr.2005.03.028

Jess, D. B., Mathioudakis, M., Erdélyi, R., Crockett, P. J., Keenan, F. P., and Christian, D. J. (2009). Alfvén waves in the lower solar atmosphere. *Science* 323, 1582–1585. doi: 10.1126/science.1168680

Judge, P. G. (1998). Spectral lines for polarization measurements of the coronal magnetic field. I. theoretical intensities. *Astrophys. J.* 500, 1009–1022. doi: 10.1086/305775

Judge, P. G., Habbal, S., and Landi, E. (2013). From forbidden coronal lines to meaningful coronal magnetic fields. *Solar Phys.* 288, 467–480. doi: 10.1007/s11207-013-0309-5

Kigure, H., Takahashi, K., Shibata, K., Yokoyama, T., and Nozawa, S. (2010). Generation of alfvén waves by magnetic reconnection. *PASJ,* 62, 993–1004. doi: 10.1093/pasj/62.4.993

Klimchuk, J. A. (2000). Cross-sectional properties of coronal loops. *Solar Phys.* 193, 53–75. doi: 10.1023/A:1005210127703

Ko, Y.-K., Raymond, J. C., Lin, J., Lawrence, G., Li, J., and Fludra, A. (2003). Dynamical and physical properties of a post-coronal mass ejection current sheet. *Astrophys. J.* 594, 1068–1084. doi: 10.1086/376982

Koza, J., Ambroz, J., Gomory, P., Habaj, P., Kozak, M., Kucera, A., et al. (2013). "The CoMP-S instrument at the Lomnicky Peak Observatory and possible synergy aspects with the space-born observations," in

Synergies between Ground and Space Based Solar Research - 1st SOLARNET - 3rd EAST/ATST Meeting (Oslo). Available online at: http://www.astro.sk/~choc/open/apvv_081611/output/presentations/2013_solarnet_conference_oslo/koza_comp-s_at_LSO_oslo_east_conference.pdf

Kramar, M., Inhester, B., Lin, H., and Davila, J. (2013). Vector tomography for the coronal magnetic field. II. hanle effect measurements. *Astrophys. J.* 775, 25–36. doi: 10.1088/0004-637X/775/1/25

Kramar, M., Inhester, B., and Solanki, S. K. (2006). Vector tomography for the coronal magnetic field. I. Longitudinal Zeeman effect measurements. *Astron. Astrophys.* 456, 665–673. doi: 10.1051/0004-6361:20064865

Krieger, A. S., Timothy, A. F., and Roelof, E. C. (1973). A coronal hole and its identification as the source of a high velocity solar wind stream. *Solar Phys.* 29, 505–525. doi: 10.1007/BF00150828

Kucera, A., Ambroz, J., Gomory, P., Kozak, M., and Rybak, J. (2010). CoMP-S - the coronal multi-channel polarimeter for Slovakia. *Contrib. Astron. Obs. Skalnate Pleso.* 40, 135–138.

Lambour, R. L., Coster, A. J., Clouser, R., Thornton, L. E., Sharma, J., and Cott, T. A. (2003). Operational impacts of space weather. *Geophys. Res. Lett.* 30, 1136. doi: 10.1029/2002gl015168

Laming, J. M. (2004). A unified picture of the fip and inverse fip effects. *Astrophys. J.* 614, 1063–1072. doi: 10.1086/423780

Laming, J. M. (2009). Non-WKB models of the FIP effect: implications for solar coronal heating and the coronal helium and neon abundances. *Astrophys. J.* 695, 954–969. doi: 10.1088/0004-637X/695/2/954

Laming, J. M. (2012). Non-WKB models of the FIP effect: the role of slow mode waves. *Astrophys. J.* 744, 115–127. doi: 10.1088/0004-637X/744/2/115

Laming, J. M. (2015). The FIP and inverse FIP effects in solar and stellar coronae. *Living Rev. Solar Phys.* 12:2. doi: 10.1007/lrsp-2015-2

Laming, J. M., Moses, J. D., Ko, Y.-K., Ng, C. K., Rakowski, C. E., and Tylka, A. J. (2013). On the remote detection of suprathermal ions in the solar corona and their role as seeds for solar energetic particle production. *Astrophys. J.* 770, 73–84. doi: 10.1088/0004-637X/770/1/73

Lin, H., and Casini, R. (2000). A classical theory of coronal emission line polarization. *Astrophys. J.* 542, 528–534. doi: 10.1086/309499

Lin, H., Kuhn, J. R., and Coulter, R. (2004). Coronal magnetic field measurements. *Astrophys. J.* 613, L177–L180. doi: 10.1086/425217

Lin, H., Penn, M. J., and Tomczyk, S. (2000). A new precise measurement of the coronal magnetic field strength. *Astrophys. J.* 541, L83–L86. doi: 10.1086/312900

Liu, J., McIntosh, S. W., De Moortel, I., Threlfall, J., and Bethge, C. (2014). Statistical evidence for the existence of alfvénic turbulence in solar coronal loops. *Astrophys. J.* 797, 7–16. doi: 10.1088/0004-637X/797/1/7

Liu, Y.-H., Drake, J. F., and Swisdak, M. (2011). The effects of strong temperature anisotropy on the kinetic structure of collisionless slow shocks and reconnection exhausts. part I. particle-in-cell simulations. *Phys. Plasmas* 18, 062110. doi: 10.1063/1.3627147

Longcope, D. W., Guidoni, S. E., and Linton, M. G. (2009). Gas-Dynamic shock heating of post-flare loops due to retraction following localized impulsive reconnection. *Astrophys. J.* 690, L18–L22. doi: 10.1088/0004-637X/690/1/L18

Lynch, B. J., Antiochos, S. K., MacNeice, P. J., Zurbuchen, T. H., and Fisk, L. A. (2004). Observable properties of the breakout model for coronal mass ejections. *Astrophys. J.* 617, 589–599. doi: 10.1086/424564

Malanushenko, A., and Schrijver, C. J. (2013). On anisotropy in expansion of magnetic flux tubes in the solar corona. *Astrophys. J.* 775, 120–135. doi: 10.1088/0004-637X/775/2/120

Mann, G., Klassen, A., Aurass, H., and Classen, H.-T. (2003). Formation and development of shock waves in the solar corona and the near-sun interplanetary space. *Astron. Astrophys.* 400, 329–336. doi: 10.1051/0004-6361:20021593

McIntosh, S. W., and De Pontieu, B. (2012). Estimating the "Dark" energy content of the solar corona. *Astrophys. J.* 761, 138–145. doi: 10.1088/0004-637X/761/2/138

McIntosh, S. W., De Pontieu, B., Carlsson, M., Hansteen, V., Boerner, P., and Goossens, M. (2011). Alfvénic waves with sufficient energy to power the quiet solar corona and fast solar wind. *Nature* 475, 477–480. doi: 10.1038/nature10235

Morton, R. J., Tomczyk, S., and Pinto, R. (2015). Investigating alfvénic wave propagation in coronal open-field regions. *Nat. Commun.* 6, 7813. doi: 10.1038/ncomms8813

Nakariakov, V. M., Ofman, L., DeLuca, E. E., Roberts, B., and Davila, J. M. (1999). TRACE observation of damped coronal loop oscillations: implications for coronal heating. *Science* 285, 862–864. doi: 10.1126/science.285.5429.862

Ng, C. K., and Reames, D. V. (2008). Shock acceleration of solar energetic protons: the first ten minutes. *Astrophys. J.* 686, L123–L126. doi: 10.1086/592996

Okamoto, T. J., and De Pontieu, B. (2011). Propagating waves along spicules. *Astrophys. J.* 736, L24–L29. doi: 10.1088/2041-8205/736/2/L24

Orozco Suárez, D., Asensio Ramos, A., and Trujillo Bueno, J. (2014). The magnetic field configuration of a solar prominence inferred from spectropolarimetric observations in the He I 10830 Å triplet. *Astron. Astrophys.* 566, A46. doi: 10.1051/0004-6361/201322903

Patsourakos, S., Vourlidas, A., and Stenborg, G. (2013). Direct evidence for a fast coronal mass ejection driven by the prior formation and subsequent destabilization of a magnetic flux rope. *Astrophys. J.* 764, 125–137. doi: 10.1088/0004-637X/764/2/125

Penn, M. J., H., Lin, S., Tomczyk, D., and Elmore, P., (Judge) (2004). Background-induced measurement errors of the coronal intensity, density, velocity and magnetic field. *Solar Phys.* 222, 61–78. doi: 10.1023/B:SOLA.0000036850.34404.5f

Pesnell, W. D., Thompson, B. J., and Chamberlain, P. C. (2012). The Solar Dynamics Observatory (SDO). *Solar Phys.* 275, 3–15. doi: 10.1007/s11207-011-9841-3

Peter, H. (2001). On the nature of the transition region from the chromosphere to the corona of the sun. *Astron. Astrophys.* 374, 1108–1120. doi: 10.1051/0004-6361:20010697

Peter, H. (2010). Asymmetries of solar coronal extreme ultraviolet emission lines. *Astron. Astrophys.* 521, A51. doi: 10.1051/0004-6361/201014433

Peter, H., Bingert, S., Klimchuk, J. A., de Forest, C., Cirtain, J. W., Golub, L., et al. (2013). Structure of solar coronal loops: from miniature to large-scale. *Astron. Astrophys.* 556, A104. doi: 10.1051/0004-6361/201321826

Rachmeler, L. A., Gibson, S. E., Dove, J. B., DeVore, C. R., and Fan, Y. (2013). Polarimetric properties of flux ropes and sheared arcades in coronal prominence cavities. *Solar Phys.* 288, 617–636. doi: 10.1007/s11207-013-0325-5

Rachmeler, L. A., Platten, S. J., Bethge, C., Seaton, D. B., and Yeates, A. R. (2014). Observations of a hybrid double-streamer/pseudostreamer in the solar corona. *Astrophys. J.* 787, L3–L8. doi: 10.1088/2041-8205/787/1/L3

Rakowski, C. E., and Laming, J. M. (2012). On the origin of the slow speed solar wind: helium abundance variations. *Astrophys. J.* 754, 65–74. doi: 10.1088/0004-637X/754/1/65

Régnier, S. (2013). Magnetic field extrapolations in the corona: success and future improvements. *Solar Phys.* 288, 481–505. doi: 10.1007/s11207-013-0367-8

Roberts, B., Edwin, P. M., and Benz, A. O. (1983). Fast pulsations in the solar corona. *Nature* 305, 688–690. doi: 10.1038/305688a0

Rouillard, A. P., Odstřcil, D., Sheeley, N. R., Tylka, A., Vourlidas, A., Mason, G., et al. (2011). Interpreting the properties of solar energetic particle events by using combined imaging and modeling of interplanetary shocks. *ApJ*, 735, 7–17. doi: 10.1088/0004-637X/735/1/7

Rybak, J., Ambroz, J., Gomory, P., Habaj, P., Koza, J., Kozak, M., et al. (2013). *Coronal Multi-channel Polarimeter at the Lomnicky Peak Observatory: CoMP-S@LSO, Kanzelhoehe Colloquium*, (Treffen), Austria. Available online at: http://www.astro.sk/~choc/open/apvv_0816-11/output/presentations/2013_kanzelhoehe_colloquium_2013/rybak_comp-s_at_lso_kso70.pdf

Sakai, J. I., Minamizuka, R., Kawata, T., and Cramer, N. F. (2001). Nonlinear torsional and compressional waves in a magnetic flux tube with electric current near the quiet solar photospheric network. *Astrophys. J.* 550, 1075–1092. doi: 10.1086/319802

Savage, S. L., McKenzie, D. E., Reeves, K. K., Forbes, T. G., and Longcope, D. W. (2010). Reconnection outflows and current sheet observed with Hinode/XRT in the 2008 April 9 "Cartwheel CME" Flare. *Astrophys. J.* 722, 329–342. doi: 10.1088/0004-637X/722/1/329

Scherrer, P. H., Bogart, R. S., Bush, R. I., Hoeksema, J. T., Kosovichev, A. G., Schou, J., et al. (1995). The solar oscillations investigation–michelson doppler imager. *Sol. Phys.* 162, 129–188. doi: 10.1007/BF00733429

Schrijver, C. J., Title, A. M., Berger, T. E., Fletcher, L., Hurlburt, N. E., Nightingale, R. W., et al. (1999). A new view of the solar outer atmosphere by the transition region and coronal explorer. *Sol. Phys.* 187, 261–302. doi: 10.1023/A:1005194519642

Sturrock, P. A. (1999). Chromospheric magnetic reconnection and its possible relationship to coronal heating. *Astrophys. J.* 521, 451–459. doi: 10.1086/307544

Testa, P., De Pontieu, B., Martínez-Sykora, J., DeLuca, E., Hansteen, V., Cirtain, J., et al. (2013). Observing coronal nanoflares in active region moss. *Astrophys. J.* 770, L1–L7. doi: 10.1088/2041-8205/770/1/L1

Thurgood, J. O., and McLaughlin, J. A. (2013). 3D Alfvén wave behaviour around proper and improper magnetic null points. *Astron. Astrophys.* 558, A127. doi: 10.1051/0004-6361/201322021

Tian, H., Tomczyk, S., McIntosh, S. W., Bethge, C., de Toma, G., and Gibson, S. (2013). Observations of coronal mass ejections with the coronal multichannel polarimeter. *Solar Phys.* 288, 637–650. doi: 10.1007/s11207-013-0317-5

Tomczyk, S., Card, G. L., Darnell, T., Elmore, D. F., Lull, R., Nelson, P. G., et al. (2008). An instrument to measure coronal emission line polarization. *Solar Phys.* 247, 411–428. doi: 10.1007/s11207-007-9103-6

Tomczyk, S., and McIntosh, S. W. (2009). Time-distance seismology of the solar corona with CoMP. *Astrophys. J.* 697, 1384–1391. doi: 10.1088/0004-637X/697/2/1384

Tomczyk, S., McIntosh, S. W., Keil, S. L., Judge, P. G., Schad, T., Seeley, D. H., et al. (2007). Alfvén waves in the solar corona. *Science* 317, 1192–1196. doi: 10.1126/science.1143304

Tylka, A. J., and Lee, M. A. (2006). A model for spectral and compositional variability at high energies in large, gradual solar particle events. *Astrophys. J.* 646, 1319–1334. doi: 10.1086/505106

Ugarte-Urra, I., Warren, H. P., and Wineberger, A. R. (2007). The magnetic topology of coronal mass ejection sources. *Astrophys. J.* 662, 1293–1301. doi: 10.1086/514814

Van Doorsselaere, T., Nakariakov, V. M., and Verwichte, E. (2008). Detection of waves in the solar corona: kink or alfvén? *Astrophys. J.* 676, L73–L75. doi: 10.1086/587029

Vasheghani Farahani, S., Nakariakov, V. M., and Doorsselaere, T. (2010). Long wavelength torsional modes of solar coronal plasma structures. *Astron. Astrophys.* 517, A29. doi: 10.1051/0004-6361/201014502

von Steiger, R., Wimmer Schweingruber, R. F., Geiss, J., and Gloeckler, G. (1995). Abundance variations in the solar wind. *Adv. Space Res.* 15, 3–12. doi: 10.1016/0273-1177(94)00013-Q

Vourlidas, A., Wu, S. T., Wang, A. H., Subramanian, P., and Howard, R. A. (2003). Direct detection of a coronal mass ejection-associated shock in large angle and spectrometric coronagraph experiment white-light images. *Astrophys. J.* 598, 1392–1402. doi: 10.1086/379098

Wang, T. J., Ofman, L., and Davila, J. M. (2009). Propagating slow magnetoacoustic waves in coronal loops observed by hinode/EIS. *Astrophys. J.* 696, 1448–1460. doi: 10.1088/0004-637X/696/2/1448

Warren, H. P., Winebarger, A. R., and Brooks, D. H. (2010). Evidence for steady heating: observations of an active region core with Hinode and TRACE. *Astrophys. J.* 711, 228–238. doi: 10.1088/0004-637X/711/1/228

Wiegelmann, T., and Sakurai, T. (2012). Solar force-free magnetic fields. *Living Rev. Solar Phys.* 9:5. doi: 10.12942/lrsp-2012-5

Conflict of Interest Statement: The authors declare that the research was conducted in the absence of any commercial or financial relationships that could be construed as a potential conflict of interest.

12

Self-Gravitating Stellar Collapse: Explicit Geodesics and Path Integration

Jayashree Balakrishna[1], Ruxandra Bondarescu[2] and Christine C. Moran[3*]

[1] Department of Mathematics and Natural Sciences, College of Arts and Sciences, Harris-Stowe State University, St. Louis, MO, USA, [2] Department of Physics, University of Zurich, Zurich, Switzerland, [3] TAPIR, Department of Theoretical Astrophysics, California Institute of Technology, Pasadena, CA, USA

We extend the work of Oppenheimer and Synder to model the gravitational collapse of a star to a black hole by including quantum mechanical effects. We first derive closed-form solutions for classical paths followed by a particle on the surface of the collapsing star in Schwarzschild and Kruskal coordinates for space-like, time-like, and light-like geodesics. We next present an application of these paths to model the collapse of ultra-light dark matter particles, which necessitates incorporating quantum effects. To do so we treat a particle on the surface of the star as a wavepacket and integrate over all possible paths taken by the particle. The waveform is computed in Schwarzschild coordinates and found to exhibit an ingoing and an outgoing component, where the former contains the probability of collapse, while the latter contains the probability that the star will disperse. These calculations pave the way for investigating the possibility of quantum collapse that does not lead to black hole formation as well as for exploring the nature of the wavefunction inside $r = 2M$.

Keywords: ultra-light particles, black hole formation, quantum effects, self-gravitating stellar collapse, geodesics inside black holes

Edited by:
Martin Anthony Hendry,
University of Glasgow, UK

Reviewed by:
Kazuharu Bamba,
Fukushima University, Japan
Siamak Akhshabi,
Golestan University, Iran

*Correspondence:
Christine C. Moran
corbett@tapir.caltech.edu

Specialty section:
This article was submitted to
Cosmology,
a section of the journal
Frontiers in Astronomy and Space
Sciences

1. INTRODUCTION

Black holes play a pivotal role in the evolution of the universe providing an important test laboratory for general relativity. Within classical general relativity, Oppenheimer and Synder modeled the gravitational collapse of star to a black hole by approximating the star with a uniform sphere of dust (hereafter O-S model) (Oppenheimer and Snyder, 1939). This model provides an analytic solution for stellar collapse that connects the Schwarzschild exterior of a star to a contracting Friedmann-Robertson-Walker (FRW) interior. Once the surface has passed within $r = 2M$, no internal pressures can halt the collapse and all configurations collapse to a point-like singularity at $r = 0$. The general features of this toy collapse model have been examined by many authors (Vaidya, 1951; Misner et al., 1973; Singh and Witten, 1997; Goswami and Joshi, 2004; Joshi, 2007).

In the standard formalism from Misner et al. (1973), the stellar surface is considered to be initially at rest. Here we consider configurations with all possible initial velocities. We derive closed-form solutions for the equations of motion in Schwarzschild and Kruskal coordinates for space-like, time-like, and light-like geodesics.

As an example application of our closed-form solutions for the classical O-S model with non-zero initial velocities, we consider the macroscopic collapse of a spherically symmetric sphere of dust composed of ultra-light particles. To approximate quantum effects, the radius of the star is approximated by a Gaussian wavepacket that is initially centered far from $r = 2M$. Its evolution is then followed via a simple path integral approach that extends the results of Redmount and Suen (1993) from a relativistic free particle to a particle constrained by non-trivial gravity. Gravitational collapse incorporating approximate treatments of quantum mechanics has been considered by a variety of authors (Hájíček et al., 1992; Hawkins, 1994; Casadio and Venturi, 1996; Ansoldi et al., 1997; Berezin, 1997; Zloshchastiev, 1998; Alberghi et al., 1999; Vaz and Witten, 2001; Corichi et al., 2002; Ortíz and Ryan, 2007). They involve non-equivalent ways of quantization that often produce physically different results (Dolgov and Khriplovich, 1997). Our path-integral approach approach is simpler, and can be easily compared to the assumption that the particle obeys the relativistic Schrödinger equation.

In the classical model a star is idealized as a collapsing self-gravitating dust sphere of uniform density and zero pressure where the constituent particles have the attributes of classical dust: each particle is assumed to be infinitesimal in size and to interact only gravitationally with other matter. The inclusion of quantum mechanical effects lifts some of these assumptions allowing for the possibility that some configurations will not collapse to black holes but will disperse or even form stable new configurations. Quantitatively, quantum treatment is necessary in macroscopic stellar collapse when the action S is of the order $\hbar$ (Narlikar, 1977). In our case, the $S/\hbar \sim 1$ condition corresponds to $mM \sim \hbar$. For macroscopic black holes of masses $M = 1M_\odot - 10^9 M_\odot$, this implies that quantum treatment is necessary for ultra-light constituent particles of $m \sim 10^{-10} - 10^{-19}$ eV. In nature, such ultralight particle could be dark matter. Dark matter comes close to the attributes of classical dust, and some dark matter clouds may be dense enough to collapse to black holes. Particles as light as $10^{-22} - 10^{-23}$ eV have been proposed as constituents of dark matter halos (Hu et al., 2000; Lesgourgues et al., 2002; Lundgren et al., 2010; Khmelnitsky and Rubakov, 2014).

To approximate quantum effects of the collapse of such a halo comprised of ultra-light particles, we start with an initial wavefunction that represents the position of the particle on the stellar surface, and is at first far from $2M$. To study the propagation of the wavefunction, we integrate over all possible paths taken by the particle. At a given time, the outgoing wavefunction comprises the probability that the star disperses, and the ingoing wavefunction the probability that it collapses. We compute the propagator in analytical form for a particle on the surface of the star in the WKB approximation, and compare this to the limited assumption that the particle obeys the relativistic Schrödinger equation. Our equations reduce to a free particle case when the mass of the star is zero; then the solution to the relativistic Schrödinger equation is the exact representation of the wave function (Redmount and Suen, 1993). In the more general case of a particle on the surface of a star, this representation is no longer correct. However, the comparison is instructive. As expected, we observed that the WKB and relativistic Schrödinger approximations are out of step at early and late times,

and appear to converge toward one another at intermediate times.

In Schwarzschild coordinates, we can follow the evolution of the surface of the star only until the formation of an apparent horizon due to the $r = 2M$ coordinate singularity. In Kruskal coordinates, one can continue to study the evolution of the star inside the apparent horizon ($r < 2M$). The paths we have derived here include time-reversing space-like paths, which allow for the possibility of extraction of information from inside the horizon to the outside. These paths turn inside the horizon $r > 0$, and head toward $r = 2M$. In future work, explicit geodesic equations in Kruskal coordinates may be used as a stepping stone to model behavior inside $r = 2M$.

The rest of the paper is structured as follows: in Section 2 we describe the Oppenheimer–Snyder model. Section 3 includes an overview of the classical action and paths, a computation of the Oppenheimer–Snyder limit and the derivation of the classical paths in Schwarzschild coordinates. Section 4 details the classical paths in Kruskal coordinates, which comprise a starting point for future work that explores the collapse inside $r = 2M$. As an example application of the closed form geodesics, Section 5 describes the quantum treatment of the dust collapse applicable to ultra-light particles in Schwarzschild coordinates. The conclusions follow in Section 6.

2. THE OPPENHEIMER–SNYDER MODEL

In General Relativity, a first approximation to the exterior space-time of any star, planet or black hole is a spherically symmetric space-time modeled by the Schwarzschild metric. This is a consequence of Birkhoff's theorem (Misner et al., 1973). The Schwarzschild line element thus takes the usual form

$$ds^2 = -\left(1 - \frac{2M}{r}\right) dt^2 + \frac{dr^2}{1 - 2M/r} + r^2 d\Omega^2, \quad (1)$$

where

$$d\Omega^2 = d\theta^2 + \sin^2\theta \, d\phi^2. \quad (2)$$

Throughout this paper we use geometric units with $G = c = 1$.

The Oppenheimer–Snyder (O-S) model follows the collapse of a star that is idealized as a dust sphere with uniform density and zero pressure from the perspective of an observer located on the surface of the star. The motion of the collapsing surface initially at radius r_i can be parametrized by

$$r = \frac{r_i}{2}(1 + \cos\eta) \quad (3)$$

$$\tau = \frac{r_i}{2}\sqrt{\frac{r_i}{2M}}(\eta + \sin\eta) \quad (4)$$

$$t = 2M \log \left| \frac{\sqrt{r_i/2M - 1} - \tan\eta/2}{\sqrt{r_i/2M - 1} + \tan\eta/2} \right| \quad (5)$$

$$+ 2M\sqrt{\frac{r_i}{2M} - 1}\left[\eta + \frac{r_i}{4M}(\eta + \sin\eta)\right]$$

These equations correspond to Equations (32.10a–32.10c) of Misner et al. (1973). The star collapses to a singularity in finite proper time. However, it takes an infinite Schwarzschild t to reach the apparent horizon at $r = 2M$ and thus an external observer will never see the star passing its gravitational radius ($r = 2M$).

In Misner et al. (1973) the stellar surface is considered to be initially at rest. Here we integrate the equations of motion for configurations with all possible initial velocities.

3. CLASSICAL ACTION AND RADIAL GEODESICS IN SCHWARZSCHILD COORDINATES

The relativistic action S and the Lagrangian $\mathcal{L}$ for this system are

$$S_{cl} = -m \int d\tau = -m \int dt \sqrt{\left(1 - \frac{2M}{r}\right) - \frac{\dot{r}^2}{1 - 2M/r}}, \quad (6)$$

where τ is the proper time, and $\dot{r} = dr/dt$. The Lagrangian is

$$\mathcal{L} = -m \sqrt{\left(1 - \frac{2M}{r}\right) - \frac{\dot{r}^2}{1 - 2M/r}} \quad (7)$$

The momentum

$$p = \frac{\partial \mathcal{L}}{\partial \dot{r}} = \frac{m\dot{r}}{\sqrt{(1 - 2M/r)\left[(1 - 2M/r)^2 - \dot{r}^2\right]}} \quad (8)$$

and the Hamiltonian is

$$H = p\dot{r} - \mathcal{L} = \sqrt{1 - \frac{2M}{r}} \sqrt{m^2 + p^2 \left(1 - \frac{2M}{r}\right)}, \quad (9)$$

which reduces to the free particle Hamiltonian in the $M \to 0$ limit. The equation of motion derived from the Euler-Lagrange equations is

$$-\frac{3M}{r^2}\left(\frac{dr}{dt}\right)^2 + \left(1 - \frac{2M}{r}\right)\frac{d^2r}{dt^2} + \frac{M}{r^2}\left(1 - \frac{2M}{r}\right)^2 = 0, \quad (10)$$

which can be integrated to

$$\dot{r}^2 = \left(1 - \frac{2M}{r}\right)^2 \left[c_1 + \frac{2M}{r}(1 - c_1)\right]. \quad (11)$$

From this the classical paths are determined

$$\int_{t_i}^{t_f} dt = t_f - t_i = \pm \int_{r_i}^{r_f} dr \frac{r^2}{x_r(r - 2M)}, \quad (12)$$

where

$$x_r = \sqrt{c_1 r^2 + 2Mr(1 - c_1)}. \quad (13)$$

The + sign represents motions from r_i to $r_f > r_i$, and − sign represents the star collapsing from r_i to $r_f < r_i$. The parameter c_1 can be used to separate the space-time regions

$$ds^2 > 0 \iff c_1 > 1 \quad \text{outside the light cone} \quad (14)$$
$$ds^2 = 0 \iff c_1 = 1 \quad \text{on the light cone}$$
$$ds^2 < 0 \iff c_1 < 1 \quad \text{inside the light cone} \quad (15)$$

In the free particle case ($M \to 0$ in Equation 11), $c_1 = \dot{r}^2$ represents the velocity squared of the particle.

3.1. Stellar Surface at Rest Limit

When the surface of the star is initially at rest, initial velocity

$$\left.\frac{dr}{dt}\right|_{r=r_i} = 0.$$

From Equation (11), this corresponds to

$$c_1 = -\frac{2M}{r_i - 2M}. \quad (16)$$

We can recover our equations of motion for this c_1 value from the O-S model. We proceed by taking the derivative of Equations (3) and (5) w.r.t. η

$$\frac{dr}{d\eta} = -\frac{r_i}{2}\sin\eta = -\sqrt{r(r_i - r)} \quad (17)$$

$$\frac{dt}{d\eta} = \frac{r_i^2}{2M}\sqrt{\frac{r_i}{2M} - 1}\left[\frac{\cos^4(\eta/2)}{(r_i/2M)\cos^2(\eta/2) - 1}\right]$$
$$= \frac{r^2}{r - 2M}\sqrt{\frac{r_i}{2M} - 1},$$

where we used that $r = r_i \cos^2 \eta/2$. We then divide $dr/d\eta$ by $dt/d\eta$, and use $r_i = 2M(c_1 - 1)/c_1$ from Equation (16) to recover the equation of motion

$$\frac{dr}{dt} = -\frac{r - 2M}{r^2}\sqrt{c_1 r^2 + 2Mr(1 - c_1)}. \quad (18)$$

The negative sign − fits in with our convention for a collapsing star. While some paths will have low probability, a path integral approach that includes quantum effects requires the consideration of space-like, time-like, and light-like paths with all possible initial velocities.

3.2. Classical Paths in Schwarzschild Coordinates

Schwarzschild coordinates allow us to explore the space-time outside $r = 2M$. The possible paths taken by a particle on the surface of a collapsing star are written explicitly in **Table 1**. Every point (r_f, t_f) may be reached either via a direct path or via an indirect path. **Table 2** contains the boundary regions that determine whether a path is direct or indirect, and whether it lies inside or outside the light cone.

The table uses

$$x(r_s) = x_s = \sqrt{c_1 r_s^2 + 2Mr_s(1 - c_1)}, \quad (19)$$

$$\beta_s = \sqrt{c_1}x_s + c_1 r_s + M(1 - c_1), \quad (20)$$

$$\alpha_s = \frac{x_s}{c_1} + 2M\log(r_s - 2M) - 2M\log\left(\frac{(r_s + x_s)^2}{r_s}\right), \quad (21)$$

where $s = i, f, a$.

TABLE 1 | Closed-form solution to Equation (12) in Schwarzschild coordinates.

Range of r_i	c_1	Equations of motion
$2M - b_2$	$c_1 > 0$	$t_f - t_i = \alpha_f - \alpha_i + M\frac{(3c_1 - 1)}{c_1\sqrt{c_1}} \log\left(\frac{\beta_f}{\beta_i}\right)$
$b_2 - b_3$	$-\frac{2M}{r_f - 2M} < c_1 < 0$	$t_f - t_i = \alpha_f - \alpha_i + \frac{M(3c_1 - 1)}{c_1\sqrt{-c_1}}\left[\arcsin\left(\frac{-c_1 r_f}{M(1 - c_1)} - 1\right) - \arcsin\left(\frac{-c_1 r_i}{M(1 - c_1)} - 1\right)\right]$
$b_3 - b_4$	$-\frac{2M}{r_a - 2M}$	$t_f - t_i = 2\alpha_a - \alpha_f - \alpha_i + M\frac{(3c_1 - 1)}{c_1\sqrt{-c_1}}\left[2\arcsin(1) - \arcsin\left(\frac{-c_1 r_f}{M(1 - c_1)} - 1\right) - \arcsin\left(\frac{-c_1 r_i}{M(1 - c_1)} - 1\right)\right]$
$b_4 - b_5$	$-\frac{2M}{b_4 - 2M} < c_1 < 0$	$t_i - t_f = \alpha_f - \alpha_i + \frac{M(3c_1 - 1)}{c_1\sqrt{-c_1}}\left[\arcsin\left(\frac{-c_1 r_f}{M(1 - c_1)} - 1\right) - \arcsin\left(\frac{-c_1 r_i}{M(1 - c_1)} - 1\right)\right]$
$b_5 - \infty$	$c_1 > 0$	$t_i - t_f = \alpha_f - \alpha_i + M\frac{(3c_1 - 1)}{c_1\sqrt{c_1}} \log\left(\frac{\beta_f}{\beta_i}\right)$

*At each region boundary b_j ($j = 2 - 5$) the equations converge to the form given in **Table 2**. All discontinuities cancel resulting in smooth transitions between regions without singularities.*

TABLE 2 | Region boundaries for the equation of motion (Equation 12).

r_i	c_1	Regional boundaries
b_1	1	$t_f - t_i = r_f - b_1 + 2M\log\left(\frac{r_f - 2M}{b_1 - 2M}\right)$
b_2	0	$t_f - t_i = \frac{\sqrt{2}}{3\sqrt{M}}\left(r_f^{3/2} - b_2^{3/2}\right) + 2\sqrt{2M}\left(\sqrt{r_f} - \sqrt{b_2}\right) + 2M\log\left[\frac{\left(1 - \sqrt{\frac{2M}{r_f}}\right)\left(1 + \sqrt{\frac{2M}{b_2}}\right)}{\left(1 + \sqrt{\frac{2M}{r_f}}\right)\left(1 - \sqrt{\frac{2M}{b_2}}\right)}\right]$
b_3	$-\frac{2M}{r_f - 2M}$	$t_f - t_i = 2M\log\left(1 - \frac{2M}{r_f}\right) - \frac{x_{b_3}}{c_1} + 2M\log\left[\frac{(x_{b_3} + b_3)^2}{b_3(b_3 - 2M)}\right] + M\frac{(3c_1 - 1)}{c_1\sqrt{-c_1}}\left[\arcsin(1) - \arcsin\left(\frac{-b_3 c_1}{M(1 - c_1)} - 1\right)\right]$
b_4	$-\frac{2M}{b_4 - 2M}$	$t_i - t_f = \frac{x_f}{c_1} - 2M\log(1 - \frac{2M}{b_4}) - 2M\log\left[\frac{(x_f + r_f)^2}{r_f(r_f - 2M)}\right] + M\frac{(3c_1 - 1)}{c_1\sqrt{-c_1}}\left[-\arcsin(1) + \arcsin\left(\frac{-r_f c_1}{M(1 - c_1)} - 1\right)\right]$
b_5	0	$t_i - t_f = \frac{\sqrt{2}}{3\sqrt{M}}\left(r_f^{3/2} - b_5^{3/2}\right) + 2\sqrt{2M}\left(\sqrt{r_f} - \sqrt{b_5}\right) + 2M\log\left[\frac{\left(1 - \sqrt{\frac{2M}{r_f}}\right)\left(1 + \sqrt{\frac{2M}{b_5}}\right)}{\left(1 + \sqrt{\frac{2M}{r_f}}\right)\left(1 - \sqrt{\frac{2M}{b_5}}\right)}\right]$
b_6	1	$t_i - t_f = r_f - b_6 + 2M\log\left(\frac{r_f - 2M}{b_6 - 2M}\right)$

The first column is the border value of r_i for each region given the r_f and c_1 values. It is calculated using the equation of motion in the third column. The light cone boundary corresponds to $c_1 = 1$.

The direct paths include a logarithmic term when $c_1 > 0$ and an arc sin term when $c_1 < 0$. The indirect paths occur for r_i in the $b_3 - b_4$ region. They each contain a zero velocity point where the path turns around. Mathematically, the velocity passes through zero ($dr/dt = 0$ at some radius r_a) when $x(r_a) = 0$, which corresponds to $c_1 = -2M/(r_a - 2M)$. The indirect paths reach the final point (r_f, t_f) after going through a zero velocity point at $r_a(t_a < t_f)$ with $r_a > r_f$ and $r_a > r_i$. The paths from r_i to r_a correspond to the positive sign in Equation (12), while the path from r_a to r_f corresponds to the negative sign.

To understand these paths we make an analogy with the vertical motion of a ball under gravity (See **Figure 1**). This analogy arises naturally due to the correspondence between the Friedmann equation and the equation of motion for a test particle in the gravitational field of a point mass in Newtonian gravity. It can be made only in the Schwarzschild case because then the interior of the star can be represented by an FRW model, which provides the link to Newtonian gravity.

Consider a ball traveling from point r_i (initial radius) to point r_f (final radius) with r_f below r_i. It can either go down directly from r_i to r_f (direct path) or it can go up from r_i, reach its highest point, where its velocity will be zero, and then fall back down to r_f (indirect path). This is shown schematically in **Figure 1A**. Analogously, if a ball has to travel from r_i to r_f with r_f above r_i, it can go directly from r_i to r_f or it can pass above r_f, reach its zero velocity point above r_f and return down to r_f. This case can be seen in **Figure 1C**. The middle figure (**Figure 1B**) shows the special case when $r_i = r_f$. Then the ball can either stay put or it can be thrown up, reach its higher point, and fall back down. Each of these paths takes a different amount of time to complete. For a fixed travel time Δt, the two points r_i and r_f are connected by a single path that is either a direct path or a turn-around path. At the turning point of a turn-around path (also its highest point) the speed of the particle is always zero.

If we keep the time interval constant, we find that for a given r_f there is a point $r_i = b_4 > r_f$, where the velocity dr/dt at $b_4 = 0$. Similarly, there is a point $r_i = b_3 < r_f$ such that the velocity dr/dt at $r_f = 0$. For any point r_i between b_3 and b_4, the only way a particle can reach r_f in the same time Δt is if it goes through a turning point $r_a > r_f$ and $r_a > r_i$. The particle cannot reach $r_f > r_i$ if it's velocity vanishes at any point between r_i and r_f.

The motion toward or away from $r = 2M$ can be compared to the vertical motion of a ball under gravity when thrown straight down or up, respectively. Classically, the star will always collapse to a black hole. Quantum mechanically, we include all possible initial velocities of the surface of the star toward and away from $2M$. The wavefunction of a particle on the surface of the star determines the spread of the trajectory of the surface of the star itself.

3.3. Escape Velocity

For a particle of mass m that is thrown up from the surface of the Earth, an initial upward velocity $v_{escape} \geq 11.2$ km/s ensures that the particle escapes Earth gravity. At this speed

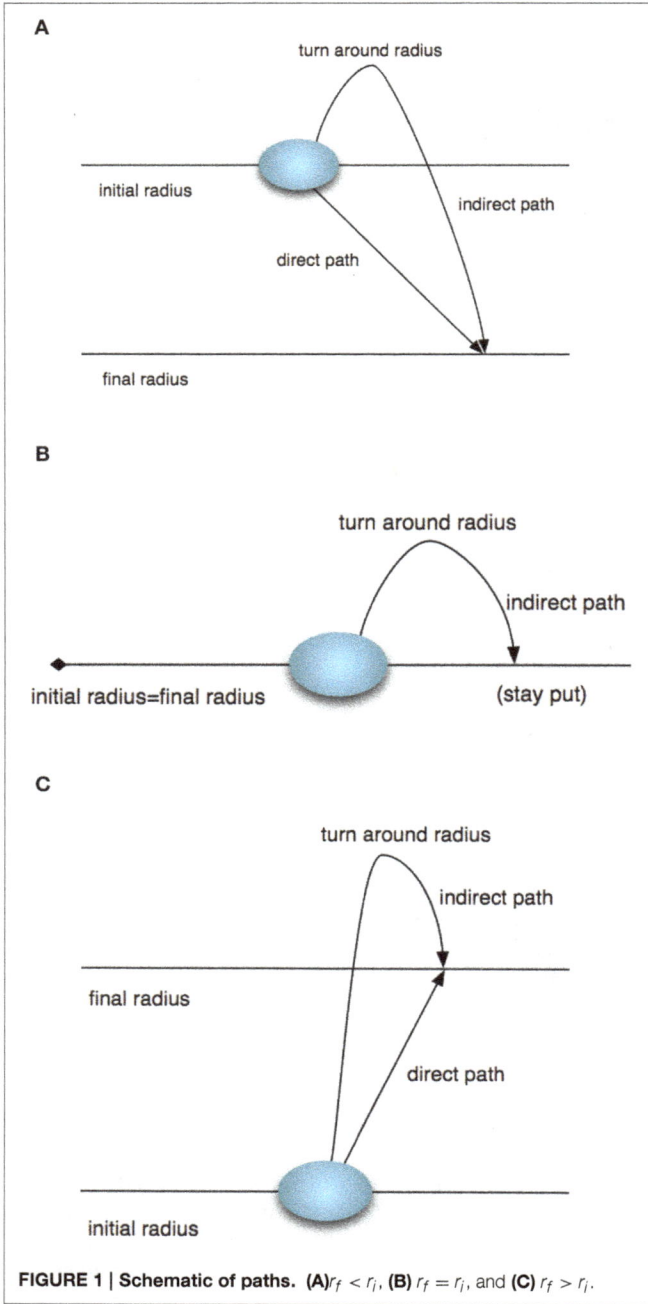

FIGURE 1 | Schematic of paths. (A)$r_f < r_i$, (B) $r_f = r_i$, and (C) $r_f > r_i$.

$$\frac{GM_E m}{R_E} = \frac{mv_i^2}{2},\qquad (22)$$

where M_E, and R_E are the Earth mass and radius.

We continue with the ball analogy to estimate the escape velocity for our particle. In our case, R_E is replaced by the initial radius of the star r_i, and $v_i = dr/d\tau$. The kinetic energy of the particle on the surface of the star equals its potential energy when

$$\frac{Mm}{r_i} = \frac{m}{2}\left|\frac{dr}{d\tau}\right|_{r=r_i}^2.\qquad (23)$$

Here

$$\frac{dr}{d\tau} = \left(1 - \frac{2M}{r}\right)^{-1}\frac{dr}{dt}\qquad (24)$$

with $dr/dt = \dot{r}$ given by Equation (11). This results in

$$\frac{Mm}{r_i} = \frac{mc_1}{2} + \frac{Mm}{r_i}(1 - c_1),\qquad (25)$$

whose solution is $c_1 = 0$. So, while all outgoing paths with $c_1 < 0$ will turn-around, outgoing paths with $c_1 \geq 0$ will not have turning points. When $c_1 = 0$, the turning point ($dr/d\tau|_{r_a} = 0$, $x_a = 0$) occurs at $r_a \to \infty$ as expected for an escaped particle.

The different paths (**Table 1**) and their various regional boundaries (**Table 2**) are summarized below for a given (r_f, t_f) outside $r = 2M$:

- **$c_1 = 1$**: the paths are light-like ($ds^2 = 0$) with $r_i = b_1 < r_f$ and $r_i = b_6 > r_f$. The paths with $c_1 > 1$ are space-like, while those with $c_1 < 1$ are time-like.
- **$c_1 = 0$**: $r_i = b_2 < r_f$ and $r_i = b_5 > r_f$. This path defines the escape velocity. All paths with $c_1 > 0$ that are outgoing will not reverse, allowing the particle to escape the gravity of the star.
- **$c_1 = -2M/(r_f - 2M)$**: $r_i = b_3 < r_f$. The minimum value of r_i where r_f is reached directly with $x(r_f) = 0$ and null final velocity

$$\frac{dr}{dt}\Big|_{r=r_f} = 0.$$

Paths with $r_i \leq b_3$ are direct. Paths with r_i between b_3 and b_4 are indirect.
- **$c_1 = -2M/(b_4 - 2M)$**: $r_i = b_4 > r_f$. This path starts with zero initial velocity. All paths with $r_i > b_4$ are direct.

4. CLASSICAL PATHS IN KRUSKAL COORDINATES

Our study is the first to consider space-like paths in Kruskal coordinates for a particle on the surface of a collapsing star. They have not been explicitly considered before because they are believed to be of low probability. We conjecture that they play an important role in quantum mechanical collapse in the same way the probability of passing through the potential barier is important in tunneling. Classically, it can never happen, and yet this probability cannot be ignored in quantum mechanics.

In this section we derive and discuss the closed-form solutions to the equation of motion for the space-like, light-like and time-like paths in Kruskal coordinates. We find that space-like geodesics have the interesting property that they can turn around outside $r = 2M$ and move back in time. Unlike the Schwarzschild t and r, which are time and space coordinates, respectively, outside $r = 2M$ but switch roles for $r < 2M$, the Kruskal coordinate v is always a time coordinate and the coordinate u is always a space coordinate. The relation between the Schwarzschild r and t and the Kruskal variables u and v is given by

$$u = \begin{cases} \sqrt{\frac{r}{2M} - 1}\, e^{r/4M} \cosh \frac{t}{4M} & r > 2M \\ \sqrt{1 - \frac{r}{2M}}\, e^{r/4M} \sinh \frac{t}{4M} & r < 2M \end{cases}$$

$$v = \begin{cases} \sqrt{\frac{r}{2M} - 1}\, e^{r/4M} \sinh \frac{t}{4M} & r > 2M \\ \sqrt{1 - \frac{r}{2M}}\, e^{r/4M} \cosh \frac{t}{4M} & r < 2M \end{cases}$$

with

$$u^2 - v^2 = \left(\frac{r}{2M} - 1\right) e^{r/2M} \tag{26}$$

and

$$\frac{v}{u} = \tanh \frac{t}{4M}, \quad r > 2M \tag{27}$$

$$\frac{u}{v} = \tanh \frac{t}{4M}, \quad r < 2M.$$

Thus the line element in Kruskal coordinates takes the form

$$ds^2 = \frac{32M^3}{r} e^{-r/2M} \left(du^2 - dv^2\right) + r^2 d\Omega^2. \tag{28}$$

The singularity at $r = 0$ occurs at $u^2 - v^2 = -1$. In this paper we consider only the first quadrant $u > 0$ and $v > 0$. The horizon $r = 2M$ occurs at $u = v$. Outside the horizon $u > v$, whereas inside the horizon $u < v$.

The classical equation of motion

$$\frac{\ddot{u}}{1 - \dot{u}^2} = \left(1 - \frac{4M^2}{r^2}\right) \frac{u - v\dot{u}}{u^2 - v^2} \tag{29}$$

can be derived from the Euler-Lagrange equations

$$\frac{d}{dv}\frac{\partial \mathcal{L}}{\partial \dot{u}} - \frac{\partial \mathcal{L}}{\partial u} = 0, \tag{30}$$

for the Lagrangian

$$\mathcal{L}(u, \dot{u}, v) = -4M \sqrt{\frac{(1 - \dot{u}^2)(1 - 2M/r)}{(u^2 - v^2)}}. \tag{31}$$

Note that here $\dot{u} = du/dv$.

The equation of motion can be reduced to

$$\frac{du}{dv} = \frac{v/u \pm x_r/r}{1 \pm (v/u)(x_r/r)}, \tag{32}$$

where x_r is given by Equation (13). Direct paths from r_i to $r_f > r_i$ have the "+"-sign in the numerator and denominator with $du/dv > 0$. Direct paths from r_i to r_f with $r_f < r_i$ will have the "−"-sign in the numerator and denominator.

The classical paths describing the evolution of the star outside $r = 2M$ have been discussed in Sec. 3.2 in Schwarzschild

coordinates. It was shown that there were direct space-like and light-like paths as well as direct and indirect time-like paths. Any two points in (r, t) were connected by a unique classical path.

What is new in Kruskal space-time are the turning points in the $u - v$ plane for all space-like, light-like and time-like paths. Spacelike paths ($c_1 > 1$, $|du/dv| > 1$) turn in time at points when $dv/du = 0$, whereas timelike paths ($c_1 < 1$, $|du/dv| < 1$) turn in space when $du/dv = 0$. Additionally, there are indirect space-like and light-like paths that have turning points in r inside the apparent horizon (here we refer to paths as being indirect when they turn in r; the light-like paths turn at $r = 0$). In contrast to the time-like paths, the in-going space-like paths are *not* unique. A point outside $r = 2M$ might be connected to a point inside the apparent horizon via a direct and an indirect space-like path or via two indirect space-like paths.

We determine the regions (**Table 3**) and the light-cone boundaries (**Table 4**) that delimit the different kinds of paths that start at $(u_i, v_i = 0)$ outside the horizon and reach (u_f, v_f) inside the apparent horizon. Outside the horizon, the light cone boundary values of "u_i" for a given u_f and v_f can be determined by replacing $t_f - t_i$ in **Table 2** by $\tanh^{-1}(v_f/u_f)$. The boundary values of r_i ($v_i = 0$) with $r_f < 2M$, are calculated from the equations in **Table 4**. Note that u_f is a function of v_f and r_f and u_i depends on v_i, which is typically taken to be zero, and r_i. Thus if one finds r_i, this determines u_i. The point $r_i = b'_4$ corresponds to $x_i = 0$ with $c_1 = -2M/(b'_4 - 2M)$. The $c_1 = 0$ and $c_1 = 1$ (lightlike) paths that reach points (u_f, v_f) inside the horizon are used to determine the r_i values b'_5 and b'_{56}, respectively. The $r_i = b'_6$ value corresponds to $x_f = 0$ with $c_1 = -2M/(r_f - 2M)$ (turning point at $r = r_f < 2M$, space-like path, $c_1 > 1$). The value of $r_i = b'_7$ is to be determined by using $c_1 = -2M/(r_a - 2M)$ ($x_a = 0, r_a < r_f < 2M$) in the equation of motion. The two unknowns b'_7 and r_a are determined by using the equation of motion and its partial derivative with respect to c_1. Both equations are given in **Table 4**. In the region outside b'_7 ($r_i > b'_7$), there exist no classical paths to (u_f, v_f).

In **Figures 2A,B** we trace direct time-like, direct space-like, and indirect space-like trajectories that start at $v_i = 0$ and pass through a fixed point $\mathbf{P}(u_P = 9.995, v_P = 10.0)$ with $r_\mathbf{P} \approx 0.8M$, which is located inside the apparent horizon. **Path 1** and **path 2** are time-like, and originate at different points outside $r = 2M$. **Path 1** ($u_i = 2.59$, $c_1 = -1.05$) lies in the $b'_4 - b'_5$ region of **Table 3**. **Path 2** ($u_i = 16.15$, $c_1 = 0.8$) lies in the $b'_5 - b'_{56}$ region, which contains time-like paths with a turning point in the $u - v$ plane ($du/dv = 0$) that lies outside the apparent horizon.

The next region is $b'_{56} - b'_6$, where each point outside the horizon is connected to points inside the horizon by two space-like paths: one direct and one indirect. **Path 3a** and **path 3b** originate at the same $u_i = 27.6$ ($r_i = 10.5M$). **Path 3a** reaches P directly. However, after having passed through P, it turns at $r_a = 0.48M$ before reaching the horizon like all space-like paths. **Path 3b** is indirect turning at $r_a = 0.3M$ ($u_a = 11.3, v_a = 11.35$), and then reaching P on its way back toward the apparent horizon. Both paths also have a $v - u$ turning point outside the apparent horizon where $dv/du = 0$. After having passed through its $v - u$ turning point, each path moves back in time (the space

TABLE 3 | The equation of motion with $r_f < 2M$.

Range of r_i	c_1	Equations of motion for paths with $r_i > 2M$ and $r_f < 2M$.
$b'_4 - b'_5$	$-\frac{2M}{b'_4 - 2M} < c_1 < 0$	$-4M\tanh^{-1}\left(\frac{u_f}{v_f}\right) = \alpha_f - \alpha_i + M\frac{(3c_1-1)}{c_1\sqrt{-c_1}}\left[\arcsin\left(\frac{-c_1 r_f}{M(1-c_1)} - 1\right) - \arcsin\left(\frac{-c_1 r_i}{M(1-c_1)} - 1\right)\right]$
$b'_5 - b'_6$	$0 < c_1 < -\frac{2M}{r_f - 2M}$	$-4M\tanh^{-1}\left(\frac{u_f}{v_f}\right) = \alpha_f - \alpha_i + M\frac{(3c_1-1)}{c_1\sqrt{c_1}}\log\left(\frac{\beta_f}{\beta_i}\right)$
$b'_{56} - b'_7$	$c_1 = -\frac{2M}{r_a - 2M}$	$-4M\tanh^{-1}\left(\frac{u_f}{v_f}\right) = 2\alpha_a - \alpha_f - \alpha_i + M\frac{(3c_1-1)}{c_1\sqrt{c_1}}\log\left(\frac{\beta_a^2}{\beta_i\beta_f}\right)$

*At the light cone boundaries they converge to the form given in **Table 4**. All discontinuities cancel each other and there are no singularities anywhere in the equations of motion.*

TABLE 4 | Equations of motion at light cone boundaries for paths with $r_f < 2M$.

r_i	c_1	Light cone boundaries
b'_4	$-\frac{2M}{b'_4 - 2M}$	$-4M\tanh^{-1}\left(\frac{u_f}{v_f}\right) = \alpha_f - 2M\log\left|1 - \frac{2M}{b'_4}\right| + M\frac{(3c_1-1)}{c_1\sqrt{-c_1}}\left[-\arcsin(1) + \arcsin\left(\frac{-r_f c_1}{M(1-c_1)} - 1\right)\right]$
b'_5	0	$-4M\tanh^{-1}\left(\frac{u_f}{v_f}\right) = \frac{\sqrt{2}}{3\sqrt{2M}}\left(r_f^{3/2} - b_5'^{3/2}\right) + 2\sqrt{2M}\left(\sqrt{r_f} - \sqrt{b'_5}\right) + 2M\log\left[\frac{\left(\sqrt{\frac{2M}{r_f}} - 1\right)\left(1 + \sqrt{\frac{2M}{b'_5}}\right)}{\left(1 + \sqrt{\frac{2M}{r_f}}\right)\left(1 - \sqrt{\frac{2M}{b'_5}}\right)}\right]$
b'_{56}	1	$-4M\tanh^{-1}\left(\frac{u_f}{v_f}\right) = 2M\log\left|\frac{4M^2}{(r_f - 2M)(b'_{56} - 2M)}\right| - r_f - b'_{56}$
b'_6	$-\frac{2M}{r_f - 2M}$	$-4M\tanh^{-1}\left(\frac{u_f}{v_f}\right) = 2M\log\left(\frac{2M}{r_f} - 1\right) - \alpha_{b'_6} + M\frac{(3c_1-1)}{c_1\sqrt{c_1}}\log\left(\frac{c_1 r_f + M(1-c_1)}{\beta_{b'_6}}\right)$
b'_7	$-\frac{2M}{r_a - 2M}$	$-4M\tanh^{-1}\left(\frac{u_f}{v_f}\right) = 2\alpha_a - \alpha_f - \alpha_{b7'} + M\frac{(3c_1-1)}{c_1\sqrt{c_1}}\log\left(\frac{\beta_a^2}{\beta_{b'_7}\beta_f}\right)$, $\quad\frac{x^2_{b'_7} + 4Mb'_7(1-c_1)}{2c_1^2 x_{b'_7}} + \frac{x_f^2 + 4Mr_f(1-c_1)}{2c_1^2 x_f} - \frac{3M(1-c_1)}{2c_1^2\sqrt{c_1}}\log\frac{\beta_{b'_7}\beta_f}{\beta_a} = 0$

*Outside 2M the equations for **Table 1** hold with $t_f - t_i$ replaced by $4\,M\tanh^{-1}(v_f/u_f)$. Like before, the value $c_1 = 1$ corresponds to a light-like path. Note that b'_7 is the caustic point where both the path $f(-2M/(r_a - 2M),u,v)$ and its partial derivative $\partial f/\partial c_1$ vanish.*

coordinate u continuously decreases, while the time coordinate v increases up to the turning point and then decreases).

In the $r_i = b'_6 - b'_7$ region there are two indirect paths that connect the same point outside the horizon to a point inside $r = 2M$. **Paths 4a** and **4b** connect the point of $u_i = 46.3$ to P. They each have a $dv/du = 0$ turning point outside the horizon, and also turn in r inside the horizon before reaching P. **Path 4a** turns at $r_a = 0.77147 < r_P$ and **path 4b** turns at $r_a = 0.79969 < r_P$. It can be seen that the paths are very close together. As u_i increases, the paths in this region become progressively closer until they merge at $r_i = b'_7$. **Path 5** shows the single indirect path that originates at $u_i = 46.7$ ($r_i = b'_7 = 12.13M$). Beyond this point, there are no real solutions and hence no way to reach point P.

Figure 2C displays the c_1 values of space-like paths reaching point **P** ($u_P = 9.995, v_P = 10.0$) from $u_i \geq 46.1$ at $v_i = 0$. At first, for each u_i in the figure there are two c_1 values with which the final point **P** can be reached. The figure clearly shows the roots (c_1 values) coming closer and closer together until they merge at the caustic point $r = b'_7$. Beyond this point there are no paths that reach point **P**. Extensions to this work to determine the Kruskal-WKB wavefunction will require an analysis of the caustic at $r = b'_7$ to remove any potential divergences.

The indirect paths connecting (u_i, v_i) with $r_i > 2M$ to (u_P, v_P) with $r_P < 2M$ penetrate the horizon deeper than r_P turning at $r_a < r_P$ before reaching r_P (e.g., see the green path in **Figure 2A**). After turning, all paths must continue toward the apparent horizon reaching it at $u = v = 0$. These paths move back in time from a point outside the horizon, where $dv/du = 0$,

after having traveled to that point forward in Kruskal time from $(u_i, v_i = 0)$ with $r_i > 2M$.

As discussed before, the turning points in r correspond to $x_r = 0$ (in the Kruskal paths x_r appears in Equation 58). They occur at $r = r_a$ when $x_a = 0$, where $c_1 = -2M/(r_a - 2M)$. Since $c_1 > 1$ for space-like paths, clearly $r_a < 2M$ (the turning points are inside the horizon). For any r on a path with r_a as the turning point, x_r can be written as $\sqrt{2Mr(r - r_a)/(2M - r_a)}$. To ensure the non-negativity of the term under the square root all points on this path must have $r > r_a$. By the same token, for time-like paths $r_a > 2M$ and $r < r_a$. This case was described in the Schwarzschild analysis.

All space-like paths that reach points inside the horizon subsequently head toward the horizon $r = 2M$ ($u = v$). Classically, they end at $u_f = v_f = 0$. However, quantum mechanically there may be paths close to the classical path that bring information from within the black hole horizon to the outside.

5. QUANTUM TREATMENT IN SCHWARZSCHILD COORDINATES

We consider a collapsing cloud of ultra-light particles with $mM \sim \hbar$. Such particles are not localized and thus neither is the surface of the star. The position of a particle on the surface of the star is then approximated by a wavefunction. In a relativistic path integral approach, this wavefunction can be computed in

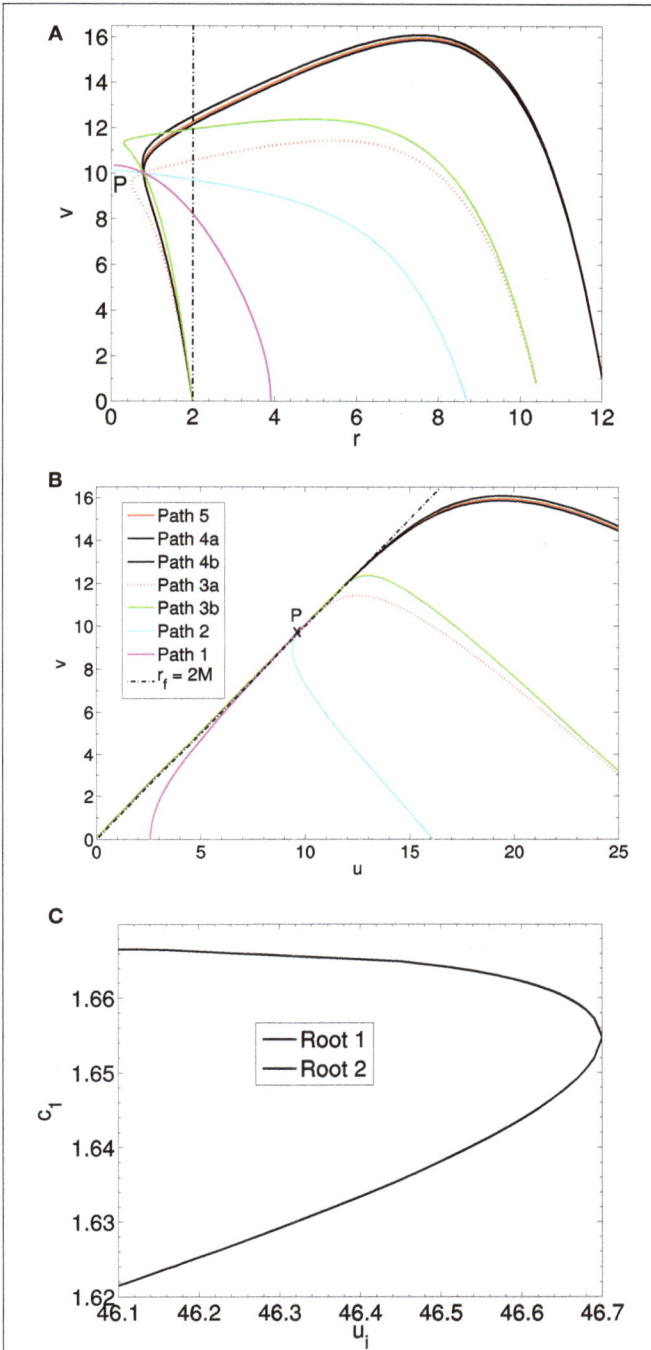

FIGURE 2 | (A) v vs. r and **(B)** v vs. u are shown for space-like and time-like paths with $v_i = 0$ that pass through $P(u_P = 9.995, v_P = 10.0)$ of $r_P = 0.8$, a point located inside the apparent horizon. **(C)** The c_1 values for indirect space-like paths in the region between b'_6 and b'_7 from which point P can be reached are displayed for each u_i. No paths exist beyond $r = b'_7$, which is where the two roots merge into one.

paths derived above are needed. In this paper, we only perform the WKB analysis in Schwarzschild coordinates, where we can only use the paths to $r_f > 2M$.

5.1. WKB Approximation: Closed-Form Propagator, Numerical Wavefunction

In order to study the wavefunction we use a WKB approximation of the propagator

$$G(r, t; r_i, t_i) = \int_C \mathcal{D}r \exp[iS/\hbar], \qquad (33)$$

where S is the action associated with each path. The set of paths C include space-like, time-like and light-like paths.

The WKB approximation involves expanding about the classical action S_{cl} to include paths that slightly deviate from the classical paths. The propagator is dominated by paths near the classical trajectory between (r_i, t_i) and (r, t) and is approximated by the WKB expression (Schulman, 2005)

$$G_{WKB}(r, t; r_i, t_i) = \sqrt{\frac{i}{2\pi\hbar} \frac{\partial^2 S_{cl}}{\partial r_i \partial r_f}} \exp\left(i\frac{S_{cl}}{\hbar}\right). \qquad (34)$$

The WKB wave function is obtained from the integration of the propagator

$$\Psi_{WKB}(r_f, t_f) = \int_0^\infty G_{WKB}(r_f, t_f; r_i, t_i)\Psi(r_i, t_i)dr_i. \qquad (35)$$

The initial wavefunction that describes a particle on the surface of a star of radius r_i is taken to be a Gaussian centered about r_c

$$\Psi(r_i, t_i = 0) = \exp\left[-\frac{(r_i - r_c)^2 m^2}{\hbar^2}\right] \qquad (36)$$

with $r_c >> 2M$.

The action from Equation (6) is rewritten using Equations (11) and (13) as

$$S_{cl}(r_i, t_i; r_f, t_f) = \pm im\sqrt{c_1 - 1} \int_{(r_i, t_i)}^{(r_f, t_f)} dr \frac{r}{x(r)} \qquad (37)$$

for direct paths connecting (r_i, t_i) to (r_f, t_f). The $+$ sign corresponds to $r_f > r_i$ and the $-$ sign corresponds to $r_f < r_i$.

Indirect paths connecting (r_i, t_i) to (r_f, t_f) through the turning point (r_a, t_a) with $r_a > r_i$ and $r_a > r_f$ are described by the classical action

$$S_{cl} = im\sqrt{c_1 - 1}\left[\int_{(r_i, t_i)}^{(r_a, t_a)} dr \frac{r}{x(r)} - \int_{(r_a, t_a)}^{(r_f, t_f)} dr \frac{r}{x(r)}\right]. \qquad (38)$$

In Schwarzschild coordinates, the wavefunction is continuous across all regions outside $r = 2M$. The presence of the $r = 2M$ coordinate singularity prohibits the inclusion of paths at or beyond the horizon region. Closed-form expressions for S_{cl} and $\partial^2 S_{cl}/(\partial r_i \partial r_f)$ are given in **Table 5**. Both functions are continuous across all regions outside $r = 2M$. Once c_1 is known for every

the WKB approximation via an integral over the classical action (Schulman, 2005). It is necessary to include configurations will all possible initial velocities. For each point (r_f, t_f) all classical point (r, t), we have all the ingredients to find the propagator in Equation (34) using the S_{cl}, and $\partial^2 S_{cl}/\partial r_i \partial r_f$ expressions from **Table 5**. The propagator is then integrated to find the wavefunction.

5.2. Relativistic Schrödinger Solution

5.2.1. Free Particle

The $M \to 0$ limit describes a relativistic free particle that is no longer confined by the gravity of the sphere of dust. This case was first studied by Redmount and Suen (1993). We revisit this limit to investigate whether the wavefunction in the WKB approximation converges to the relativistic Schrödinger solution, which is the exact solution for this case. In Redmount and Suen (1993) some disagreement was observed, but while they attempted a physical interpretation, it is likely due to numerical error.

When $M \to 0$, the Hamiltonian from Equation (9) reduces to

$$H = \sqrt{\mathbf{p}^2 + m^2}, \tag{39}$$

where $\mathbf{p}$ is now the momentum operator. The Hamiltonian is nonpolynomial in $\mathbf{p}$. The square root term thus corresponds to a non-local momentum operator that is interpreted as acting on any wavefunction (Redmount and Suen, 1993)

$$\Psi(x, t) = \int_{-\infty}^{\infty} dk e^{ikx} \phi(k, t). \tag{40}$$

to give

$$H\Psi(x, t) = \int_{-\infty}^{\infty} dk e^{ikx} \sqrt{\hbar^2 k^2 + m^2} \phi(k, t). \tag{41}$$

This integrates to

$$\phi(k, t) = A(k, x_0) \exp\left(-\frac{i\Delta t}{\hbar}\sqrt{\hbar^2 k^2 + m^2}\right). \tag{42}$$

When $\Delta t = 0$, we recover the original wavefunction, which is chosen to be a Gaussian centered around the origin,

$$\Psi(x_0, t_0) = \mathcal{N} \exp\left(-\frac{m^2 x_0^2}{\hbar^2}\right),$$

and so

$$\Psi(x, t) = \frac{\mathcal{N}}{2\pi} \int_{-\infty}^{\infty} dk \int_{-\infty}^{\infty} dx_0 \exp\left(-m^2 x_0^2/\hbar^2\right) \tag{43}$$

$$\times \exp\left(ik\Delta x\right) \exp\left(-\frac{i\Delta t}{\hbar}\sqrt{\hbar^2 k^2 + m^2}\right).$$

The normalization $|\mathcal{N}|^2 = m\sqrt{2}/(\hbar\sqrt{\pi})$ is time independent.

In terms of the propagator

$$\Psi(x_f, t_f) = \int_{-\infty}^{\infty} dx_0 G(x_f, t_f; x_0, t_0)\Psi(x_0, t_0), \tag{44}$$

where

$$G(x, t; x_0, t_0) = \int_{-\infty}^{\infty} \frac{dk}{2\pi} \exp\left(ik\Delta x\right)$$

$$\times \exp\left[-\frac{i\Delta t}{\hbar}\sqrt{\hbar^2 k^2 + m^2}\right],$$

$$= \lim_{\epsilon \to 0^+} \frac{m(i\Delta t + \epsilon)}{\pi \hbar \sqrt{\lambda_\epsilon}} K_1(m\lambda_\epsilon^{1/2}/\hbar). \tag{45}$$

Here K_1 is the modified Bessel function and $\lambda_\epsilon \equiv \Delta x^2 + (i\Delta t + \epsilon)^2$. When $\Delta t = 0$ the propagator reduces to $\delta(\Delta x)$ as expected.

TABLE 5 | The classical action in various regions.

$c_1 > 0$ (direct path)	S_{cl}	$\pm im\sqrt{c_1 - 1}\left[\frac{x_f - x_i}{c_1} - \frac{M(1 - c_1)}{c_1\sqrt{c_1}}\log\left(\frac{\beta_f}{\beta_i}\right)\right]$
	$\frac{\partial^2 S_{cl}}{\partial r_i \partial r_f}$	$\frac{\mp imr_i^2 r_f^2\left\{\left(\frac{x_i + 4Mr_i(1 - c_1)}{2c_1^2 x_i}\right) - \left(\frac{x_f + 4Mr_f(1 - c_1)}{2c_1^2 x_f}\right) - \frac{3M(1 - c_1)}{2c_1^{5/2}}\log\left(\frac{\beta_i}{\beta_f}\right)\right\}^{-1}}{2X_i X_f(r_i - 2M)(r_f - 2M)(c_1 - 1)^{3/2}}$
$c_1 < 0$ (direct path)	S_{cl}	$\pm im\sqrt{c_1 - 1}\left[\frac{x_f - x_i}{c_1} - \frac{x_f - x_i}{c_1}\left(\arcsin\left(\frac{-c_1 r_f}{M(1 - c_1)} - 1\right) - \arcsin\left(\frac{-c_1 r_i}{M(1 - c_1)} - 1\right)\right)\right]$
	$\frac{\partial^2 S_{cl}}{\partial r_i \partial r_f}$	$\frac{\mp imr_i^2 r_f^2\left\{\left(\frac{x_i + 4Mr_i(1 - c_1)}{2c_1^2 x_i}\right) - \left(\frac{x_f + 4Mr_f(1 - c_1)}{2c_1^2 x_f}\right) + \frac{3M(1 - c_1)}{2c_1^2\sqrt{-c_1}}\left[\arcsin\left(\frac{c_1 r_f}{M(1 - c_1)} - 1\right) - \arcsin\left(\frac{c_1 r_i}{M(1 - c_1)} - 1\right)\right]\right\}^{-1}}{2X_i X_f(r_i - 2M)(r_f - 2M)(c_1 - 1)^{3/2}}$
$c_1 = -\frac{2M}{r_a - 2M}$ (indirect path)	S_{cl}	$\pm im\sqrt{c_1 - 1}\left[-\frac{x_f - x_i}{c_1} - \frac{M(1 - c_1)}{c_1\sqrt{c_1}}\left(2\arcsin(1) - \arcsin\left(\frac{2r_i}{r_a} - 1\right) - \arcsin\left(\frac{2r_f}{r_a} - 1\right)\right)\right]$
	$\frac{\partial^2 S_{cl}}{\partial r_i \partial r_f}$	$\frac{+ imr_i^2 r_f^2\left\{\left(\frac{x_i + 4Mr_i(1 - c_1)}{2c_1^2 x_i}\right) - \left(\frac{x_f + 4Mr_f(1 - c_1)}{2c_1^2 x_f}\right) - \frac{3M(1 - c_1)}{2c_1^2\sqrt{-c_1}}\left[\arcsin\left(\frac{2r_f}{r_a} - 1\right) - \arcsin\left(\frac{2r_i}{r_a} - 1\right) - 2\arcsin(1)\right]\right\}^{-1}}{2X_i X_f(r_i - 2M)(r_f - 2M)(c_1 - 1)^{3/2}}$

The + sign is chosen when $r_f > r_i$, and the − sign when $r_f < r_i$.

The WKB propagator for the relativistic free particle is then computed using Equation (34) Redmount and Suen (1993)

$$G_{WKB} = \sqrt{\frac{m}{2\pi} \frac{(i\Delta t + \epsilon)^2}{\hbar \lambda_\epsilon^{3/2}}} \exp\left[-\frac{m\lambda_\epsilon^{1/2}}{\hbar}\right], \quad (46)$$

where WKB wavefunction is given by

$$\Psi_{WKB}(x,t) = \int_{-\infty}^{\infty} dx_0 G_{WKB}(x,t;x_0,t_0)\Psi(x_0,t_0). \quad (47)$$

The integrals from Equations (44) and (47) are evaluated numerically. The results are shown in **Figure 3**. It can be seen that the wavefunction in the WKB approximation converges to the Schrödinger solution in all parts of the light cone. When no star M is present, the relativistic Schrödinger is exact as originally discussed in Redmount and Suen (1993). However, when they performed the same numerical comparison, they found some disagreement between the two solutions. This we believe to be due to numerical error of Redmount and Suen (1993). We observe convergence (see **Figure 3**). of the wavefunction in the WKB approximation to the Schrödinger solution in the $M \rightarrow 0$ limit.

5.2.2. Non-zero Mass
While in the free particle case the Schrödinger solution was exact, when $M \neq 0$ it becomes a rough approximation, which fails as we approach $r = 2M$. Comparing the two solutions is still instructive. If we follow the same procedure as for the free particle above, and re-write the Hamiltonian from Equation (9) as

$$H = \sqrt{B^2 \mathbf{p}^2 + Bm^2}, \quad (48)$$

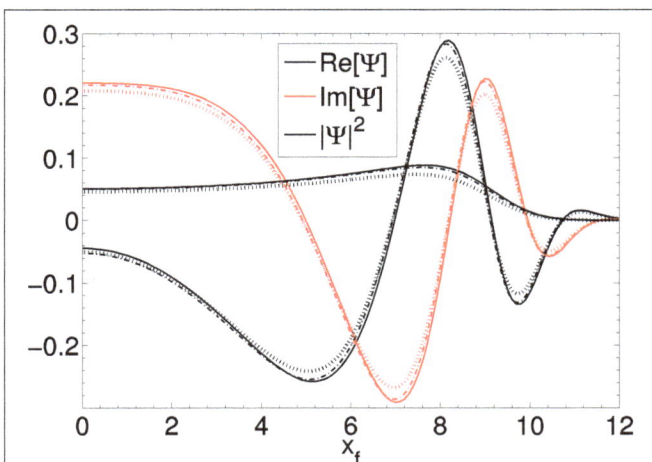

FIGURE 3 | Re[Ψ], Im[Ψ], and |Ψ|² are shown for the free particle case at $t = 10\hbar/m$ as a function of x_f, which also has units of $\hbar/m$. The solid lines show the Schrödinger solution, while dotted lines and dashed lines show the WKB approximation for $\epsilon = 0.05$, $\epsilon = 0.005$, respectively. It can be seen that as $\epsilon \rightarrow 0$ the WKB approximation converges to the Schrödinger solution.

where $\mathbf{p}$ is the momentum operator and

$$B = B(M,r) = 1 - \frac{2M}{r}. \quad (49)$$

The whole wavefunction can then be written as

$$\Psi(r,t) = \int dr_i G(r,t;r_i,t_i)\Psi(r_i,\tau_i), \quad (50)$$

where the propagator is given by

$$G(r,t;r_i,t_i) = \int_{-\infty}^{\infty} \frac{dk}{2\pi} \exp(ik\Delta r) \quad (51)$$
$$\times \exp\left[-i\frac{\Delta t}{\hbar}\sqrt{\hbar^2 k^2 B^2 + m^2 B}\right].$$

Like before, the propagator can then be integrated exactly to obtain

$$G(x,t;x_i,t_i) = \lim_{\epsilon \to 0} \frac{m(i\Delta t + \epsilon)}{\pi \hbar B^{1/2} \lambda_\epsilon^{1/2}} K_1(mB^{1/2}\lambda_\epsilon^{1/2}/\hbar), \quad (52)$$

where K_1 is the modified Bessel function and

$$\lambda_\epsilon = \left(\frac{\Delta r}{B}\right)^2 + (i\Delta t + \epsilon)^2. \quad (53)$$

When $r \rightarrow 2M$ the propagator vanishes. Thus this solution is inaccurate at late times and cannot model the final stages of the collapse of the dust sphere.

5.3. Numerical Results
We integrate Equation (35) numerically for the Gaussian wavefunction from Equation (36) centered about $r_c = 30M$. Classically, the star collapses to $r = 2M$ in infinite Schwarzschild time. However, quantum mechanically there is a possibility for the star expanding and dispersing as well. **Figure 4** shows the time evolution from $t = 0$ to $t = 45M$ of $|\Psi_{WKB}(r_f,t_f)|^2$ as a function of r_f. An ingoing peak and an outgoing peak appear with the amplitude of the ingoing peak always remaining larger than that of the outgoing peak for the course of the evolution, which is expected for a collapsing star. For numerical computations we scale all variables to be dimensionless. The only parameter is $mM/\hbar$. We note that **Figure 4** is qualitatively similar to the WKB results of Corichi et al. (2002).

Figures 5A,B show the ingoing and outgoing peaks of $|\Psi_{WKB}(r,t_f)|^2$ for different values of $mM/\hbar$ at times of $t_f = 5M$ and $t_f = 10M$, respectively. For a given particle number $N = M/m$, the speed of the collapse increases slightly with the particle mass m as does the value of the ingoing peak. This is consistent with our expectation that for higher particle mass the star is more classical. For lower masses, the star resists collapse longer until probability of dispersion exceeds the probability of collapse.

In **Figure 6** we compare the WKB and Schrödinger solutions. It can be seen that the radial position of the ingoing peak drifts out of step at late times, while the outgoing peak continues to evolve at about the same position. As the star

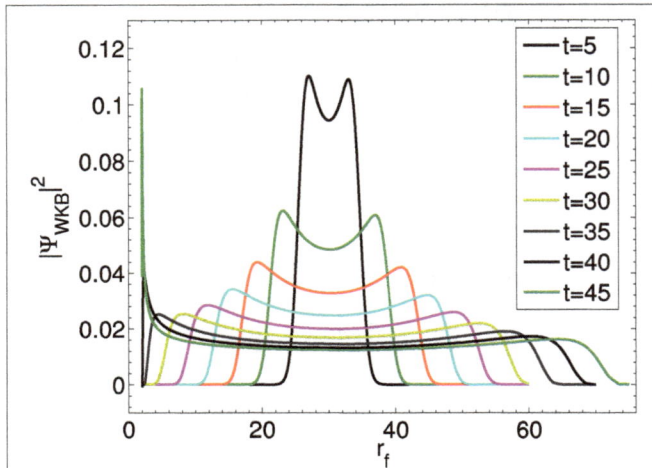

FIGURE 4 | The time evolution of WKB $|\Psi(r_f, t_f)|^2$ is shown from $t = 5M$ to $t = 45M$ in the case of $mM/h = 1$. We can see that the star collapses to a black hole.

approaches $r_f = 2M$, the ingoing peak amplitude for the WKB approximation increases developing a numerical singularity that indicates the formation of an apparent horizon, while the ingoing peak amplitude of the Schrödinger solution decreases since the propagator in Equation (52) vanishes at $r_f \to 2M$. The latter behavior reinforces that the relativistic Schrödinger equation is not the correct representation for a particle in the gravitational field of a collapsing star.

6. CONCLUSION

Closed form solutions for the classical paths taken by a particle on the surface of a collapsing star have been determined for all initial configurations in Schwarzschild and Kruskal coordinates.

We found that

(i) all time-like paths are unique. For a given time interval, a path between an initial and final point can be either direct or indirect, where it turns around in space. Thus some particles that initially move away from the star, can return and contribute to the collapse.
(ii) all space-like paths are unique outside $r = 2M$.
(iii) Kruskal space-like paths can turn around in Kruskal time, but cannot turn in Kruskal space. Multiple paths can connect a given point outside the horizon to a final point that lies inside $r = 2M$.
(iv) Classical Kruskal space-like paths connect points outside $r = 2M$ with points inside $r = 2M$ up to a critical value of $r = r_a$. Spacelike paths from $(r_i, v_i = 0)$ to any given point r inside the horizon, where $r_a < r < 2M$, are non-unique (two paths exist). These paths get less and less separated and merge into a single space-like path at $r = r_a$ and any point $r < r_a$ is unreachable from r_i. Upon reaching the critical value $r = r_a$, the space-like paths turn back toward $r = 2M$, reaching it at $u = v = 0$. Therefore, classically, no information from $r < 2M$ can exit a black

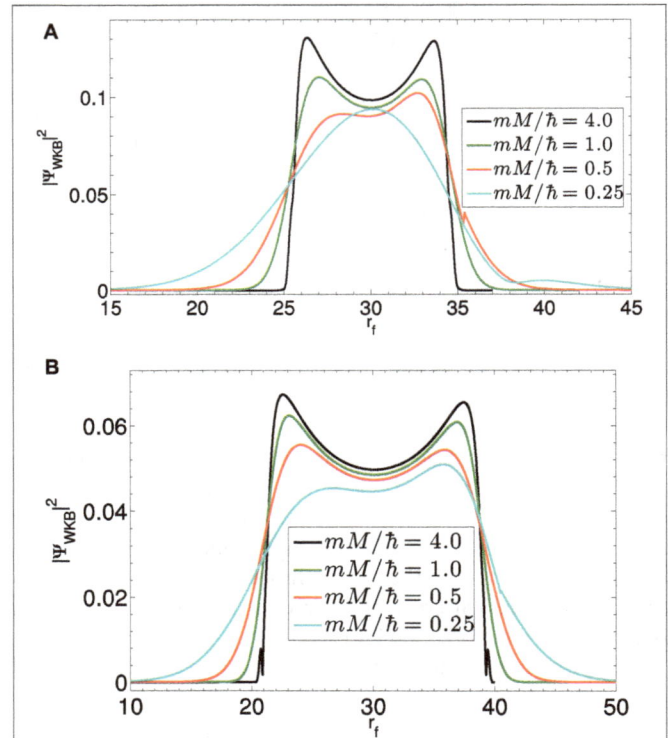

FIGURE 5 | (A) $|\Psi(r_f, t_f)|^2$ at $t = 5M$ as a function of r_f for $mM/h = 4.0$, $mM/h = 1$, $mM/h = 0.5$, and $mM/h = 0.25$. It can be seen that that lower masses behave more quantum mechanically, and are less likely to collapse to a black hole. **(B)** $|\Psi(r_f, t_f)|^2$ at $t = 10M$ as a function of r_f for the same masses.

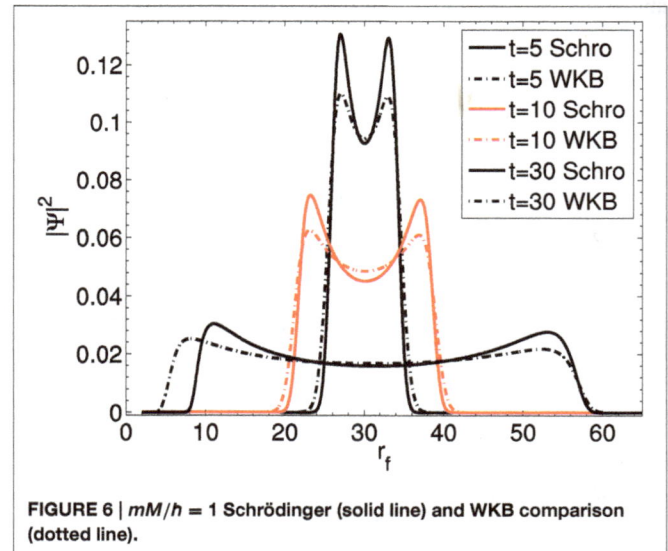

FIGURE 6 | $mM/h = 1$ Schrödinger (solid line) and WKB comparison (dotted line).

hole. However, by taking into account paths close to the classical paths, one might be able to extract information from $r < 2M$.

If we make an analogy to tunneling, the particle has a low probability of passing through a potential barrier. Classically, it never happens, and yet this probability cannot be ignored in quantum mechanics. Similarly, it may be that the low probability space-like paths will play a crucial role in our understanding of quatum mechanical collapse.

The collapse of a self-gravitating star was next modeled as a ball of dust using a WKB approximation. We integrated around the classical paths with all possible initial velocities to include quantum effects. This extended the analysis of Redmount and Suen (1993) from the relativistic free particle to the case of non-trivial gravity. In practice, quantum mechanical effects are important in macroscopic stellar collapse when the sphere is composed of ultra-light particles. A number of ultralight dark matter particles have been proposed, which could physically motivate such stars.

The evolution of the wavefunction of a particle on the surface of a collapsing star was followed numerically. We showed that in the case of a star collapsing with zero initial velocity, our path equations reduce to the Oppenheimer-Schneider equations of motion. In the $M \rightarrow 0$ limit, the relativistic free particle wavefunction of Redmount and Suen (1993) is obtained, and the WKB and relativistic Schrödinger solutions match. Since in this limit the Schrödinger solution is exact (Kiefer and Singh, 1991), the covergence of the WKB to the Schrödinger solution enforces the validity of the WKB approximation.

Our results for this self-gravitating collapse are summarized as follows:

(i) The wavefunction representing a particle on the surface of a collapsing star typically exhibits an outgoing and an ingoing component, where the former contains the probability that the star will disperse and the latter the collapse probability. For a given particle number $N = M/m$, we find that the rate at which collapse occurs increases with particle mass. This is consistent with the expectation that for higher particle mass the star is more classical. As the particle mass is lowered, the star resists collapse until the probability that it disperses exceeds the collapse probability. Note that some part of the star always disperses even when a black hole forms.

(ii) In the case of the collapsing star the relativistic Schrödinger solution is not a good approximation for the wave function. On comparing the WKB and relativistic Schrödinger solutions, we find that the ingoing wavefunction gets more and more out of step at late times. The outgoing component of the wavefunction for the two solutions is in better agreement because it is further from the coordinate singularity at $r = 2M$. As expected, the Schrödinger and WKB solutions are out of step at early times when WKB solution is more inaccurate, they come closer together at intermediate times, and fall out of step again at late times, when the Schrödinger approximation fails to model the singularity formation. The presence of the coordinate singularity at $r = 2M$, motivate a potential investigation beyond $r = 2M$ via the WKB approximation deployed in Kruskal coordinates where no analytical relativistic Schrödinger solution is available.

The traditional Oppenheimer Schneider model describes the zero pressure collapse of a star starting from rest. We use a path integral approach that requires including paths of all velocities.

For a particle on the surface of a star, we explicitly parametrically describe all such paths of different initial velocities. This is done in Schwarzschild coordinates until $r = 2M$ and in Kruskal coordinates until the $r = 0$ singularity. We include ingoing and outgoing velocities, and time-like, light-like and space-like paths. In the process we have generated the paths shown in Fuller and Wheeler (1962) and more. We have determined equations for the space-like caustics (degenerate paths) which have not been described before in this manner. We have used a WKB path integral approach and determined the Schwarzschild wave function for the collapse. Using a similar approach one can in the future use the parametric path equations we have determined in Kruskal coordinates to explore the space-time inside the horizon. It will require expansion around the caustic. The possibility of paths moving back in time might have important ramifications.

It has long been argued whether quantum gravity should be causal or not (Teitelboim, 1983; Hartle and Kucha, 1986; Hartle, 1988; Mattingly, 2005) and as of today there is no accepted theory of quantum gravity. We evaluate the propagator both for a free relativistic particle, where we have an exact solution, and for a collapsing ball of dust. Since the propagator does not vanish outside the light cone, we include all paths in our integration. In the free particle case, the WKB and exact solutions match. Space-like classical paths give the dominant contribution in the construction of the WKB approximation far outside the light cone, where the approximation is valid. This propagation will be acausal (backward in time) in some Lorentz frames. Whether or not this proves to be admissible in the ultimate theory of quantum gravity is beyond the purpose of our work. Our work's exploration of the consequence of an acausal propogator in the collapse of a zero pressure star might nevertheless contribute to the dialogue in the search of a consistent theory of quantum gravity.

In addition, we propose that an understanding of these low probability paths that are traditionally ignored could shed some light in theories of black hole formation. Furthermore, if the black holes observed by LIGO are formed from scalar particles (Abbott et al., 2016), the particles would be very light; then quantum effects might play a role in their formation, constraining analog gravity theorems (Barceló et al., 2005).

AUTHOR CONTRIBUTIONS

All authors listed, have made substantial, direct and intellectual contribution to the work, and approved it for publication.

ACKNOWLEDGMENTS

JB acknowledges Prof. Wai-Mo Suen for the initial impetus to approach this research and for subsequent guidance and useful discussions. We also thank Prof. Ian Redmount for his inital work on this topic. Further, we are particularly grateful to Prof. Mihai Bondarescu and Prof. Philippe Jetzer for useful discussions and advice. RB has received support from the Dr. Tomalla Foundation and the Swiss National Science Foundation. CM is supported by the NSF Astronomy and Astrophysics Postdoctoral Fellowship under award AST-1501208.

REFERENCES

Abbott, B., Abbott, R., Abbott, T. D., Abernathy, M. R., Acernese, F. et al. (2016). Astrophysical implications of the binary black hole merger gw150914. *Astrophys. J. Lett.* 818:L22. doi: 10.3847/2041-8205/818/2/L22

Alberghi, G., Casadio, R., Vacca, G., and Venturi, G. (1999). Gravitational collapse of a shell of quantized matter. *Class. Quantum Grav.* 16, 131.

Ansoldi, S., Aurilia, A., Balbinot, R., and Spallucci, E. (1997). Classical and quantum shell dynamics, and vacuum decay. *Class. Quantum Grav.* 14, 2727. doi: 10.1088/0264-9381/14/10/004

Barceló, C., Liberati, S., and Visser, M. (2005). Analogue gravity. *Living Rev. Relativ.* 8, 214. doi: 10.12942/lrr-2005-12

Berezin, V. A. (1997). Quantum black hole model and Hawking's radiation. *Phys. Rev. D* 55, 2139.

Casadio, R., and Venturi, G. (1996). Semiclassical collapse of a sphere of dust. *Class. Quantum Grav.* 13, 2715.

Corichi, A., Cruz-Pacheco, G., Minzoni, A., Padilla, P., Rosenbaum, M., Ryan, M. Jr., et al. (2002). Quantum collapse of a small dust shell. *Phys. Rev. D* 65:064006. doi: 10.1103/PhysRevD.65.064006

Dolgov, A., and Khriplovich, I. (1997). Properties of the quantized gravitating dust shell. *Phys. Lett. B* 400, 12–14. doi: 10.1016/S0370-2693(97)00352-3

Fuller, R. W. and Wheeler, J. A. (1962). Causality and multiply connected space-time. *Phys. Rev.* 128:919. doi: 10.1103/PhysRev.128.919

Goswami, R., and Joshi, P. S. (2004). Spherical dust collapse in higher dimensions. *Phys. Rev. D* 69:044002. doi: 10.1103/PhysRevD.69.044002

Hájíček, P., Kay, B. S., and Kuchar, K. V. (1992). Quantum collapse of a self-gravitating shell: equivalence to coulomb scattering. *Phys. Rev. D* 46:5439. doi: 10.1103/PhysRevD.46.5439

Hartle, J. B. (1988). Quantum kinematics of spacetime. II. a model quantum cosmology with real clocks. *Phys. Rev. D* 38:2985.

Hartle, J. B., and Kucha, K. V. (1986). Path integrals in parametrized theories: the free relativistic particle. *Phys. Rev. D* 34:2323.

Hawkins, E. (1994). Quantum gravitational collapse of a charged dust shell. *Phys. Rev. D* 49:6556.

Hu, W., Barkana, R., and Gruzinov, A. (2000). Fuzzy cold dark matter: the wave properties of ultralight particles. *Phys. Rev. Lett.* 85, 1158–1161. doi: 10.1103/PhysRevLett.85.1158

Joshi, P. S. (2007). *Gravitational Collapse and Spacetime Singularities.* Cambridge, UK: Cambridge University Press.

Khmelnitsky, A., and Rubakov, V. (2014). Pulsar timing signal from ultralight scalar dark matter. *J. Cosmol. Astropart. Phys.* 2014:019. doi: 10.1088/1475-7516/2014/02/019

Kiefer, C., and Singh, T. P. (1991). Quantum gravitational corrections to the functional schrödinger equation. *Phys. Rev. D* 44:1067.

Lesgourgues, J., Arbey, A., and Salati, P. (2002). A light scalar field at the origin of galaxy rotation curves. *New Astron. Rev.* 46, 791–799. doi: 10.1016/S1387-6473(02)00247-6

Lundgren, A. P., Bondarescu, M., Bondarescu, R., and Balakrishna, J. (2010). Lukewarm dark matter: bose condensation of ultralight particles. *Astrophys. J. Lett.* 715:L35. doi: 10.1088/2041-8205/715/1/L35

Mattingly, D. (2005). Modern tests of lorentz invariance. *Liv. Rev. Relativ.* 8:5. doi: 10.12942/lrr-2005-5

Misner, C. W., Thorne, K. S., Wheeler, J. A., and Gravitation, W. (1973). *Freeman and Company.* San Francisco, CA: W.H. Freeman.

Narlikar, J. (1977). Quantum uncertainty in the final state of gravitational collapse. *Nature* 269:129.

Oppenheimer, J. R., and Snyder, H. (1939). On continued gravitational contraction. *Phys. Rev.* 56:455.

Ortíz, L., and Ryan, M. Jr. (2007). Quantum collapse of dust shells in 2 + 1 gravity. *Gen. Relativ. Gravit.* 39, 1087–1107. doi: 10.1007/s10714-007-0458-7

Redmount, I. H., and Suen, W.-M. (1993). Path integration in relativistic quantum mechanics. *Int. J. Mod. Phys. A* 8, 1629–1635. doi: 10.1142/S0217751X93000667

Schulman, L. S. (2005). *Techniques and Applications of Path Integration.* Mineola, NY: Courier Dover Publications.

Singh, T., and Witten, L. (1997). Spherical gravitational collapse with tangential pressure. *Class. Quantum Grav.* 14:3489.

Teitelboim, C. (1983). Causality versus gauge invariance in quantum gravity and supergravity. *Phys. Rev. Lett.* 50:705.

Vaidya, P. (1951). Nonstatic solutions of einstein's field equations for spheres of fluids radiating energy. *Phys. Rev.* 83:10.

Vaz, C., and Witten, L. (2001). Quantum black holes from quantum collapse. *Phys. Rev. D* 64:084005. doi: 10.1103/PhysRevD.64.084005

Zloshchastiev, K. G. (1998). Monopole and electrically charged dust thin shells in general relativity: classical and quantum comparison of hollow and atomlike configurations. *Phys. Rev. D* 57, 4812–4820.

Conflict of Interest Statement: The authors declare that the research was conducted in the absence of any commercial or financial relationships that could be construed as a potential conflict of interest.

mxCSM: A 100-slit, 6-Wavelength Wide-Field Coronal Spectropolarimeter for the Study of the Dynamics and the Magnetic Fields of the Solar Corona

Haosheng Lin*

Institute for Astronomy, University of Hawaii, Pukalani, HI, USA

Tremendous progress has been made in the field of observational coronal magnetometry in the first decade of the Twenty-First century. With the successful construction of the Coronal Multichannel Magnetometer (CoMP) instrument, observations of the linear polarization of the coronal emission lines (CELs), which carry information about the azimuthal direction of the coronal magnetic fields, are now routinely available. However, reliable and regular measurements of the circular polarization signals of the CELs remain illusive. The CEL circular polarization signals allow us to infer the magnetic field strength in the corona, and is critically important for our understanding of the solar corona. Current telescopes and instrument can only measure the coronal magnetic field strength over a small field of view. Furthermore, the observations require very long integration time that preclude the study of dynamic events even when only a small field of view is required. This paper describes a new instrument concept that employs large-scale multiplexing technology to enhance the efficiency of current coronal spectropolarimeter by more than two orders of magnitude. This will allow for the instrument to increase the integration time at each spatial location by the same factor, while also achieving a large field of view coverage. We will present the conceptual design of a 100-slit coronal spectropolarimeter that can observe six CELs simultaneously. Instruments based on this concept will allow us to study the evolution of the coronal magnetic field even with coronagraphs with modest aperture.

Keywords: corona, magnetic fields, spectropolarimetry, instrumentation

Edited by:
Sarah Gibson,
National Center for Atmospheric
Research, USA

Reviewed by:
Gordon James Duncan Petrie,
National Solar Observatory, USA
Jonathan Wesley Cirtain,
NASA, USA

*Correspondence:
Haosheng Lin
lin@ifa.hawaii.edu

Specialty section:
This article was submitted to
Stellar and Solar Physics,
a section of the journal
Frontiers in Astronomy and Space
Sciences

1. INTRODUCTION

In the low density, high temperature and highly ionized coronal plasma, magnetic fields suppress cross-field-line motion of charged particles, thereby creating an atmosphere with highly anisotropic local thermodynamic properties, thus shaping the appearance of the corona to closely resemble that of the magnetic field lines. However, large scale flows of charged particles (electric currents) in turn alter the large scale structure of the magnetic fields. Therefore, detailed observations of the behavior of magnetic fields and plasmas during major eruptions, as well as during quiet periods, are the crucial data needed for understanding the interaction between the fields and the coronal plasmas,

and the physics of solar eruptions. While it has been more than
a century since the ground-breaking work of Hale (1908) that
revealed the magnetic nature of the sun, direct measurement
of magnetic fields in the outer layer of the solar atmosphere
remains difficult. Much progress has been made in the past
two decades in the field of coronal polarimetry to directly
measure the polarization of coronal emission lines (CELs)
that responds directly to coronal magnetic fields (Casini and
Judge, 1999; Lin and Casini, 2000; Lin et al., 2000, 2004). The
Coronal Multichannel Polarimeter (CoMP), in particular, can
now provide direct measurements of the FeXIII line linear
polarization on a daily basis (Tomczyk et al., 2008). Nevertheless,
direct inference of the coronal magnetic fields from these
measurements remains a challenging task due to the low optical
density of the coronal atmosphere.

Recent advancements in scalar and vector tomography based
on space EUV intensity and ground intensity and linear
polarization data of CELs have further allowed us to make direct
inference of the 3D magnetic and thermodynamic structure of
the corona from these observations (Kramar et al., 2016). As
shown by Kramar et al. (2016), coronal magnetic fields derived
from tomographic reconstruction of EUV and white light data for
Carrington Rotation 2112, revealed several possible deficiencies
in MHD simulated models of the same period, and demonstrated
the importance of direct 'observations' for research of coronal
magnetic fields.

While we have finally attained the capability to derive the
3D magnetic and thermodynamic structures of the corona
from direct observations, their accuracy and applicability are
still subject to the limitations of existing instrumentation. For
examples, CEL polarimetry at present is only possible from
ground-based instrumentation, and with a single sight line from
Earth. True tomographic observations that sample the corona
from multiple lines of sight (LOS) *simultaneously* is currently
not possible. Therefore, we have relied on the rotation of
the Sun to provide the multiple LOS measurements for use
with tomographic inversion tools. Accordingly the results from
these tomographic inversions are the static component of the
coronal fields during the period of the observations. *To resolve
the dynamic time scales of solar eruptions, true tomography
with simultaneous observations from multiple sight lines is
needed.* Next, tomography with only linear polarization input
is insensitive to certain magnetic field configurations (Kramar
et al., 2013). However, the amplitudes of the circular polarization
signals of the CELs are two orders of magnitude lower than that
of the linear polarization. Therefore, *the sensitivity of our synoptic
coronal polarimeters needs to be greatly improved for tomographic
inversion to provide coverage of the full range of coronal magnetic
field configurations.* Finally, **Figure 1** shows the solar corona
observed in six different CELs with excitation temperatures
ranging from 0.5 to 2.2 MK (Habbal et al., 2011). The dramatic
difference in the appearance of the corona between ions with
low and high ionization temperatures is a manifestation of the
(spatial) non-uniformity of the coronal temperature. Therefore,
polarization measurements at only one CEL sample only coronal
plasma with temperature within a narrow range around the
ionization temperature of the spectral line. In order to "see" the

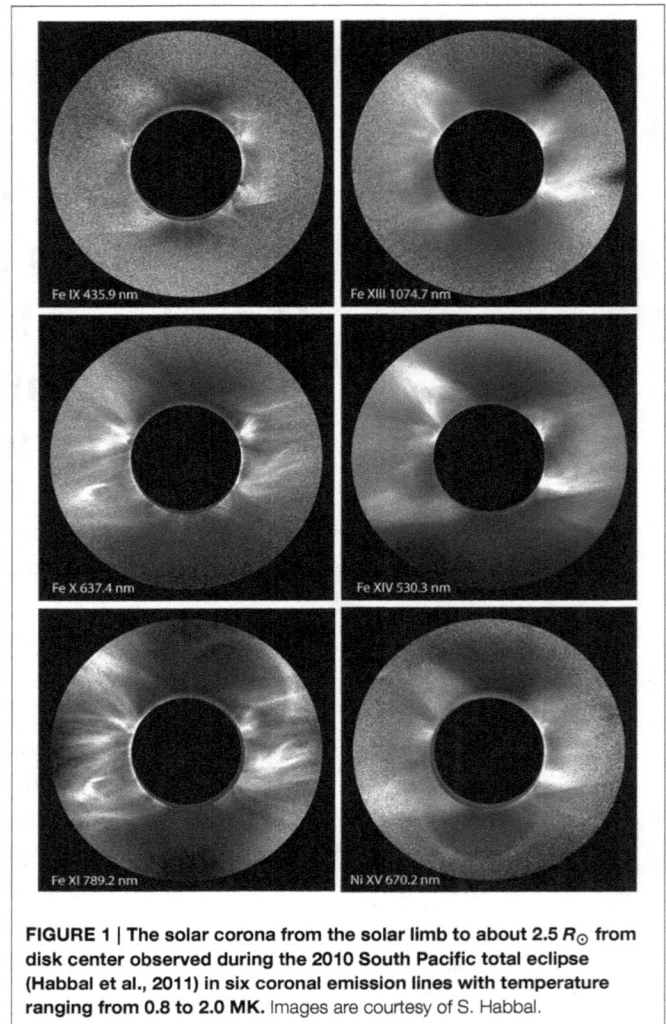

FIGURE 1 | The solar corona from the solar limb to about 2.5 $R_\odot$ from
disk center observed during the 2010 South Pacific total eclipse
(Habbal et al., 2011) in six coronal emission lines with temperature
ranging from 0.8 to 2.0 MK. Images are courtesy of S. Habbal.

entire corona, *observations at multiple spectral lines spanning a
broad range of the coronal temperature is needed.*

Observational coronal magnetometry is a new field that
will eventually provide solar physicists the magnetic and
thermodynamic structure of the solar corona needed for every
aspect of coronal research. Even with its current limitations, this
new capability to directly derive the 3D corona magnetic field
structure from measurements of signals originating from the
corona, as opposed to extrapolations or MHD simulations that
rely on information of magnetic fields at the lower boundary
layers, is an important new capability that will provide new
insight into the physical processes that define the corona.
Therefore, it is critically important that we continue to develop
and improve our observing capability to provide data with
comprehensive coverage in spatial, temporal, polarization, and
temperature domains to allow tomographic inversion techniques
to yield a complete picture of the corona.

In this paper, we present the conceptual design of a
new coronal magnetometer, called the **m**assively-multiple**x**ed
Coronal **S**pectropolarimetric **M**agnetometer (mxCSM), for the

measurement of the intensities, velocities, and polarizations of the six CELs shown in **Figure 1** *simultaneously*. This instrument will provide CEL polarization data with sufficient spatial, spectral, and temporal resolution, and spatial, temperature, and velocity field coverage to enable high precision vector tomographic inversion of the coronal magnetic fields. mxCSM adapts the large-scale multiplexing strategy demonstrated by the **m**assively-multiplexed **SPEC**troheliograph (mxSPEC, Lin, 2014) proof-of-concept instrument. Its optical system consists of a new catadioptric off-axis Gregorian coronagraph and two100-slit, 3-line spectrographs. It uses six 4096 × 4096 format CCDs and/or IR cameras to observe the six spectral lines simultaneously. With this new design, mxCSM can deliver close to three orders of magnitude improvement in capability over current generation of coronal magnetometers without the use of large aperture telescopes. This conceptual design that will be described in the following sections serves to showcase the potential of the large-scale multiplexing strategy for instrumentations for coronal magnetism research in particular, and for future solar spectroscopic instrumentations in general. More importantly, due to its compact design, mxCSM is ideally suited for deployment in space. We envision that future missions with two or more mxCSMs deployed in circumsolar orbits similar to the Solar Terrestrial Relations Observatory (STEREO) mission will provide multiple LOS measurements to enable true tomographic inversion of the coronal magnetic and thermodynamic structures.

2. THE DEVELOPMENT OF MODERN MULTIPLE-SLIT SPECTROGRAPHS

2.1. FIRS: The Facility IR Spectropolarimeter

Multi-slit spectroscopy is a simple, yet very effective method of multiplexing spectra from multiple slices of a 2D spatial field onto a 2D array detector. It was first conceived 40 years ago (Martin et al., 1974; Livingston et al., 1980) but had not been widely adapted in the digital age until it was revived by Srivastava and Mathew (1999). Lin (2003), unaware of the historical developments and the new effort in India, independently conceived the design for a multi-slit, multi-wavelength IR spectropolarimeter in late 1990s. Its design goals were to take advantage the large multiplexing capability of modern large-format, high-performance visible and IR focal plane arrays (FPAs) and new high-efficiency narrow passband Dense Wavelength Division Multiplexing (DWDM) filter technology developed for telecommunication applications to improve the imaging capability and operational efficiency of grating-based spectropolarimeters. This design was first realized through the development and construction of the Facility IR Spectropolarimeter (FIRS, Jaeggli et al., 2010) for the Dunn Solar Telescope (DST) of the National Solar Observatory (NSO) in Sunspot, New Mexico. In addition to realizing the potential of the multi-slit design, FIRS also advanced an achromatic reflecting spectrograph design that uses coarsely-ruled echelle grating to allow for observations of multiple spectral lines simultaneously

on the same spectrograph to further increase the operational efficiency of the instrument. FIRS supports observations with a maximum of 4 slits. The primary constraints on the number of slits for FIRS are (1) the relatively modest size of the 1024 × 1024 format IR camera available, (2) the requirements for high spectral resolution (with a spectral sampling size $\Delta\lambda \approx \lambda/300,000$), and (3) substantial Doppler velocity field coverage (e.g., ≥ 200 kms). In order to cover a spectral window of approximately ±120 km/s around the spectral lines, 256 pixels per slit were dedicated to record the spectra at each spatial sampling point. Nevertheless, even with a relatively small number of slits, it takes only 190 scan steps to observe a $150'' \times 75''$ field with $0.3''$ spatial sampling size, a factor of 4 faster than a single slit instrument.

2.2. mxSPEC: The Massively Multiplex Spectroheliograph Concept

For science that requires only low or medium spectral resolution, a large field of view coverage with moderate spatial resolution, multi-slit spectrograph can be reconfigured with low dispersion gratings to accommodate a large number of slits without the reduction of the spectral window (and Doppler velocity field) coverage, thereby greatly reducing the number of scan steps needed to scan the full 2D field. This enables a scanning grating spectrograph to perform 3D imaging spectroscopy with very high temporal resolution. The proof-of-concept instrument, mxSPEC, was assembled at the full-disk port of the DST in 2014 using DST inventory optics, a He I 1083 nm DWDM bandpass isolation filter (BIF) with 1.4 nm bandpass, and a 10 frame per second (fps) Raytheon Virgo 2K 2048 × 2048 IR camera. In this setup, the DST aperture was reduced to 135 mm, and mxSPEC observed the full solar disk with $1''$ pixel^{-1} spatial sampling. mxSPEC was equipped with a photolithographically etched 49-slit mask. The slits are separated by 750 μm distance, with a slit width of 12.5 μm. The Sun illuminates 34 to 35 of the 49 slits at any given time, while the IR camera sees 41 of the 49 slits. Thus, only 60 scan steps, or less than 8 seconds (including processing overhead) are required to obtain a $2460 \times 2048 \times 50$ (x, y, λ) hyper-spectral data cube. The spectrograph yields a 250 mÅ /pixel spectral sampling size ($\lambda/\Delta\lambda = 40,000$), and a ±225 km/s Doppler velocity coverage centered on the nearby Si I 1072.7 nm line. **Figure 2** shows a sample multi-slit full-Sun spectral image from mxSPEC, detailed spectra around a sunspot near the east limb and a full-disk He I 1083 nm line core image constructed from a full disk scan. Image sequences of the whole sun in He I 1083 nm line showing many aspects of chromosphere dynamics are available at http://www.ifa.hawaii.edu/users/lin/default/mxSPEC.html.

3. THE MASSIVELY-MULTIPLEX CORONAL SPECTROPOLARIMETRIC MAGNETOMETER

In order to observe six spectral lines spanning over two octaves in wavelength simultaneously through a single telescope, an optical system that can project an achromatic image of the sun at the full-disk occulter, as well as at the entrance slits of the spectrographs is needed. Furthermore, the spectrograph also needs to be

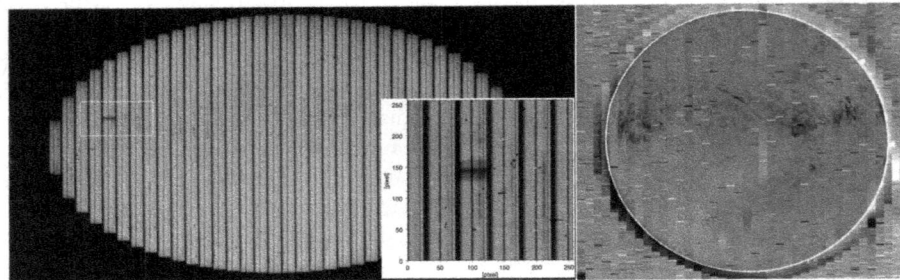

FIGURE 2 | Left: 34-slit spectra of the full Sun obtained with the mxSPEC proof-of-concept instrument. Each vertical slice in the image is a horizontally dispersed spectrum. The 2048 × 2048 image was rescaled to a 2408 × 1024 image to better display the slit spectra. Close-up of spectra of a sunspot in the rectangular box in the full-disk spectral image are shown in the insert. **Right:** He I 1083 nm line core image constructed from a full-disk scan. The black and white short horizontal lines are data reduction artifacts due to defective pixels.

achromatic to support multi-wavelength observations. We have designed a new wide-field catadioptric coronagraph based on an off-axis Gregorian telescope, and a 100-slit refractive Czerny-Turner spectrograph that can observe three spectral lines simultaneously. The system is equipped with two spectrographs to support simultaneous observations of six spectral lines. The following sections describe the designs of the coronagraph and the spectrographs.

3.1. The Catadioptric Off-Axis Gregorian Coronagraph

3.1.1. The Imaging System of mxCSM Coronagraph

Classical Lyot coronagraphs employ low-scatter, super polished singlet objective lens to minimize the scattered light at the prime focus, and a prime focus occulter to prevent the disk light from traveling further downstream. However, due to the dispersion of the index of reflection, the size and location the solar image formed by the objective lens vary as functions of wavelength. It is therefore difficult to observe more than one spectral line at the same time with the Lyot coronagraphs. The catadioptric optical system of mxCSM overcomes this limitation of the classical Lyot coronagraph. **Figures 3–6** shows the ZEMAX layouts of the optical system of mxCSM projected in the Y-Z, X-Y, Z-X plane, and an isometric 3D shaded model, respectively. This optical design is based on a CCD camera with 4096 × 4096 format and 9 μm pixels. The catadioptric off-axis Gregorian coronagraph consists of an aspheric aperture corrector (AC, aperture $\Phi = 300$ mm) followed by an off-axis parabolic primary mirror (M1, $\Phi = 300$ mm, Focal Length FL = 800 mm, Off-Axis Distance OAD = 300 mm) and a concave off-axis elliptical secondary mirror (M2, $\Phi = 120$ mm, FL = 244.6 mm, OAD = 160 mm, conic $C = -0.158$). A full-disk occulter is placed at the M1 focus to block disk light. The effective focal length at the Gregorian focus is 1850 mm. An entrance aperture stop (AS) placed between AC and M1, and a Lyot stop (LS) placed on the image of AS formed by M2 limit the effective aperture of the telescope. The location of AS is chosen such that the Lyot stop can be located at an accessible location outside of the region with overlapping beams near M2. The diameter of AS is oversized to 275 mm. Edge diffractions of AS will be blocked by the Lyot stop, which

also limits the effective telescope aperture to 250 mm. This design achieves better than 1″ spatial resolution within a 0.375 degree field (1.5 $R_\odot$) from Fe XIV 530 nmto Si X 1430 nm, as demonstrated by the spot diagram in **Figure 7**.

To support simultaneous observations of six spectral lines, a Gregorian beam splitter (GBS) splits the beam coming down from M2 into two arms. Light with wavelengths shorter than 750 nm is reflected again with the spectrograph 1 fold and scan mirror (SG1FSM) to feed the first spectrograph SG1. The transmitted light is reflected by a Gregorian fold mirror (GFM) and then SG2FSM to feed the second spectrograph. SG1FSM and SG2FSM also serve as the field scanning mirrors of the spectrographs.

The optical system of mxCSM is designed for use with CCDs with 9 μm pixels. With an effective focal length of 1850 mm at the Gregorian foci, and a 1:1 magnification between the entrance and exit slit focal planes, the spatial sampling size is 1″ per pixel. The slit masks, SG1SM and SG2SM of the spectrographs have 100 parallel slits with width of 9 μm, separated by a distance of 432 μm . It takes only 48 scan steps to complete the scanning of the 1.2 × 1 degree field of view (FOV).

For polarization measurements, a simple polarimeter consists of a rotating achromatic $\lambda/3$ waveplate polarization modulator PM followed by a linear polarizer PA are shown between M2 and GBS for single-beam polarimetry. For dual beam polarimetry, the linear polarization analyzer PA can be replaced by Wollaston prisms placed near the focal planes of the spectrograph (§ 3.2).

3.1.2. Scattered Light Considerations

The most important aspect of a coronagraph design is the scattered light performance of the optical system. The primary source of instrumental scattered light of a classical Lyot coronagraph is the scattering of disk light off the dusts and imperfections on the surfaces and in the substrate of the objective lens. For ground-based observations, Elmore (2007) has found that the scattered light of the MK4 coronagraph located at Mauna Loa Solar Observovatory is dominated by dust particle accumulation on the objective lens of the coronagraphs even immediately after cleaning the objective. As Mauna Loa is one of the best ground-based coronal sites, this finding strongly suggests that for ground-based coronal observations dust control

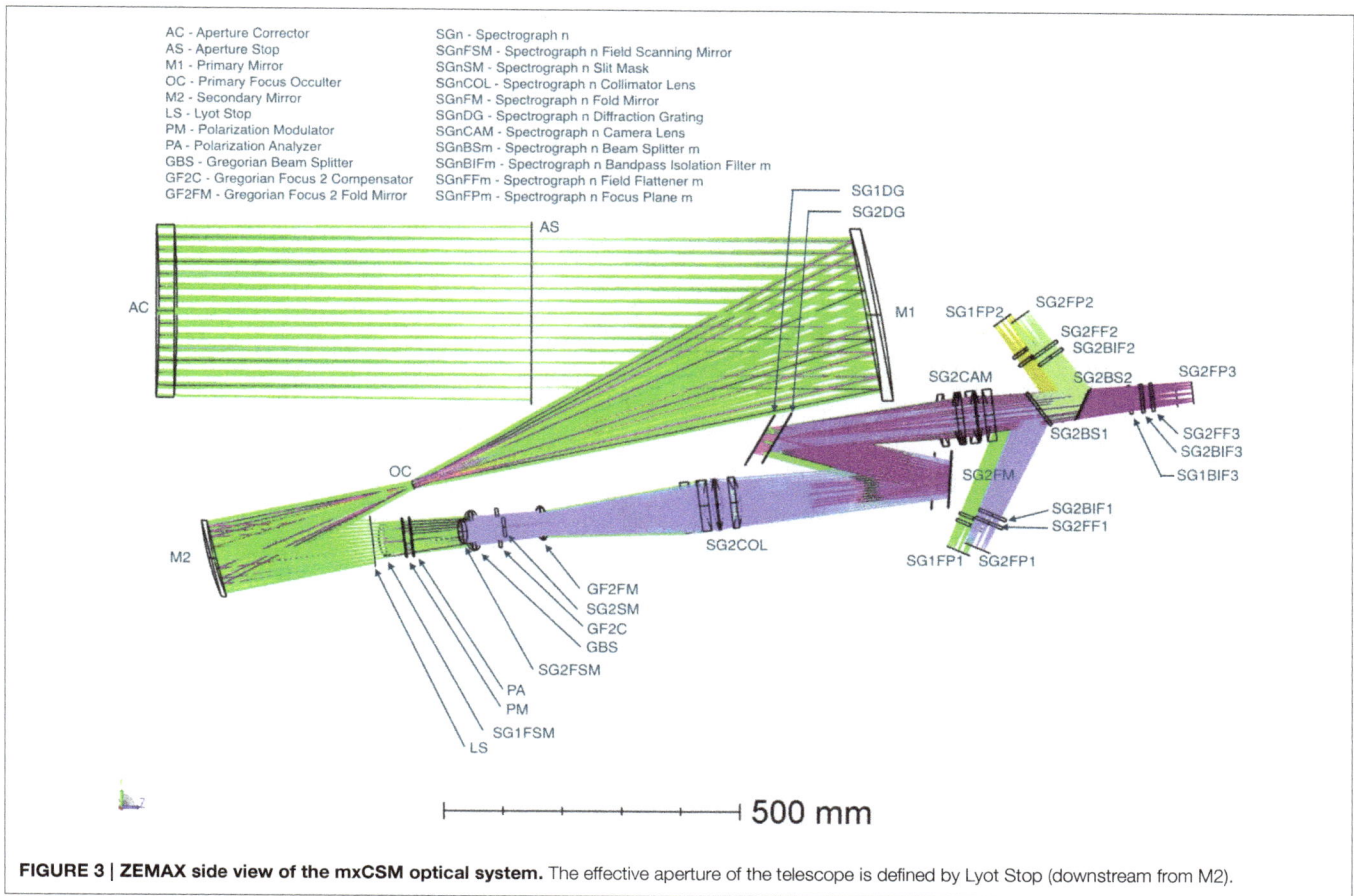

FIGURE 3 | ZEMAX side view of the mxCSM optical system. The effective aperture of the telescope is defined by Lyot Stop (downstream from M2).

is the most critical element for achieving low scattered light level, provided that the objective of the coronagraph is of sufficient quality. Indeed, a recent scattered light study performed for the COSMO Large Coronagraph project at HAO (Gallagher, 2015) showed that a super polished coronagraph singlet lens with a 0.7 nm root-mean-square (RMS) micro roughness polish over the spatial period from 3 to 0.4 mm produces a scattered light of approximately $4 \times 10^{-6} I_\odot$ at a distance of 1.1 $R_\odot$ from disk center and at the wavelength of of the Fe XIII 1075 nm coronal emission line. Whereas, the total scattered light produced by the 0.7 nm RMS objective plus contribution from dust accumulation on the objective surface equivalent to a cleanliness level (CL) of CL220 is approximately $10 \times 10^{-6} I_\odot$ at 1.1 μm and 1.1 $R_\odot$. Here $I_\odot$ and $R_\odot$ denote the intensity of the Sun at disk center, and the radius of the Sun, respectively.

For a catadioptric Gregorian telescope, the scattering off the surface of the primary mirror M1 is an additional source of instrumental scattered light. Given identical surface quality, a mirror produces approximately two times the scattered light than a lens Gallagher (2015). We estimated that for a catadioptric Gregorian coronagraph, the combined scattered light off the aperture corrector and the primary mirror with 0.4 nm RMS micro roughness on all three optical surface is equal to that of a classical Lyot coronagraph with 0.7 nm RMS surface quality on

the two surfaces of the objective lens. Modern optical fabrication techniques can now produce mirrors with RMS micro roughness well below 0.5 nm. Therefore, while a catadioptric Gregorian coronagraph will need to be polished with a surface micro roughness specification that is almost two times more stringent than that for a lens coronagraph in order to achieve the same instrumental scattered light performance, the cost differential for the additional polishing (which should be a small fraction of the total cost required to build a functional system) is more than compensated by the multi-spectral-line observing capability that the catadioptric optical system enables, and the multitude of improvements in the operational efficiency that it will provide.

With properly controlled instrumental scattered light using super-polished surfaces, the primary source of the instrumental scattered light is the accumulation of dust particles on the optical surfaces upstream of the occulter at the primary focus of the system. Decades of operations of Lyot coronagraphs at the Evans Facility of the National Solar Observatory in Sunspot, New Mexico, Mauna Loa Solar Observatory on the Big Island of Hawaii, Mees Solar Observatory on Haleakala, and other coronal observatories around the world had demonstrated the effectiveness of lens cleaning techniques and the robustness of the super polished lens surfaces to withstand repeated

FIGURE 4 | ZEMAX front view of mxCSM.

cleaning without long-term degradation of the scattered light performance of the coronagraphs. On the other hand, while mirror coronagraphs like SOLARC (Kuhn et al., 2003) offer unparalleled wavelength coverage and simultaneous multi-line observations, our experience from operation of SOLARC was that cleaning of its primary mirror is difficult and inevitably scratches the mirror surface and degrades its scattered light performance over time. Therefore, it is important that the primary mirror surface of mxCSM be kept clean to minimize dust contamination. With a catadioptric optical system, the aperture corrector can serve as the window of an air-tight or semi-air-tight enclosure to enclose M1 to eliminate or minimize the need of periodic cleaning of the M1 surface. With comprehensive dust control measure, such as (slightly) clean air over-pressurized dome, the semi-air-tight M1 enclosure, and extended lens tube in front of AC, supplemented with cleaning of the external surface of AC when necessary, low total scattered light can be achieved on a regular basis.

3.2. Compact 3-Line Spectrographs

Spectroscopy and spectropolarimetry of CELs are particularly well suited for the mxSPEC concept. In the million-degree corona, the Doppler width of the CELs are of the order of $\lambda/10,000$, where λ is the wavelength of the spectral line. Therefore, a spectrograph with only moderate resolution that samples the spectra with about $\lambda/40,000$ sample size is sufficient. However, in order to observe multiple spectral lines over a large wavelength range simultaneously, an achromatic spectrograph is needed. mxCSM uses two identical (except for the grating angles)

3-line spectrographs for observation of a total of 6 spectral lines simultaneously. The spectrographs are folded refractive Czerny-Turner spectrographs based on a pair of air-spaced achromatic triplet lenses and a medium-resolution diffraction grating with 7.9 line/mm ruling blazed at 26.7 degree. The optical layout of the first spectrograph is shown again in **Figure 8** without the rest of the optical system for clarity. The effective focal length of the air-spaced triplet lenses is 381 mm. The nominal spectrograph angle $\phi(\equiv \alpha - \beta)$ is set to 20 degree **Table 1** shows the spectrograph configuration parameters and performance characteristics of the six spectral lines shown in **Figure 1**. The collimator SG1COL forms an image of the pupil approximately 650 mm away from the collimator, where the diffraction grating SG1DG is located. The fold mirror SG1FM and the grating redirect the beam away from the direction of the entrance slit. Due to the coarse-ruling and the moderate blaze angle, most of the wavelengths in the visible and near-IR wavelengths are diffracted in approximately the same angular direction. This allows all the spectra to be formed by a single camera lens with minimal image quality degradation. The camera lens (SG1CAM) is placed one focal length away from the grating to produce a telecentric beam. Two dichroic beam splitters (SG1BS1 and SG1BS2) split the exit beams into three arms with different wavelength bands, allowing for observation of three spectral lines simultaneously. Three ultra-narrow bandpass isolation filters (BIFs) with bandwidth equal to $\lambda/1000$ centered at the wavelength of the spectral lines are placed in the telecentric beams in the three arms, each followed by a field flattener optimized for each spectral line.

The slit masks of mxCSM will be configured with 100 parallel slits with slit width of 9 μm and slit separation of 432 μm between neighboring slits. Therefore, it will take only 48 scan steps to cover a 1.2×1 degree FOV. Due to the anamorphic demagnification of the spectrographs, the images of the entrance slits are separated by 360 μm, or 40 pixels on the CCDs at the exit slit planes. With the 380 mm focal length, 20 degree spectrograph angle, and the 7.9 line/mm grating blazed at 26.7 degree, the 9 μm pixel of the CCD samples the spectra with a sample size of $\Delta\lambda \approx \lambda/39,000$ for all wavelengths. The width of $\lambda/1000$ of the 40-pixel spectral windows of each slit thus cover a Doppler velocity window of $\pm$ 150 km/s.

The ultra-narrow bandpass isolation filters will have a flat top transmission profile with a minimum 90% (-0.5 dB) transmission bandwidth of $\lambda/2000$, and a maximum 0.1% (-30 dB) transmission bandwidth of $\lambda/1000$ to eliminate crosstalk of spectra from neighboring slits and overlapping orders. The size of the BIFs will be approximately 50×50 mm.

3.3. Polarimetry Sensitivity Estimates

With the simple optical system, mxCSM will have very high photon throughput. The maximum number of optical surfaces (Spectrograph 2, Arm 3) is 35, excluding the diffraction grating and the bandpass isolation filters. Assuming a nominal 99.5% efficiency for each optical surface using high-performance anti-reflection coating, 70% efficiency for the diffraction grating, 90% efficiency for the BIFs, and 75% quantum efficiency for the

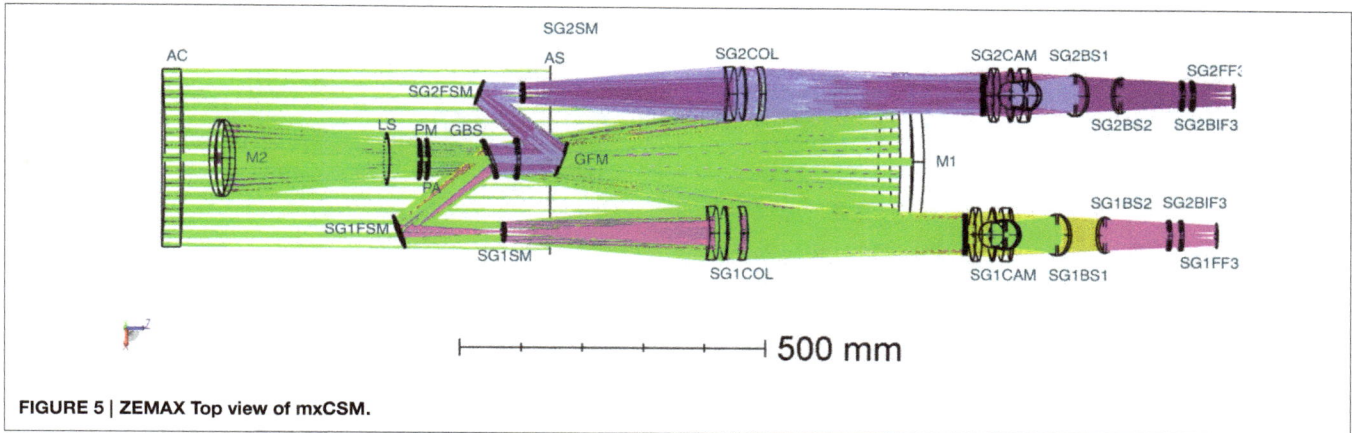

FIGURE 5 | ZEMAX Top view of mxCSM.

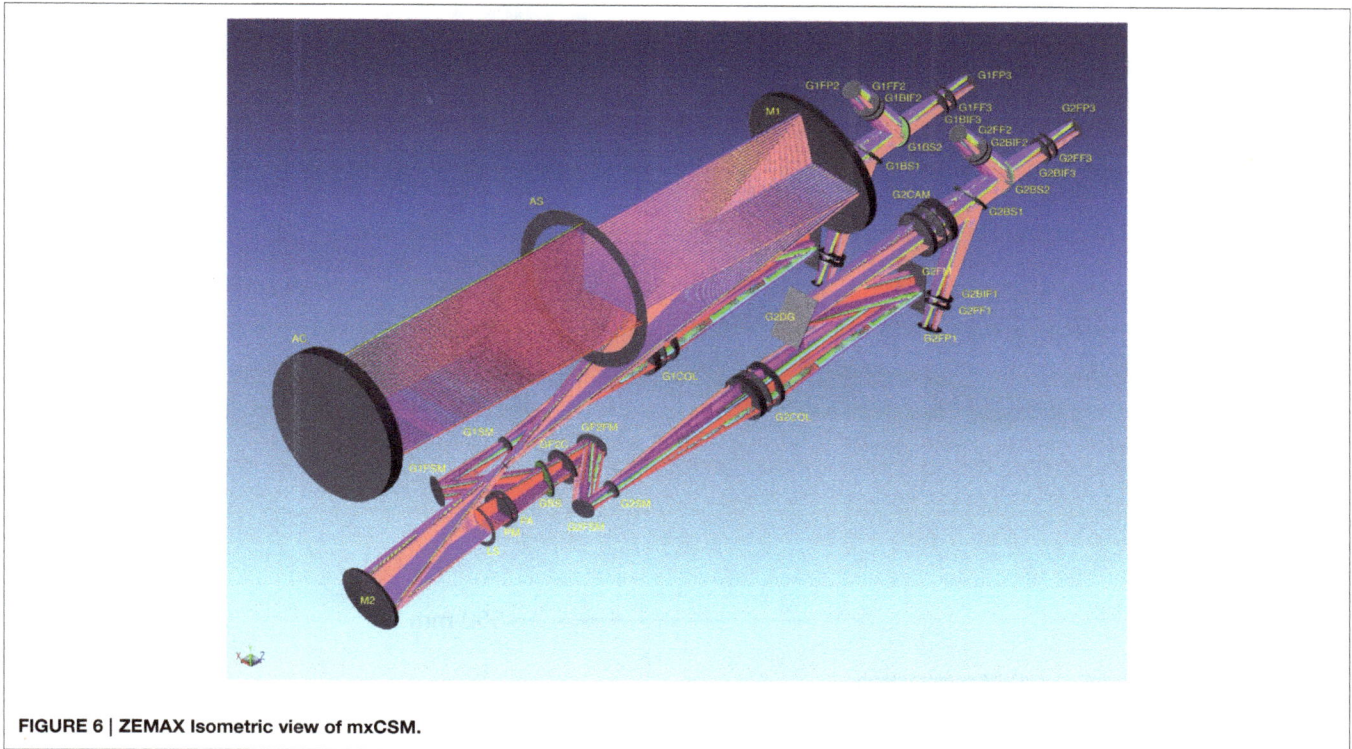

FIGURE 6 | ZEMAX Isometric view of mxCSM.

detector, the overall system throughput of the system is about 40%. The expected photon flux in the continuum spectra near the Fe XIII 1075 nm line with a total scattered light (including sky, dust, and instrumental contribution) of $10 \times 10^{-6} I_\odot$, with different spatial and temporal binning is listed in **Table 2**. Using a circular polarization amplitude of 1×10^{-3} for a 10 G magnetic field, the estimated 3σ detection limits of the line-of-sight component of the coronal magnetic field $B_{3\sigma}$ of mxCSM for the Fe XIII 1075 nm line are 35 G, 12 G, and 4 G, respectively, with spatial resolution of $1''$, $3''$ and $10''$ and temporal resolution of 2 h per map if the scattered light background is $10 \times 10^{-6} I_\odot$. This is sufficient to measure the coronal magnetic field in most active regions up to $1.4 R_\odot$, based on the experience from the SOLARC

coronagraph. In comparison, it would take a conventional single slit spectropolarimeter a minimum of 26.7 h of observation to obtain one full-polarization map of the 1.2 × 1.0 degree field.

The high polarization sensitivity of the hourly and daily averaged data will also allow us to measure the orientation of the coronal magnetic fields from the orientation of the linear polarization of the CEL at very large distance from the limb. It will also allow us to explore the possibility of measuring the line-of-sight magnetic field strength in quiet regions. These observations will provide the data necessary for tomographic inversion of the coronal magnetic fields.

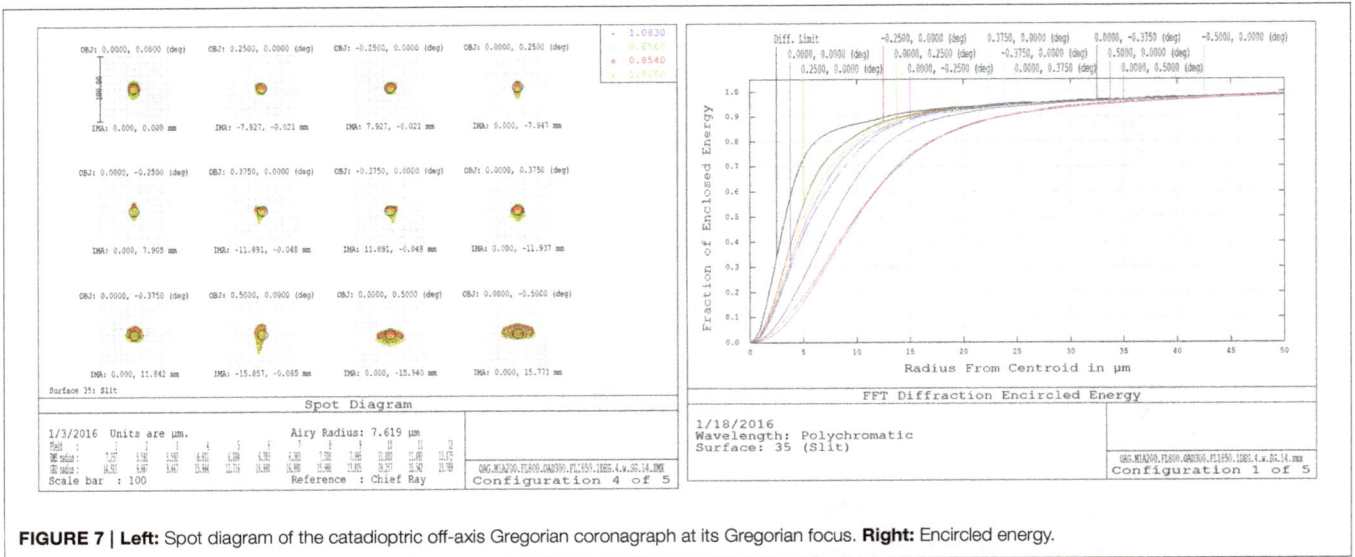

FIGURE 7 | Left: Spot diagram of the catadioptric off-axis Gregorian coronagraph at its Gregorian focus. **Right:** Encircled energy.

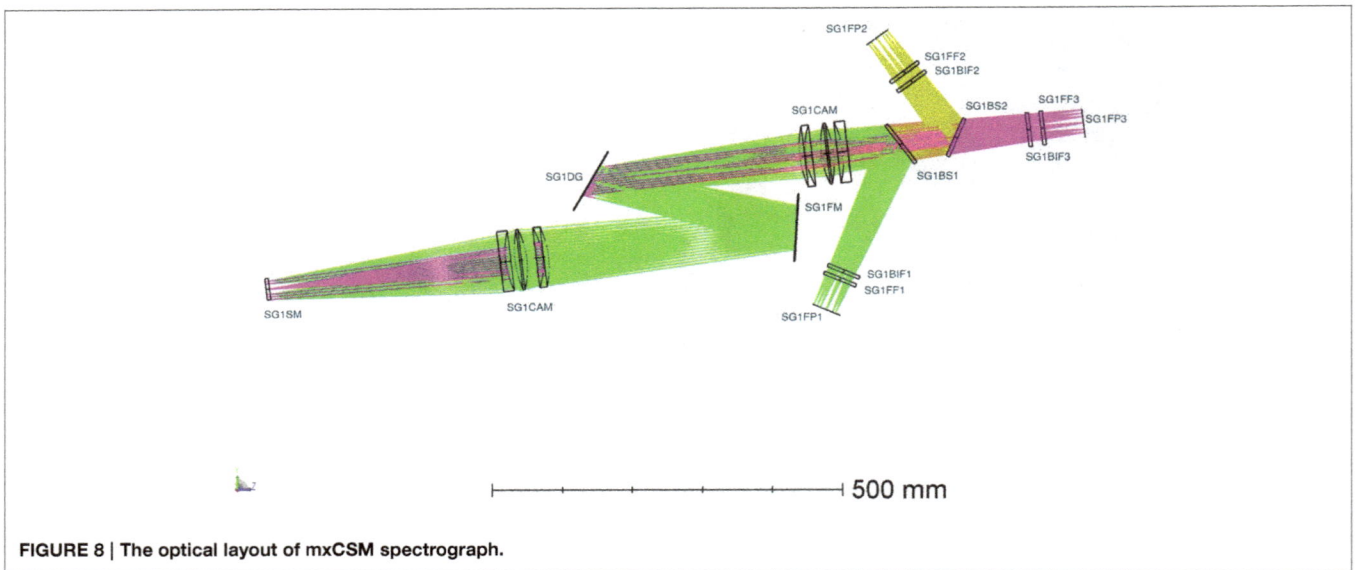

FIGURE 8 | The optical layout of mxCSM spectrograph.

TABLE 1 | Instrument characteristics of the mxCSM spectrographs.

Ion	λ [nm]	m	α	β	φ	ε	Δλ [pm]	λ/Δλ	ΔV [km/s]
Fe IX	436	257	36.701	16.701	20.000	1.00	11	39,114	±150
Fe XIV	530	211	36.701	16.638	20.063	0.87	13	39,053	±150
Fe X	637	176	36.701	16.775	19.926	0.86	16	39,185	±150
Ni XIV	670	167	36.674	16.674	20.000	0.94	17	39,072	±150
Fe XI	789	142	36.674	16.742	19.932	0.99	20	39,135	±150
Fe XIII	1075	104	36.674	16.601	20.070	0.87	27	39,034	±150

m denotes the blazing order of the wavelength, and α and β are the incident and exit angles of the line with respect to the grating normal, respectively. The first three lines are observed with Spectrograph 1 (SG1) and therefore have identical grating incident angle α. The second set of lines are observed by SG2. The nominal spectrograph angle φ ≡ α − β of the spectrographs is 20 degree ε is the amplitude of the blazing function for each spectral line, Δλ denotes the spectral sampling size per pixel, and ΔV is the Doppler velocity coverage of each slit for the spectrograph.

TABLE 2 | Estimates of continuum photon flux N_v, normalized photon noise level σ_P, 3σ longitudinal magnetic field detection limit $B_{3\sigma}$, and total time ΔT_{map} required to scan a 1.2 × 1.0 degree FOV of mxCSM for the Fe XIII 1075 nm line polarimetric observations, assuming a 20% system efficiency, which includes the 50% transmission of the exit linear polarizer of the polarimeter, and a $10 \times 10^{-6} I_\odot$ scattered light background for coronal observation.

No. Slit	No. Scan step	Δx	Δt [s]	CoAdd	ΔT	N_v	σ_P	$3\sigma_B$	ΔT_{map}
1	4800	$1''$	20 (2.5 × 8)	1	20 s	66,350	3.8×10^{-3}	100 G	26.7 h
100	48	$1''$	20 (2.5 × 8)	1	20 s	66,350	3.8×10^{-3}	100 G	16 min
				8	160 s	531,000	1.4×10^{-3}	35 G	2 h
				30	600 s	2,000,000	7.1×10^{-4}	18 G	8 h
100	48	$3''$	20 (2.5 × 8)	1	20 s	600,000	1.3×10^{-3}	33 G	16 min
				8	160 s	4,800,000	4.6×10^{-4}	12 G	2 h
				30	600 s	18,000,000	2.4×10^{-4}	6 G	8 h
100	48	$10''$	20 (2.5 × 8)	1	20 s	6,600,000	3.8×10^{-4}	10 G	16 min
				8	160 s	53,000,000	1.4×10^{-4}	4 G	2 h
				30	600 s	200,000,000	7.1×10^{-5}	2 G	8 h

The first row of the table shows the estimates for a single-slit instrument for comparison. The spectrograph samples the corona with a 1" × 1" sample size. Δx denotes the size of the spatial sampling element, which can be increased by binning the data. The total integration time at each spatial sampling element ΔT is calculated from the 2.5 s individual integration, an 8-state modulation sequence, and the number of **averaged polarization sequences** (CoAdd). ΔT_{map} is estimated assuming a camera with high speed readout that operates with near 100% duty cycle. N_v is the total number of photons that each spatial sampling element collects with integration time ΔT for each polarimetry measurement. $\sigma_P(= 1/\sqrt{N_v})$ denotes the amplitude of noise in the continuum of the polarized spectra if the intensity of the continuum spectra is equal to $10 \times 10^{-6} I_\odot$.

4. SUMMARIES AND DISCUSSIONS

We have presented a conceptual design for a new instrument, optimized for high-temporal resolution spectroscopic measurements of the intensity and polarization of multiple CELs over a very large field of view for research in coronal magnetism. Key technologies that enable this new instrument configuration are (1) large-format focal plane arrays, (2) high-efficiency ultra narrow bandpass isolation filters, and (3) new catadioptric wide-field coronagraph designs. Integration of these technologies enables the implementation of large-scale multiplexing technique to efficiently project the spectra of a very large number of slices of the image plane of the telescope onto multiple focal plane arrays to be recorded simultaneously. The large-scale multiplexing design greatly enhances the capability of the telescope with moderate apertures for observations that require a large field of view coverage. For comparison, the time required for the 25 cm aperture, 6-line, 100-slit coronal spectropolarmeter coronagraph presented in this paper to observe the 1 degree FOV is comparable to that of a 6-m coronagraph equipped with current single-slit, single-wavelength spectropolarimeter. But mxCSM can be constructed with only a fraction of the cost required for the construction of a 6-m class coronagraph.

The high system throughput of this design also makes it an ideal design for future space missions where size and weight of the instruments are severely limited. Finally, we note that although large-scale multiplexing strategy can greatly enhance the capabilities of a current generation of small aperture telescopes, it should also be implemented for future large ground-based telescope projects. For example, a 1-m class mcCSM would yield a magnetic field sensitivity of 10 G with a 3" spatial sampling and 15 min temporal resolution. This will directly enable study of the evolution of active region coronal magnetic fields during solar flares and coronal mass ejections, and finally allow us to test theoretical models of solar eruptions.

AUTHOR CONTRIBUTIONS

The author confirms being the sole contributor of this work and approved it for publication.

FUNDING

The hardware used for the demonstration of the mxSPEC proof-of-concept instrument were acquired with two NSF Major Research Instrument grants, NSF ATM#0421582, and NSF ATM#0923560.

ACKNOWLEDGMENTS

The author thanks Shadia Habbal for helpful comments and review of the paper, and for graciously providing the 2010 Eclipse data for inclusion in this paper. The author also thanks John W. Harvey and Sarah F. Martin for information on early multiple-slit spectroscopy works.

REFERENCES

Casini, R., and Judge, P. G. (1999). Spectral lines for polarization measurements of the coronal magnetic field. II. Consistent treatment of the stokes vector for magnetic-dipole transitions. *Astrophys. J.* 522, 524–539. doi: 10.1086/307629

Elmore, D. (2007). *TMk4 Scattered Light Analysis.* High Altitude Observatory Coronal Solar Magnetism Observatory Technical Note 10.

Gallagher, D. (2015). *COSMO Stray Light Analysis*. High Altitude Observatory Coronal Solar Magnetism Observatory Document 16-COSMOLC-DE-7003.

Habbal, S. R., Druckmüller, M., Morgan, H., Ding, A., Johnson, J., Druckmüllerová, H., et al. (2011). Thermodynamics of the solar corona and evolution of the solar magnetic field as inferred from the total solar eclipse observations of 2010 july 11. *Astrophys. J.* 734, 120. doi: 10.1088/0004-637X/734/2/120

Hale, G. E. (1908). On the probable existence of a magnetic field in sun-spots. *Astrophys. J.* 28, 315. doi: 10.1086/141602

Jaeggli, S. A., Lin, H., Mickey, D. L., Kuhn, J. R., Hegwer, S. L., Rimmele, T. R., et al. (2010). FIRS: a new instrument for photospheric and chromospheric studies at the DST. *Mem. Soc. Astronom. Ital.* 81, 763.

Kramar, M., Inhester, B., Lin, H., and Davila, J. (2013). Vector tomography for the coronal magnetic field. II. Hanle effect measurements. *Astrophys. J.* 775, 25. doi: 10.1088/0004-637X/775/1/25

Kramar, M., Lin, H., and Tomczyk, S. (2016). Direct observation of coronal magnetic fields by vector tomography of the coronal emission line polarizations. *Astrophys. J. Lett.* 819:36. doi: 10.3847/2041-8205/819/2/L36

Kuhn, J. R., Coulter, R., Lin, H., and Mickey, D. L. (2003). "The SOLARC off-axis coronagraph," in *Innovative Telescopes and Instrumentation for Solar Astrophysics*, Vol. 4853 of *Proc. SPIE*, eds S. L. Keil and S. V. Avakyan, 318–326. doi: 10.1117/12.460296

Lin, H. (2003). "ATST near-IR spectropolarimeter," in *Innovative Telescopes and Instrumentation for Solar Astrophysics*, Vol. 4853 of *Proc. SPIE*, eds S. L. Keil and S. V. Avakyan, 215–222. doi: 10.1117/12.460374

Lin, H. (2014). "mxSPEC: a massively multiplexed full-disk spectroheliograph for solar physics research," in *Ground-Based and Airborne Instrumentation for Astronomy V*, Vol. 9147 of *Proc. SPIE*, eds S. K. Ramsay, I. S. McLean and H. Takami (Montréal, QC), 914712.

Lin, H., and Casini, R. (2000). A classical theory of coronal emission line polarization. *Astrophys. J.* 542, 528–534. doi: 10.1086/309499

Lin, H., Kuhn, J. R., and Coulter, R. (2004). Coronal magnetic field measurements. *Astrophys. J. Lett.* 613, L177–L180. doi: 10.1086/425217

Lin, H., Penn, M. J., and Tomczyk, S. (2000). A new precise measurement of the coronal magnetic field strength. *Astrophys. J. Lett.* 541, L83–L86. doi: 10.1086/312900

Livingston, W., Harvey, J., Doe, L. A., Gillespie, B., and Ladd, G. (1980). The kitt-peak coronal velocity experiment. *Bull. Astronom. Soc. Ind.* 8, 43.

Martin, S. F., Ramsey, H. E., Carroll, G. A., and Martin, D. C. (1974). Multi-slit spectrograph and H alpha Doppler system. *Solar Phys.* 37, 343–350. doi: 10.1007/BF00152493

Srivastava, N., and Mathew, S. K. (1999). A digital imaging multi-slit spectrograph for measurement of line-of-Sight velocities on the sun. *Solar Phys.* 185, 61–68. doi: 10.1023/A:1005189319845

Tomczyk, S., Card, G. L., Darnell, T., Elmore, D. F., Lull, R., Nelson, P. G., et al. (2008). An instrument to measure coronal emission line polarization. *Solar Phys.* 247, 411–428. doi: 10.1007/s11207-007-9103-6

Conflict of Interest Statement: The author declares that the research was conducted in the absence of any commercial or financial relationships that could be construed as a potential conflict of interest.

14

Reference Ellipsoid and Geoid in Chronometric Geodesy

Sergei M. Kopeikin [1,2]*

[1] Department of Physics and Astronomy, University of Missouri, Columbia, MO, USA, [2] Department of Physical Geodesy and Remote Sensing, Siberian State University of Geosystems and Technologies, Novosibirsk, Russia

Edited by:
Agnes Fienga,
Observatoire de la Côte d'Azur, France

Reviewed by:
Alberto Vecchiato,
Osservatorio Astrofisico di Torino -
INAF, Italy
Ramesh Jaga Govind,
University of Cape Town, South Africa

*Correspondence:
Sergei M. Kopeikin
kopeikins@missouri.edu

Specialty section:
This article was submitted to
Fundamental Astronomy,
a section of the journal
Frontiers in Astronomy and Space
Sciences

Chronometric geodesy applies general relativity to study the problem of the shape of celestial bodies including the earth, and their gravitational field. The present paper discusses the relativistic problem of construction of a background geometric manifold that is used for describing a reference ellipsoid, geoid, the normal gravity field of the earth and for calculating geoid's undulation (height). We choose the perfect fluid with an ellipsoidal mass distribution uniformly rotating around a fixed axis as a source of matter generating the geometry of the background manifold through the Einstein equations. We formulate the post-Newtonian hydrodynamic equations of the rotating fluid to find out the set of algebraic equations defining the equipotential surface of the gravity field. In order to solve these equations we explicitly perform all integrals characterizing the interior gravitational potentials in terms of elementary functions depending on the parameters defining the shape of the body and the mass distribution. We employ the coordinate freedom of the equations to choose these parameters to make the shape of the rotating fluid configuration to be an ellipsoid of rotation. We derive expressions of the post-Newtonian mass and angular momentum of the rotating fluid as functions of the rotational velocity and the parameters of the ellipsoid including its bare density, eccentricity and semi-major axes. We formulate the post-Newtonian Pizzetti and Clairaut theorems that are used in geodesy to connect the parameters of the reference ellipsoid to the polar and equatorial values of force of gravity. We expand the post-Newtonian geodetic equations characterizing the reference ellipsoid into the Taylor series with respect to the eccentricity of the ellipsoid, and discuss the small-eccentricity approximation. Finally, we introduce the concept of relativistic geoid and its undulation with respect to the reference ellipsoid, and discuss how to calculate it in chronometric geodesy by making use of the anomalous gravity potential.

Keywords: physical geodesy, reference ellipsoid, geoid, relativistic geodesy, reference frames, general relativity and gravitation

1. INTRODUCTION

Accurate definition, determination and realization of terrestrial reference frame is essential for fundamental astronomy, celestial mechanics, geophysics as well as for precise satellite and aircraft navigation, positioning, and mapping. The International Terrestrial Reference Frame (ITRF) is materialized by coordinates of a number of geodetic points (stations) located on the Earth's surface, and spread out across the globe. Definition of the coordinates is derived from the International Terrestrial Reference System (ITRS) which is a theoretical concept. ITRF is used to measure plate tectonics, regional subsidence or loading and to describe the Earth's rotation in space by measuring its rotational parameters (Petit and Luzum, 2010).

Nowadays, four main geodetic techniques are used to compute the accurate terrestrial coordinates and velocities of stations—GPS, VLBI, SLR, and DORIS, for the realization of ITRF. The observations are so accurate that geodesists have to model and include to the data processing the secular Earth's crust changes to reach self-consistency between the various ITRF realizations referred to different epoch. It is recognized that in order to maintain the up-to-date ITRF realization as accurate as possible the development of the most precise theoretical model and relationships between parameters of the model is of a paramount importance.

This model is not of a pure kinematic origin because the gravity field plays an essential role in geodetic network computations (Heiskanen and Moritz, 1967; Hofmann-Wellenhof and Moritz, 2006) making a particular theory of gravity a part of definition of the ITRS. Until recently, such a theory was the Newtonian theory of gravity. However, current SLR/GPS/VLBI techniques allow us to determine the transformation parameters between the coordinates and velocities of the collocation points of the ITRF realization with the precision approaching to ~1 mm and ~1 mm/year, respectively (Petit and Luzum, 2010, Table 4.1). At the same time, general relativity predicts that the post-Newtonian effects in measuring the ITRF coordinates, as compared with the Newtonian theory of gravity, is of the order of the Earth's gravitational radius that is about 1 cm (Kopejkin, 1991; Müller et al., 2008). This general-relativistic effect distorts the coordinate grid of ITRF at the noticeable level which can be detected with the currently available geodetic techniques and, hence, should be taken into account.

ITRS is specified by the Cartesian equatorial coordinates rotating rigidly with the Earth. For the purposes of geodesy and gravimetry the Cartesian coordinates are often converted to a geographical system of elliptical coordinates (h, θ, λ) referred to an international reference ellipsoid which is a non-trivial solution of the Newtonian gravity field equations found by Maclaurin (Chandrasekhar, 1967). It describes the figure of a fluid body with a homogeneous mass density rotating around a fixed z-axis with a constant angular velocity ω. The post-Newtonian effects deforms the shape of Maclaurin's ellipsoid (Chandrasekhar, 1965; Bardeen, 1971) and modify the basic equations of classic geodesy (Müller et al., 2008; Kopeikin et al., 2011, 2016). These relativistic effects must be properly calculated to ensure the adequacy of the geodetic coordinate transformations at the millimeter level of accuracy. A pioneering study of relativistic effects in geodesy have been carried out by Bjerhammar (1985).

Post-Newtonian equilibrium configurations of uniformly rotating fluids have been discussed in literature by researchers from USA (Chandrasekhar, 1965, 1967a,b,c, 1971a,b; Bardeen, 1971; Chandrasekhar and Elbert, 1974, 1978; Chandrasekhar and Miller, 1974), Lithuania (Bondarenko and Pyragas, 1974; Pyragas et al., 1974, 1975), USSR (Tsirulev and Tsvetkov, 1982a,b; Tsvetkov and Tsirulev, 1983; Galtsov et al., 1984) and, the most recently, scientists from the Fridrich Schiller

University of Jena in Germany (Petroff, 2003; Ansorg et al., 2004; Meinel et al., 2008; Gürlebeck and Petroff, 2010, 2013). These papers focused primarily on studying the astrophysical aspects of the problem like stability of the rotating stars, the points of bifurcations, exact axially-symmetric spacetimes, emission of gravitational waves, etc. The present paper focuses on the post-Newtonian effects in physical geodesy. More specifically, we extend the research on the figures of equilibrium into the realm of relativistic geodesy and pay attention to the possible geodetic applications for the adequate processing of the high-precise data obtained by various geodetic techniques that include but not limited to SLR, LLR, VLBI, DORIS, and GNSS (Drewes, 2009; Plag and Pearlman, 2009).

Important stimulating factor for pursuing more advanced theoretical research on relativistic effects in geodesy and the Earth figure of equilibrium is related to the recent breakthrough in manufacturing ring laser gyroscopes (Schreiber, 2013; Beverini et al., 2014; Hurst et al., 2015). Earth rotation and orientation provide the link between the terrestrial (ITRF) and celestial reference frames (ICRF). Traditionally, the Earth orientation parameters (EOPs) are observed by the International Earth Rotation Service (IERS) that operates the global VLBI network. VLBI technique determines EOPs off-line after accumulating the data over sufficiently long period of time. Moreover, VLBI depends on suitable models of Earth's troposphere and geophysical factors which are difficult to predict. Ring laser gyroscopes provide direct access to the Earth rotation axis and a high resolution in the short-term. Currently, the modern quantum optics has matured to a point where the ring laser gyroscopes can resolve rotation rates of 10^{-12} rad/s after 1 h of integration and demonstrate an impressive stability over several month (Schreiber, 2013). The combination of VLBI and ring laser measurements offers an improved sensitivity for the EOPs in the short-term and the direct access to the Earth rotation axis. Another application of the ring laser gyroscopes is to measure the Lense-Thirring effect predicted by Einstein's general relativity (Ciufolini and Wheeler, 1995), in a terrestrial laboratory environment (Beverini et al., 2014).

Another, rapidly emerging branch of relativistic geodesy is called *chronometric geodesy* (Petit et al., 2014). This new line of development is pushed forward by the fascinating progress in making up quantum clocks (Poli et al., 2013), ultra-precise time-scale dissemination over the globe (Kómár et al., 2014), and geophysical applications of the clocks (Bondarescu et al., 2012, 2015). Clocks at rest in a gravitational potential tick slower than clocks outside of it. On Earth, this translates to a relative frequency change of 10^{-16} per meter of height difference (Falke et al., 2014). Comparing the frequency of a probe clock with a reference clock provides a direct measure of the gravity potential difference between the two clocks. This novel technique has been dubbed *chronometric leveling*. It is envisioned as one of the most promising application of the relativistic geodesy in a near future (Mai and Müller, 2013; Mai, 2014; Petit et al., 2014). Optical frequency standards have recently reached stability of 2.2 × 10^{-16} at 1 s, and demonstrated an overall fractional frequency uncertainty of 2.1 × 10^{-18} (Nicholson et al., 2015) which enables their use for *chronometric geodesy* at the absolute level of one centimeter.

The chronometric leveling directly measures the equipotential surface of gravity field (geoid) without conducting a complicated gravimetric survey and solving the differential equations for the anomalous gravity potentials (Stoke's or Molodensky's problem Hofmann-Wellenhof and Moritz, 2006; Torge and Müller, 2012). Combining the data of the chronometric leveling with those of the conventional geodetic techniques will allow us to determine the normal heights of reference points with an unprecedented accuracy (Bondarescu et al., 2012). An adequate physical interpretation of this type of measurements requires more precise theoretical definition of geoid (Kopeikin et al., 2015; Oltean et al., 2015) and is inconceivable without an accompanying development of the corresponding mathematical algorithms accounting for the major relativistic effects in the description of the equipotential level surface and its, so-called, *normal*) gravity field (Hofmann-Wellenhof and Moritz, 2006; Torge and Müller, 2012).

The objective of the present paper is to suggest a solution of Einstein's gravity field equations that gives rise to the post-Newtonian equipotential surface described by the second-order polynomial which exactly coincides with the shape of the Maclaurin ellipsoid in the Newtonian gravity. Our previous paper (Kopeikin et al., 2016) has explored the post-Newtonian case of a homogeneous density distribution of a uniformly rotating perfect fluid. We have shown that the post-Newtonian effects inevitably distort the Maclaurin ellipsoid of rotation to a surface described by the fourth-order polynomial which cannot be reduced to the ellipsoidal surface of the second order by doing coordinate transformations. In the present paper we assume that the mass distribution is not homogeneous but an ellipsoidal one with a free parameter, and prove that under this assumption the equipotential level surface can be made as an exact ellipsoid of the second order which may be found more convenient for doing the reduction of high-precise geodetic measurements with taking into account relativistic corrections.

The paper consists of several sections, bibliography and appendix. Section 2 explains briefly the principles of the post-Newtonian approximations and describes the post-Newtonian metric tensor. In Section 3 we give definition of the post-Newtonian geoid. Section 4 discusses the post-Newtonian ellipsoid which generalizes the Maclaurin ellipsoid and is, in general, the surface of the fourth order. It also introduces the reader to the concept of the post-Newtonian coordinate freedom and shows how this freedom can be used to simplify the mathematical description of the PN-ellipsoid. Sections 5–7 calculate respectively the Newtonian and post-Newtonian gravitational potentials inside the rotating PN-ellipsoid. Sections 8, 9 give the post-Newtonian definitions of the conserved mass and angular momentum. Section 10 derives the post-Newtonian equations of the equipotential level surface. Sections 11, 12 provide the reader with the relativistic generalization of the Pizzetti and Clairaut theorem of classical geodesy (Pizzetti, 1913). Section 13 approximates the relativistic formulas which can be used in practical applications of relativistic geodesy and shows how to build the exact reference-ellipsoid in the post-Newtonian

approximation. Finally, Section 14 introduces the concept of the post-Newtonian geoid's undulation and its relation to the current problem of the measurement of the global average sea level. Appendix contains technical details of calculations of the mathematical integrals which appear in the main part of the paper.

The following notations are used throughout the paper:

- the Greek indices α, β, ... run from 0 to 3,
- the Roman indices i, j, ... run from 1 to 3,
- repeated Greek indices mean Einstein's summation from 0 to 3,
- repeated Roman indices mean Einstein's summation from 1 to 3,
- the unit matrix (also known as the Kroneker symbol) is denoted by $\delta_{ij} = \delta^{ij}$,
- the fully antisymmetric symbol Levi-Civita is denoted as $\varepsilon_{ijk} = \varepsilon^{ijk}$ with $\varepsilon_{123} = +1$,
- the bold letters $\boldsymbol{a} = (a^1, a^2, a^3) \equiv (a^i)$, $\boldsymbol{b} = (b^1, b^2, b^3) \equiv (b^i)$, and so on, denote spatial 3-dimensional vectors,
- a dot between two spatial vectors, for example $\boldsymbol{a} \cdot \boldsymbol{b} = a^1 b^1 + a^2 b^2 + a^3 b^3 = \delta_{ij} a^i b^j$, means the Euclidean dot product,
- the cross between two vectors, for example $(\boldsymbol{a} \times \boldsymbol{b})^i \equiv \varepsilon^{ijk} a^j b^k$, means the Euclidean cross product,
- we use a shorthand notation for partial derivatives $\partial_\alpha = \partial/\partial x^\alpha$,
- covariant derivative with respect to a coordinate x^α is denoted as ∇_α;
- the Minkowski (flat) space-time metric $\eta_{\alpha\beta} = \text{diag}(-1, +1, +1, +1)$,
- $g_{\alpha\beta}$ is the physical spacetime metric,
- $\bar{g}_{\alpha\beta}$ is the background spacetime metric describing the gravitational field of the reference ellipsoid,
- the Greek indices are raised and lowered with the metric $\bar{g}_{\alpha\beta}$,
- G is the universal gravitational constant,
- c is the speed of light in vacuum,
- ω is a constant rotational velocity,
- ρ is a bare constant density of the fluid,
- a is a semi-major axis of the ellipsoid of revolution,
- b is a semi-minor axis of the ellipsoid of revolution,
- ρ_0 is a constant bare density of matter,
- $\kappa \equiv \pi G \rho_0 a^2/c^2$ is a dimensional parameter characterizing the strength of gravitational field,
- a bar above any quantity indicates that it belongs to the background spacetime manifold with the metric $\bar{g}_{\alpha\beta}$.

Other notations are explained in the text as they appear.

2. POST-NEWTONIAN SPACETIME AND METRIC TENSOR

Discussion of the chronometric geodesy starts from the construction of the spacetime manifold for the case of a rotating Earth. We shall employ Einstein's general relativity to build such a manifold though some other alternative theories of gravity discussed, for example in Will (1993), can be used as well. Einstein's field equations represent a system of ten non-linear differential equations in partial derivatives for the ten components of the metric tensor, $g_{\alpha\beta}$, which represent gravitational potentials. Because the equations are difficult to solve exactly due to their non-linearity, we apply the post-Newtonian approximations (PNA) for their solution (Chandrasekhar, 1965).

The PNA are applied in case of slowly-moving matter having a weak gravitational field. This is exactly the situation in the solar system which makes PNA highly appropriate for constructing relativistic theory of reference frames (Soffel et al., 2003) and relativistic celestial mechanics in the solar system (Soffel, 1989; Brumberg, 1991; Kopeikin et al., 2011). The PNA are based on the assumption that a Taylor expansion of the metric tensor can be done in the inverse powers of the speed of gravity c that is equal to the speed of light in vacuum in general relativity. Exact mathematical formulation of a set of basic axioms required for doing the post-Newtonian expansion was given by Rendall (1990). Practically, it requires to have several small parameters characterizing the source of gravity. They are: $\varepsilon_i \sim v_i/c$, $\varepsilon_e \sim v_e/c$, and $\eta_i \sim U_i/c^2$, $\eta_e \sim U_e/c^2$, where v_i is a characteristic velocity of motion of matter inside a body, v_e is a characteristic velocity of the relative motion of the bodies with respect to each other, U_i is the internal gravitational potential of each body, and U_e is the external gravitational potential between the bodies. If one denotes a characteristic radius of a body as L and a characteristic distance between the bodies as R, the internal and external gravitational potentials will be $U_i \simeq GM/L$ and $U_e \simeq GM/R$, where M is a characteristic mass of the body. Due to the virial theorem of the Newtonian gravity (Chandrasekhar, 1965) the small parameters are not independent. Specifically, one has $\varepsilon_i^2 \sim \eta_i$ and $\varepsilon_e^2 \sim \eta_e$. Hence, parameters ε_i and ε_e are sufficient in doing post-Newtonian approximations. Because within the solar system these parameters do not significantly differ from each other, we shall not distinguished them. Quite often we shall use notation $\kappa \equiv \pi G\rho a^2/c^2$ to mark the powers of the fundamental speed c in the post-Newtonian terms.

We assume that physical spacetime has the metric tensor denoted as $g_{\alpha\beta}$. This spacetime is well-approximated by a background manifold with the metric tensor denoted $\bar{g}_{\alpha\beta}$. The perturbation of gravitational field is denoted with $\varkappa_{\alpha\beta}$ and is called the anomalous gravity potential. We, first, build the metric tensor of the background manifold by solving Einstein's equations in harmonic coordinates $x^\alpha = (x^0, x^i)$, where $x^0 = ct$, and t is the coordinate time. The class of the harmonic coordinates is defined by imposing the de Donder gauge condition on the metric tensor (Fock, 1964; Weinberg, 1972),

$$\partial_\alpha \left(\sqrt{-\bar{g}}\bar{g}^{\alpha\beta} \right) = 0 . \qquad (1)$$

Einstein equations for the metric tensor are a complicated non-linear system of differential equations in partial derivatives. Because gravitational field of the solar system is weak and motion of matter is slow, we focus on the first post-Newtonian approximation of general relativity. We also assume that the source of the background metric is a perfect fluid. Under these assumptions the post-Newtonian background metric has the

following form (Kopeikin et al., 2011)

$$\bar{g}_{00} = -1 + \frac{2\bar{V}}{c^2} + \frac{2}{c^4}\left(\bar{\Phi} - \bar{V}^2\right) + O\left(c^{-6}\right) , \qquad (2)$$

$$\bar{g}_{0i} = -\frac{4\bar{V}^i}{c^3} + O\left(c^{-5}\right) , \qquad (3)$$

$$\bar{g}_{ij} = \delta_{ij}\left(1 + \frac{2\bar{V}}{c^2}\right) + O\left(c^{-4}\right) , \qquad (4)$$

where the gravitational potentials entering the metric, satisfy the Poisson equations,

$$\Delta\bar{V} = -4\pi G\bar{\rho} , \qquad (5)$$

$$\Delta\bar{V}^i = -4\pi G\bar{\rho}v^i , \qquad (6)$$

$$\Delta\bar{\Phi} = -4\pi G\bar{\rho}\left(2v^2 + 2\bar{V} + \bar{\Pi} + \frac{3\bar{p}}{\bar{\rho}}\right) , \qquad (7)$$

with $\bar{p}$ and v^i being pressure and velocity of matter respectively, and $\bar{\Pi}$ is the internal energy of matter per unit mass. We emphasize that $\bar{\rho}$ is the local mass density of baryons per a unit of invariant (3-dimensional) volume element $dV = \sqrt{-\bar{g}}u^0 d^3x$, where u^0 is the time component of the 4-velocity of matter. The local mass density, $\bar{\rho}$, relates in the post-Newtonian approximation to the invariant mass density $\bar{\rho}^* = \sqrt{-\bar{g}}u^0\bar{\rho}$, namely (Kopeikin et al., 2011)

$$\bar{\rho}^* = \bar{\rho} + \frac{\bar{\rho}}{c^2}\left(\frac{1}{2}v^2 + 3\bar{V}\right) . \qquad (8)$$

The internal energy, $\bar{\Pi}$, is related to pressure, $\bar{p}$, and the local density, $\bar{\rho}$, through the thermodynamic equation

$$d\bar{\Pi} + \bar{p}d\left(\frac{1}{\bar{\rho}}\right) = 0 , \qquad (9)$$

and the equation of state, $\bar{p} = \bar{p}(\bar{\rho})$.

We shall further assume that the background matter rotates rigidly around fixed z axis with a constant angular velocity ω. This makes the background spacetime stationary with the background metric being independent on time. In the stationary spacetime, the mass density ρ^* obeys the *exact* equation of continuity

$$\partial_i\left(\bar{\rho}^*v^i\right) = 0 . \qquad (10)$$

Velocity of rigidly rotating fluid is

$$v^i = \varepsilon^{ijk}\omega^j x^k , \qquad (11)$$

where $\omega^i = (0, 0, \omega)$ is the constant angular velocity. Replacing velocity v^i in Equation (10) with Equation (11), and differentiating, reveals that

$$v^i\partial_i\bar{\rho} = 0 , \qquad (12)$$

which means that velocity of the fluid is tangent to the surfaces of a constant density $\bar{\rho}$.

3. RELATIVISTIC GEOID

The Newtonian concept of Earth's geoid was extended to the post-Newtonian approximation of general relativity in Soffel (1989) and Kopejkin (1991). *Exact* concept of the relativistic geoid in general relativity that is not limited to the first post-Newtonian approximation has been discussed in Kopeikin et al. (2015) and Oltean et al. (2015).

We consider the Earth as made up of the background matter with small perturbations which reflect the actual mass distribution, stresses, and velocity flow of Earth's matter with respect to the rigidly rotating frame of reference. The angular velocity, ω, of Earth's rotation also slowly changes because of precession, nutation, polar motion and variations in the length-of-day. The physical metric, $g_{\alpha\beta}$, depends on time and spatial coordinates. It is decomposed into an algebraic sum of the background metric $\bar{g}_{\alpha\beta}$, and its perturbation, $\varkappa_{\alpha\beta}$,

$$g_{\alpha\beta} = \bar{g}_{\alpha\beta} + \varkappa_{\alpha\beta} . \quad (13)$$

In the present paper, we shall neglect dependence of the perturbation $\varkappa_{\alpha\beta}$ on time because it produces very tiny relativistic effects that are currently unobservable.

Terrestrial reference frame is formed by the world lines of observers being fixed with respect to the crust of the Earth. Each observer moves in spacetime with four-velocity $u^\alpha = c^{-1}dx^\alpha/d\tau$ where $x^\alpha = \{x^0, x^1, x^2, x^3\}$ are coordinates of the observer, and τ is its proper time defined in terms of the metric tensor Equation (13) as follows,

$$c^2 d\tau^2 = -g_{\alpha\beta}dx^\alpha dx^\beta . \quad (14)$$

Physical space of observers at each instant of time is represented by a three-dimensional hypersurface of constant proper time that is orthogonal everywhere to the world lines of the observers. The metric tensor on this hypersurface is denoted as $h_{\alpha\beta}$ and is given by Zelmanov et al. (2006) and Landau and Lifshitz (1975)

$$h_{\alpha\beta} \equiv g_{\alpha\beta} + u_\alpha u_\beta . \quad (15)$$

It is used to measure the spatial distances. Rising and lowering of Greek indices of geometric objects residing on the perturbed manifold are done with the help of the full metric $g_{\alpha\beta}$.

Similarly to classic geodesy, general relativity offers two definitions of relativistic geoid (Soffel, 1989; Kopejkin, 1991)

Definition 1: The relativistic *u*-geoid represents a two-dimensional surface at any point of which the rate of the proper time, τ, of an ideal clock carried out by a set of fiducial terrestrial observers is constant.

The *u*-geoid is determined by equation $W = $ const., where the physical gravity potential

$$W(x) = c^2 \left(1 - \frac{d\tau}{dt}\right) . \quad (16)$$

Because $d\tau/dt = \sqrt{-g_{\alpha\beta}\dot{x}^\alpha \dot{x}^\beta}$, where the dot means the total time derivative with respect to the coordinate time t, Equation (16) becomes,

$$W(x) \equiv c^2 \left[1 - \left(-g_{00} - \frac{2}{c}g_{0i}v^i - \frac{1}{c^2}g_{ij}v^i v^j\right)^{1/2}\right] . \quad (17)$$

This matches the Newtonian definition of the geoid after decomposition of the metric tensor in the post-Newtonian series and discarding all post-Newtonian terms.

Definition 2: The relativistic *a*-geoid represents a two-dimensional surface at any point of which the direction of a plumb line measured by the terrestrial observer, is orthogonal to the tangent plane of geoid's surface.

In order to derive equation of *a*-geoid, we notice that the direction of the plumb line is given by a four-vector of the physical acceleration of gravity, $g_\alpha \equiv -c^2 a_\alpha$ where $a_\alpha = u^\beta \nabla_\beta u_\alpha$ is a four-acceleration of the terrestrial observer being orthogonal to u^α. In case of a stationary rotating configuration of matter, we get (Lightman et al., 1975, Problems 10.14 and 16.17)

$$g_\alpha = -c^2 \partial_\alpha \ln\left(1 - \frac{W}{c^2}\right) , \quad (18)$$

where W is calculated in Equation (17). We consider an arbitrary displacement, $dx^\alpha_\perp \equiv h^\alpha{}_\beta dx^\beta$, on the spatial hypersurface being orthogonal to u^α everywhere, and make a scalar product of $dx^\alpha_\perp$ with the direction of the plumb line. It gives,

$$dx^\alpha_\perp g_\alpha = dx^\alpha g_\alpha = -c^2 d\ln\left(1 - \frac{W}{c^2}\right) . \quad (19)$$

From definition of *a*-geoid the left-hand side of Equation (19) must vanish due to the condition of orthogonality of the two vectors, $dx^\alpha_\perp$ and g_α. Therefore, it makes $d\ln\left(1 - W/c^2\right) = 0$, which means the constancy of the gravity potential W on the three-dimensional surface of the *a*-geoid. Thus, the surface of *a*-geoid coincides with that of *u*-geoid.

Calculation of the geoid is a difficult mathematical problem of geodesy. It is solved by introducing a reference surface of reference ellipsoid. Geoid's surface is defined in terms of the geoid undulation (height) with respect to the reference ellipsoid.

4. POST-NEWTONIAN REFERENCE ELLIPSOID

In classical geodesy the reference figure for calculation of geoid's undulation is the Maclaurin ellipsoid which is a surface of the second order formed by a rigidly rotating fluid of constant density ρ. Maclaurin's ellipsoid is a surface of the second order (Chandrasekhar, 1969)

$$\frac{x^2 + y^2}{a^2} + \frac{z^2}{b^2} = 1 , \quad (20)$$

where a and b are semi-major and semi-minor axes of the ellipsoid. We also assume $a > b$, and define the eccentricity of the Maclaurin ellipsoid as

$$e \equiv \frac{\sqrt{a^2 - b^2}}{a^2} \,. \qquad (21)$$

Physically, the ellipsoidal shape of rotating, self-gravitating fluid is formed because the Newtonian gravity potential is a scalar function represented by a polynomial of the second order with respect to the Cartesian spatial coordinates, and the differential Euler equation defining the equilibrium of the gravity and pressure is of the first order partial different equation which leads to the quadratic (w.r.t. the coordinates) equation of the level surface.

We shall demonstrate in the following sections that in the post-Newtonian approximation the gravity potential, W, of the rotating fluid is a polynomial of the fourth order as was first noticed by Chandrasekhar (1965). Hence, the post-Newtonian equation of the level surface of a rigidly-rotating fluid is expected to be a surface of the fourth order. We shall assume that the surface remains axisymmetric in the post-Newtonian approximation and dubbed the body with such a surface as PN-ellipsoid.

We shall denote all quantities taken on the surface of the PN-ellipsoid with a bar to distinguish them from the coordinates outside of the surface. Let the barred coordinates $\bar{x}^i = \{\bar{x}, \bar{y}, \bar{z}\}$ denote a point on the surface of the PN-ellipsoid with the axis of symmetry directed along the rotational axis and with the origin located at its post-Newtonian center of mass [1]. Let the rotational axis coincide with the direction of z axis. Then, the most general equation of the PN-ellipsoid is

$$\frac{\bar{\sigma}^2}{a^2} + \frac{\bar{z}^2}{b^2} = 1 + \kappa F(\bar{\mathbf{x}}) \,, \qquad (22)$$

where $\kappa \equiv \pi G \rho a^2 / c^2$, $\bar{\sigma}^2 \equiv \bar{x}^2 + \bar{y}^2$, is the post-Newtonian parameter which is convenient in the calculations that follows,

$$F(\bar{\mathbf{x}}) \equiv \mathcal{K}_1 \frac{\bar{\sigma}^2}{a^2} + \mathcal{K}_2 \frac{\bar{z}^2}{b^2} + \mathcal{B}_1 \frac{\bar{\sigma}^4}{a^4} + \mathcal{B}_2 \frac{\bar{z}^4}{b^4} + \mathcal{B}_3 \frac{\bar{\sigma}^2 \bar{z}^2}{a^2 b^2} \,, \qquad (23)$$

and $\mathcal{K}_1, \mathcal{K}_2, \mathcal{B}_1, \mathcal{B}_2, \mathcal{B}_3$ are arbitrary numerical coefficients.

Each cross-section of the PN-ellipsoid being orthogonal to the rotational axis, represents a circle. The equatorial cross-section has an equatorial radius, $\bar{\sigma} = r_e$, being determined from Equation (22) by the condition $\bar{z} = 0$. It yields

$$r_e = a \left(1 + \frac{1}{2} \kappa \mathcal{B}_1 \right) \,. \qquad (24)$$

The meridional cross-section of the PN-ellipsoid is no longer an ellipse (as it was in case of the Maclaurin ellipsoid) but a curve of the fourth order. Nonetheless, we can define the polar radius,

[1] Post-newtonian definitions of mass, center of mass, and other multipole moments can be found, for example, in Kopeikin et al. (2011).

$\bar{z} = r_p$, of the PN-ellipsoid by the condition, $\bar{\sigma} = 0$. Equation (22) yields

$$r_p = b \left(1 + \frac{1}{2} \kappa \mathcal{B}_2 \right) \,. \qquad (25)$$

The equatorial and polar radii of the PN-ellipsoid should be used in the post-Newtonian approximation instead of the equatorial and polar radii of the Maclaurin reference-ellipsoid for calculation of observable physical effects like the force of force, etc. We characterize the "oblateness" of the PN-ellipsoid by the post-Newtonian eccentricity

$$\epsilon \equiv \frac{\sqrt{r_e^2 - r_p^2}}{r_e} \,. \qquad (26)$$

It differs from the eccentricity Equation (21) of the Maclaurin ellipsoid by relativistic correction

$$\epsilon = e - \kappa \frac{1 - e^2}{2e} (\mathcal{B}_2 - \mathcal{B}_1) \,. \qquad (27)$$

The PN-ellipsoid vs. Maclaurin's ellipsoid is visualized in **Figure 1**.

Theoretical formalism for calculation of the post-Newtonian level surface can be worked out in arbitrary coordinates. For mathematical and historic reasons the most convenient are harmonic coordinates which are also used by the IAU (Soffel et al., 2003) and IERS (Petit and Luzum, 2010). The class of the harmonic coordinates is selected by the gauge condition Equation (1). Different harmonic coordinates are interrelated by coordinate transformations which are not violating the gauge condition 1. This freedom is known as a residual gauge (or coordinate) freedom. The field Equations (5)–(7) and their solutions are form-invariant with respect to the residual gauge transformations.

The residual gauge freedom is described by a post-Newtonian coordinate transformation,

$$x'^\alpha = x^\alpha + \kappa \xi^\alpha(x) \,, \qquad (28)$$

where functions, ξ^α, obey the Laplace equation,

$$\Delta \xi^\alpha = 0 \,. \qquad (29)$$

Solution of the Laplace equation which is convergent at the origin of the coordinate system, is given in terms of the harmonic polynomials which are selected by the conditions imposed by the statement of the problem. In our case, the problem is to determine the shape of the PN-ellipsoid which has the surface described by the polynomial of the fourth order Equation (32) with yet unknown coefficients $\mathcal{B}_1, \mathcal{B}_2, \mathcal{B}_3$. The form of the Equation 32 does not change (in the post-Newtonian approximation) if the functions ξ^α in Equation (28) are polynomials of the third order. It is straightforward to show

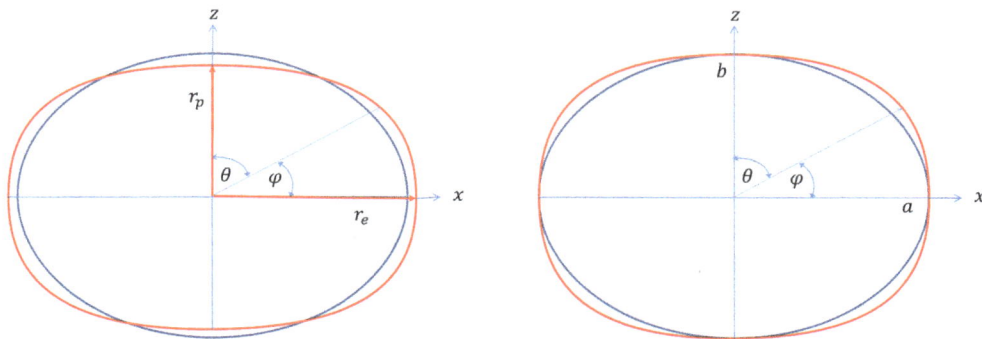

FIGURE 1 | Meridional cross-section of the PN-ellipsoid (a red curve in on-line version) vs. the Maclaurin ellipsoid (a blue curve in on-line version). The left panel represents the most general case with arbitrary values of the shape parameters $\mathcal{B}_1, \mathcal{B}_2, \mathcal{B}_3$ when the equatorial, r_e, and polar, r_p, radii of the PN-ellipsoid differ from the semi-major, a, and semi-minor, b, axes of the Maclaurin ellipsoid, $r_e \neq a, r_p \neq b$. The right panel shows the most important physical case of $\mathcal{B}_1 = \mathcal{B}_2 = 0$ when the equatorial and polar radii of the PN-ellipsoid and the Maclaurin ellipsoid are equal. The angle φ is the geographic latitude ($-90° \leq \varphi \leq 90°$), and the angle θ is a complementary angle (co-latitude) used for calculation of integrals in appendix of the present paper ($0 \leq \theta \leq \pi$). In general, when $\mathcal{B}_1 \neq \mathcal{B}_2 \neq 0$, the maximal radial difference (the "height" difference) between the surfaces of the PN-ellipsoid and the Maclaurin ellipsoid depends on the choice of the post-Newtonian coordinates, and may amount to a few cm (see Section 13). Carefully operating with the residual gauge freedom of the post-Newtonian theory allows us to make the difference between the two surfaces much less than that.

that the admissible harmonic polynomials of the third order have the following form

$$\xi^1 = hx + p\frac{x}{a^2}\left(\sigma^2 - 4z^2\right), \tag{30a}$$

$$\xi^2 = hy + p\frac{y}{a^2}\left(\sigma^2 - 4z^2\right), \tag{30b}$$

$$\xi^3 = kz + q\frac{z}{b^2}\left(3\sigma^2 - 2z^2\right), \tag{30c}$$

where h, k, p, and q are arbitrary constant parameters. Polynomials Equations (30a)–(30c) represent solutions of the Laplace Equation (29). We choose $\xi^0 = 0$ because we consider stationary spacetime so all functions are time-independent.

Coordinate transformation Equation (28) with ξ^i taken from Equations (30a)–(30c) does not violate the harmonic gauge condition Equation (1) but it changes the numerical post-Newtonian coefficients in the mathematical form of Equations (22) and Equation (23)

$$\mathcal{K}_1 \rightarrow \mathcal{K}_1 + 2h, \tag{31a}$$

$$\mathcal{K}_2 \rightarrow \mathcal{K}_2 + 2k, \tag{31b}$$

$$\mathcal{B}_1 \rightarrow \mathcal{B}_1 + 2p, \tag{31c}$$

$$\mathcal{B}_2 \rightarrow \mathcal{B}_2 - 4q, \tag{31d}$$

$$\mathcal{B}_3 \rightarrow \mathcal{B}_3 - 8p\frac{b^2}{a^2} + 6q\frac{a^2}{b^2}, \tag{31e}$$

Thus, it makes evident that only one out of the five coefficients $\mathcal{K}_1, \mathcal{K}_2, \mathcal{B}_1, \mathcal{B}_2, \mathcal{B}_3$ is algebraically independent while the four others can be chosen arbitrary. To simplify our calculations and eliminate the gauge-dependent terms from mathematical equations we choose $\mathcal{K}_1 = \mathcal{K}_2 = \mathcal{B}_1 = \mathcal{B}_2 = 0$. The constant $\mathcal{B}_3 \equiv \mathcal{B}$ is left free. It will be fixed later on. With the accepted choice of the coordinates the polar radius $r_p = b$, the equatorial

radius $r_e = a$, the eccentricity of the PN-ellipsoid $\epsilon = e$, and function

$$F(\bar{x}) \equiv \mathcal{B}\frac{\bar{\sigma}^2\bar{z}^2}{a^2b^2}. \tag{32}$$

Let $x^i = \{x, y, z\}$ is any point inside the PN-ellipsoid. We introduce a quadratic polynomial

$$C(x) \equiv \frac{\sigma^2}{a^2} + \frac{z^2}{b^2} - 1. \tag{33}$$

This polynomial vanishes on the boundary surface of the PN-ellipsoid in the Newtonian approximation. However, in the post-Newtonian approximation we have on the boundary the following condition

$$C(\bar{x}) = \kappa F(\bar{x}), \tag{34}$$

as a consequence of Equation (22). In terms of the polynomial $C(\bar{x})$ function $F(\bar{x})$ in Equation (32) can be formally recast to

$$F(\bar{x}) = \mathcal{B}\frac{\bar{z}^2}{b^2}\left[1 + C(\bar{x}) - \frac{\bar{z}^2}{b^2}\right]. \tag{35}$$

The term being proportional to $C(\bar{x})$ can be discarded on the boundary of the PN-ellipsoid. Now we shall calculate the gravitational potentials of the rotating PN ellipsoid which enter the metric tensor Equations (2)–(4).

5. NEWTONIAN POTENTIAL $\bar{V}$

Newtonian gravitational potential $\bar{V}$ satisfies the inhomogeneous Poisson's equation

$$\Delta \bar{V}(x) = -4\pi G\bar{\rho}(x), \tag{36}$$

inside the mass. Its particular solution is given by

$$\bar{V}(\boldsymbol{x}) = G \int_{\mathcal{V}} \frac{\bar{\rho}(\boldsymbol{x}')d^3x'}{|\boldsymbol{x} - \boldsymbol{x}'|} , \qquad (37)$$

where $\mathcal{V}$ is the volume occupied by the matter distribution. We shall assume that the density has an ellipsoidal distribution inside the body which differs from the homogeneous density, ρ_0, only in the post-Newtonian approximation. Moreover, we shall consider the most simple, linearized case of the ellipsoidal distribution

$$\bar{\rho}(\boldsymbol{x}) = \rho_0 \left[1 + \kappa \mathcal{A} \left(\frac{\sigma^2}{a^2} + \frac{z^2}{b^2} \right) \right] , \qquad (38)$$

where ρ_0 is a constant bare density, and $\mathcal{A}$ is as yet undetermined numerical constant. This assumption allows us to perform integration in Equation (37) explicitly. Inside the mass the integral Equation (37) can be calculated by making use of spherical coordinates. The procedure is as follows (Chandrasekhar, 1969).

Let us consider a point $x^i = \{x, y, z\}$ inside the PN-ellipsoid Equation (22). It is connected to the point $\bar{x}^i$ on the surface of the ellipsoid by vector $R^i = \bar{x}^i - x^i$ where $R^i = R\ell^i$, $R = \sqrt{\delta_{ij}R^iR^j}$, and a unit vector, $\ell^i = \{\sin\theta\cos\lambda, \sin\theta\sin\lambda, \cos\theta\}$. We have

$$\bar{x}^i = x^i + \ell^i R , \qquad (39)$$

Substituting Equations (39)–(22) yields a quadratic equation

$$AR^2 + 2BR + C = \kappa F (\boldsymbol{x} + \boldsymbol{\ell}R) , \qquad (40)$$

where

$$A \equiv \frac{\sin^2\theta}{a^2} + \frac{\cos^2\theta}{b^2} ,$$
$$B \equiv \frac{\sin\theta \left(x\cos\lambda + y\sin\lambda\right)}{a^2} + \frac{z\cos\theta}{b^2} ,$$
$$C \equiv \frac{\sigma^2}{a^2} + \frac{z^2}{b^2} - 1 . \qquad (41)$$

We solve Equation (40) by iterations by expanding $R = \hat{R} + c^{-2}\Delta R$, where $\hat{R}$ is either $\hat{R}_+$ or $\hat{R}_-$ corresponding to two solutions of the quadratic Equation (40) with the right hand side being nil. The bare solution $\hat{R}$ is used, then, to calculate the right side of Equation (40) for performing the second iteration. We obtain

$$R_\pm = -\frac{B}{A} \pm \frac{\sqrt{B^2 - AC + \kappa AF_\pm}}{A} , \qquad (42)$$

where

$$F_\pm \equiv \mathcal{B} \left(\frac{z + \cos\theta\hat{R}_\pm}{b} \right)^2 - \mathcal{B} \left(\frac{z + \cos\theta\hat{R}_\pm}{b} \right)^4 , \qquad (43)$$

where the term $C(\bar{\boldsymbol{x}}) = AR^2 + 2BR + C$ vanishes because of Equation (40).

We make replacement of variable $\boldsymbol{x}'$ in Equation (37) to $\boldsymbol{r} = \boldsymbol{x} - \boldsymbol{x}'$, and use spherical coordinates to perform integration with respect to the radial coordinate $r = |\boldsymbol{x} - \boldsymbol{x}'|$ from the point $r = 0$ to the point lying on the surface of the PN ellipsoid, $r = R_-(\theta, \lambda)$ or $r = R_+(\theta, \lambda)$. For the internal point, the angular integration in the remaining integral over the surface of the PN ellipsoid is performed over the solid angle 4π, and the integration with the point $R_-(\theta, \lambda)$ equals to that with the radial direction $R_+(\theta, \lambda)$. This observation makes the angular integration easier because the integral Equation (37) can be written down in the following form (Chandrasekhar, 1969)

$$\bar{V} = \frac{1}{4}G\rho_0 \oint_{S^2} \left(R_+^2 + R_-^2 \right) d\Omega + \kappa\mathcal{A}\bar{I}_1 , \qquad (44)$$

where R_+ and R_- are defined in Equation (42), and the integral

$$\bar{I}_1 = G\rho_0 \int_{\mathcal{V}} \frac{d^3x'}{|\boldsymbol{x} - \boldsymbol{x}'|} \left(\frac{\sigma'^2}{a^2} + \frac{z'^2}{b^2} \right) \qquad (45)$$

takes into account the ellipsoidal distribution of the density. After making use of Equation (42) and expanding the integrand in Equation (44) with respect to the post-Newtonian parameter κ, the Newtonian potential takes on the following form

$$\bar{V} = \frac{1}{2}G\rho_0 \oint_{S^2} \left\{ \frac{2B^2 - AC}{A^2} + \frac{\kappa}{2A} \left[F_+ + F_- \right. \right. \\ \left. \left. - \frac{B}{\sqrt{B^2 - AC}} \left(F_+ - F_- \right) \right] \right\} d\Omega + \kappa\mathcal{A}\bar{I}_1 , \qquad (46)$$

where all post-Newtonian terms of the higher order in κ have been discarded, and integration is performed over a unit sphere S^2 with respect to the angles λ and θ, $d\Omega \equiv \sin\theta d\theta d\lambda$ is the element of the solid angle.

Now, we expand $F_\pm$ in polynomial w.r.t. $R_\pm$,

$$F_\pm = \alpha_0 + \alpha_1 \cos\theta R_\pm + \alpha_2 (\cos\theta R_\pm)^2 + \alpha_3 (\cos\theta R_\pm)^3 \\ + \alpha_4 (\cos\theta R_\pm)^4 , \qquad (47)$$

where the coefficients

$$\alpha_0 = \mathcal{B}\frac{z^2}{b^2} \left(1 - \frac{z^2}{b^2} \right) , \qquad (48)$$

$$\alpha_1 = \frac{2\mathcal{B}z}{b^2} \left(1 - 2\frac{z^2}{b^2} \right) \qquad (49)$$

$$\alpha_2 = \frac{\mathcal{B}}{b^2} \left(1 - 6\frac{z^2}{b^2} \right) \qquad (50)$$

$$\alpha_3 = -\frac{4\mathcal{B}z}{b^4} , \qquad (51)$$

$$\alpha_4 = -\frac{\mathcal{B}}{b^4} , \qquad (52)$$

are the polynomials of z only. We also notice that on the surface of the PN-ellipsoid, $F(\bar{\boldsymbol{x}}) = \alpha_0$, as follows from Equation (35) where we can use, $C(\bar{\boldsymbol{x}}) = 0$, in the post-Newtonian terms.

Replacing Equations (47) in (46) transforms it to

$$\bar{V} = \bar{V}_N + \kappa\bar{V}_{pN} + \kappa\mathcal{A}\bar{I}_1 , \qquad (53)$$

where

$$\bar{V}_{\mathrm{N}} = G\rho_0 \oint_{S^2} \left(\frac{B^2}{A^2} - \frac{C}{2A}\right) d\Omega \tag{54}$$

$$\bar{V}_{\mathrm{pN}} = G\rho_0 \oint_{S^2} \left[\frac{\alpha_0}{2A} - \alpha_1 \cos\theta \frac{B}{A^2} + \alpha_2 \cos^2\theta \left(\frac{2B^2}{A^3} - \frac{C}{2A^2}\right)\right.$$
$$-2\alpha_3 \cos^3\theta \left(\frac{2B^3}{A^4} - \frac{BC}{A^3}\right)$$
$$\left. + \alpha_4 \cos^4\theta \left(\frac{8B^4}{A^5} - \frac{6B^2C}{A^4} + \frac{C^2}{2A^3}\right)\right] d\Omega , \tag{55}$$

Equations (53)–(55) describe the Newtonian potential *exactly* both on the surface of the PN-ellipsoid and inside it. These equations cannot be extended to the external space which requires a separate integration which will be discussed somewhere else.

The integrals entering V in Equations (54), (55) are discussed in Appendix A in Supplementary Material. The integrals I_1 appears besides Equation (53) also in the post-Newtonian potentials of the gravitational field in Equation (80), and its discussion is postponed to section 7. After evaluating the integrals in the internal point and reducing similar terms, potentials V_{N} and V_{pN} take on the following form:

$$\bar{V}_{\mathrm{N}} = \pi G\rho_0 a^2 \left[\left(1 - \frac{z^2}{b^2}\right) \beth_0 - \left(1 - 3\frac{z^2}{b^2}\right) \beth_1 - C(\boldsymbol{x})\beth_1\right], \tag{56}$$

$$\bar{V}_{\mathrm{pN}} = \pi G\rho_0 a^2 \left[F_1(z) + b^2 F_2(z)C(\boldsymbol{x}) + b^4 F_3(z)C^2(\boldsymbol{x})\right], \tag{57}$$

where

$$F_1(z) = \alpha_0 \beth_0 - 2\alpha_1 z\beth_1 + \tag{58}$$
$$2\alpha_2 b^2 \left[\left(1 - \frac{z^2}{b^2}\right) \beth_1 - \left(1 - \frac{3z^2}{b^2}\right) \beth_2\right]$$
$$-4\alpha_3 b^2 z \left[3\left(1 - \frac{z^2}{b^2}\right) \beth_2 - \left(3 - \frac{5z^2}{b^2}\right) \beth_3\right] +$$
$$6\alpha_4 b^4 \left[\left(1 - \frac{z^2}{b^2}\right)^2 \beth_2 - 2\left(1 - 6\frac{z^2}{b^2} + 5\frac{z^4}{b^4}\right) \beth_3\right.$$
$$\left. + \left(1 - 10\frac{z^2}{b^2} + \frac{35}{3}\frac{z^4}{b^4}\right) \beth_4\right],$$

$$F_2(z) = \alpha_2 \left(\beth_1 - 2\beth_2\right) - 4\alpha_3 z \left(2\beth_2 - 3\beth_3\right) + \tag{59}$$
$$6\alpha_4 b^2 \left[\left(1 - \frac{z^2}{b^2}\right) \beth_2 - 3\left(1 - \frac{3z^2}{b^2}\right) \beth_3\right.$$
$$\left. + 2\left(1 - \frac{5z^2}{b^2}\right) \beth_4\right],$$

$$F_3(z) = \alpha_4 \left(\beth_2 - 6\beth_3 + 6\beth_4\right) , \tag{60}$$

and the polynomial coefficients $\alpha_0, \alpha_1, \alpha_2, \alpha_3, \alpha_4$ are given in Equations (48)–(52). It is worth noticing that V_{pN} obeys the Laplace equation

$$\Delta \bar{V}_{\mathrm{pN}} = 0 , \tag{61}$$

and Equation (57) represents the harmonic polynomial of the fourth order.

6. POST-NEWTONIAN VECTOR POTENTIAL $\bar{V}^i$

Vector potential V^i obeys the Poisson equation

$$\Delta \bar{V}^i = -4\pi G\bar{\rho}(\boldsymbol{x})v^i(\boldsymbol{x}) , \tag{62}$$

which has a particular solution

$$\bar{V}^i = G \int_{\mathcal{V}} \frac{\bar{\rho}(\boldsymbol{x}')v^i(\boldsymbol{x}')}{|\boldsymbol{x} - \boldsymbol{x}'|} d^3x' . \tag{63}$$

For a rigidly rotating configuration, $v^i(x) = \varepsilon^{ijk}\omega^j x^k$ so that

$$\bar{V}^i = \varepsilon^{ijk}\omega^j \bar{\mathcal{D}}^k , \tag{64}$$

where

$$\bar{\mathcal{D}}^i = G \int \frac{\bar{\rho}(\boldsymbol{x}')x'^i d^3x'}{|\boldsymbol{x} - \boldsymbol{x}'|} . \tag{65}$$

It can be recast to the following form

$$\mathcal{D}^i = x^i \bar{V}_{\mathrm{N}} + G \int \frac{\bar{\rho}(\boldsymbol{x}')(x'^i - x^i)}{|\boldsymbol{x} - \boldsymbol{x}'|} d^3x' , \tag{66}$$

where $\bar{V}_{\mathrm{N}}$ is the Newtonian potential given in Equation (56). Because the potential $\bar{V}^i$ appears only in the post-Newtonian terms we can consider the density $\bar{\rho}$ as constant, $\bar{\rho} = \rho_0$. In this case the second term in the right hand side of Equation (66) can be integrated over the radial coordinate, yielding

$$\int_{\mathcal{V}} \frac{\bar{\rho}(\boldsymbol{x}')(x'^i - x^i)}{|\boldsymbol{x} - \boldsymbol{x}'|} d^3x' = \frac{\rho_0}{6} \oint_{S^2} \left(R_+^3 + R_-^3\right) l^i d\Omega . \tag{67}$$

After making use of Equation (42) to replace R_+ and R_-, we obtain

$$\int_{\mathcal{V}} \frac{\bar{\rho}(x'^i - x^i)}{|\boldsymbol{x} - \boldsymbol{x}'|} d^3x' = \rho_0 \oint_{S^2} \left(-\frac{4}{3}\frac{B^3}{A^3} + \frac{BC}{A^2}\right) l^i d\Omega , \tag{68}$$

where we have omitted the post-Newtonian terms (with κ) since the vector potential $\bar{V}^i$ itself appears only in the post-Newtonian terms. Integrals entering Equation (68) are given in Appendix A in Supplementary Material. Calculation reveals

$$\int_{\mathcal{V}} \frac{\bar{\rho}(\boldsymbol{x}')(x' - x)}{|\boldsymbol{x} - \boldsymbol{x}'|} d^3x' = -\pi\rho_0 a^2 x \left[\left(1 - \frac{z^2}{b^2}\right) \beth_0\right.$$
$$-2\left(1 - \frac{3z^2}{b^2}\right) \beth_1 + \left(1 - \frac{5z^2}{b^2}\right) \beth_2\right]$$
$$\left. + x\pi\rho_0 a^2 C(\boldsymbol{x}) \left(\beth_1 - \beth_2\right) , \tag{69}\right.$$

$$\int_{\mathcal{V}} \frac{\bar{\rho}(\boldsymbol{x}')(y' - y)}{|\boldsymbol{x} - \boldsymbol{x}'|} d^3x' = -\pi\rho_0 a^2 y \left[\left(1 - \frac{z^2}{b^2}\right) \beth_0\right.$$
$$-2\left(1 - \frac{3z^2}{b^2}\right) \beth_1 + \left(1 - \frac{5z^2}{b^2}\right) \beth_2\right]$$
$$\left. + y\pi\rho_0 a^2 C(\boldsymbol{x}) \left(\beth_1 - \beth_2\right) , \tag{70}\right.$$

$$\int_{\mathcal{V}} \frac{\bar{\rho}(\boldsymbol{x}')(z'-z)}{|\boldsymbol{x}-\boldsymbol{x}'|} d^3x' = -4\pi\rho_0 a^2 z \left[\left(1-\frac{z^2}{b^2}\right)\beth_1 \right.$$
$$\left. -\left(1-\frac{5z^2}{3b^2}\right)\beth_2\right] - 2z\pi\rho_0 a^2 C(\boldsymbol{x})(\beth_1 - 2\beth_2) \,. \tag{71}$$

Substituting this result to Equation (66) and making use of Equation (56) yields

$$\bar{\mathcal{D}}^i \equiv \left(\bar{\mathcal{D}}^x, \bar{\mathcal{D}}^y, \bar{\mathcal{D}}^z\right) = (x\bar{D}_1, y\bar{D}_1, z\bar{D}_2) \,, \tag{72}$$

where

$$\bar{D}_1 \equiv \pi G\rho_0 a^2 \left[\left(1-\frac{3z^2}{b^2}\right)\beth_1 - \left(1-\frac{5z^2}{b^2}\right)\beth_2 - C(\boldsymbol{x})\beth_2\right] \tag{73}$$

$$\bar{D}_2 \equiv \pi G\rho_0 a^2 \left[\left(1-\frac{z^2}{b^2}\right)\beth_0 - \left(5-7\frac{z^2}{b^2}\right)\beth_1 + 4\left(1-\frac{5z^2}{3b^2}\right)\beth_2 \right.$$
$$\left. + C(\boldsymbol{x})\left(4\beth_2 - 3\beth_1\right)\right] \,, \tag{74}$$

7. POST-NEWTONIAN SCALAR POTENTIAL $\bar{\Phi}$

Potential $\bar{\Phi}$ is defined by equation

$$\Delta\bar{\Phi} = -4\pi G\bar{\rho}(\boldsymbol{x}')\bar{\phi}(\boldsymbol{x}') \,, \tag{75}$$

where function

$$\bar{\phi}(\boldsymbol{x}') \equiv 2\omega^2\sigma^2 + 3\frac{\bar{p}}{\bar{\rho}} + 2\bar{V}_N \,. \tag{76}$$

Potential $\bar{\Phi}$ enters only the post-Newtonian equations. Therefore, everywhere in calculation of $\bar{\Phi}$ we can assume the density $\bar{\rho}$ being approximated as constant. The pressure $\bar{p}$ inside the massive body can be calculated from the hydrostatic equilibrium equation with a constant value of the bare density $\bar{\rho} = \rho_0$. The solution is Chandrasekhar (1969)

$$\frac{\bar{p}}{\rho_0} = -\pi G\rho_0 a^2 C(\boldsymbol{x})(\beth_0 - 2\beth_1) \,. \tag{77}$$

Making use of Equation (77) and Equation (56) we can write down function $\bar{\phi}(\boldsymbol{x}')$ as

$$\bar{\phi}(\boldsymbol{x}') = a^2 \left[2\omega^2 - \pi G\rho_0 (3\beth_0 - 4\beth_1)\right]\left(\frac{\sigma^2}{a^2} + \frac{z^2}{b^2}\right)$$
$$- 2a^2\left[\omega^2 - \pi G\rho_0(\beth_0 - 3\beth_1)\right]\frac{z^2}{b^2} + \pi G\rho_0 a^2 (5\beth_0 - 6\beth_1) \,. \tag{78}$$

Particular solution of Equation (75) can be written as

$$\bar{\Phi} = 2\omega^2 a^2 \left(\bar{I}_1 - \bar{I}_2\right) - \pi G\rho_0 a^2 \left[\left(3\bar{I}_1 + 2\bar{I}_2 - 5\bar{V}_N\right)\beth_0 \right.$$
$$\left. + 2\left(2\bar{I}_1 + 3\bar{I}_2 - 3\bar{V}_N\right)\beth_1\right] \,, \tag{79}$$

where we have introduced two integrals

$$\bar{I}_1 = G\rho_0 \int_{\mathcal{V}} \frac{d^3x'}{|\boldsymbol{x}-\boldsymbol{x}'|}\left(\frac{\sigma'^2}{a^2} + \frac{z'^2}{b^2}\right) \,, \tag{80}$$

$$\bar{I}_2 = \frac{G\rho_0}{b^2} \int_{\mathcal{V}} \frac{z'^2}{|\boldsymbol{x}-\boldsymbol{x}'|}d^3x' \,. \tag{81}$$

Notice that the integral $\bar{I}_1$ has already appeared in the calculation of the Newtonian potential in Equation (45) while the integral $\bar{I}_2$ is a new one.

The integrals can be split in several algebraic pieces,

$$\bar{I}_1 = \left(\frac{\sigma^2}{a^2} + \frac{z^2}{b^2}\right)\left(2\bar{D}_1 - \bar{V}_N\right) + 2\frac{z^2}{b^2}\left(\bar{D}_2 - \bar{D}_1\right)$$
$$+ \frac{G\rho_0}{8}\oint_{S^2}\left(R_+^4 + R_-^4\right)\left(\frac{\sin^2\theta}{a^2} + \frac{\cos^2\theta}{b^2}\right)d\Omega \,, \tag{82}$$

$$\bar{I}_2 = \frac{z^2}{b^2}\left(2\bar{D}_2 - \bar{V}_N\right) + \frac{G\rho_0}{8b^2}\oint_{S^2}\left(R_+^4 + R_-^4\right)\cos^2\theta d\Omega \,, \tag{83}$$

where the integrals

$$\frac{1}{8}\oint_{S^2}\left(R_+^4 + R_-^4\right)\left(\frac{\sin^2\theta}{a^2} + \frac{\cos^2\theta}{b^2}\right)d\Omega$$
$$= \oint_{S^2}\left(\frac{2B^4}{A^3} - 2\frac{B^2C}{A^2} + \frac{C^2}{4A}\right)d\Omega \,, \tag{84}$$

$$\frac{1}{8}\oint_{S^2}\left(R_+^4 + R_-^4\right)\cos^2\theta d\Omega$$
$$= \oint_{S^2}\left(\frac{2B^4}{A^4} - 2\frac{B^2C}{A^3} + \frac{C^2}{4A^2}\right)\cos^2\theta d\Omega \,, \tag{85}$$

We use the results of Appendix A in Supplementary Material to calculate these integrals, and obtain

$$\oint_{S^2}\left(\frac{2B^4}{A^3} - 2\frac{B^2C}{A^2} + \frac{C^2}{4A}\right)d\Omega \tag{86}$$
$$= \frac{3}{2}\pi a^2\left[\left(1-\frac{z^2}{b^2}\right)^2\beth_0 - 2\left(1-6\frac{z^2}{b^2} + 5\frac{z^4}{b^4}\right)\beth_1\right.$$
$$\left. + \left(1-10\frac{z^2}{b^2} + \frac{35}{3}\frac{z^4}{b^4}\right)\beth_2\right] + \pi a^2\left[\left(1-\frac{z^2}{b^2}\right)\beth_0\right.$$
$$\left. -4\left(1-3\frac{z^2}{b^2}\right)\beth_1 + 3\left(1-5\frac{z^2}{b^2}\right)\beth_2\right]C(\boldsymbol{x})$$
$$-\pi a^2\left[\beth_1 - \frac{3}{2}\beth_2\right]C^2(\boldsymbol{x})$$

$$\oint_{S^2}\left(\frac{2B^4}{A^4} - 2\frac{B^2C}{A^3} + \frac{C^2}{4A^2}\right)\frac{\cos^2\theta}{b^2}d\Omega \tag{87}$$
$$= \frac{3}{2}\pi a^2\left[\left(1-\frac{z^2}{b^2}\right)^2\beth_1 - 2\left(1-6\frac{z^2}{b^2} + 5\frac{z^4}{b^4}\right)\beth_2\right.$$
$$\left. + \left(1-10\frac{z^2}{b^2} + \frac{35}{3}\frac{z^4}{b^4}\right)\beth_3\right]$$

$$+\pi a^2\left[\left(1-\frac{z^2}{b^2}\right)\mathfrak{I}_1 - 4\left(1-3\frac{z^2}{b^2}\right)\mathfrak{I}_2\right.$$
$$\left.+3\left(1-5\frac{z^2}{b^2}\right)\mathfrak{I}_3\right]C(x) + \pi a^2\left[\mathfrak{I}_2 - \frac{3}{2}\mathfrak{I}_3\right]C^2(x)$$

Substituting these results in Equations (82) and (83) yields

$$\bar{I}_1 = \frac{1}{2}\pi G\rho_0 a^2\left[\left(1-\frac{z^4}{b^4}\right)\mathfrak{I}_0 - 6\frac{z^2}{b^2}\left(1-\frac{5}{3}\frac{z^2}{b^2}\right)\mathfrak{I}_1\right. \tag{88}$$
$$\left.-\left(1-10\frac{z^2}{b^2}+\frac{35}{3}\frac{z^4}{b^4}\right)\mathfrak{I}_2\right]$$
$$-\pi G\rho_0 a^2\left[\frac{3z^2}{b^2}\mathfrak{I}_1 + \left(1-5\frac{z^2}{b^2}\right)\mathfrak{I}_2 + \frac{1}{2}\mathfrak{I}_2 C(x)\right]C(x),$$

$$\bar{I}_2 = \pi G\rho_0 a^2\left[\left(1-\frac{z^2}{b^2}\right)\frac{z^2}{b^2}\mathfrak{I}_0 + \left(\frac{3}{2}-12\frac{z^2}{b^2}+\frac{25}{2}\frac{z^4}{b^4}\right)\mathfrak{I}_1\right. \tag{89}$$
$$\left.-\left(3-26\frac{z^2}{b^2}+\frac{85}{3}\frac{z^4}{b^4}\right)\mathfrak{I}_2 + \left(\frac{3}{2}-15\frac{z^2}{b^2}+\frac{35}{2}\frac{z^4}{b^4}\right)\mathfrak{I}_3\right]$$
$$+\pi G\rho a^2\left[\left(1-6\frac{z^2}{b^2}\right)\mathfrak{I}_1 - 4\left(1-5\frac{z^2}{b^2}\right)\mathfrak{I}_2\right.$$
$$\left.+3\left(1-5\frac{z^2}{b^2}\right)\mathfrak{I}_3 + \left(\mathfrak{I}_2-\frac{3}{2}\mathfrak{I}_3\right)C(x)\right]C(x).$$

Replacing these expressions to Equation (79) results in

$$\bar{\Phi} = \Phi_0 + \Phi_1 C(x) + \Phi_2 C^2(x), \tag{90}$$

where

$$\Phi_0 = \frac{1}{2}\pi^2 G^2\rho_0^2 a^4\left[7\mathfrak{I}_0^2 - 3\mathfrak{I}_0(8\mathfrak{I}_1 - 5\mathfrak{I}_2 + 2\mathfrak{I}_3)\right. \tag{91}$$
$$\left.+2\mathfrak{I}_1(15\mathfrak{I}_1 - 20\mathfrak{I}_2 + 9\mathfrak{I}_3)\right] - \pi^2 G^2\rho_0^2 a^4\left[7\mathfrak{I}_0^2\right.$$
$$+\mathfrak{I}_0(-60\mathfrak{I}_1 + 67\mathfrak{I}_2 - 30\mathfrak{I}_3) + 2\mathfrak{I}_1(51\mathfrak{I}_1 - 88\mathfrak{I}_2$$
$$\left.+45\mathfrak{I}_3)\right]\frac{z^2}{b^2} + \frac{1}{6}\pi^2 G^2\rho_0^2 a^4\left[21\mathfrak{I}_0^2 + \mathfrak{I}_0(-288\mathfrak{I}_1 + 445\mathfrak{I}_2\right.$$
$$\left.-210\mathfrak{I}_3) + 10\mathfrak{I}_1(57\mathfrak{I}_1 - 116\mathfrak{I}_2 + 63\mathfrak{I}_3)\right]\frac{z^4}{b^4}$$
$$+\pi G\rho_0 a^4\omega^2\left[\mathfrak{I}_0 - 3\mathfrak{I}_1 + 5\mathfrak{I}_2 - 3\mathfrak{I}_3 - 2(\mathfrak{I}_0 - 9\mathfrak{I}_1 + 21\mathfrak{I}_2\right.$$
$$\left.-15\mathfrak{I}_3)\frac{z^2}{b^2} + (\mathfrak{I}_0 - 5(3\mathfrak{I}_1 - 9\mathfrak{I}_2 + 7\mathfrak{I}_3))\frac{z^4}{b^4}\right],$$

$$\Phi_1 = \pi^2 G^2\rho_0^2 a^4\left[\mathfrak{I}_0(-7\mathfrak{I}_1 + 11\mathfrak{I}_2 - 6\mathfrak{I}_3) + 2\mathfrak{I}_1(6\mathfrak{I}_1\right. \tag{92}$$
$$\left.-14\mathfrak{I}_2 + 9\mathfrak{I}_3)\right] + \pi^2 G^2\rho_0^2 a^4\left[\mathfrak{I}_0(21\mathfrak{I}_1 - 55\mathfrak{I}_2 + 30\mathfrak{I}_3)\right.$$
$$\left.-2\mathfrak{I}_1(24\mathfrak{I}_1 - 70\mathfrak{I}_2 + 45\mathfrak{I}_3)\right]\frac{z^2}{b^2} - 2\pi G\rho_0 a^4\omega^2\left[(\mathfrak{I}_1\right.$$
$$\left.-3\mathfrak{I}_2 + 3\mathfrak{I}_3) - 3(\mathfrak{I}_1 - 5\mathfrak{I}_2 + 5\mathfrak{I}_3)\frac{z^2}{b^2}\right]$$

$$\Phi_2 = \frac{1}{2}\pi G\rho_0 a^4\left[(-\mathfrak{I}_0\mathfrak{I}_2 + 8\mathfrak{I}_1\mathfrak{I}_2 + 6\mathfrak{I}_0\mathfrak{I}_3 - 18\mathfrak{I}_1\mathfrak{I}_3)\pi G\rho_0\right.$$
$$\left.-6(\mathfrak{I}_2 - \mathfrak{I}_3)\omega^2\right], \tag{93}$$

This finalizes the calculation of the post-Newtonian potentials inside the mass.

8. POST-NEWTONIAN MASS

The post-Newtonian conservation laws have been discussed by a number of researchers the most notably in textbooks (Fock, 1964; Will, 1993; Kopeikin et al., 2011). General relativity predicts that the integrals of energy, linear momentum, angular momentum and the center of mass of an isolated system are conserved in the post-Newtonian approximation. In the present paper we deal with a single isolated body so that the integrals of the center of mass and the linear momentum are trivial and we can always choose the origin of the coordinate system at the center of mass of the body with the linear momentum being nil. The integrals of energy and angular momentum are less trivial and requires detailed calculations which are given below.

The law of conservation of energy yields the post-Newtonian mass of a rotating fluid ball that is defined as follows (Fock, 1964; Will, 1993; Kopeikin et al., 2011)

$$M = M_{\mathrm{N}} + \frac{1}{c^2}M_{\mathrm{pN}}, \tag{94}$$

where $\mathcal{V}$ is the volume of PN-ellipsoid, and

$$M_{\mathrm{N}} = \int_{\mathcal{V}}\bar{\rho}(x)d^3x, \tag{95}$$

is the rest mass of baryons comprising the body, and

$$M_{\mathrm{pN}} = \int_{\mathcal{V}}\bar{\rho}(x)\left(v^2 + \Pi + \frac{5}{2}V_{\mathrm{N}}\right)d^3x, \tag{96}$$

is the post-Newtonian correction.

In order to calculate the rest mass, M_{N}, we introduce normalized spherical coordinates r, θ, ϕ related to the Cartesian (harmonic) coordinates x, y, z as follows,

$$x = ar\sin\theta\cos\phi, \qquad y = ar\sin\theta\sin\phi, \qquad z = br\cos\theta. \tag{97}$$

In these coordinates the integral Equation (95) is given by

$$M_{\mathrm{N}} = a^2 b\rho_0\int_0^{r(\theta)}\int_0^{\pi}\int_0^{2\pi}\left(1 + \kappa\mathcal{A}r^2\right)r^2\sin\theta\,dr d\theta\,d\phi, \tag{98}$$

where $r(\theta)$ describes the surface of the PN-ellipsoid defined above in Equations (22), (32)

$$r^2(\theta) = 1 + \kappa\mathcal{B}\sin^2\theta\cos^2\theta. \tag{99}$$

Integration in Equation (98) results in

$$M_{\mathrm{N}} = \frac{4\pi}{3}\rho_0 a^2 b\left[1 + \frac{\kappa}{5}(\mathcal{A}+\mathcal{B})\right]. \tag{100}$$

The post-Newtonian contribution, M_{pN}, to the rest mass reads

$$M_{\mathrm{pN}} = \rho_0 a^4 b\int_0^1\int_0^{\pi}\int_0^{2\pi}\left\{\omega^2 r^2\sin^2\theta + \frac{5}{2}\pi G\rho_0\left[\mathfrak{I}_0\left(1 - r^2\cos^2\theta\right)\right.\right.$$

$$- \beth_1 r^2 \left(1 - 3\cos^2\theta\right)\Big]\Big\} \, r^2 \sin\theta \, dr d\theta d\phi$$

$$= \frac{8\pi}{15}\rho_0 a^4 b \left(\omega^2 + 5\pi G\rho_0\beth_0\right) \,. \tag{101}$$

The total mass Equation (94) becomes

$$M = \frac{4\pi}{3}\rho_0 a^2 b \left[1 + \frac{\kappa}{5}\left(\frac{2\omega^2}{\pi G\rho_0} + \mathcal{A} + \mathcal{B} + 10\beth_0\right)\right] \,. \tag{102}$$

The inverse relation is used to convert the density ρ_0 to the total mass,

$$\rho_0 = \frac{3M}{4\pi a^2 b}\left[1 - \frac{\kappa}{5}\left(\mathcal{A} + \mathcal{B} + 10\beth_0 + \frac{2\omega^2}{\pi G\rho_0}\right)\right] \,. \tag{103}$$

The total post-Newtonian mass, M, depends on the sum of the parameters, $\mathcal{A} + \mathcal{B}$.

9. POST-NEWTONIAN ANGULAR MOMENTUM

Vector of the post-Newtonian angular momentum, $S^i = (S^x, S^y, S^z)$, is defined by Fock (1964), Will (1993), and Kopeikin et al. (2011)

$$S^i = S_N^i + \frac{1}{c^2}S_{pN}^i \,, \tag{104}$$

where S_N^i and S_{pN}^i are the Newtonian and post-Newtonian contributions respectively,

$$S_N^i = \int \bar{\rho}(\boldsymbol{x}) \left(\boldsymbol{x} \times \boldsymbol{v}\right)^i d^3x \,, \tag{105}$$

$$S_{pN}^i = \int \bar{\rho}(\boldsymbol{x}) \left(v^2 + 6\bar{V} + \frac{\bar{p}}{\bar{\rho}}\right) \left(\boldsymbol{x} \times \boldsymbol{v}\right)^i d^3x$$
$$- 4 \int \bar{\rho}(\boldsymbol{x}) \left(\boldsymbol{x} \times \bar{\boldsymbol{V}}\right)^i d^3x \,, \tag{106}$$

and vector-potential $\bar{V} \equiv \bar{V}^i$ has been given in Equations (64) and (72).

It can be checked up by inspection that in case of an axisymmetric mass distribution with the ellipsoidal density distribution Equation (38), the only non-vanishing component of the angular momentum is $S^3 = S^z \equiv S$. Indeed, $\boldsymbol{v} = (\boldsymbol{\omega} \times \boldsymbol{x})$, and $(\boldsymbol{x} \times \boldsymbol{v})^i = (\boldsymbol{x} \times (\boldsymbol{\omega} \times \boldsymbol{x}))^i = \omega^i(x^2 + y^2 + z^2) - x^i \omega z$. It results in

$$S_N^x = -\omega \int \bar{\rho} x z d^3x \,, \quad S_N^y = -\omega \int \bar{\rho} y z d^3x \,,$$

$$S_N^z = \omega \int \bar{\rho}(x^2 + y^2) d^3x \,. \tag{107}$$

By making use of the coordinates Equation (97) the reader can easily confirm that $S_N^x = S_N^y = 0$, and $S_N^z \equiv S_N$ reads

$$S_N = a^4 b\omega \int_0^{r(\theta)} \int_0^\pi \int_0^{2\pi} \bar{\rho}(\boldsymbol{x}) r^4 \sin^3\theta \, dr d\theta d\phi \,, \tag{108}$$

where $r(\theta)$ is defined in Equation (99). After integration we obtain,

$$S_N = \frac{8\pi}{15}a^4 b\rho_0\omega \left[1 + \frac{\kappa}{7}\left(5\mathcal{A} + 2\mathcal{B}\right)\right] \,. \tag{109}$$

Replacing the density ρ_0 by the total mass M with the help of Equation (102), makes the Newtonian part of the angular momentum take on the following form,

$$S_N = \frac{2}{5}Ma^2\omega \left[1 - \frac{\kappa}{35}\left(2\mathcal{A} - 3\mathcal{B} + 70\beth_0 + \frac{14\omega^2}{\pi G\rho_0}\right)\right] \,. \tag{110}$$

It is straightforward to prove that x and y components of S_{pN}^i also vanish due to the axial symmetry, and only its z component, $S_{pN}^z \equiv S_{pN}$, remains. We notice that

$$\int \bar{\rho}(\boldsymbol{x}) \left(\boldsymbol{x} \times \bar{\boldsymbol{V}}\right)^z d^3x = \int \bar{\rho}(\boldsymbol{x})\bar{D}_1 \left(\boldsymbol{x} \times \boldsymbol{v}\right)^z d^3x \,, \tag{111}$$

where D_1 is taken from Equation (73). Therefore,

$$S_{pN} = \int \bar{\rho}(\boldsymbol{x}) \left(v^2 + 6\bar{V}_N + \frac{\bar{p}}{\bar{\rho}} - 4\bar{D}_1\right) \left(\boldsymbol{x} \times \boldsymbol{v}\right)^z d^3x \,. \tag{112}$$

Making transformation to the coordinates Equation (97) and approximating $\bar{\rho} = \rho_0$, yields

$$S_{pN} = \omega a^6 b\rho_0 \int_0^1 \int_0^\pi \int_0^{2\pi} \Big\{\omega^2 r^2 \sin^2\theta + 7\beth_0 - 6\beth_1 - (\beth_0 + 4\beth_1$$
$$- 4\beth_2)\, r^2 - 2\left(3\beth_0 - 15\beth_1 + 10\beth_2\right) r^2 \cos^2\theta\Big\}$$
$$\quad r^4 \sin^3\theta \, dr d\theta d\phi \tag{113}$$
$$= \frac{4}{35}Ma^2\omega \left[2\omega^2 + \pi G\rho_0 \left(19\beth_0 - 16\beth_1\right)\right] \,.$$

The total angular momentum becomes

$$S = \frac{2}{5}Ma^2\omega \left\{1 - \frac{\kappa}{35}\left[2\mathcal{A} - 3\mathcal{B} - 40\left(3\beth_0 - 4\beth_1\right) - \frac{6\omega^2}{\pi G\rho_0}\right]\right\} \,. \tag{114}$$

10. POST-NEWTONIAN EQUATION OF THE LEVEL SURFACE

The figure of the rotating fluid body is defined by the boundary condition of vanishing pressure, $p = 0$. The surface $p = 0$ is called the level surface. Relativistic Euler equation derived for the rigidly rotating fluid body tells us that the level surface coincides with the equipotential surface of the post-Newtonian gravitational potential $\bar{W}$ which is expressed in terms of the centrifugal and gravitational potentials by the following expression (Kopejkin, 1991; Kopeikin et al., 2011, 2015)

$$\bar{W} = \frac{1}{2}\omega^2\sigma^2 + \bar{V}_N + \kappa\left(\bar{V}_{pN} + \mathcal{A}\bar{I}_1\right) + \frac{1}{c^2}\left(\frac{1}{8}\omega^4\sigma^4\right.$$

$$+ \frac{3}{2}\omega^2\sigma^2\bar{V}_N - 4\omega^2\sigma^2\bar{D}_1 - \frac{1}{2}\bar{V}_N^2 + \bar{\Phi}\Big) , \tag{115}$$

where $\kappa \equiv \pi G\rho_0 a^2/c^2$, and the potentials $\bar{V}_N$, $\bar{V}_{pN}$, $\bar{I}_1$, $\bar{D}_1$, $\bar{\Phi}$ have been explained in Sections 5–7. After substituting these potentials to equation Equation (115) it can be presented as a quadratic polynomial with respect to function $C(x)$,

$$\bar{W}(x) = \bar{W}_0 + \bar{W}_1 C(x) + \bar{W}_2 C^2(x) , \tag{116}$$

where the coefficients of the expansion are the polynomials of the z coordinate only. In particular, the coefficient $\bar{W}_0$ is a polynomial of the fourth order,

$$\bar{W}_0 = K_0 + K_1\frac{z^2}{b^2} + K_2\frac{z^4}{b^4} , \tag{117}$$

where

$$K_0 = \frac{1}{2}\omega^2 a^2 + \pi G\rho_0 a^2 (\mathrm{J}_0 - \mathrm{J}_1) \tag{118}$$
$$+ \frac{1}{8c^2}\omega^4 a^4 + \frac{1}{2}\kappa\omega^2 a^2 (5\mathrm{J}_0 - 17\mathrm{J}_1 + 18\mathrm{J}_2 - 6\mathrm{J}_3)$$
$$+ \frac{1}{2}\kappa\pi G\rho_0 a^2 \big[6\mathrm{J}_0^2 - \mathrm{J}_0(22\mathrm{J}_1 - 15\mathrm{J}_2 + 6\mathrm{J}_3)$$
$$+ \mathrm{J}_1(29\mathrm{J}_1 - 40\mathrm{J}_2 + 18\mathrm{J}_3)\big] + \kappa\pi G\rho_0 a^2 [2(\mathrm{J}_1 - 4\mathrm{J}_2$$
$$+ 6\mathrm{J}_3 - 3\mathrm{J}_4)\mathcal{B} + \frac{1}{2}(\mathrm{J}_0 - \mathrm{J}_2)\mathcal{A}] ,$$

$$K_1 = -\frac{1}{2}\omega^2 a^2 - \pi G\rho_0 a^2 (\mathrm{J}_0 - 3\mathrm{J}_1) \tag{119}$$
$$- \frac{1}{4c^2}\omega^4 a^4 - \kappa\omega^2 a^2 (5\mathrm{J}_0 - 40\mathrm{J}_1 + 66\mathrm{J}_2 - 30\mathrm{J}_3)$$
$$- \kappa\pi G\rho_0 a^2 \big(6\mathrm{J}_0^2 - 56\mathrm{J}_0\mathrm{J}_1 + 99\mathrm{J}_1^2 + 67\mathrm{J}_0\mathrm{J}_2 - 176\mathrm{J}_1\mathrm{J}_2$$
$$- 30\mathrm{J}_0\mathrm{J}_3 + 90\mathrm{J}_1\mathrm{J}_3\big) + \kappa\pi G\rho_0 a^2 [(\mathrm{J}_0 - 18\mathrm{J}_1 + 78\mathrm{J}_2$$
$$- 120\mathrm{J}_3 + 60\mathrm{J}_4)\mathcal{B} - (3\mathrm{J}_1 - 5\mathrm{J}_2)\mathcal{A}] ,$$

$$K_2 = \frac{1}{8c^2}\omega^4 a^4 + \frac{1}{2}\kappa\omega^2 a^2 (5\mathrm{J}_0 - 63\mathrm{J}_1 + 130\mathrm{J}_2 - 70\mathrm{J}_3) \tag{120}$$
$$+ \frac{1}{6}\kappa\pi G\rho_0 a^2 \big[18\mathrm{J}_0^2 - 5\mathrm{J}_0(54\mathrm{J}_1 - 89\mathrm{J}_2$$
$$+ 42\mathrm{J}_3) + \mathrm{J}_1(543\mathrm{J}_1 - 1160\mathrm{J}_2 + 630\mathrm{J}_3)\big]$$
$$- \kappa\pi G\rho_0 a^2 [(\mathrm{J}_0 - 20\mathrm{J}_1 + 90\mathrm{J}_2 - 140\mathrm{J}_3 + 70\mathrm{J}_4)\mathcal{B}$$
$$+ \frac{1}{2}\left(\mathrm{J}_0 - 10\mathrm{J}_1 + \frac{35}{3}\mathrm{J}_2\right)\mathcal{A}] .$$

Coefficient $\bar{W}_1$ in Equation (116) is a polynomial of the second order,

$$\bar{W}_1 = P + P_1\frac{z^2}{b^2} , \tag{121}$$

where

$$P = \frac{1}{2}\omega^2 a^2 \left[1 + \frac{1}{2c^2}\omega^2 a^2 + \kappa (3\mathrm{J}_0 - 18\mathrm{J}_1 + 28\mathrm{J}_2 - 12\mathrm{J}_3)\right]$$
$$- \pi G\rho_0 a^2\mathrm{J}_1 + \kappa\pi G\rho_0 a^2 [(\mathrm{J}_1 - 8\mathrm{J}_2 + 18\mathrm{J}_3 - 12\mathrm{J}_4)\mathcal{B}$$
$$- \mathrm{J}_2\mathcal{A}] - \kappa\pi G\rho_0 a^2 (6\mathrm{J}_0\mathrm{J}_1 - 11\mathrm{J}_1^2 - 11\mathrm{J}_0\mathrm{J}_2 + 28\mathrm{J}_1\mathrm{J}_2$$

$$+ 6\mathrm{J}_0\mathrm{J}_3 - 18\mathrm{J}_1\mathrm{J}_3) , \tag{122}$$
$$P_1 = -\frac{1}{4c^2}\omega^2 a^4 - \frac{3}{2}\kappa\omega^2 a^2 (\mathrm{J}_0 - 16\mathrm{J}_1 + 36\mathrm{J}_2 - 20\mathrm{J}_3) \tag{123}$$
$$+ \kappa\pi G\rho_0 a^2 [20\mathrm{J}_0\mathrm{J}_1 - 45\mathrm{J}_1^2 - 55\mathrm{J}_0\mathrm{J}_2 + 140\mathrm{J}_1\mathrm{J}_2$$
$$+ 30\mathrm{J}_0\mathrm{J}_3 - 90\mathrm{J}_1\mathrm{J}_3] - \kappa\pi G\rho_0 a^2 [2(3\mathrm{J}_1 - 25\mathrm{J}_2 + 51\mathrm{J}_3$$
$$- 30\mathrm{J}_4)\mathcal{B} + (3\mathrm{J}_1 - 5\mathrm{J}_2)\mathcal{A}] .$$

Coefficient $\bar{W}_2$ in Equation (116) is constant,

$$\bar{W}_2 = \frac{1}{8c^2}\omega^4 a^4 - \frac{1}{2}\kappa\omega^2 a^2 (3\mathrm{J}_1 - 2\mathrm{J}_2 - 6\mathrm{J}_3) \tag{124}$$
$$- \kappa\pi G\rho_0 a^2 \left[(\mathrm{J}_2 - 6\mathrm{J}_3 + 6\mathrm{J}_4)\mathcal{B} + \frac{1}{2}\mathrm{J}_2\mathcal{A}\right.$$
$$\left. + \frac{1}{2}(\mathrm{J}_1^2 + \mathrm{J}_0\mathrm{J}_2 - 8\mathrm{J}_1\mathrm{J}_2 - 6\mathrm{J}_0\mathrm{J}_3 + 18\mathrm{J}_1\mathrm{J}_3)\right] .$$

On the level surface of the PN-ellipsoid[2] we have all three coordinates interconnected by Equation (34) of the PN-ellipsoid, $C(\bar{x}) = \kappa\alpha_0(\bar{z})$, so that Equation (116) becomes

$$\bar{W} \equiv \bar{W}_0 + \kappa\bar{W}_1\alpha_0(\bar{z}) , \tag{125}$$

and the term with $\bar{W}_2 \sim O(\kappa^2)$, is discarded as negligibly small. After reducing similar terms, the potential $\bar{W}$ on the level surface is simplified to the polynomial of the fourth order,

$$\bar{W} = \bar{K}_0 + \bar{K}_1\frac{\bar{z}^2}{b^2} + \bar{K}_2\frac{\bar{z}^4}{b^4} , \tag{126}$$

where

$$\bar{K}_0 = K_0 , \tag{127}$$
$$\bar{K}_1 = K_1 + \kappa\left(\frac{1}{2}\omega^2 a^2 - \pi G\rho_0 a^2\mathrm{J}_1\right)\mathcal{B} , \tag{128}$$
$$\bar{K}_2 = K_2 - \kappa\left(\frac{1}{2}\omega^2 a^2 - \pi G\rho_0 a^2\mathrm{J}_1\right)\mathcal{B} . \tag{129}$$

Because the potential $\bar{W}$ is to be constant on the level surface (Kopeikin et al., 2015), the numerical coefficients $\bar{K}_1$ and $\bar{K}_2$ must vanish. The first condition, $\bar{K}_1 = 0$, yields a relation between the angular velocity of rotation, ω, and the eccentricity, e, of the rotating fluid ellipsoid:

$$\frac{\omega^2}{2\pi G\rho_0}\left[1 + \frac{\omega^2 a^2}{2c^2} + 2\kappa (5\mathrm{J}_0 - 40\mathrm{J}_1 + 66\mathrm{J}_2 - 30\mathrm{J}_3 - \mathcal{B})\right]$$
$$= 3\mathrm{J}_1 - \mathrm{J}_0 \quad - \kappa\left[6\mathrm{J}_0^2 - 56\mathrm{J}_0\mathrm{J}_1 + 99\mathrm{J}_1^2 + 67\mathrm{J}_0\mathrm{J}_2 - 176\mathrm{J}_1\mathrm{J}_2\right.$$
$$- 30\mathrm{J}_0\mathrm{J}_3 + 90\mathrm{J}_1\mathrm{J}_3 + (3\mathrm{J}_1 - 5\mathrm{J}_2)\mathcal{A} - (\mathrm{J}_0 - 18\mathrm{J}_1 + 78\mathrm{J}_2$$
$$\left. - 120\mathrm{J}_3 + 60\mathrm{J}_4)\mathcal{B}\right] . \tag{130}$$

Equation (130) generalizes the famous result that was first obtained by Colin Maclaurin in 1742, from the Newtonian theory of gravity to the realm of general relativity.

---

[2] We remind the reader that the coordinates on the surface of the PN-ellipsoid are denoted as $\bar{x}, \bar{y}, \bar{z}$.

The second condition, $\bar{K}_2 = 0$, yields an algebraic equation for the two coefficients $\mathcal{A}$, $\mathcal{B}$,

$$
\frac{1}{2}\left(\mathrm{J}_0 - 10\mathrm{J}_1 + \frac{35}{3}\mathrm{J}_2\right)\mathcal{A} + \left(\frac{\omega^2}{2\pi G\rho_0} + \mathrm{J}_0 - 21\mathrm{J}_1 + 90\mathrm{J}_2\right.
$$
$$
\left. - 140\mathrm{J}_3 + 70\mathrm{J}_4\right)\mathcal{B} = \frac{\omega^4}{8\pi^2 G^2 \rho^2} + \frac{\omega^2}{4\pi G\rho}\left(5\mathrm{J}_0 - 63\mathrm{J}_1\right.
$$
$$
+ 130\mathrm{J}_2 - 70\mathrm{J}_3) + \frac{1}{6}\left[18\mathrm{J}_0^2 - 5\mathrm{J}_0(54\mathrm{J}_1 - 89\mathrm{J}_2 + 42\mathrm{J}_3)\right.
$$
$$
\left. + \mathrm{J}_1(543\mathrm{J}_1 - 1160\mathrm{J}_2 + 630\mathrm{J}_3)\right] . \tag{131}
$$

Equation (131) imposes one constraint on two coefficients $\mathcal{A}$, $\mathcal{B}$ defining the shape of the PN-ellipsoid and the law of distribution of mass density. One more constraint is required to fix these coefficients. We can use, for example, the Maclaurin relation Equation (130) to set an additional (geodetic) constraint on the parameters by demanding that the post-Newtonian part of Equation (130) vanishes.

We explored another possibility to impose self-consistent constraints on the shape parameters of the PN-ellipsoid directly related to the gravimetric measurements in geodesy. Newtonian theory of rotating reference-ellipsoid connects the shape parameters of the Maclaurin ellipsoid with the measured values of its gravity force at the pole and equator of the ellipsoid through the Pizzetti and Clairaut theorems (Pizzetti, 1913; Torge and Müller, 2012). Let us denote the force of gravity measured at the pole of the ellipsoid by $\gamma_p \equiv \gamma_i(x = 0, y = 0, z = b)$, and the force of gravity measured on equator by $\gamma_e \equiv \gamma_i(x = 0, y = a, z = 0)$. Due to the rotational symmetry of the ellipsoid the equatorial point can be, in fact, chosen arbitrary. The classic form of the Pizetti theorem is

$$
2\frac{\gamma_e}{a} + \frac{\gamma_p}{b} = \frac{3GM}{a^2 b} - 2\omega^2 , \tag{132}
$$

while the Clairaut theorem states

$$
\frac{\gamma_e}{a} - \frac{\gamma_p}{b} = \frac{3GM}{2a^2 b}\frac{3e - e^3 - 3\sqrt{1 - e^2}\arcsin e}{e^3} + \omega^2 . \tag{133}
$$

These two theorems were crucial in geodesy of XIX-th century for deeper understanding that the geometric shape of Earth's figure can be determined not only from the measurements of the geodetic arcs but, independently, from rendering the intrinsic measurements of the force of gravity on its surface. The gravity-geometry correspondence led to the pioneering idea that the gravity force and geometry of space must be always interrelated—the idea which paved a way in XX-th century to the development of the general theory of relativity by A. Einstein. We derive the post-Newtonian analogs of the Pizzetti and Clairaut theorems in the next two sections and explore what kind of natural limitations (if any) they can set on the shape parameters of the post-Newtonian ellipsoid.

11. POST-NEWTONIAN PIZZETTI'S THEOREM

The force of gravity on the equipotential surface of a massive rotating body is defined by equation (Kopeikin et al., 2011)

$$
\gamma_i = \left[\Lambda^j{}_i\partial_j\bar{W}\right]_{x=\bar{x}} , \tag{134}
$$

where $\partial_i \equiv \partial/\partial x^i$, the post-Newtonian gravity potential $\bar{W}$ is given in Equation (116), and

$$
\Lambda^j{}_i = \delta^{ij}\left(1 - \frac{1}{c^2}\bar{V}_N\right) - \frac{1}{2c^2}v^i v^j , \tag{135}
$$

is the matrix of transformation from the global coordinates to the local inertial frame of observer, $v^i = (\omega \times x)^i$ is velocity of the observer with respect to the global coordinates, and V_N is the Newtonian potential Equation (56). We emphasize that we, first, take a derivative in Equation (134), and then, take the spatial coordinates, x, on the equipotential surface described by the equation of the post-Newtonian ellipsoid Equation (22).

Velocity v^i is orthogonal to the gradient $\partial_i\bar{W}$ everywhere, that is

$$
v^i\partial_i\bar{W} = 0 . \tag{136}
$$

Indeed, it is easy to prove that

$$
v^i\partial_i\bar{W} = \omega\left(x\partial_y\bar{W} - y\partial_x\bar{W}\right) . \tag{137}
$$

Partial derivatives of W are calculated from Equation (116),

$$
\partial_x\bar{W} = \frac{d\bar{W}}{dC}\partial_x C(x) = \frac{d\bar{W}}{dC}\frac{2x}{a^2} , \quad \partial_y\bar{W} = \frac{d\bar{W}}{dC}\partial_y C(x) = \frac{d\bar{W}}{dC}\frac{2y}{a^2} . \tag{138}
$$

Replacing the partial derivative from Equations (138) to (137) yields Equation (136). After accounting for Equations (136), (134) is simplified to

$$
\gamma_i(\bar{x}) = \left[\left(1 - \frac{1}{c^2}\bar{V}_N\right)\partial_i\bar{W}\right]_{x=\bar{x}} . \tag{139}
$$

We shall denote the force of gravity on the pole by $\gamma_p \equiv \gamma_i(x = 0, y = 0, z = r_p)$, and the force of gravity on the equator by $\gamma_e \equiv \gamma_i(x = 0, y = r_e, z = 0)$ where the equatorial r_e and polar r_p radii are defined in Equations (24) and (25) respectively. Taking the partial derivative from $\bar{W}$ in Equation (139), yields

$$
\gamma_p = \frac{2\pi G\rho_0 a^2}{b}(\mathrm{J}_0 - 2\mathrm{J}_1) + 16\frac{\omega^2 a^2}{b}\kappa(\mathrm{J}_1 - 3\mathrm{J}_2 + 2\mathrm{J}_3) \tag{140}
$$
$$
+ \frac{2\pi G\rho_0 a^2}{b}\kappa\left[(\mathrm{J}_0 - 17\mathrm{J}_1 + 60\mathrm{J}_2 - 76\mathrm{J}_3 + 32\mathrm{J}_4)\mathcal{B}\right.
$$
$$
\left. + \left(\mathrm{J}_0 - 4\mathrm{J}_1 + \frac{8}{3}\mathrm{J}_2\right)\mathcal{A}\right] + \frac{4\pi G\rho_0 a^2}{3b}\kappa\left[\mathrm{J}_0(27\mathrm{J}_1 - 56\mathrm{J}_2\right.
$$
$$
\left. + 24\mathrm{J}_3) - 2\mathrm{J}_1(33\mathrm{J}_1 - 74\mathrm{J}_2 + 36\mathrm{J}_3)\right] ,
$$
$$
\gamma_e = a\left(2\mathrm{J}_1 G\pi\rho_0 - \omega^2\right) - \frac{\omega^4 a^3}{2c^2} + \kappa\left[-3\mathrm{J}_0 + 18\mathrm{J}_1 - 28\mathrm{J}_2\right. \tag{141}
$$

$$+ 12\mathcal{J}_3 + (\mathcal{J}_0 - \mathcal{J}_1)]\,\omega^2 a - 2\kappa\pi G\rho_0 a\,[(\mathcal{J}_1 - 8\mathcal{J}_2 + 18\mathcal{J}_3$$
$$- 12\mathcal{J}_4)\,\mathcal{B} + \mathcal{J}_2\mathcal{A}] + 2\kappa\pi G\rho_0 a\,[\mathcal{J}_0\,(5\mathcal{J}_1 - 11\mathcal{J}_2 + 6\mathcal{J}_3)$$
$$- 2\mathcal{J}_1\,(5\mathcal{J}_1 - 14\mathcal{J}_2 + 9\mathcal{J}_3)]\,.$$

We form a linear combination

$$2\frac{\gamma_e}{a} + \frac{\gamma_p}{b} = 2\pi G\rho_0\left[2\mathcal{J}_1 + \frac{a^2}{b^2}(\mathcal{J}_0 - 2\mathcal{J}_1)\right] - 2\omega^2 - \frac{1}{c^2}\omega^4 a^2$$

$$- 2\kappa\left[2\mathcal{J}_0 - 17\mathcal{J}_1 + 28\mathcal{J}_2 - 12\mathcal{J}_3 - 8\frac{a^2}{b^2}(\mathcal{J}_1\right.\quad(142)$$
$$- 3\mathcal{J}_2 + 2\mathcal{J}_3)]\,\omega^2 - 2\kappa\pi G\rho_0\left[2\mathcal{J}_1 - 16\mathcal{J}_2 + 36\mathcal{J}_3\right.$$
$$\left.- 24\mathcal{J}_4 - \frac{a^2}{b^2}(\mathcal{J}_0 - 17\mathcal{J}_1 + 60\mathcal{J}_2 - 76\mathcal{J}_3 + 32\mathcal{J}_4)\right]\mathcal{B}$$
$$- 2\kappa\pi G\rho_0\left[2\mathcal{J}_2 + \frac{a^2}{b^2}\left(\mathcal{J}_0 - 4\mathcal{J}_1 + \frac{8}{3}\mathcal{J}_2\right)\right]\mathcal{A}$$
$$+ 4\kappa\pi G\rho_0\,[\mathcal{J}_0\,(5\mathcal{J}_1 - 11\mathcal{J}_2 + 6\mathcal{J}_3) - 2\mathcal{J}_1\,(5\mathcal{J}_1$$
$$- 14\mathcal{J}_2 + 9\mathcal{J}_3)] + \frac{4a^2}{3b^2}\kappa\pi G\rho_0\,[\mathcal{J}_0(27\mathcal{J}_1 - 56\mathcal{J}_2$$
$$+ 24\mathcal{J}_3) - 2\mathcal{J}_1(33\mathcal{J}_1 - 74\mathcal{J}_2 + 36\mathcal{J}_3)]\,.$$

Making use of integrals given in Appendix A in Supplementary Material, we can check that Equation (142) is simplified to

$$2\frac{\gamma_e}{a} + \frac{\gamma_p}{b} = 4\pi G\rho_0 - 2\omega^2 - \frac{1}{c^2}\omega^4 a^2\qquad(143)$$
$$- \frac{\kappa}{3e^7}\left[e(1 - e^2)(105 - 104e^2 + 42e^4)\right.$$
$$- 3\sqrt{1 - e^2}(5 - 4e^2)(7 - 6e^2 + 2e^4)\arcsin e\left]\omega^2\right.$$
$$- \frac{\kappa\pi G\rho_0}{12e^9}(7 - 4e^2)\left[5e(21 - 31e^2 + 10e^4)\right.$$
$$- 3\sqrt{1 - e^2}(35 - 40e^2 + 8e^4)\arcsin e\left]\mathcal{B}\right.$$
$$+ \frac{\kappa\pi G\rho_0}{e^5}\left[e\left(7 - \frac{19}{3}e^2 - 2e^4\right)\right.$$
$$- \sqrt{1 - e^2}(7 - 4e^2)\arcsin e\left]\mathcal{A}\right.$$
$$+ \frac{\kappa\pi G\rho_0}{3e^{10}}(1 - e^2)\left[-315e^2 + 621e^4 - 250e^6\right.$$
$$+ 24e^8 + 2e\sqrt{1 - e^2}(315 - 516e^2 + 169e^4 - 18e^6)$$
$$\arcsin e - 3(105 - 242e^2 + 178e^4 - 40e^6)\arcsin^2 e].$$

In order to compare Equation (143) with its classic counterpart Equation (132), we convert, then, the density ρ_0, to the total relativistic mass of the PN-ellipsoid by making use of Equation (103). It allows us to recast Equation (143) to

$$2\frac{\gamma_e}{a} + \frac{\gamma_p}{b} = \frac{3GM}{a^2 b} - 2\omega^2 - \frac{a^2\omega^4}{c^2}\qquad(144)$$
$$+ \frac{3\sqrt{1 - e^2}}{16e^9 a^2 b^2}\frac{G^2 M^2}{c^2}\left[e\sqrt{1 - e^2}\,(-315 + 621e^2\right.$$
$$- 250e^4 + 24e^6) + 2\,(315 - 831e^2 + 685e^4$$
$$- 187e^6 + 42e^8)\arcsin e]$$

$$+ \frac{1}{20e^7 b}\frac{GM\omega^2}{c^2}\left[e(-525 + 1045e^2 - 730e^4 + 234e^6)\right.$$
$$+ 15\,\frac{35 - 93e^2 + 92e^4 - 42e^6 + 8e^8}{\sqrt{1 - e^2}}\arcsin e\left]\right.$$
$$- \frac{9}{16}\frac{G^2 M^2}{c^2}\frac{105 - 347e^2 + 420e^4 - 218e^6 + 40e^8}{e^{10}a^2 b^2}\arcsin^2 e$$
$$- \frac{3}{320e^8 a^2 b^2}\frac{G^2 M^2}{c^2}e\left[3675 - 7525e^2 + 4850e^4\right.$$
$$- 1000e^6 - 48e^8 - 15\frac{\sqrt{1 - e^2}}{e}(7 - 4e^2)\,(35 - 40e^2$$
$$+ 8e^4)\arcsin e]\mathcal{B}$$
$$+ \frac{9}{16}\frac{G^2 M^2}{a^2 b^2 e^5 c^2}\left[e\left(7 - \frac{19}{3}e^2 - 2e^4\right)\right.$$
$$- \sqrt{1 - e^2}\,(7 - 4e^2)\arcsin e\left]\mathcal{A}.\right.$$

This equation represents the post-Newtonian extension of the classical Pizzetti theorem (132).

12. POST-NEWTONIAN CLAIRAUT'S THEOREM

In order to derive the post-Newtonian analog of the Clairaut theorem Equation (133) we follow its classic derivation given, for example, in Pizzetti (1913). To this end we form a linear difference between the forces of gravity measured at the pole and on the equator of the PN-ellipsoid by making use of EquationS (139) and (140). We get

$$\frac{\gamma_e}{a} - \frac{\gamma_p}{b} = 2\pi G\rho_0\left[-\mathcal{J}_1 + \frac{a^2}{b^2}(\mathcal{J}_0 - 2\mathcal{J}_1)\right]\qquad(145)$$
$$+ \omega^2 + \frac{1}{2c^2}\omega^4 a^2$$
$$+ \kappa\,[2\mathcal{J}_0 - 17\mathcal{J}_1 + 28\mathcal{J}_2 - 12\mathcal{J}_3$$
$$+ 16\frac{a^2}{b^2}(\mathcal{J}_1 - 3\mathcal{J}_2 + 2\,\mathcal{J}_3)]\,\omega^2$$
$$- \kappa\pi G\rho_0\,[-2\mathcal{J}_1 + 16\mathcal{J}_2 - 36\mathcal{J}_3 + 24\mathcal{J}_4$$
$$- \frac{2a^2}{b^2}(\mathcal{J}_0 - 17\mathcal{J}_1 + 60\mathcal{J}_2 - 76\mathcal{J}_3 + 32\mathcal{J}_4)\Big]\mathcal{B}$$
$$- 2\kappa\pi G\rho_0\left[2\mathcal{J}_2 - \frac{a^2}{b^2}\left(\mathcal{J}_0 - 4\mathcal{J}_1 + \frac{8}{3}\mathcal{J}_2\right)\right]\mathcal{A}$$
$$+ 2\kappa\pi G\rho_0\,[\mathcal{J}_0(-5\mathcal{J}_1 + 11\mathcal{J}_2 - 6\mathcal{J}_3)$$
$$+ 2\mathcal{J}_1(5\mathcal{J}_1 - 14\mathcal{J}_2 + 9\mathcal{J}_3)]$$
$$+ \frac{4a^2}{3b^2}\kappa\pi G\rho_0\,[\mathcal{J}_0(27\mathcal{J}_1 - 56\mathcal{J}_2 + 24\mathcal{J}_3)$$
$$- 2\mathcal{J}_1(33\mathcal{J}_1 - 74\mathcal{J}_2 + 36\mathcal{J}_3)]\,.$$

We use the results of Appendix A in Supplementary Material to replace the integrals entering the right side of Equation (145), with their explicit expressions given in terms of the eccentricity e of the Maclaurin ellipsoid Equation (20). It yields

$$\frac{\gamma_e}{a} - \frac{\gamma_p}{b} = \frac{6\pi G\rho_0}{e^3}\left(e - \sqrt{1 - e^2}\arcsin e\right)\qquad(146)$$

$$- 2\pi G\rho + \omega^2 + \frac{1}{2c^2}\omega^4 a^2$$

$$+ \frac{\kappa}{6e^7}\left(-75e - 5e^3 + 122e^5 - 42e^7\right.$$

$$+ 3\sqrt{1-e^2}(25 + 10e^2 - 34e^4 + 8e^6)\arcsin e\Big)\,\omega^2$$

$$- \frac{\kappa}{24e^9}\pi G\rho_0\left[e\left(525 - 1075e^2 + 662e^4 - 112e^6\right)\right.$$

$$- 3\sqrt{1-e^2}(175 - 300e^2 + 144e^4 - 16e^6)\arcsin e\Big]\mathcal{B}$$

$$- \frac{\kappa}{6e^5}\pi G\rho_0\left[e\left(15 - 29e^2 + 6e^4\right)\right.$$

$$- 3\sqrt{1-e^2}\left(5 - 8e^2\right)\arcsin e\Big]\mathcal{A}$$

$$- \frac{\kappa}{6e^9}\pi G\rho_0(1-e^2)\left[e(225 - 351e^2 - 58e^4 + 24e^6)\right.$$

$$- 2\sqrt{1-e^2}(225 - 276e^2 - 85e^4 + 18e^6)\arcsin e\Big]$$

$$- \frac{\kappa}{2e^{10}}\pi G\rho_0(1-e^2)\left(75 - 142e^2 + 38e^4 + 40e^6\right)$$

$$\arcsin^2 e.$$

This form of the Clairaut theorem apparently depends on the shape parameters $\mathcal{A}$, $\mathcal{B}$ in different combinations and can be used to impose a constraints on one of them in addition to the constraint given by the level surface Equation (131). For example, we could demand that all post-Newtonian terms in Equation (146) vanish.

The last step is to replace the density ρ_0 in Equation (146) with the total mass of the PN-ellipsoid by making use of the gauge-invariant expression Equation (103). We get

$$\frac{\gamma_e}{a} - \frac{\gamma_p}{b} = \frac{3GM}{2a^2 b}\frac{3e - e^3 - 3\sqrt{1-e^2}}{e^3}\arcsin e + \omega^2 + \frac{a^2\omega^4}{2c^2}$$

$$- \frac{\begin{array}{c}e(375 + 25e^2 - 682e^4 + 234e^6)\\ -3\sqrt{1-e^2}(125 + 50e^2 - 194e^4 + 40e^6)\\ \arcsin e\end{array}}{40be^7}\frac{GM}{c^2}\omega^2$$

$$- \frac{3}{640}\frac{\begin{array}{c}e(2625 - 5375e^2 + 3310e^4 -\\ 704e^6 + 48e^8) - \sqrt{1-e^2}(875\\ -1500e^2 + 720e^4 - 128e^6)\arcsin e\end{array}}{a^3 be^9}\frac{G^2M^2}{c^2}\mathcal{B}$$

$$- \frac{3}{32}\frac{e\left(15 - 29e^2 + 6e^4\right) - 3\sqrt{1-e^2}\left(5 - 8e^2\right)\arcsin e}{a^2 b^2 e^5}\frac{G^2M^2}{c^2}\mathcal{A}$$

$$- \frac{3}{160}\frac{\begin{array}{c}e\sqrt{1-e^2}(225 - 351e^2 - 58e^4\\ +24e^6) - 2(225 - 501e^2 + 191e^4 +\\ 175e^6 - 42e^8)\arcsin e\end{array}}{a^4 e^9}\frac{G^2M^2}{c^2}$$

$$- \frac{9}{32}\frac{G^2M^2}{c^2}\frac{75 - 142e^2 + 38e^4 + 88e^6}{a^4 e^{10}}\arcsin^2 e \quad (147)$$

This is the post-Newtonian extension of the classical Clairaut theorem Equation (133).

13. REFERENCE ELLIPSOID IN SMALL-ECCENTRICITY APPROXIMATION

Let us apply the formalism of previous sections to derive practically useful post-Newtonian relationships used for processing geodetic and gravimetric measurements on the surface of Earth and in space. To this end we notice that the flattening, f of the terrestrial reference ellipsoid is about $f \simeq e^2/2 \simeq 1/298 = 0.0034$ (Petit and Luzum, 2010, Section 1) that can be used as a small parameter for expanding all post-Newtonian terms in the convergent Taylor series with respect to f. We shall keep the Newtonian expressions as they are without expansion, and take into account only terms of the order of e^2 in the post-Newtonian parts of equations by systematically discarding terms of the order of $\sim e^4$, and higher. Because, according to Maclaurin's relation Equation (130), the square of the angular velocity, $\omega^2 \simeq e^2$, (see Pizzetti, 1913; Chandrasekhar, 1969; Torge and Müller, 2012 for more detail), we shall also discard terms of the order of ω^4 and $\omega^2 e^2$. This procedure shall dubbed as a *small-eccentricity approximation*.

Now, we have to make a decision about what kind of constraints on the model parameters of the PN ellipsoid $\mathcal{A}$, $\mathcal{B}$ would be more preferable for practical applications. We have considered the case of $\mathcal{A} = 0$, $\mathcal{B} \neq 0$ in our previous work (Kopeikin et al., 2016). We have proved that $\mathcal{B} \simeq O\left(e^4\right)$, and can be ignored in all geodetic equations of the small-eccentricity approximation under assumption of the homogeneous density distribution. In the present paper we explore the complementary case $\mathcal{A} \neq 0$, $\mathcal{B} = 0$. This constraint makes the level surface of the post-Newtonian ellipsoid coinciding exactly with the classic ellipsoid Equation (20) which means that the ellipsoidal coordinate system, having been ubiquitously used in geodesy, is not deformed by the relativistic corrections to the gravity field at all. This requires to adopt that the mass density inside the ellipsoid has an inhomogeneous ellipsoidal distribution given by Equation (38) depending on the parameter $\mathcal{A}$ which is determined from the equation of the level surface Equation (131). We solve Equation (131) under this constraint with respect to $\mathcal{A}$, and find out that it can be approximated with

$$\frac{8}{315}e^4\mathcal{A} = \frac{472}{4725}e^4 + \frac{11152}{363825}e^6 + O\left(e^8\right), \quad (148)$$

that yields

$$\mathcal{A} = \frac{59}{15} + \frac{1394e^2}{1155}, \quad (149)$$

Post-Newtonian correction to the mass of the Earth, $M_\oplus$, can be evaluated from Equation (102). Because the mass couples with the universal gravitational constant G, it contributes to the numerical value of the geocentric gravitational constant $GM_\oplus = 3.986004418 \times 10^{14}$ m^{3}s^{-2}. Numerical values for the geopotential and the semi-major axis, $a_\oplus$, of reference ellipsoid are given in Petit and Luzum (2010, Table 1.1.) They yield, $GM_\oplus/c^2 a_\oplus = 6.7 \times 10^{-10}$. After expansion of the right side of Equation (102) with respect to the eccentricity e, the relativistic variation in the value of $GM_\oplus$ is

$$\frac{\delta(GM_\oplus)_{\text{pN}}}{GM_\oplus} \simeq 2\kappa J_0 \simeq 3\frac{GM_\oplus}{c^2 a_\oplus} \simeq 2.08 \times 10^{-9}\,, \qquad (150)$$

where the numerical value of κ was calculated on the basis of relationship $\kappa = \pi G \rho_0 a_\oplus^2 \simeq (3/4)GM_\oplus/c^2 a_\oplus$. The current uncertainty in the numerical value of $GM_\oplus$ is $\delta GM_\oplus = 8 \times 10^5$ m^{3}s^{-2} (Petit and Luzum, 2010, Table 1.1) which gives the fractional uncertainty

$$\frac{\delta(GM_\oplus)}{GM_\oplus} \simeq 2.0 \times 10^{-9}\,. \qquad (151)$$

This is comparable with the relativistic contribution Equation (150) that must be taken into account in the reduction of precise geodetic data processing which are currently based on the Newtonian theory.

The Maclaurin relation Equation (130) takes on the following approximate form

$$\frac{\omega^2}{2\pi G \rho_0} = \frac{\sqrt{1-e^2}(3-2e^2)\arcsin e - 3e(1-e^2)}{e^3} + \frac{99}{175}\frac{GM}{ac^2}e^2\,, \qquad (152)$$

which, after replacing ρ_0 with the help of Equation (103) and expansion, can be recast to

$$\omega^2 = \frac{3GM}{2a^3}\left[\frac{(3-2e^2)\arcsin e - 3e\sqrt{1-e^2}}{e^3} + \frac{99}{175}\frac{GM}{ac^2}e^2\right]\,. \qquad (153)$$

This can be further decomposed and presented as follows

$$\omega = \omega_\oplus\left(1 + \frac{297}{280}\frac{GM}{ac^2}\right)\,, \qquad (154)$$

where

$$\omega_\oplus = \left(\frac{3GM}{2a^3}\frac{(3-2e^2)\arcsin e - 3e\sqrt{1-e^2}}{e^3}\right)^{1/2}\,, \qquad (155)$$

is the currently used value for the Earth angular velocity of rotation, $\omega_\oplus = 729211.5 \times 10^{-10}$ rad·s^{-1} (Petit and Luzum, 2010). Contribution from the post-Newtonian term in the right side of Equation (154) amounts to 7.1×10^{-10} which makes it evident that the currently adopted value for $\omega_\oplus$ should be corrected by the IUGG.

The approximate version of the post-Newtonian Pizzetti theorem Equation (144) reads

$$2\frac{\gamma_e}{a} + \frac{\gamma_p}{b} = -2\omega^2 + \frac{3GM}{a^2b} + \frac{4GM}{ac^2}\left[\left(3 + \frac{47}{25}e^2\right)\frac{GM}{a^3} + \frac{1}{4}\omega^2\right]\,. \qquad (156)$$

The post-Newtonian corrections in the Clairaut theorem Equation (147), after they are expanded with respect to the eccentricity, yield

$$\frac{\gamma_e}{a} - \frac{\gamma_p}{b} = \omega^2 + \frac{3GM}{2a^2b}\frac{e(3-e^2) - 3\sqrt{1-e^2}\arcsin e}{e^3}$$

$$+ \frac{GM}{ac^2}\left(\frac{59}{25}\frac{GM}{a^3}e^2 + \frac{11}{14}\omega^2\right)\,. \qquad (157)$$

The post-Newtonian corrections to the gravitational field entering the the Pizzetti and Clairaut theorems Equations (156), (157) are not so negligibly small, amount to the magnitude of approximately 3 μGal (1 Gal = 1 cm/s^2), and are to be taken into account in calculation of the parameters of the reference-ellipsoid from astronomical and gravimetric data in a foreseeable future.

Equations (148)–(151) extrapolates the Newtonian relations adopted by IERS and IUGG for definition of the Earth reference ellipsoid and for connecting its parameters to the angular velocity ω and gravity field, to the realm of general relativity.

14. POST-NEWTONIAN GEOID'S UNDULATION

We define the anomalous (disturbing) gravity potential $\mathcal{T} \equiv \mathcal{T}(\boldsymbol{x})$ as the difference between the physical gravity potential, $W \equiv W(\boldsymbol{x})$, and the background gravity potential, $\bar{W}(\boldsymbol{x})$, of the PN reference ellipsoid

$$\mathcal{T} = W - \bar{W}\,. \qquad (158)$$

The gravity potential W is defined by the Einstein equations with the sources that take into account the real distribution of density, pressure, etc. of the real Earth. We don't need to solve the full Einstein equations again in order to find out $\mathcal{T}$. Because the difference between W and $\bar{W}$ is very small, we can consider merely a linear perturbation of the background metric tensor [see Equation (13)]

$$\varkappa_{\alpha\beta} = g_{\alpha\beta}(\boldsymbol{x}) - \bar{g}_{\alpha\beta}(\boldsymbol{x})\,. \qquad (159)$$

The gravity potential W is constant on the perturbed equipotential surface, and is defined by

$$W = c^2\left(1 - \frac{d\tau}{dt}\right)\,, \qquad (160)$$

where $d\tau = c^{-1}\sqrt{-g_{\alpha\beta}dx^\alpha dx^\beta}$ is the proper time of observer in the real physical spacetime. Making use of Equations (160) and (16) allows us to recast Equation (158) to

$$\mathcal{T} = c^2\left(\frac{d\bar{\tau}}{dt} - \frac{d\tau}{dt}\right)\,, \qquad (161)$$

which can be further simplified by noticing that

$$\left(\frac{d\tau}{dt}\right)^2 = \left(1 - \bar{u}^\alpha \bar{u}^\beta \varkappa_{\alpha\beta}\right)\left(\frac{d\bar{\tau}}{dt}\right)^2\,, \qquad (162)$$

where $\bar{u}^\alpha$ is the unperturbed four-velocity of observer which coincides with the four-velocity of matter of the rotating reference ellipsoid. Accounting for definition Equation (161), we get the anomalous gravity potential in the form,

$$\mathcal{T} = \frac{c^2}{2}\bar{u}^\alpha \bar{u}^\beta \varkappa_{\alpha\beta}\,, \qquad (163)$$

which is fully sufficient for practical applications. We emphasize that $\varkappa_{\alpha\beta}$ is the difference between the actual physical metric, $g_{\alpha\beta}$, and the metric $\bar{g}_{\alpha\beta}$ of the background manifold, and as such, should not be confused with the post-Newtonian expansion of the metric $g_{\alpha\beta}$ around a flat spacetime with the Minkowski metric $\eta_{\alpha\beta} = \text{diag}(-1, 1, 1, 1)$ like it is shown in Equations (2)–(4). Our next task is to derive the differential equation for the anomalous gravity potential $\mathcal{T}$.

Let us assume that inside the Earth the deviation of the real matter distribution from its unperturbed value is described by the symmetric energy-momentum tensor

$$c^2 \mathfrak{T}^{\alpha\beta} = \mathfrak{e} \, u^\alpha u^\beta + \mathfrak{s}^{\alpha\beta} \,, \tag{164}$$

where u^α is four-velocity, $\mathfrak{e}$ is the energy density, and $\mathfrak{s}^{\alpha\beta}$ is the symmetric stress tensor of the perturbing matter. The stress tensor includes the isotropic pressure (diagonal components) and shear (off-diagonal components), and is orthogonal to u^α, that is $\mathfrak{s}_{\alpha\beta} u^\alpha = 0$. The energy density of the matter perturbation

$$\mathfrak{e} = \rho \left(c^2 + \mathfrak{P} \right) \,, \tag{165}$$

where ρ is the mass density of the perturbation, and $\mathfrak{P}$ is the internal (compression) energy of the perturbation. For further calculations, a more convenient metric variable is

$$l_{\alpha\beta} \equiv -\varkappa_{\alpha\beta} + \frac{1}{2}\bar{g}_{\alpha\beta}\varkappa \,, \tag{166}$$

where $\varkappa \equiv \bar{g}^{\alpha\beta}\varkappa_{\alpha\beta}$. The dynamic field theory of manifold perturbations leads to the following equation for $l_{\alpha\beta}$ (Kopeikin and Petrov, 2013, 2014),

$$l_{\alpha\beta}{}^{|\mu}{}_{|\mu} + \bar{g}_{\alpha\beta}A^\mu{}_{|\mu} - 2A_{\alpha|\beta} - \bar{R}^\mu{}_\alpha l_{\beta\mu} - \bar{R}^\mu{}_\beta l_{\alpha\mu}$$
$$- 2\bar{R}_{\alpha\mu\nu\beta}l^{\mu\nu} + 2F^{\text{m}}_{\alpha\beta} = \frac{16\pi G}{c^4}\mathfrak{T}_{\alpha\beta} \,, \tag{167}$$

where $A^\alpha \equiv l^{\alpha\beta}{}_{|\beta}$ is the gauge vector function, depending on the choice of the coordinates, $\bar{R}_{\alpha\mu\nu\beta}$ is the Riemann (curvature) tensor of the background manifold depending on the metric tensor $\bar{g}_{\alpha\beta}$, its first and second derivatives, $\bar{R}_{\alpha\beta} = \bar{g}^{\mu\nu}\bar{R}_{\mu\alpha\nu\beta}$— the Ricci tensor, and $F^{\text{m}}_{\alpha\beta}$ is the tensorial perturbation of the background matter induced by the presence of the perturbation $\mathfrak{T}^{\alpha\beta}$ (see Kopeikin and Petrov, 2013, Equations 148–150 for particular details).

In what follows, we focus on derivation of the master equation for the anomalous gravity potential $\mathcal{T}$ in the exterior space that is outside of the background matter of the reference ellipsoid. This assumption is normally used in geodesy (Torge and Müller, 2012). To achieve our goal, we introduce two auxiliary scalars,

$$\mathfrak{q} \equiv \bar{u}^\alpha \bar{u}^\beta l_{\alpha\beta} + \frac{l}{2} \,, \tag{168}$$

$$\mathfrak{p} \equiv \left(\bar{g}^{\alpha\beta} + \bar{u}^\alpha \bar{u}^\beta \right) l_{\alpha\beta} \,, \tag{169}$$

where

$$l \equiv \bar{g}^{\alpha\beta} l_{\alpha\beta} = 2(\mathfrak{p} - \mathfrak{q}) \,. \tag{170}$$

In terms of the scalar $\mathfrak{q}$ the anomalous gravity potential Equation (163) reads

$$\mathcal{T} = -\frac{c^2}{2}\mathfrak{q} \,, \tag{171}$$

where we have used the property $\varkappa = l$. Taking from both sides of Equation (171) the covariant Laplace operator yields

$$\Box \mathcal{T} \equiv -\frac{c^2}{2}\Box \mathfrak{q} \,, \tag{172}$$

where $\Box \mathfrak{q} \equiv \mathfrak{q}^{|\mu}{}_{|\mu}$ is to be calculated from Equation (167).

We notice that according to Kopeikin and Petrov (2013, Equations 148–150) $F^{\text{m}}_{\alpha\beta}$ is directly proportional to the thermodynamic quantities of the background matter and, thus, vanishes in the exterior (with respect to the background matter) space. Hence, we can drop off $F^{\text{m}}_{\alpha\beta}$ in (167) in the exterior-to-matter domain. After contracting Equation (167) with $\bar{g}^{\alpha\beta}$, and accounting for Equation (170) we obtain

$$\Box \mathfrak{q} - \mathfrak{p}^{|\alpha}{}_{|\alpha} - A^\alpha{}_{|\alpha} = -\frac{8\pi G}{c^2}\mathfrak{T} \,, \tag{173}$$

where all terms depending on the Ricci tensor $\bar{R}_{\alpha\beta}$ cancel out, $\mathfrak{T} \equiv \bar{g}^{\alpha\beta}\mathfrak{T}_{\alpha\beta}$, and we still have terms with the gauge field A^α. Now, we use the gauge freedom of general relativity to simplify (173). More specifically, we impose the gauge condition

$$A^\alpha = -\mathfrak{p}^\alpha \,, \tag{174}$$

where $\mathfrak{p}^\alpha \equiv \bar{g}^{\alpha\beta}\partial_\beta\mathfrak{p}$. This gauge allows us to eliminate function $\mathfrak{p}$ from Equation (173) and, after making use of Equation (172), reduce it to to a simple form of a covariant Poisson equation

$$\Box \mathcal{T} = 4\pi G\mathfrak{T} \,. \tag{175}$$

In the Newtonian approximation the trace of the energy-momentum tensor is reduced to the negative value of the matter density of the perturbation, $\mathfrak{T} \simeq -\rho$, while $\Box \mathcal{T} \simeq \Delta_{\text{N}}\mathcal{T}$. Hence, Equation (175) matches its Newtonian counterpart. Outside the mass distribution the master equation for the anomalous gravity potential is reduced to the covariant Laplace equation

$$\Box \mathcal{T} = 0 \,. \tag{176}$$

Equations (175), (176) extend equations of classic geodesy to the realm of the post-Newtonian chronometric geodesy. The main difference is that the covariant Laplace operator in Equations (175), (176) is taken in curved space with the metric $\bar{g}_{\alpha\beta}$. The explicit form of the covariant Laplace operator applied to a scalar $\mathcal{T}$ in coordinates $x^i = \{x^1, x^2, x^3\}$, reads (Lightman et al., 1975, Problem 7.7.)

$$\Box \mathcal{T} \equiv \frac{1}{\sqrt{-\bar{g}}}\partial_i \left(\sqrt{-\bar{g}}\,\bar{g}^{ij}\partial_j\mathcal{T} \right) \,, \tag{177}$$

where the repeated indices mean the Einstein summation, $\partial_i \equiv \partial/\partial x^i$ is the partial derivative with respect to the spatial coordinates, and we omitted all time derivatives because the background spacetime is stationary. For the geodetic purposes it is convenient to choose a spherical coordinate system, $\{x^1, x^2, x^3\} = \{r, \theta, \lambda\}$ co-rotating rigidly along with the reference ellipsoid and with z-axis directed along the axis of the rotation. With a sufficient post-Newtonian accuracy the metric tensor of the background spacetime in this coordinate system reads

$$\bar{g}_{00} = -1 + \frac{1}{c^2}\left(\omega^2 r^2 \sin^2\theta + 2\bar{V}_{\mathrm{N}}\right) + O\left(c^{-3}\right) , \quad (178)$$

$$\bar{g}_{0\lambda} = \frac{1}{c}\omega r^2 \sin^2\theta + O\left(c^{-3}\right) , \quad (179)$$

$$\bar{g}_{rr} = 1 + \frac{2\bar{V}_{\mathrm{N}}}{c^2} + O\left(c^{-3}\right) , \quad (180)$$

$$\bar{g}_{\theta\theta} = \left(1 + \frac{2\bar{V}_{\mathrm{N}}}{c^2}\right)r^2 + O\left(c^{-3}\right) , \quad (181)$$

$$\bar{g}_{\lambda\lambda} = \left(1 + \frac{2\bar{V}_{\mathrm{N}}}{c^2}\right)r^2 \sin^2\theta + O\left(c^{-3}\right) , \quad (182)$$

where $\bar{V}$ is the Newtonian gravitational potential defined in Equation (56). Determinant of the metric

$$\bar{g} \equiv \det[\bar{g}_{\alpha\beta}] = -r^4 \sin^2\theta \left(1 + \frac{2\bar{V}}{c^2}\right)^2 , \quad (183)$$

After calculating the inverse metric $\bar{g}^{\alpha\beta}$, substituting it to Equation (177), and discarding all terms od the order of $O(c^{-3})$, the equation for the anomalous gravity potential takes on the following form

$$\Delta\mathcal{T} = \frac{\omega^2}{c^2}\frac{\partial^2\mathcal{T}}{\partial\lambda^2} , \quad (184)$$

where

$$\Delta \equiv \partial_{rr} + \frac{2}{r}\partial_r + \frac{1}{r^2}\partial_{\theta\theta} + \frac{1}{r^2\tan\theta}\partial_\theta + \frac{1}{r^2\sin^2\theta}\partial_{\lambda\lambda} , \quad (185)$$

is the Laplace operator in the spherical coordinates. The post-Newtonian equation Equation (184) can be solved by iterations by expanding the distrubing potential in the post-Newtonian series

$$\mathcal{T} = T_{\mathrm{N}} + \frac{1}{c^2}T_{\mathrm{pN}} + O(c^{-4}) , \quad (186)$$

where T_{N} is the Newtonian disturbing potential expressed, for example, by the Stokes integral formula (Heiskanen and Moritz, 1967, Equations 2–163a), and T_{pN} is the post-Newtonian correction which we are searching for. Substituting Equation (186) in Equation (184) yields the inhomogeneous differential equation for T_{pN},

$$\Delta T_{\mathrm{pN}} = \omega^2 \frac{\partial^2 T_{\mathrm{N}}}{\partial\lambda^2} , \quad (187)$$

which can be solved by means of standard mathematical techniques for known T_{N}.

We introduce the relativistic geoid height, $\mathcal{N}$, by making use of relativistic generalization of Bruns' formula (Heiskanen and Moritz, 1967, Equations 2–144). We consider a set of equipotential surfaces defined by constant values of the gravity potential W. Let a point Q lie on an equipotential reference surface $\mathcal{S}_1$ and has coordinates x_Q^α, and a point $\mathcal{P}$ lie on another equipotential surface $\mathcal{S}_2$, and has coordinates $x_\mathcal{P}^\alpha$. The height difference, $\mathcal{N}$, between the two surfaces is defined as the absolute value of the integral taken along the direction of the plumb line passing through the points Q and $\mathcal{P}$,

$$\mathcal{N} = \int_Q^\mathcal{P} n_\alpha \frac{dx^\alpha}{d\ell}d\ell , \quad (188)$$

where $n_\alpha \equiv g_\alpha/|\boldsymbol{g}|$ is the unit (co)vector along the plumb line, g_α is the relativistic acceleration of gravity Equation (18), $|\boldsymbol{g}| \equiv (\bar{h}^{\alpha\beta}g_\alpha g_\beta)^{1/2}$, and ℓ is the proper length defined in space by Zelmanov et al. (2006) and Landau and Lifshitz (1975)

$$d\ell^2 = \bar{h}_{\alpha\beta}dx^\alpha dx^\beta , \quad (189)$$

where $\bar{h}_{\alpha\beta} \equiv \bar{g}_{\alpha\beta} + \bar{u}_\alpha\bar{u}_\beta$ is the metric tensor in three-dimensional space. In case, when the height difference is small enough, we can use the second mean value theorem for integration (Hobson, 1909) and approximate the integral in Equation (188) as follows

$$\mathcal{N} = \int_Q^\mathcal{P} \frac{g_\alpha(x)dx^\alpha}{|\boldsymbol{g}(x)|} = -\frac{c^2}{g_Q}\int_Q^\mathcal{P} \partial_\alpha \ln\left(1 - \frac{W}{c^2}\right)dx^\alpha$$

$$= \frac{c^2}{g_Q}\ln\left|\frac{1 - \dfrac{W(Q)}{c^2}}{1 - \dfrac{W(\mathcal{P})}{c^2}}\right| , \quad (190)$$

where $g_Q = |\boldsymbol{g}(Q)|$ denotes the magnitude of the relativistic acceleration of gravity taken on the equipotential surface $\mathcal{S}_1$. Equation (190) is exact. Separation of the height $\mathcal{N}$ in the Newtonian part and the post-Newtonian corrections depends on how we define the reference equipotential surface $\mathcal{S}_1$.

Let us choose the reference surface by equation $W(Q) = \bar{W}$ where $\bar{W}$ is the normal (relativistic) gravity field of the reference ellipsoid. Then, expanding the logarithm in Equation (190) with respect to the ratio W/c^2 and making use of definition Equation (158) of the anomalous gravity potential $\mathcal{T}$, we obtain from (190)

$$\mathcal{N} = \frac{|\mathcal{T}(\mathcal{P})|}{\gamma_Q} , \quad (191)$$

where the disturbing potential, $\mathcal{T}$, is measured at the point $\mathcal{P}$ on the geoid surface W, and the acceleration of gravity $\gamma_Q \equiv g_Q$ is measured at point Q on the reference surface $\bar{W}$.

Relativistic Bruns' formula Equation (191) yields geoid's undulation with respect to the unperturbed reference level surface, $\bar{W} = \mathrm{const.}$, in general relativity. Because we have defined this surface as an equipotential surface $\bar{W}$ which is the post-Newtonian solution of the Einstein equations, the height

$\mathcal{N}$ does not represent the undulation of the relativistic geoid W with respect to the Newtonian equipotential surface defined by the surface of the constant Newtonian gravity potential, $\bar{V}_N =$ const. Expansion of the height $\mathcal{N}$ in Equation (191) in the post-Newtonian series around the value of the surface $\bar{W}$, yields

$$\mathcal{N} = \mathfrak{N} + \frac{1}{c^2}\mathfrak{N}_{pN} + O\left(c^{-4}\right) , \qquad (192)$$

where $\mathfrak{N}$ is the classic definition of the geoid height given in terms of the Newtonian disturbing potential, $V_N - \bar{V}_N$ as explained, for example, in textbook (Vaníček and Krakiwsky, 1986, chapter 21). The post-Newtonian correction $\mathfrak{N}_{pN}$ to the height $\mathfrak{N}$ is caused by the difference between the post-Newtonian terms in W and $\bar{W}$, and has a magnitude of the order $\mathfrak{N}_{pN} \simeq (V_N/c^2) \times \mathfrak{N}$, where V_N is the Newtonian gravitational potential of the Earth. Because the largest undulation of the Newtonian geoid of the Earth does not exceed 100 m (Torge and Müller, 2012), the *explicit* post-Newtonian correction $\mathfrak{N}_{pN}$ to the undulation is exceedingly small, $\mathfrak{N}_{pN} \simeq 7 \times 10^{-6}$ cm.

Thus, the main relativistic effects in terrestrial geodesy enter through the difference between the equipotential surface of the relativistic potential of the gravity force, $\bar{W}$, and one of the Newtonian gravity potential V_N. This difference amounts to $(V_N/c^2) \times R_\oplus \simeq 1$ cm which looks pretty small but crucially important in the current study of the changes in the global average sea level which is now rising twice as fast as it did over the past century, providing clear evidence of global warming on a short time scale. The measurements are done with the help of the the satellite laser altimetery which is so precise that allows us to measure the change in the global average sea level with the uncertainty ≤ 1 mm/year (Fu and Haines, 2013; Ablain et al., 2015) which is a factor of 10 larger than the post-Newtonian effects in the determination of the equipotential level surface!

ACKNOWLEDGMENTS

We thank the referees for valuable comments and suggestions for improving the presentation of the manuscript. The work of SK has been supported by the grant No 14-27-00068 of the Russian Science Foundation (RSF).

REFERENCES

Ablain, M., Cazenave, A., Larnicol, G., Balmaseda, M., Cipollini, P., Faugère, Y., et al. (2015). Improved sea level record over the satellite altimetry era (1993-2010) from the Climate Change Initiative project. *Ocean Sci.* 11, 67–82. doi: 10.5194/os-11-67-2015

Ansorg, M., Fischer, T., Kleinwächter, A., Meinel, R., Petroff, D., and Schöbel, K. (2004). Equilibrium configurations of homogeneous fluids in general relativity. *Mon. Not. Roy. Astron. Soc.* 355, 682–688. doi: 10.1111/j.1365-2966.2004.08371.x

Bardeen, J. M. (1971). A Reexamination of the Post-Newtonian Maclaurin Spheroids. *Astrophys. J.* 167, 425. doi: 10.1086/151040

Beverini, N., Allegrini, M., Beghi, A., Belfi, J., Bouhadef, B., Calamai, M., et al. (2014). Measuring general relativity effects in a terrestrial lab by means of laser gyroscopes. *Laser Phys.* 24:074005. doi: 10.1088/1054-660X/24/7/074005

Bjerhammar, A. (1985). On a relativistic geodesy. *Bull. Géodésique* 59, 207–220. doi: 10.1007/BF02520327

Bondarenko, N. P., and Pyragas, K. A. (1974). On the Equilibrium figures of an ideal rotating liquid in the post-newtonian approximation of general relativity. II: Maclaurin's P-Ellipsoid. *Astrophys. Space Sci.* 27, 453–466. doi: 10.1007/BF00643890

Bondarescu, R., Bondarescu, M., Hetényi, G., Boschi, L., Jetzer, P., and Balakrishna, J. (2012). Geophysical applicability of atomic clocks: direct continental geoid mapping. *Geophys. J. Int.* 191, 78–82. doi: 10.1111/j.1365-246X.2012.05636.x

Bondarescu, R., Schärer, A., Lundgren, A., Hetényi, G., Houlié, N., Jetzer, P., et al. (2015). Ground-based optical atomic clocks as a tool to monitor vertical surface motion. *Geophys. J. Int.* 202, 1770–1774. doi: 10.1093/gji/ggv246

Brumberg, V. A. (1991). *Essential Relativistic Celestial Mechanics.* Bristol: Adam Hilger.

Chandrasekhar, S. (1965). The post-newtonian effects of general relativity on the equilibrium of uniformly rotating bodies. I. The maclaurin spheroids and the virial theorem. *Astrophys. J.* 142, 1513–1518. doi: 10.1086/148433

Chandrasekhar, S. (1967). Ellipsoidal figures of equilibrium - a historical account. *Commun. Pure Appl. Math.* 20, 251–265. doi: 10.1002/cpa.3160200203

Chandrasekhar, S. (1967a). The post-newtonian effects of general relativity on the equilibrium of uniformly rotating bodies. II. The deformed figures of the maclaurin spheroids. *Astrophys. J.* 147, 334–352. doi: 10.1086/149003

Chandrasekhar, S. (1967b). The post-newtonian effects of general relativity on the equilibrium of uniformly rotating bodies. III. The deformed figures of the jacobi ellipsoids. *Astrophys. J.* 148, 621–644. doi: 10.1086/149183

Chandrasekhar, S. (1967c). The post-newtonian effects of general relativity on the equilibrium of uniformly rotating bodies.IV. The roche model. *Astrophys. J.* 148, 645–649. doi: 10.1086/149184

Chandrasekhar, S. (1969). *Ellipsoidal Figures of Equilibrium.* New Haven, CT: Yale University Press.

Chandrasekhar, S. (1971a). The post-newtonian effects of general relativity on the equilibrium of uniformaly rotating bodies.VI. The deformed figures of the jacobi ellipsoids. *Astrophys. J.* 167, 455–463. doi: 10.1086/151042

Chandrasekhar, S. (1971b). The post-newtonian effects of general relativity on the equilibrium of uniformly rotating bodies. V. The deformed figures of the maclaurin spheroids. *Astrophys. J.* 167, 447–453. doi: 10.1086/151041

Chandrasekhar, S., and Elbert, D. D. (1974). The deformed figures of the Dedekind ellipsoids in the post-Newtonian approximation to general relativity. *Astrophys. J.* 192, 731–746. doi: 10.1086/153111

Chandrasekhar, S., and Elbert, D. D. (1978). The deformed figures of the Dedekind ellipsoids in the post-Newtonian approximation to general relativity - Corrections and amplifications. *Astrophys. J.* 220, 303–313. doi: 10.1086/155906

Chandrasekhar, S., and Miller, J. C. (1974). On slowly rotating homogeneous masses in general relativity. *Mon. Not. Roy. Astron. Soc.* 167, 63–80. doi: 10.1093/mnras/167.1.63

Ciufolini, I., and Wheeler, J. A. (1995). *Gravitation and Inertia.* Princeton, NJ: Princeton University Press.

Drewes, H. (ed.). (2009). *Geodetic Reference Frames.* Berlin: Springer. doi: 10.1007/978-3-642-00860-3

Falke, S., Lemke, N., Grebing, C., Lipphardt, B., Weyers, S., Gerginov, V., et al. (2014). A strontium lattice clock with 3×10^{-17} inaccuracy and its frequency. *New J. Phys.* 16:073023. doi: 10.1088/1367-2630/16/7/073023

Fock, V. A. (1964). *The Theory of Space, Time and Gravitation, 2nd Edn. (Trans. N. Kemmer).* New York, NY: Macmillan.

Fu, L.-L., and Haines, B. J. (2013). The challenges in long-term altimetry calibration for addressing the problem of global sea level change. *Adv. Space Res.* 51, 1284–1300. doi: 10.1016/j.asr.2012.06.005

Galtsov, D. V., Tsvetkov, V. P., and Tsirulev, A. N. (1984). The spectrum and polarization of the gravitational radiation of pulsars. *Sov. Phys. JETP* 59, 472–477.

Gürlebeck, N., and Petroff, D. (2010). The axisymmetric case for the post-newtonian dedekind ellipsoids. *Astrophys. J.* 722, 1207–1215. doi: 10.1088/0004-637X/722/2/1207

Gürlebeck, N., and Petroff, D. (2013). A generalized family of post-newtonian dedekind ellipsoids. *Astrophys. J.* 777, 1–16. doi: 10.1088/0004-637X/777/1/60

Heiskanen, W. A., and Moritz, H. (1967). *Physical Geodesy*. San Francisco, CA: W. H. Freeman.

Hobson, E. W. (1909). On the second mean-value theorem of the integral calculus. *Proc. Lond. Math. Soc.* s2-7, 14–23. doi: 10.1112/plms/s2-7.1.14

Hofmann-Wellenhof, B., and Moritz, H. (2006). *Physical Geodesy*. Berlin: Springer.

Hurst, R. B., Rabeendran, N., Wells, J.-P. R., and Schreiber, K. U. (2015). "Large ring laser gyroscopes: towards absolute rotation rate sensing," in *Society of Photo-Optical Instrumentation Engineers (SPIE) Conference Series, volume 9444 of Society of Photo-Optical Instrumentation Engineers (SPIE) Conference Series* (Bellingham, WA), 944407.

Kómár, P., Kessler, E. M., Bishof, M., Jiang, L., Sórensen, A. S., Ye, J., et al. (2014). A quantum network of clocks. *Nat. Phys.* 10, 582–587. doi: 10.1038/nphys3000

Kopeikin, S., Efroimsky, M., and Kaplan, G. (2011). *Relativistic Celestial Mechanics of the Solar System*. Berlin: Wiley. doi: 10.1002/9783527634569

Kopeikin, S., Han, W., and Mazurova, E. (2016). Post-Newtonian reference-ellipsoid for relativistic geodesy. *Phys. Rev. D*. eprint arXiv:1510.03131

Kopeikin, S. M., Mazurova, E. M., and Karpik, A. P. (2015). Towards an exact relativistic theory of Earth's geoid undulation. *Phys. Lett. A* 379, 1555–1562. doi: 10.1016/j.physleta.2015.02.046

Kopeikin, S. M., and Petrov, A. N. (2013). Post-newtonian celestial dynamics in cosmology: field equations. *Phys. Rev. D* 87:044029. doi: 10.1103/PhysRevD.87.044029

Kopeikin, S. M., and Petrov, A. N. (2014). Dynamic field theory and equations of motion in cosmology. *Ann. Phys.* 350, 379–440. doi: 10.1016/j.aop.2014.07.029

Kopejkin, S. M. (1991). Relativistic manifestations of gravitational fields in gravimetry and geodesy. *Manuscripta Geodaetica* 16, 301–312.

Landau, L. D., and Lifshitz, E. M. (1975). *The Classical Theory of Fields*. Oxford: Pergamon Press.

Lightman, A. P., Press, W. H., Price, R. H., and Teukolsky, S. A. (1975). *Problem Book in Relativity and Gravitation*. Princeton, NJ: Princeton University Press.

Mai, E. (2014). Time, atomic clocks, and relativistic geodesy. *Report No 124, Deutsche Geodʹatische Kommission der Bayerischen Akademie der Wissenschaften (DGK)*, 128. Available online at: http://dgk.badw.de/fileadmin/docs/a-124.pdf

Mai, E., and Müller, J. (2013). General remarks on the potential use of atomic clocks in relativistic geodesy. *ZFV Zeitschrift Geodasie Geoinformation Landmanagement* 138, 257–266.

Meinel, R., Ansorg, M., Kleinwächter, A., Neugebauer, G., and Petroff, D. (2008). *Relativistic Figures of Equilibrium*. Cambridge: Cambridge University Press. doi: 10.1017/CBO9780511535154

Müller, J., Soffel, M., and Klioner, S. A. (2008). Geodesy and Relativity. *J. Geod.* 82, 133–145. doi: 10.1007/s00190-007-0168-7

Nicholson, T. L., Campbell, S. L., Hutson, R. B., Marti, G. E., Bloom, B. J., McNally, R. L., et al. (2015). Systematic evaluation of an atomic clock at 2×10^{-18} total uncertainty. *Nat. Commun.* 6, 1–8. doi: 10.1038/ncomms7896

Oltean, M., Epp, R. J., McGrath, P. L., and Mann, R. B. (2015). Geoids in general relativity: geoid quasilocal frames. eprint arXiv:1510.02858

Petit, G., and Luzum, B. (2010). IERS conventions. *IERS Tech. Note* 36, 179.

Petit, G., Wolf, P., and Delva, P. (2014). "Atomic time, clocks, and clock comparisons in relativistic spacetime: a review," in *Frontiers in Relativistic Celestial Mechanics. Vol. 2 Applications and Experiments*, ed S. M. Kopeikin (Berlin: W. de Gruyter), 249–279.

Petroff, D. (2003). Post-Newtonian Maclaurin spheroids to arbitrary order. *Phys. Rev. D* 68, 104029. doi: 10.1103/physrevd.68.104029

Pizzetti, P. (1913). *Principii Della Teoria Meccanica Della Figura dei Pianeti*. Pisa: E. Spoerri.

Plag, H.-P., and Pearlman, M. (ed.). (2009). *Global Geodetic Observing System*. Dordrecht: Springer. doi: 10.1007/978-3-642-02687-4

Poli, N., Oates, C. W., Gill, P., and Tino, G. M. (2013). Optical atomic clocks. *Nuovo Cimento Riv. Ser.* 36, 555–624. doi: 10.1393/ncr/i2013-10095-x

Pyragas, K. A., Bondarenko, N. P., and Kravtsov, O. V. (1974). On the equilibrium figures of an ideal rotating liquid in the post-newtonian approximation of general relativity. I: equilibrium conditions. *Astrophys. Space Sci.* 27, 437–452. doi: 10.1007/BF00643889

Pyragas, K. A., Bondarenko, N. P., and Kryshtal, A. N. (1975). On the equilibrium figures of an ideal rotating fluid in the post-Newtonian approximation of general relativity. III - Stability of the forms of equilibrium. *Astrophys. Space Sci.* 33, 75–97. doi: 10.1007/BF00646009

Rendall, A. D. (1990). Convergent and divergent perturbation series and the post-minkowskian approximation scheme. *Class. Quant. Grav.* 7, 803–812. doi: 10.1088/0264-9381/7/5/010

Schreiber, K. U. (2013). Variations of earth rotation from ring laser gyroscopes: one hundred years of rotation sensing with optical interferometry (invited). *AGU Fall Meet. Abstr.* Available online at: http://abstractsearch.agu.org/meetings/2013/FM/G11C-06.html

Soffel, M., Klioner, S. A., Petit, G., Wolf, P., Kopeikin, S. M., Bretagnon, P., et al. (2003). The IAU 2000 resolutions for astrometry, celestial mechanics, and metrology in the relativistic framework: explanatory supplement. *Astron. J. (USA)* 126, 2687–2706. doi: 10.1086/378162

Soffel, M. H. (1989). *Relativity in Astrometry, Celestial Mechanics and Geodesy*. Berlin: Springer-Verlag. doi: 10.1007/978-3-642-73406-9

Torge, W., and Müller, J. (2012). *Geodesy, 4th Edn*. Berlin: De Gruyter. doi: 10.1515/9783110250008

Tsirulev, A. N., and Tsvetkov, V. P. (1982a). Rotating post-newtonian near ellipsoidal configurations of a magnetized homogeneous fluid - Part I. *Sov. Astron.* 26, 289–292.

Tsirulev, A. N., and Tsvetkov, V. P. (1982b). Rotating post-newtonian near ellipsoidal configurations of a magnetized homogeneous fluid - Part II. *Sov. Astron.* 26, 407–412.

Tsvetkov, V. P., and Tsirulev, A. N. (1983). Gravitational waves emitted by a spinning magnetized blob of homogeneous post-newtonian gravitating fluid. *Sov. Astron.* 27, 66–69.

Vaníček, P., and Krakiwsky, E. J. (1986). *Geodesy, the Concepts, 2nd Edn*. Amsterdam: North Holland.

Weinberg, S. (1972). *Gravitation and Cosmology: Principles and Applications of the General Theory of Relativity*. New York, NY: John Wiley & Sons, Inc.

Will, C. M. (1993). *Theory and Experiment in Gravitational Physics*. Cambridge: Cambridge University Press. doi: 10.1017/CBO9780511564246

Zelmanov, A., Rabounski, D., Crothers, S. J., and Borissova, L. (2006). *Chronometric Invariants: on Deformations and the Curvature of Accompanying Space*. Rehoboth, NM: American Research Press. Available online at: http://www.ptep-online.com/index_files/books/zelmanov1944.pdf

Conflict of Interest Statement: The author declares that the research was conducted in the absence of any commercial or financial relationships that could be construed as a potential conflict of interest.

15

Baldwin Effect and Additional BLR Component in AGN with Superluminal Jets

Víctor M. Patiño-Álvarez[1], Janet Torrealba[1], Vahram Chavushyan[1]*, Irene Cruz-González[2], Tigran Arshakian[3,4,5], Jonathan León-Tavares[1] and Luka Popović[6]

[1] Instituto Nacional de Astrofísica, Óptica y Electrónica, Puebla, Mexico, [2] Instituto de Astronomía, Universidad Nacional Autónoma de México, Mexico City, Mexico, [3] Physikalisches Institut, Universität zu Köln, Köln, Germany, [4] Byurakan Astrophysical Observatory, Byurakan, Armenia, [5] Isaac Newton Institute of Chile in Estern Europe and Eurasia, Armenian Branch, Santiago, Chile, [6] Astronomical Observatory, Belgrade, Serbia

Edited by:
Jirong Mao,
RIKEN, Japan

Reviewed by:
Daniela Bettoni,
INAF - Osservatorio Astronomico di Padova, Italy
Giovanna Maria Stirpe,
INAF - Osservatorio Astronomico di Bologna, Italy

*Correspondence:
Vahram Chavushyan
chavushyanv@gmail.com

Specialty section:
This article was submitted to Milky Way and Galaxies, a section of the journal Frontiers in Astronomy and Space Sciences

We study the Baldwin Effect (BE) in 96 core-jet blazars with optical and ultraviolet spectroscopic data from a radio-loud AGN sample obtained from the MOJAVE 2 cm survey. A statistical analysis is presented of the equivalent widths (W_λ) of emission lines Hβ λ4861, Mg II λ2798, C IV λ1549, and continuum luminosities at 5100 Å, 3000 Å, and 1350 Å. The BE is found statistically significant (with confidence level $c.l. \geq 95\%$) in Hβ and C IV emission lines, while for Mg II the trend is slightly less significant ($c.l. = 94.5\%$). The slopes of the BE in the studied samples for Hβ and Mg II are found steeper and with statistically significant difference than those of a comparison radio-quiet sample. We present simulations of the expected BE slopes produced by the contribution to the total continuum of the non-thermal boosted emission from the relativistic jet, and by variability of the continuum components. We find that the slopes of the BE between radio-quiet and radio-loud AGN should not be different, under the assumption that the broad line is only being emitted by the canonical broad line region around the black hole. We discuss that the BE slope steepening in radio AGN is due to a jet associated broad-line region.

Keywords: galaxies:active, galaxies:jets, quasars:emission lines

1. INTRODUCTION

Baldwin (1977) discovered that quasars follow a relation between the rest frame equivalent widths for the ultraviolet lines (e.g., C IV, Lyα, etc.) and the continuum luminosity at 1350 Å (L_{1350}), known as the *Baldwin Effect* (Carswell and Smith, 1978), hereafter BE. This relation became quite important and has been the subject of many investigations because it allows to study the physics of the diverse emitting regions present in active galactic nuclei (AGN). The BE is well established for broad emission lines in the ultraviolet and optical regions (e.g., Shields, 2007), and it is also found that steepens with increasing ionization potential (Zheng and Malkan, 1993; Dietrich et al., 2002). Most recently, the BE has also been found in narrow emission lines (e.g., Croom et al., 2002; Dietrich et al., 2002; Netzer et al., 2004, 2006; Netzer and Trakhtenbrot, 2007; Kovačević et al., 2010; Popović and Kovačević, 2011; Zhang et al., 2013). It has been suggested that the BE could be used to probe the model predictions of the spectral energy distributions (SED) as a function of luminosity (Dietrich et al., 2002), or to test cosmological models at high redshifts (Shields, 2007).

Despite the advances made in this subject during the last three decades, the physical mechanisms driving the observed BE remain unclear (see Shields, 2007, and references therein for a complete review on the BE). The most widely accepted driving mechanism is that the ionization continuum softens as the luminosity increases (e.g., Zheng and Malkan, 1993), so that high-luminosity AGN decrease the fraction of ionizing photons for broad emission line formation. This is consistent with Scott et al. (2004) that the low-luminosity AGN show harder spectral continuum in the extreme-UV. Some theoretical studies support that the BE is driven, at least in part, on both the continuum shape and the metallicity of the gas (e.g., Korista et al., 1998).

Other fundamental parameters have been proposed as the principal drivers of the BE: the Eddington ratio (Baskin and Laor, 2004; Bachev et al., 2004; Dong et al., 2009), or the black hole mass (e.g., Warner et al., 2003; Xu et al., 2008). Nevertheless, a consensus on these issues has not yet emerged.

Moreover, it is now well established that emission lines originated from higher ionization species display steeper slopes in the $W_\lambda - L_c$ diagram. This means that the intensity of the correlation, traced by the slope, seems to be dependent on the emission line ionizing energy, as was shown by several authors (e.g., Zheng and Malkan, 1993; Zheng et al., 1995; Espey and Andreadis, 1999; Dietrich et al., 2002).

The aim of this paper is to investigate the BE in radio-loud AGN possessing relativistic jets. This is of great interest because, through the past decades, several studies have shown that AGN spectral properties differentiate depending on radio-loudness[1]. For example, Brotherton et al. (2001, and references therein) found that the composite spectrum of radio loud (RL; $\log R > 1$) AGN, compared to that of radio-quiet (RQ) AGN, shows a redder SED, broader Balmer lines, stronger [O III] emission, and stronger red wing/weaker blue wing asymmetry of the C IV λ1549 emission line. Other authors found that RL and RQ AGN have remarkably similar low-ionization emission lines (Mg II and C III]), while high-ionization lines are clearly stronger in RL composite spectrum (e.g., C IV, Francis et al., 1993; Zheng et al., 1997). Motivated by these spectral differences in RL and RQ AGN, in this work the authors investigate the difference of the BE between the population of RL AGN having the beamed continuum emission due to relativistic jet and the sample of RQ AGN.

The paper is presented as follows. The characteristics of the RL AGN sample and spectroscopic observational data are presented in Section 2. The comparison sample of RQ AGN is described in Section 3. The BE and simulations of the contribution of non-thermal emission to the BE is presented in Section 4. The statistical results and comparison with the RQ samples are presented in Section 5, including line-luminosity relations, and BE slope differences. The jet contribution to the total non-thermal continuum emission and the non-thermal dominance dependence on viewing angle and equivalent width for Flat Spectrum Radio Quasars (FSRQ) is discussed in Section 6. Finally, discussion and conclusions are presented in Section 7.

Throughout the paper a flat cosmology model is used with parameters $\Omega_m = 0.3$ ($\Omega_\Lambda + \Omega_m = 1$) and $H_0 = 70$ km s^{-1} Mpc^{-1}.

2. SAMPLE AND SPECTROSCOPIC DATA OF RADIO-LOUD AGN

2.1. Sample

The sample of 96 RL AGN studied here is a part of 250 compact extragalactic sources with radio jets (15 GHz) compiled and described by Kovalev et al. (2005), that comprises blazars (BL Lacs and flat-spectrum radio quasars), radio galaxies, and few sources unclassified in the optical regime. AGN of this sample have a core–jet structure on miliarcsecond scales, where the radio jet is aligned close to the line-of-sight. These sources are observed with VLBA at 2 cm (Kellermann et al., 1998, 2004; Zensus et al., 2002), and roughly half of the sample is part of the MOJAVE [2] ("Monitoring of Jets in AGN with VLBA Experiments"; see Lister et al., 2009) program. Most of the sources in the sample have flat radio spectra ($\alpha > -0.5$, $F \sim \nu^{+\alpha}$, for $\nu > 500$ MHz; Kovalev et al., 1999, 2000), their total flux density at 15 GHz (obtained in the period 1994–2003) is > 1.5 Jy for Northern hemisphere sources ($\delta > 0°$) and > 2 Jy for sources with $-20° < \delta < 0°$.

Given that 97% of the sample is comprised by AGN with flat radio spectrum, and broad lines typical of quasars, hereafter the RL AGN sample will be referred to as Flat Spectrum Radio Quasars (FSRQ).

The range in radio-loudness of the FSRQ is $1.2 \leq \log R \leq 4.5$ with an average value of $\log R = 3.5$.

The core–jet structure of the 96 FSRQ makes it a unique sample to study via spectroscopic observations the influence of the jet beaming effects on the broad and narrow emission line regions (BLR and NLR), and in particular to study the BE in RL AGN.

2.2. Spectroscopic data

Optical and ultraviolet spectroscopic data of blazars are presented in full detail in the accompanying spectral atlas[3] (Torrealba et al., 2012). Spectra are available for 123 sources from the MOJAVE/2cm sample (see Torrealba et al., 2014), but for the BE analysis presented here, the sample was restricted to AGN with $S/N > 10$ spectra which involves a sample of 96 FSRQ, which are about half of the AGN in the MOJAVE sample.

As is mentioned in the spectroscopic atlas, the observations were acquired at two 2.1 m Mexican telescopes in OAGH[4] and OAN-SPM[5]. In few cases, the spectra were complemented with available databases (HST, SDSS, etc.). Our database is homogeneous in the sense that the same spectral analysis procedures are used for fitting the emission lines, de-blending of the Fe II emission and emission-line local continuum fitting. To strengthen the analysis results, the flux, line equivalent width, and

---

[1] Radio-loudness classic criteria R: the ratio between the radio (5 GHz) and optical (4400 Å) flux densities $R = F_{5\,GHz}/Fo_{4400\,A}$ (Kellermann et al., 1989).

[2] http://www.physics.purdue.edu/astro/MOJAVE/index.html.
[3] http://vizier.cfa.harvard.edu/viz-bin/VizieR?-source=J/other/RMxAA/48.9.
[4] Observatorio Astrofísico Guillermo Haro, in Cananea, Sonora, Mexico.
[5] Observatorio Astronómico Nacional en San Pedro Mártir, Baja California, Mexico.

continuum luminosity measurements have not been mixed with data obtained from literature.

Three subsamples of FSRQ were defined:

- The Hβ subsample comprises 18 quasars and 3 radio galaxies. The narrow-line sources with FWHM Hβ $\lesssim$ 1000 km s^{-1} were excluded. The redshift range is 0.033–0.751 with optical magnitude between $13.6 < B_J < 18.5$.
- The Mg II λ2798 subsample is the largest data set which comprises 69 quasars. In this case, the redshift range is 0.295–2.118 with magnitude between $14.5 < B_J < 20.6$.
- The C IV λ1549 subsample comprises 31 quasars. The redshift range is 0.295–3.396 with magnitude between $15.1 < B_J < 20.9$.

It is important to mention that due to the redshift, more than one emission line was available for some sources.

2.3. Continuum and Emission Line Parameters

The same spectral analysis procedure was used to measure spectral line parameters (flux and equivalent width) and continuum emission for all AGN in our sample. Procedures to obtain the continuum emission and the subtraction of the Fe II contribution are described in detail in Section 6 of Torrealba et al. (2012). The emission line parameters are measured after subtracting the contribution of Fe II emission and a power-law of the local continuum. The spectral range of the data only allows to fit the local continuum with a power-law, by selecting regions free of emission or absorption lines. The total emission line flux was measured by Gaussian decomposition of the spectra. The decomposition was performed using the task MPFITEXPR from the MPFIT IDL package (Markwardt, 2009).

The uncertainty of the emission-line flux is estimated from the formula given in Tresse et al. (1999) and on the average is about 15%. The continuum flux is measured from the iron free spectrum for each AGN in the range of ± 50 Å. Then the monochromatic continuum luminosities were calculated $L_c \equiv \lambda L_\lambda$ at 5100 Å, 3000 Å, or 1350 Å for the three AGN subsamples.

The luminosity results of FSRQ samples are:

- Ranges of continuum luminosities: $44.1 \leq \log L_{5100} \leq 46.8$ and $45.6 \leq \log L_{3000} \leq 48$, and $46.3 \leq \log L_{1350} \leq 48.8$.
- Mean continuum luminosities: $\log L_{5100} = 45.7 \pm 0.8$, $\log L_{3000} = 46.7 \pm 0.5$, and $\log L_{1350} = 47.6 \pm 0.9$.
- Average uncertainty for L_c: 11, 10, and 17% for L_{5100}, L_{3000}, and L_{1350}, respectively.
- Mean total line luminosities: $\log L_{H\beta} = 43.8 \pm 0.8$, $\log L_{Mg II} = 44.8 \pm 0.5$, and $\log L_{C IV} = 45.8 \pm 0.7$.

The equivalent width for each emission line was calculated using the ratio of the total line luminosity (L_{line}) and monochromatic continuum multiplying by the wavelength associated with the corresponding continuum, $W_\lambda = (L_{line}/L_c) \times \lambda$. The W_λ uncertainties are about 30–35 % depending on the mean signal-to-noise ratio of the spectrum.

Uncertainties of the equivalent widths W_λ for the emission lines near 5100 Å with mean spectral S/N $\sim$ 15 lie in the range

10–15%. Near the 3000 Å region, the uncertainties are roughly 12% with S/N $\sim$ 20, and for W_λ(C IV) the average uncertainty is ~14% with S/N $\sim$ 15.

The mean W_λ and its standard deviation of Hβ, Mg II, and C IV emission lines are (76.6 $\pm$ 23.8) Å, (42.4 $\pm$ 21.8) Å, and (27.0 $\pm$ 14.4) Å, respectively.

3. SAMPLES OF RADIO-QUIET AGN

To compare the BE in our FSRQ sample, two samples of RQ AGN were selected. For Hβ the sample used comes from Greene and Ho (2005) while for Mg II and C IV the sample comes from Shen et al. (2011). This control sample was compared to the BE in FSRQ. Both samples of RQ AGN were selected from the Sloan Digital Sky Survey (SDSS, York et al., 2000).

The RQ control sample for Hβ emission consists of 229 RQ AGN from the Third Data Release (DR3, Abazajian et al., 2005) with $z \leq 0.35$. The second sample is taken from The Seventh Data Release (DR7, Abazajian et al., 2009) with 44,000 quasars having the Mg II emission line ($0.35 \leq z \leq 2.25$), and 10,000 quasars with C IV emission line ($1.5 \leq z \leq 4.95$). Both samples are assumed to be dominated by a population of RQ AGN (e.g., Shaw et al., 2012).

Greene and Ho (2005) and Shen et al. (2011) use the following procedure to measure the spectral line and continuum characteristics. They decompose the spectrum for each source by simultaneous fitting of two-component model consisting of featureless continuum and the empirical Fe II template from Boroson and Green (1992, Hβ region) and Vestergaard and Wilkes (2001, Mg II region). Shen et al. (2011) fitted the local continuum with a single power-law in the wavelength intervals between 2200 and 2700 Å and 2900–3090 Å near the Mg II line and 1445–1465 Å and 1700–1705 Å near the C IV emission line. The featureless continuum in the region of Hβ line was approximated by a double power-law broken at 5000 Å under the requirement that the combined flux of the two components at ~5600 Å (near Hβ) be equal to the observed flux at that point (Greene and Ho, 2005). Shen et al. (2011) measured the C IV line flux without iron subtraction which may lead to an overestimation of W_λ by ~0.05 dex on average. In the RQ samples, the emission line profile is modeled as a multicomponent Gaussian taking into account both the broad and narrow components.

4. BALDWIN EFFECT

4.1. W_λ vs. L_c

The relation between the emission line equivalent width (W_λ) and the continuum emission luminosity (L_c) is given by Baldwin (1977),

$$\log W_\lambda = \alpha + \beta \log L_c. \qquad (1)$$

The slope β is found to be negative for RQ AGN (Baldwin, 1977; Shields, 2007), i.e., the equivalent width of the emission line (or the contrast between the line and continuum luminosities) decreases toward large continuum luminosities.

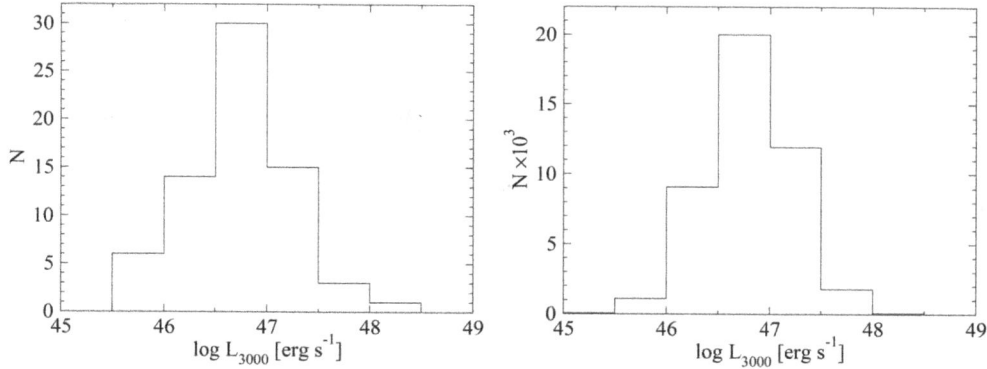

FIGURE 1 | Distributions of continuum luminosity at 3000 Å obtained for the observed FSRQ sample (left panel); and for the simulated RQ sample (right panel).

Equation (1) can be transformed to a relation between the total line luminosity L_{line} and the monochromatic continuum luminosity $\lambda L_\lambda \equiv L_c$ measured at a certain wavelength λ,

$$\log L_{\text{line}} = A + B \log L_c, \qquad (2)$$

by replacing α and β with,

$$\alpha = A + \log \lambda \qquad (3)$$
$$\beta = B - 1, \qquad (4)$$

and considering that $W_\lambda \simeq L_{\text{line}}/L_\lambda$.

4.2. Contribution of Non-thermal Emission to the Baldwin Effect

The optical continuum emission in RQ AGN is assumed to be isotropic and generated in the accretion disk, so that the continuum luminosity is $L_c = L_{\text{disk}}^{\text{RQ}}$. On the other hand, for FSRQ the optical continuum emission has two components, the thermal emission from the disk ($L_{\text{disk}}^{\text{BL}}$) and the beamed non-thermal emission from the relativistic jet (L_{jet}), i.e., $L_c = L_{\text{disk}}^{\text{BL}} + L_{jet}$. It was assumed that the main contribution to the broad line emission is attributed to the disk thermal emission, while the beamed emission from the jet is produced beyond the BLR and, hence, has no contribution to the Broad Line (BL) emission.

A simulation was performed in order to estimate the statistical properties of the RQ samples studied by Greene and Ho (2005) and Shen et al. (2011). The purpose is to compare them to the RL sample presented in this paper. First, a distribution of continuum luminosity was taken (see **Figure 1**), based on the luminosity distributions observed for our sample of FSRQ. For all three lines a gaussian distribution represents well the data. Then, using the line luminosity - continuum luminosity relations described in the afore mentioned papers, and the scatter obtained for these relationships, representative line luminosities of the sample of RQ AGN were generated. Using these line luminosities and the assumed distribution of continuum luminosities, the equivalent widths for the simulated RQ sample were calculated. The number of simulated values in each case is equal to the number of data

points in the original RQ samples. **Figure 2** shows the simulation results for the Hβ, Mg II, and CIV lines.

After generating the data, it was separated in order to match the continuum luminosity range on our sample of FSRQ. Using the simulated data that fall in our observed range, the mean and standard deviation of the equivalent width for the three lines Hβ, C IV and Mg II, were calculated. It is worth mentioning that for C IV and Mg II, all the simulated data fall inside the desired continuum luminosity ranges.

Then, a linear least-squares algorithm in one dimension was applied, using the IDL task FITEXY[6], to perform a linear fitting to the L_c - W_λ relation to obtain the Baldwin Effect of the RQ sample.

From these simulations, the next conclusions were drawn:

- The slope and uncertainty on the simulated L_c - W_λ relation remains unchanged, regardless of the input continuum luminosity distribution used.
- The changes on the intercept and uncertainty on the simulated L_c - W_λ relation, with changes on the input continuum luminosity distribution are negligible.
- The mean and standard deviation of the simulated equivalent widths can change drastically, depending on the continuum luminosity distribution used as input.

5. STATISTICAL ANALYSIS

To understand the effect of a non-thermal emission in the BE for blazars, it is necessary to analyze the difference in the $L_{\text{line}} - L_c$ and $W_\lambda - L_c$ relations between the samples of FSRQ and RQ AGN.

5.1. Comparison of Line-Luminosity Relations

The relation between the line luminosity and the continuum luminosity of RQ AGN are derived for Hβ by Greene and Ho (2005), and for Mg II and C IV by Shen et al. (2011), using the weighted linear fitting of binned data for the total flux of

---

[6]http://user.astro.columbia.edu/~williams/mpfitexy/

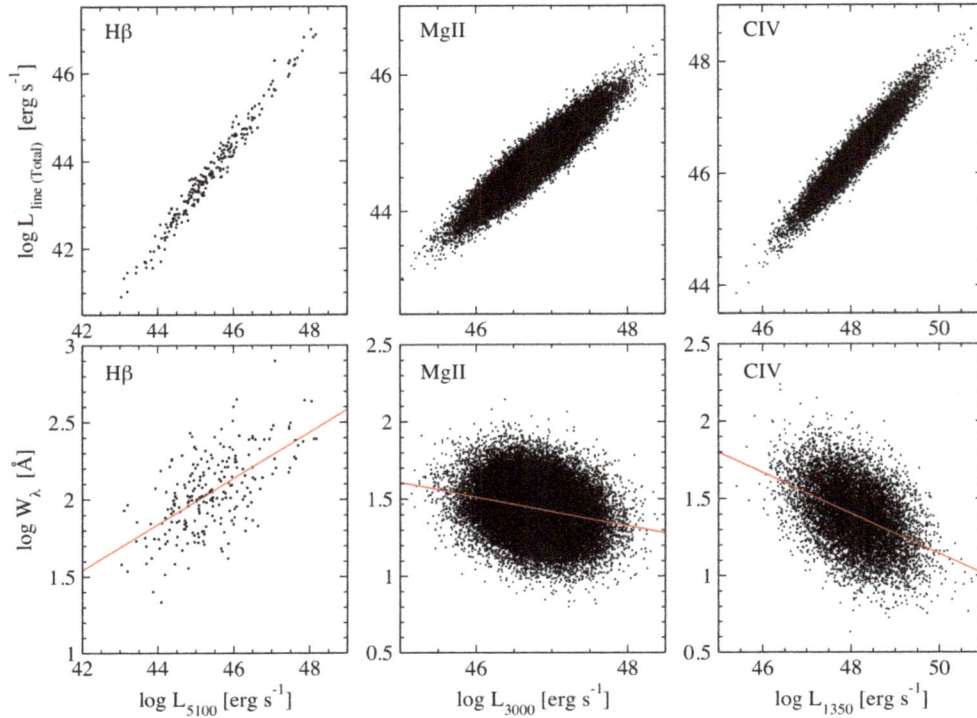

FIGURE 2 | Relations obtained for the Hβ (left panel), Mg II (middle panel), and CIV (right panel) emission lines. Top row: L_{line}-L_c plot with simulated data. **Bottom row:** EW-L_c plot with simulated data.

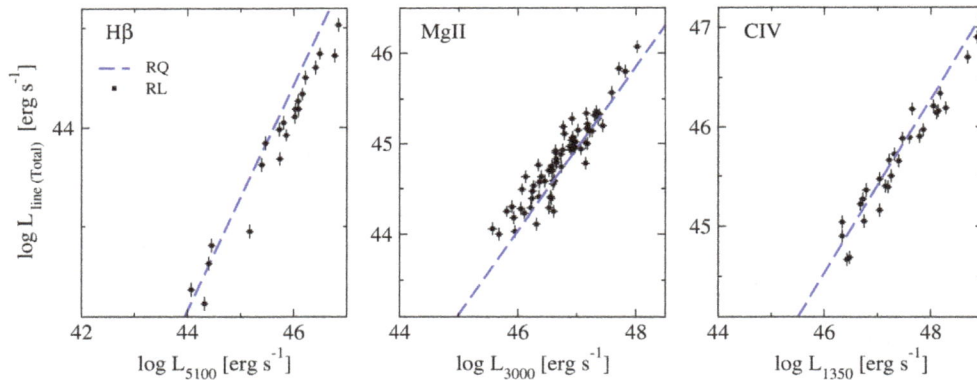

FIGURE 3 | Emission line luminosity against continuum luminosity of FSRQ (squares): $L_{H\beta}$ vs. L_{5100} **(left panel),** L_{MgII} vs. L_{3000} **(middle panel), and** L_{CIV} vs. L_{1350} **(right panel).** The dashed line reproduces the relations $L_{H\beta} - L_{5100}$ **(left panel)** from Greene and Ho (2005), $L_{MgII} - L_{3000}$ and $L_{CIV} - L_{1350}$ **(middle and right panels)** from Shen et al. (2011).

the emission lines Hβ, Mg II, and C IV, and their respective continuum luminosities at 5100 Å 3000 Å, and 1350 Å (dashed lines in **Figure 3**). The slope and intercept of their fittings are presented in the top part of **Table 1**.

The emission line and the corresponding continuum luminosity data of FSRQ are shown for Hβ, Mg II, and C IV emission lines in **Figure 3**. The same fitting procedure as in Greene and Ho (2005) and Shen et al. (2011) were followed, and

the relation defined in Section 2.2 between the line luminosity and the continuum luminosity for each subsample of FSRQ (straight lines in top panels of **Figure 4**) was derived. Fitting parameters of our subsamples and significance of correlations between line and continuum luminosities are presented in the lower part of **Table 1**.

Significant correlations for all line luminosities at the confidence level of $\geq$ 98.8% were found. It is noticeable that

TABLE 1 | Parameters of weighted linear fitting for line and continuum luminosities (log $L_{line} = A + B \log L_c$).

L_c (1)	L_{line} (2)	$A \pm \sigma_A$ (3)	$B \pm \sigma_B$ (4)	p (5)
RQ AGN				
L_{5100}	$L_{H\beta}$	-7.70 ± 0.22	1.133 ± 0.005	–
L_{3000}	$L_{Mg\,II}$	2.22 ± 0.09	0.909 ± 0.002	–
L_{1350}	$L_{C\,IV}$	4.42 ± 0.27	0.872 ± 0.006	–
FSRQ				
L_{5100}	$L_{H\beta}$	-1.32 ± 7.58	0.988 ± 0.166	0.002
L_{3000}	$L_{Mg\,II}$	7.64 ± 7.16	0.796 ± 0.153	0.012
L_{1350}	$L_{C\,IV}$	8.32 ± 7.04	0.788 ± 0.148	0.001

The first part of the Table presents the relationships between line and continuum luminosities found by Greene and Ho (2005) (Hβ) and Shen et al. (2011) (Mg II and C IV) for RQ samples. The second part shows our best weighted linear fit parameters for the L_{line} vs. L_c, and to the equivalent width of emission line luminosity and the L_c near each line in our blazar sample. Columns (1) and (2) are the continuum and emission line luminosity, respectively; Column (3) is the intercept and its error; Column (4) is the linear fit slope and its error; Column (5) is the statistical probability of the weighted linear fit, $p \leq 0.05$ means that the linear correlation is statistically significant at a c.l. $\geq 95\%$.

the slopes B of line-continuum luminosity relations measured for FSRQ have a tendency to be shallower for the three ions Hβ, Mg II, and C IV, than those found for RQ AGN, see **Figure 4** (top panels) and **Table 2**. In order to quantify the significance of the difference in the slopes of the L_c-L_{line} relations for RQ and FSRQ; an unpaired t-test[7] was applied. For the Hβ line, the two-tailed P-value is 0.0057, corresponding to a statistically significant difference. For the Mg II line, the two-tailed P-value is 0.0263, corresponding also to a statistically significant difference. However, for the C IV line, the two-tailed P-value is 0.4370, corresponding to a non-statistically significant difference. The authors suggest that the differences found for Mg II and Hβ are indicating the contribution of an extra emission line component, possibly related to the jet.

5.2. Baldwin Effect Comparison

The Baldwin Effect of FSRQ is derived by a weighted linear fitting to the binned data, taking into account uncertainties in both axes using the IDL task FITEXY. Bins of the data were set along the L_c and measure the mean and standard deviation of W_λ in each bin. It should be noted that an adaptive data bin was used, in order to get the same number of measurements in each bin. The weighted fit lines for each Hβ, Mg II, and C IV lines are presented in **Figure 4** (full lines in the lower panels) and their fitted parameters are listed at the bottom part of **Table 2**.

For each emission line, simulated values of W_λ and L_c for RQ AGN (see **Figure 2**) are used to generate the data set, which is then fitted by the weighted linear method described above (dashed lines in bottom panels of **Figure 4**). The slope, intercept, and uncertainties for each line, are presented in the top part of **Table 2**.

There is a difference in the slopes for the BE for the simulated RQ sample and the observed FSRQ sample; as can be seen in the bottom panels of **Figure 4**. The significance of these differences is

[7]http://graphpad.com/quickcalcs/ttest1/

tested by means of an unpaired t-test. For Hβ the test results in a statistically significant difference, with a P-value of 0.0007. For Mg II, the test also results in a statistically significant difference, with a P-value of 0.0016. However, for C IV, the test results in a non-statistically significant difference, with a P-value of 0.1161.

In order to explain the differences between the slopes in both the relations $L_{cont} - L_{line}$ and $L_{cont} - W_\lambda$ of the RQ (Greene and Ho, 2005; Shen et al., 2011) and the FSRQ (our sample), a simulation was designed showing the behavior of the Baldwin Effect under the accepted paradigm of RQ and RL AGN. The simulation results for three different scenarios are listed:

- First a RQ system was simulated, using as base the RQ relations for the three lines published by the above authors (solid line in **Figure 5**).
- Then a continuum component (simulating the jet) was added of the same luminosity as the disk component. The emission line is calculated using only the continuum component from the accretion disk (dashed line in **Figure 5**).
- Then both continuum components were decreased, each by a different a factor. This with the aim of probing if the differences in variation amplitudes of the disk and the jet, are responsible for the differences observed in the slopes. Many different variability factors were simulated, however, all of them just result in a Y-axis shift of the model. Again, the emission line is calculated using only the continuum component from the accretion disk (dot-dashed line in **Figure 5**).

The simulation results for the Hβ, Mg II, and C IV lines are shown in **Figure 5**.

As evidenced by the simulations presented above, the presence of an additional continuum component produced by the jet, does not explain the BE slope difference found between FSRQ and RQ AGN; the change produced is only a parallel Y-axis shift and not a change in slope.

6. NON-THERMAL CONTRIBUTION TO THE CONTINUUM EMISSION

6.1. Non-thermal Dominance for FSRQ

To quantify the contribution of the jet emission to the total optical/UV emission, the non-thermal dominance (*NTD*) introduced in Shaw et al. (2012), was estimated. They defined the *NTD* as

$$NTD = \frac{L_{obs}}{L_p}, \qquad (5)$$

where L_{obs} is the observed continuum luminosity and L_p is the predicted continuum luminosity estimated from the emission-line luminosity for a non-blazar sample. It is assumed that the broad-line emission reflects the thermal power of the accretion disk.

In this work, the authors define an alternative *NTD* for FSRQ:

$$NTD = \frac{L_{obs}}{L_p} = \frac{L_{disk} + L_{jet}}{L_p}, \qquad (6)$$

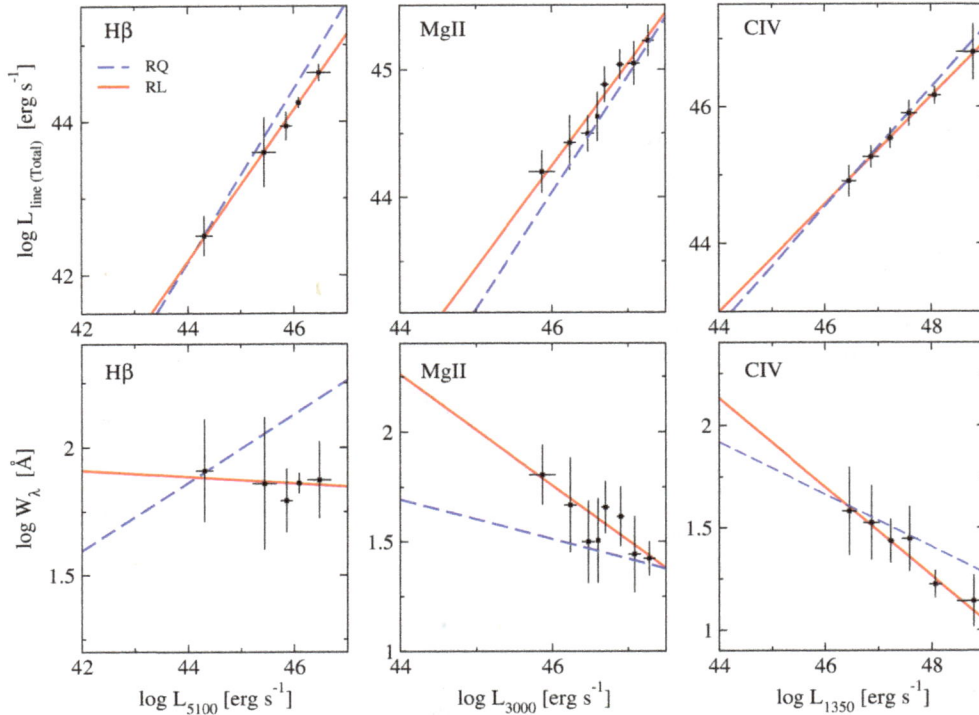

FIGURE 4 | The top panels show the continuum luminosity associated to each line (L_c) and the total line luminosity L_{line} for our binned data. The bottom panels show L_c and the equivalent widths (W_λ) estimated from the data in the former panels. The red solid lines represent the best weighted linear fits to our data. The blue dashed lines are derived from the simulations described in Section 4.2 using the relations $L_{H\beta} - L_{5100}$ from Greene and Ho (2005), $L_{MgII} - L_{3000}$ and $L_{CIV} - L_{1350}$ from Shen et al. (2011) to derive the corresponding line luminosities and equivalent widths.

TABLE 2 | Parameters of weighted linear fitting for the Baldwin Effect ($\log W_\lambda = \alpha + \beta \log L_c$).

L_c	W_λ (line)	$\alpha \pm \sigma_\alpha$	$\beta \pm \sigma_\beta$	p
(1)	(2)	(3)	(4)	(5)
RQ AGN				
L_{5100}	$W_\lambda(H\beta)$	-4.864 ± 0.708	0.149 ± 0.013	–
L_{3000}	$W_\lambda(Mg\ II)$	5.798 ± 0.099	-0.093 ± 0.002	–
L_{1350}	$W_\lambda(C\ IV)$	7.718 ± 0.154	-0.131 ± 0.003	–
FSRQ				
L_{5100}	$W_\lambda(H\beta)$	2.41 ± 2.78	-0.012 ± 0.061	0.013
L_{3000}	$W_\lambda(Mg\ II)$	13.38 ± 4.97	-0.253 ± 0.107	0.011
L_{1350}	$W_\lambda(C\ IV)$	11.66 ± 4.02	-0.216 ± 0.085	0.010

The first part of the Table presents the relationships between W_λ and continuum luminosities simulated from line-luminosity relations in Greene and Ho (2005) (Hβ) and Shen et al. (2011) (Mg II and C IV) for RQ AGN. The second part shows, for the FSRQ sample, our best weighted linear fit parameters for the relation W_λ vs. L_c, for the equivalent width of each emission line and the corresponding L_c. Columns (1) and (2) are the continuum and equivalent width, respectively; Column (3) is the intercept and its error; Column (4) is the linear fit slope and its error; Column (5) is the statistical probability of the weighted linear fit, $p \leq 0.05$ means that the linear correlation is statistically significant at a c.l. $\geq 95\%$.

where L_{obs} is the observed continuum luminosity, L_p is the predicted disk continuum luminosity estimated from the emission line, L_{disk} is the continuum luminosity emitted by the accretion disk, and L_{jet} is the jet contribution to the continuum luminosity. If the emission line is only affected by the disk $L_p = L_{disk}$, so that

$$NTD = 1 + \frac{L_{jet}}{L_{disk}}, \qquad (7)$$

which shows that $NTD \geq 1$. Note that $NTD = 1$ means that the continuum is due only to thermal emission, $NTD > 1$ shows that a superluminal jet exists that contributes to the continuum luminosity, and $NTD > 2$ means that $L_{jet} > L_{disk}$.

The emission line luminosity vs. the observed continuum luminosity for the three subsamples is presented in the top panels of **Figure 4**. It was found that the majority of blazars from the Hβ (81%) and C IV (84%) subsamples are located below the fiducial relations found by Greene and Ho (2005) and Shen et al. (2011) presented as dashed lines. Thus, these FSRQ must have a significant non-thermal contribution in the optical and UV bands, i.e., $NTD > 1$ for most sources. The excess emission is likely to be a boosted optical/UV emission from the relativistic jets, and it was expected that all quasars in our subsamples should have $NTD > 1$.

Figure 6 shows the distribution of NTD values obtained for the three emission lines of our study; it is noticeable that the peak is around $NTD = 0.5 - 1.0$. Further results concerning the NTD are discussed in the next section. Instead of the expectation,

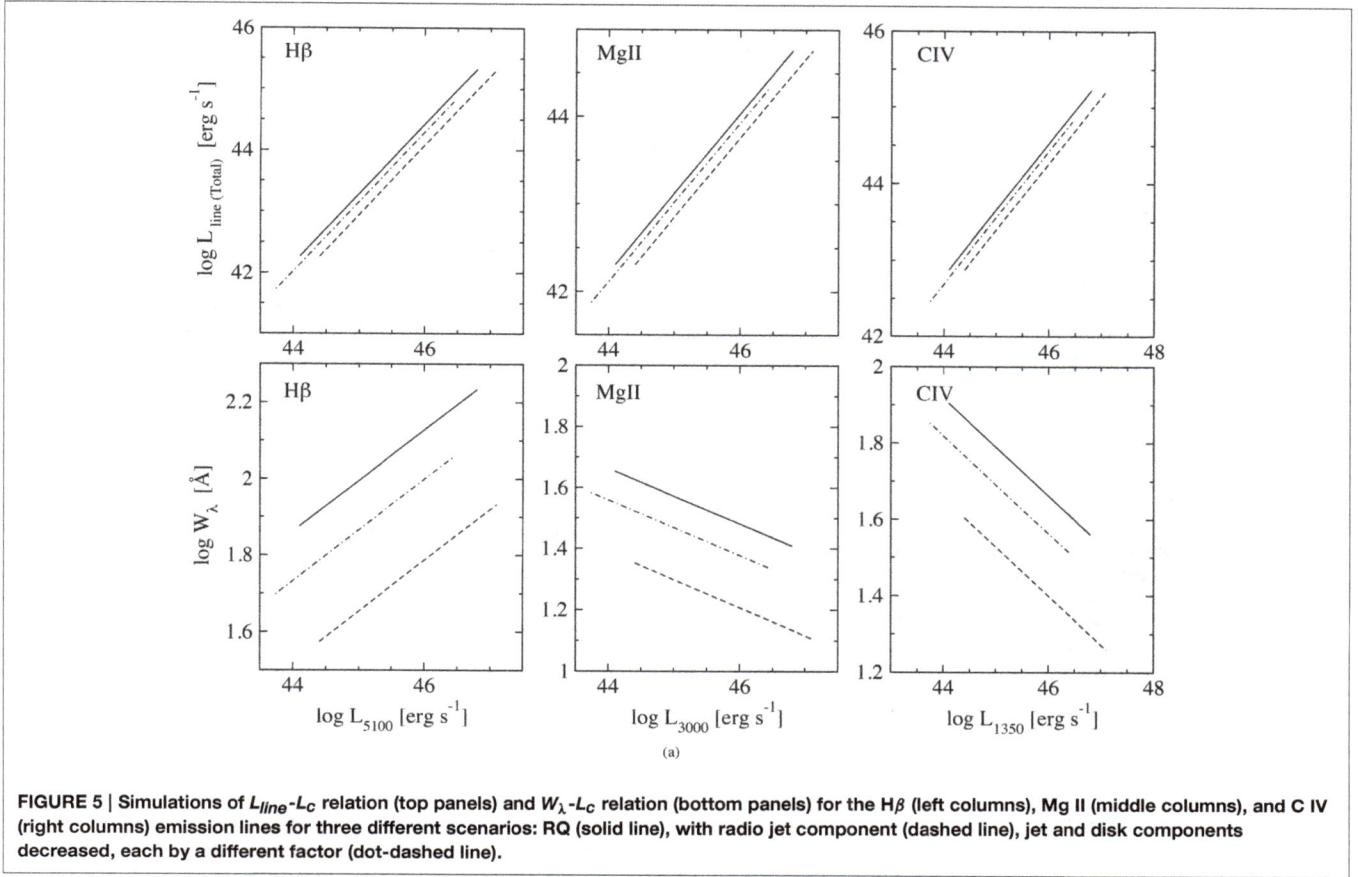

FIGURE 5 | Simulations of L_{line}-L_C relation (top panels) and W_λ-L_C relation (bottom panels) for the Hβ (left columns), Mg II (middle columns), and C IV (right columns) emission lines for three different scenarios: RQ (solid line), with radio jet component (dashed line), jet and disk components decreased, each by a different factor (dot-dashed line).

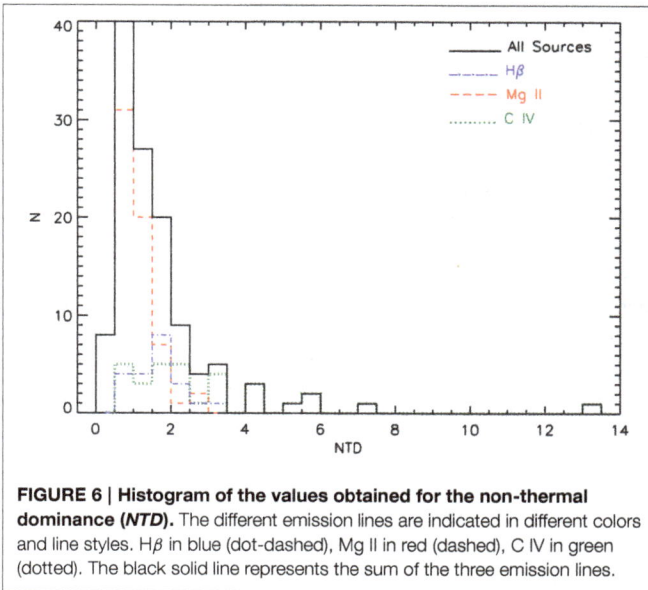

FIGURE 6 | Histogram of the values obtained for the non-thermal dominance (NTD). The different emission lines are indicated in different colors and line styles. Hβ in blue (dot-dashed), Mg II in red (dashed), C IV in green (dotted). The black solid line represents the sum of the three emission lines.

the NTD obtained for our sample of FSRQ shown in the vertical axis of **Figure 7** spans values in the regions $NTD > 2$, where $L_{jet} > L_{disk}$ in Equation (7). Few sources have $NTD > 2$. On the

other hand the region $NTD > 1$ is populated by a large number of sources, where the superluminal jet contributes to the continuum luminosity, and a number of sources have $NTD < 1$.

It was found that 56% of the Mg II subsample (filled dots in **Figure 7**) have $NTD < 1$, while only 19% of Hβ (empty squares in **Figure 7**) and 16% of C IV sources (filled triangles in **Figure 7**) have $NTD < 1$. Values of $NTD < 1$, were found as well by Shaw et al. (2012). This result probably means that, not only an additional component of the BLR (BLR2) exists together with the canonical BLR, which can be related to or activated by the jet [c.f. $L_{p(BLR2)}$ in Equation 9]; but also that it scales differently with the continuum luminosity than the canonical BLR component ($L_{p(BLR2)} > L_{jet}$). The reason why this affects more the Mg II subsample is unclear, but could be due to a larger sample than the other two ion subsamples, or a possible ion stratification in the BLR where Mg II zones are closer to the ionization source in the inner part of the jet. Observational evidence for the presence of BLR material located at parsec scales down to the radio core has been found by coordinated spectroscopic and VLBI monitoring studies. More specifically, Arshakian et al. (2010a) and León-Tavares et al. (2010) found evidence for BLR material around the radio core for the radiogalaxies 3C 390.3 and 3C 120, respectively.

In order to explore this possibility, it was assumed that there exists an emission line component related to the jet then $L_p >$

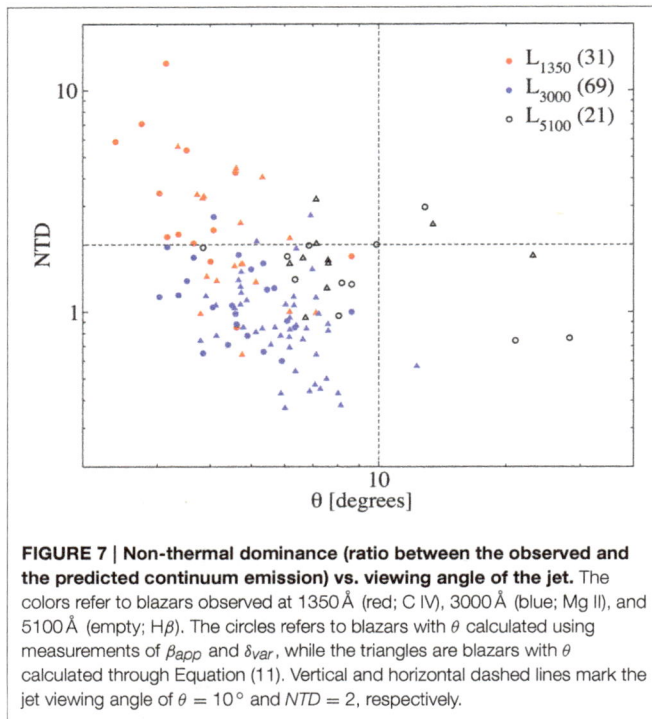

FIGURE 7 | Non-thermal dominance (ratio between the observed and the predicted continuum emission) vs. viewing angle of the jet. The colors refer to blazars observed at 1350 Å (red; C IV), 3000 Å (blue; Mg II), and 5100 Å (empty; Hβ). The circles refers to blazars with θ calculated using measurements of $β_{app}$ and $δ_{var}$, while the triangles are blazars with θ calculated through Equation (11). Vertical and horizontal dashed lines mark the jet viewing angle of θ = 10° and NTD = 2, respectively.

L_{disk} therefore:

$$NTD \neq 1 + \frac{L_{jet}}{L_{disk}}. \qquad (8)$$

If the predicted continuum luminosity obtained from the emission line component related to the disk is called $L_{p(BLR1)}$, and the predicted continuum luminosity obtained from the emission line component related to the jet is called $L_{p(BLR2)}$, then:

$$NTD = \frac{L_{disk} + L_{jet}}{L_p} = \frac{L_{disk} + L_{jet}}{L_{p(BLR1)} + L_{p(BLR2)}}, \qquad (9)$$

where $L_{p(BLR1)} = L_{disk}$. If $L_{p(BLR2)} > L_{jet}$ then it is possible to obtain values of $NTD < 1$.

The finding of values $NTD < 1$, most specially in Mg II (56% of FSRQ), seems to support the idea of a BLR component related to the jet. This analysis also suggests that the emission line component related to the jet scales differently with the continuum, than the canonical broad line region scaling.

Arshakian et al. (2010b) and Torrealba et al. (2011) showed that, for the MOJAVE blazars, optical (5100 Å) and radio VLBA total emission at 15 GHz (L_{VLBA}) are correlated on milliarcsecond scales. They suggest a synchrotron origin of radio and optical emission for quasars and BL Lacs which is boosted by the relativistic jet.

Application of partial Kendall's $τ_p$ statistical analysis[8] to the Mg II subsample shows that L_{3000} and L_{VLBA} are correlated at a $c.l. = 99.9\%$ ($τ_p = 0.27$). While, for the C IV subsample, the correlation between L_{1350} and L_{VLBA} is not significant ($τ_p =$

---

[8]Partial Kendall's $τ_p$ rank correlation removes the common dependence of luminosities on redshift.

0.21 and $c.l. = 91.4\%$), but the correlation recovers for the relation between L_{1350} and jet luminosity with $τ_p = 0.20$ and $c.l. = 97.3\%$. Note that L_{jet} is equal to the difference between total VLBA and radio core luminosities (see Arshakian et al., 2010b). These correlations indicate also that the bulk of the UV emission is non-thermal and, most likely, produced in the jet, which also supports the contribution of a jet-BLR component to the continuum luminosity.

Other evidence for a non-thermal origin of the variable optical emission comes from the link between the jet kinematics on sub-parsec scales and optical continuum flares on scales from few months to few years (Pérez et al., 1989; Arshakian et al., 2010a; León-Tavares et al., 2010, 2013). These findings suggested that the source of the non-thermal variable optical emission is located in the innermost part of the sub-parsec scale jet, which is a region that may be close to the BLR clouds and thus possibly affecting and activating it.

6.2. Dependence of Non-thermal Contribution on Jet Viewing Angle

The viewing angle of the jet ($θ_{var}$) was estimated using the variability Doppler factor ($δ_{var}$) and apparent speed of the jet $β_a$ (in units of the speed of the light; e.g., Lähteenmäki and Valtaoja, 1999):

$$θ_{var} = \arctan \frac{2β_a}{β_a^2 + δ_{var}^2 - 1}. \qquad (10)$$

Recent values of $β_a$ are taken from the MOJAVE website[9] and $δ_{var}$ from Hovatta et al. (2009). The latter parameter is available for 35 sources from the sample of 96 AGN. For the remaining objects, the empirical relation between the jet viewing angle ($θ_j$) and the total radio luminosity at 15 GHz was used (L_{VLBA}) obtained for 62 blazars from the statistically complete MOJAVE-1 sample, c.f. Equation (11) in Arshakian et al. (2010b):

$$\log(θ_j) = (7.92 \pm 0.78) + (0.26 \pm 0.03) \log L_{VLBA} \qquad (11)$$

Note that the range of L_{VLBA} of our sample ($42.1 \leq \log L_{VLBA} \leq 46.0$) is similar to the one in Arshakian et al. (2010b).

Viewing angles $θ_{var}$ of 71 blazars from the MOJAVE/2cm were estimated by Hovatta et al. (2009) using the variability Doppler factors and apparent speeds of the jets. Note that errors of viewing angles cannot be estimated because of difficulties and significant uncertainties in Doppler factor values ($δ_{var}$) associated with each source (Arshakian et al., 2010b).

Seventy one values of $θ_j$ from Equation (11) were computed and compared with independent measurements of $θ_{var}$ in Equation (10). The Spearman rank correlation between the two samples is $ρ = 0.56$ with $c.l. > 99.99\%$ ($P = 3 \times 10^{-7}$), indicating that the measurements of $θ_j$ are statistically reliable for viewing angles larger than ∼1 degree.

The viewing angles $θ_j$ of 96 FSRQ were used to analyze the $NTD - θ$ relation plane presented in **Figure 7**. There is a negative trend between NTD and viewing angle of the jet for the majority of blazars with viewing angles less than 10°. For these sources,

---

[9]http://www.physics.purdue.edu/astro/MOJAVE/index.html

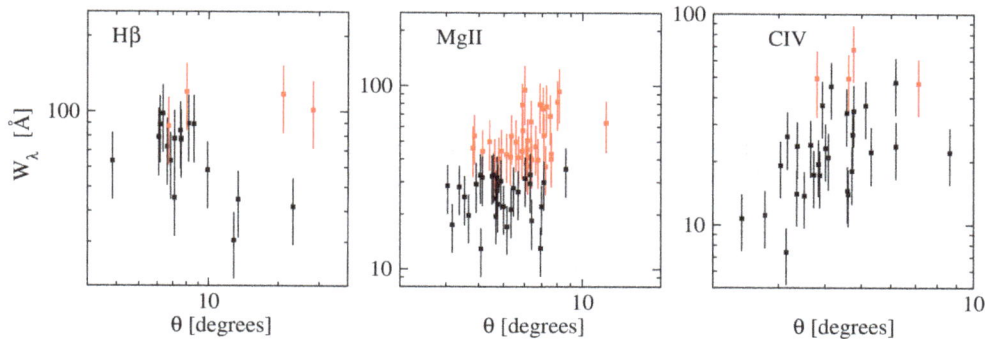

FIGURE 8 | Equivalent width vs. viewing angle of the jet. Red dots indicate sources with *NTD* < 1. The Mg II emission line shows a significant correlation at confidence level ≥ 96%. For the other lines there is no statistically significant correlation.

the Kendall's partial correlation is $\tau_p = -0.05$ with probability $P = 7.6 \times 10^{-5}$ (*c.l.* of 99.99%) indicating for a significant negative correlation between jet viewing angle and *NTD*. This correlation is mainly due to quasars of the Mg II subsample. No significant correlation is found for the Hβ and C IV subsamples, most likely, because of their smaller sampling.

In this work, the authors conclude that the Mg II subsample shows that the non-thermal dominance *NTD* of the optical and UV continuum emission decreases with viewing angles of the jet, in agreement with the prediction of the relativistic beaming theory.

6.3. Dependence of Equivalent Width on Jet Viewing Angle

Boosting of the continuum emission at smaller viewing angles should lead to a decrease of emission line equivalent widths as a result of the increase of the line-continuum contrast. W_λ and the jet viewing angle θ was compared in **Figure 8** for all three subsamples. A significant positive correlation was found between W_λ and θ for Mg II with $r = 0.25$ and *c.l.* = 96.6%, while the other lines do not show statistically significant results. It is worth noting that this correlation appears to be dominated by the points with *NTD* < 1.

In this work, the authors conclude that the equivalent width of Mg II is correlated with the jet viewing angle in the sense that increasing viewing angles produce larger values of W_λ, which is reasonable since larger viewing angles would mean less continuum boosting.

7. DISCUSSION AND CONCLUSIONS

We investigate the Baldwin effect of 96 FSRQ for which spectroscopic data are available from Torrealba et al. (2012, 2014). Our main results are the following:

- We report that a significant Baldwin Effect was found in FSRQ, shown by significant anticorrelations (at the confidence level of ≥ 95%) between equivalent widths of the Hβ, Mg II and C IV emission lines and corresponding continuum luminosities. The slopes of the BE in FSRQ seem to be steeper

than those in samples dominated by RQ quasars. Larger sampling of FSRQ is needed to confirm this result.

- The simulated toy model, shows that the difference we find in the slopes of the Baldwin Effect for RQ and FSRQ, cannot be explained by the addition of a non-thermal component to the continuum luminosity, nor by variability of the continuum components; which suggests that an extra emission line component is responsible for steepening the BE.

- We found that roughly 80% of FSRQ have a significant non-thermal contribution in optical/UV bands, i.e., *NTD* > 1, as was expected for AGN having relativistic jets. In particular, we reported that the bulk of UV emission is non-thermal and produced in the jet itself. The same evidence is corroborated for the optical continuum emission.

- We found values of *NTD* < 1 for several FSRQ, and argue that this result cannot be explained by a canonical BLR ionized only by the accretion disk, requiring the existence of an additional BLR component activated by the superluminal jet at parsec scales down to the radio-core. Probably due to their larger sampling, this is shown by sources (57%) with Mg II emission, and not in Hβ and C IV samples.

- In both optical and UV emission *NTD* increases at smaller viewing angles of the jet ($\theta \leq 10°$), which is in agreement with the prediction of the relativistic beaming theory.

- A positive correlation was found between the equivalent width W_λ and the viewing angle of the jet θ for Mg II, which is due to beaming of the continuum emission happening at small view angles that leads to a decrease in W_λ as a result of the continuum-line contrast.

It is well known that the SED of blazars are best described by beamed synchrotron emission from radio to X-ray frequencies, while the inverse Compton emission describes the SED of high frequency photons from the X-ray to TeV bands. The boosted synchrotron emission from the jet may dominate the low energy segment of the electromagnetic spectrum (radio to UV) in BL Lacs and FSRQ; while for radio galaxies (viewed at larger angles with respect to the jet) the emission from both the jet and accretion disk may significantly contribute to the total emission in optical, UV, and X-ray bands (cf. Blandford and Rees, 1978; Maraschi and Rovetti, 1994; Urry and Padovani, 1995). The thermal emission from the accretion disk may dominate during the lower state of the jet activity (when the jet power is at minimum), and vice versa, the non-thermal emission of the jet would be dominant during the high states of the jet activity (see Figure 4 in Arshakian et al., 2008).

This explains our findings presented earlier in Section 6 above, that about 10 to 25% of the MOJAVE blazars from the Hβ and C IV subsamples have a dominant thermal emission ($NTD < 1$). We would expect that roughly the same percentage of FSRQ from the Mg II subsample are thermally dominated, in disagreement with our finding that about 50% of sources have a thermal excess. For these quasars, the flux at 3000 Å coincides with the peak of the blue bump and the thermal luminosity at this frequency is higher than the luminosities at 1350 Å and 5100 Å by about 0.1 *dex*. This difference in luminosity is too small to reconcile the disagreement. Alternatively, the difference can be understood if there are two sources exciting the Mg II clouds, one is the thermal emission from the disk exciting the virialized Mg II clouds, and the other is the non-thermal jet emission which may excite both the virialized Mg II clouds and/or Mg II clouds outflowing along the jet (Pérez et al., 1989; Arshakian et al., 2010a; León-Tavares et al., 2010, 2013).

In this scenario, the observed Mg II emission is reflecting the contribution from both the jet and accretion disk, and, hence, the predicted emission line luminosity cannot only be attributed to the accretion disk in radio-loud blazars. This inevitably leads to values of $NTD < 1$ even for strongly non-thermal dominated sources. But the reason why only the Mg II line is affected by boosted jet emission remains unclear.

In the case of RQ AGN, the continuum emission radiation that ionizes the emitting gas regions comes from the thermal radiation of the accretion disk. While for FSRQ, the continuum emission also has a significant contribution from the non-thermal boosted radiation that arises from the relativistic jet (e.g., D'Arcangelo et al., 2007; Marscher et al., 2008). If the steepening of the Baldwin effect slope is inherent to FSRQ, this result could be directly related with the Doppler boosting of the continuum. If we could correct by Doppler boosting factor the continuum luminosity for all of our sources, we may find that the BE slope becomes flatter. But we need more data to do this analysis. Also a larger sample of FSRQ with optical and ultraviolet spectra is needed to confirm the BE slope steepening.

Our results listed above show that the non-thermal continuum of the jet contributes to the total continuum emission. Thus, the steep slopes of BE in blazars can be a signature of the contribution of the jet emission to the total continuum and line emission.

There is an observational evidence that optical flares and kinematics of the jet on sub-parsec scales are closely correlated: optical flare rises when superluminal component emerges into the jet. To explain the link between jet kinematics, optical continuum and emission line variability it was suggested the existence of the jet-excited BLR outflowing down the jet (Arshakian et al., 2010a; León-Tavares et al., 2010).

León-Tavares et al. (2013) reported a flare-like event of the Mg II emission line during a γ-ray outburst in 3C 454.3. They found that the highest levels of the emission line flux coincide with a superluminal jet component traversing through the radio core, which was confirmed in consequent studies by Isler et al. (2013). This is a direct observational evidence for a response of the broad emission lines to changes of the non-thermal continuum emission of the jet and, hence, the presence of the BLR material surrounding the radio core. The authors proposed an outflowing BLR which can arise from the accretion disk wind. This possibility is supported and was previously suggested by Pérez et al. (1989), and also for specific sources like 3C 273 (Paltani and Türler, 2003) and 3C 454. 3 (Finke and Dermer, 2010; León-Tavares et al., 2013).

From these previous findings combined with our results, we suggest the possibility that the jet emission greatly affects the gas that produces the emission lines and so, has an important contribution to the Baldwin Effect found in radio-loud compact AGN with superluminal jets. As a consequence, we can conclude that the relativistic plasma is tightly connected with the emitting line gas regions. The scenario we propose to explain the values of NTD< 1 and the difference in slopes of the Baldwin Effect in RQ AGN and FSRQ (steeper in RQ) consists on a second component of the Broad Line Region that is related to the jet, probably in the form of an outflow. Further work is needed aiming to quantify the contribution of the jet emission to the total continuum emission, and as well, to find out the real distribution of the gas emitting region in blazar type AGN.

AUTHOR CONTRIBUTIONS

All the authors have then contributed to its design, acquisition, interpretation of the data, and writing of the paper.

ACKNOWLEDGMENTS

This work is based on observations acquired at the Observatorio Astronómico Nacional in the Sierra San Pedro Mártir (OAN–SPM), Baja California, México, and at the Observatorio Astrofísico Guillermo Haro (OAGH), in Cananea, Sonora, México. This work is supported by CONACyT basic research grants 48484-F, 54480, and 151494 (Mexico). VA acknowledges support from the CONACyT program for Ph.D. studies. ICG acknowledges DGAPA (UNAM, Mexico) for a sabbatical scholarship and the Harvard-Smithsonian Center for Astrophysics for support as a visiting scholar. TA acknowledges support by DFG project number Os 177/2-1. LC is supported by the Ministry of Education and Science of R. Serbia through the project Astrophysical Spectroscopy of Extragalactic Objects (176001). The MOJAVE project is supported under National Science Foundation grant 0807860-AST and NASA-Fermi grant NNX08AV67G.

REFERENCES

Abazajian, K. N., Adelman-McCarthy, J. K., Agüeros, M. A., Allam, S. S., Anderson, K. S. J., Anderson, S. F., et al. (2005). The third data release of the sloan digital sky survey. *Astronom. J.* 129, 1755–1759. doi: 10.1086/427544

Abazajian, K. N., Adelman-McCarthy, J. K., Agüeros, M. A., Allam, S. S., Prieto, A. C., An, D., et al. (2009). The seventh data release of the sloan digital sky survey. *Astrophys. J. Suppl.* 182, 543–558. doi: 10.1088/0067-0049/182/2/543

Arshakian, T. G., León-Tavares, J., Lobanov, A. P., Chavushyan, V. H., Popovic, L., Shapovalova, A. I., et al. (2008). Jet-BLR connection in the radio galaxy 3C 390.3. *Mem. Soc. Astronom. Italiana* 79, 1022.

Arshakian, T. G., León-Tavares, J., Lobanov, A. P., Chavushyan, V. H., Shapovalova, A. I., Burenkov, A. N., et al. (2010a). Observational evidence for the link between the variable optical continuum and the subparsec-scale jet of the radio galaxy 3C 390.3. *Month. Notices R. Astronom. Soc.* 401, 1231–1239. doi: 10.1111/j.1365-2966.2009.15714.x

Arshakian, T. G., Torrealba, J., Chavushyan, V. H., Ros, E., Lister, M. L., Cruz-González, I., et al. (2010b). Radio-optical scrutiny of compact AGN: correlations between properties of pc-scale jets and optical nuclear emission. *Astronom. Astrophys.* 520, A62. doi: 10.1051/0004-6361/201014418

Bachev, R., Marziani, P., Sulentic, J. W., Zamanov, R., Calvani, M., and Dultzin-Hacyan, D. (2004). Average ultraviolet quasar spectra in the context of eigenvector 1: a baldwin effect governed by the eddington ratio? *Astrophys. J. Suppl.* 617, 171–183. doi: 10.1086/425210

Baldwin, J. A. (1977). Luminosity indicators in the spectra of quasi-stellar objects. *Astrophys. J.* 214, 679–684. doi: 10.1086/155294

Baskin, A., and Laor, A. (2004). On the origin of the C IV Baldwin effect in active galactic nuclei. *Month. Notices R. Astronom. Soc.* 350, L31–L35. doi: 10.1111/j.1365-2966.2004.07833.x

Blandford, R. D., and Rees, M. J. (1978). "Some comments on radiation mechanisms in Lacertids," in *BL Lac Objects*, Vol. 17, ed A. M. Wolfe (Copenhagen: Physica Scripta), 265–274.

Boroson, T. A., and Green, R. F. (1992). The emission-line properties of low-redshift quasi-stellar objects. *Astrophys. J.* 80, 109–135. doi: 10.1086/191661

Brotherton, M. S., Tran, H. D., Becker, R. H., Gregg, M. D., Laurent-Muehleisen, S. A., and White, R. L. (2001). Composite spectra from the FIRST bright quasar survey. *Astrophys. J.* 546, 775. doi: 10.1086/318309

Carswell, R. F., and Smith, M. G. (1978). Is the apparent redshift cut-off in the Tololo deep survey real ? *Month. Notices R. Astronom. Soc.* 185, 381–388. doi: 10.1093/mnras/185.2.381

Croom, S. M., Rhook, K., Corbett, E. A., Boyle, B. J., Netzer, H., Loaring, N. S., et al. (2002). The correlation of line strength with luminosity and redshift from composite quasi-stellar object spectra. *Month. Notices R. Astronom. Soc.* 337, 275–292. doi: 10.1046/j.1365-8711.2002.05910.x

D'Arcangelo, F. D., Marscher, A. P., Jorstad, S. G., Smith, P. S., Larionov, V. M., Hagen-Thorn, V. A., et al. (2007). Rapid multiwaveband polarization variability in the quasar PKS 0420-014: optical emission from the compact radio jet. *Astrophys. J.* 659, L107–L110. doi: 10.1086/517525

Dietrich, M., Hamann, F., Shields, J. C., Constantin, A., Vestergaard, M., Chaffee, F., et al. (2002). Continuum and emission-line strength relations for a large active galactic nuclei sample. *Astrophys. J.* 581, 912–924. doi: 10.1086/344410

Dong, X., Wang, T., Wang, J., Fan, X., Wang, H., Zhou, H., et al. (2009). Eddington ratio governs the equivalent width of Mg II emission line in active galactic nuclei. *Astrophys. J.* 703, L1–L5. doi: 10.1088/0004-637X/703/1/L1

Espey, B., and Andreadis, S. (1999). "Observational evidence for an ionization-dependent baldwin effect," in *Quasars and Cosmology*, Vol. 162 of *Astronomical Society of the Pacific Conference Series*, eds G. Ferland and J. Baldwin (San Francisco, CA: Astronomical Society of the Pacific), 351.

Finke, J. D., and Dermer, C. D. (2010). On the Break in the Fermi-Large Area Telescope Spectrum of 3C 454.3. *Astrophys. J.* 714, L303–L307. doi: 10.1088/2041-8205/714/2/L303

Francis, P. J., Hooper, E. J., and Impey, C. D. (1993). The ultraviolet spectra of radio-loud and radio-quiet quasars. *Astronom. J.* 106, 417–425. doi: 10.1086/116651

Greene, J. E., and Ho, L. C. (2005). Estimating Black Hole Masses in Active Galaxies Using the Hα Emission Line. *Astrophys. J.* 630, 122–129. doi: 10.1086/431897

Hovatta, T., Valtaoja, E., Tornikoski, M., and Lähteenmäki, A. (2009). Doppler factors, Lorentz factors and viewing angles for quasars, BL Lacertae objects and radio galaxies. *Astronom. Astrophys.* 494, 527–537. doi: 10.1051/0004-6361:200811150

Isler, J. C., Urry, C. M., Coppi, P., Bailyn, C., Chatterjee, R., Fossati, G., et al. (2013). A time-resolved study of the broad-line region in blazar 3C 454.3. *Astrophys. J.* 779, 100. doi: 10.1088/0004-637X/779/2/100

Kellermann, K. I., Lister, M. L., Homan, D. C., Vermeulen, R. C., Cohen, M. H., Ros, E., et al. (2004). Sub-milliarcsecond imaging of quasars and active galactic nuclei. III. Kinematics of parsec-scale radio jets. *Astrophys. J.* 609, 539. doi: 10.1086/421289

Kellermann, K. I., Sramek, R., Schmidt, M., Shaffer, D. B., and Green, R. (1989). VLA observations of objects in the Palomar Bright Quasar Survey. *Astronom. J.* 98, 1195. doi: 10.1086/115207

Kellermann, K. I., Vermeulen, R. C., Zensus, J. A., and Cohen, M. H. (1998). Sub-milliarcsecond imaging of quasars and active galactic nuclei. *Astronom. J.* 115, 1295. doi: 10.1086/300308

Korista, K., Baldwin, J., and Ferland, G. (1998). Quasars as cosmological probes: the ionizing continuum, gas metallicity, and the W_lambda-L relation. *Astrophys. J.* 507, 24–30. doi: 10.1086/306321

Kovalev, Y. A., Kovalev, Y. Y., and Nizhelsky, N. A. (2000). "Instantaneous 1-22 GHz spectra of 214 VSOP survey sources," in *Astrophysical Phenomena Revealed by Space VLBI*, Vol. 52, eds H. Hirabayashi, P. G. Edwards, and D. W. Murphy (Moscow: Publications of the Astronomical Society of Japan), L1027–L1036.

Kovalev, Y. Y., Kellermann, K. I., Lister, M. L., Homan, D. C., Vermeulen, R. C., Cohen, M. H., et al. (2005). Sub-milliarcsecond imaging of quasars and active galactic nuclei. IV. Fine-scale structure. *Astronom. J.* 130, 2473. doi: 10.1086/497430

Kovalev, Y. Y., Nizhelsky, N. A., Kovalev, Y. A., Berlin, A. B., Zhekanis, G. V., Mingaliev, M. G., et al. (1999). Survey of instantaneous 1-22 GHz spectra of 550 compact extragalactic objects with declinations from −30° to +43°. *Astronom. Astrophys. Suppl.* 139, 545. doi: 10.1051/aas:1999406

Kovačević, J., Popović, L. Č., and Dimitrijević, M. S. (2010). Analysis of optical Fe II emission in a sample of active galactic nucleus spectr a. *Astrophys. J. Suppl. Ser.* 189, 15–36. doi: 10.1088/0067-0049/189/1/15

Lähteenmäki, A., and Valtaoja, E. (1999). Total flux density variations in extragalactic radio sources. III. Doppler boosting factors, lorentz factors, and viewing angles for active galactic nuclei. *Astrophys. J.* 521, 493–501. doi: 10.1086/307587

León-Tavares, J., Chavushyan, V., Patiño-Álvarez, V., Valtaoja, E., Arshakian, T. G., Popović, L. Č., et al. (2013). Flare-like Variability of the Mg II λ2800 Emission Line in the Γ-Ray Blazar 3C 454.3. *Astrophys. J.* 763:L36. doi: 10.1088/2041-8205/763/2/L36

León-Tavares, J., Lobanov, A. P., Chavushyan, V. H., Arshakian, T. G., Doroshenko, V. T., Sergeev, S. G., et al. (2010). Relativistic plasma as the dominant source of the optical continuum emission in the broad-line radio galaxy 3C 120. *Astrophys. J.* 715, 355. doi: 10.1088/0004-637X/715/1/355

Lister, M. L., Cohen, M. H., Homan, D. C., Kadler, M., Kellermann, K. I., Kovalev, Y. Y., et al. (2009). MOJAVE: monitoring of jets in active galactic nuclei with VLBA experiments. VI. kinematics analysis of a complete sample of blazar jets. *Astronom. J.* 138, 1874. doi: 10.1088/0004-6256/138/6/1874

Maraschi, L., and Rovetti, F. (1994). A unified relativistic beaming model for BL Lacertae objects and flat spectrum radio quasars. *Astrophys. J.* 436, 79–88. doi: 10.1086/174882

Markwardt, C. B. (2009). *Astronomical Science of the Pacific Conference Series*. Québec City, QC: Astronomical Society of the Pacific.

Marscher, A. P., Jorstad, S. G., D'Arcangelo, F. D., Smith, P. S., Williams, G. G., Larionov, V. M., et al. (2008). The inner jet of an active galactic nucleus as revealed by a radio-to-γ-ray outburst. *Nature* 452, 966–969. doi: 10.1038/nature06895

Netzer, H., Mainieri, V., Rosati, P., and Trakhtenbrot, B. (2006). The correlation of narrow line emission and X-ray luminosity in active galactic nuclei. *Astronom. Astrophys.* 453, 525–533. doi: 10.1051/0004-6361:200654203

Netzer, H., Shemmer, O., Maiolino, R., Oliva, E., Croom, S., Corbett, E., et al. (2004). Near-infrared spectroscopy of high-redshift active galactic nuclei. II. Disappearing narrow-line regions and the role of accretion. *Astrophys. J.* 614, 558–567. doi: 10.1086/423608

Netzer, H., and Trakhtenbrot, B. (2007). Cosmic evolution of mass accretion rate and metallicity in active galactic nuclei. *Astrophys. J.* 654, 754–763. doi: 10.1086/509650

Paltani, S., and Türler, M. (2003). Dynamics of the Lyα and C IV Emitting Gas in 3C 273. *Astrophys. J.* 583, 659–669. doi: 10.1086/345421

Pérez, E., Penston, M. V., and Moles, M. (1989). Spectrophotometric monitoring of high luminosity active galactic nuclei - ii. first results. *Month. Notices R. Astronom. Soc.* 239, 75–90. doi: 10.1093/mnras/239.1.75

Popović, L. Č., and Kovačević, J. (2011). Optical emission-line properties of a sample of the broad-line active galactic nuclei: the baldwin effect and eigenvector 1. *Astrophys. J.* 738, 68. doi: 10.1088/0004-637X/738/1/68

Scott, J. E., Kriss, G. A., Brotherton, M., Green, R. F., Hutchings, J., Shull, J. M., et al. (2004). A composite extreme-ultraviolet QSO spectrum from FUSE. *Astrophys. J.* 615, 135–149. doi: 10.1086/422336

Shaw, M. S., Romani, R. W., Cotter, G., Healey, S. E., Michelson, P. F., Readhead, A. C. S., et al. (2012). Spectroscopy of Broad-line Blazars from 1LAC. *Astrophys. J.* 748, 49. doi: 10.1088/0004-637X/748/1/49

Shen, Y., Richards, G. T., Strauss, M. A., Hall, P. B., Schneider, D. P., Snedden, S., et al. (2011). A catalog of quasar properties from sloan digital sky survey data release 7. *Astrophys. J. Suppl.* 194, 45. doi: 10.1088/0067-0049/194/2/45

Shields, J. C. (2007). "Emission-line versus continuum correlations in active galactic nuclei," in *The Central Engine of Active Galactic Nuclei*, Vol. 373 of *Astronomical Society of the Pacific Conference Series*, eds L. C. Ho and J.-W. Wang (Astronomical Society of the Pacific), 355.

Torrealba, J., Arshakian, T. G., Chavushyan, V., and Cruz-González, I. (2011). "Correlations between radio emission of the parsec-scale jet and optical nuclear emission of host AGN," in *XIII Latin American Regional IAU Meeting, Vol. 40 of Revista Mexicana de Astronomía y Astrofísica, (Serie de Conferencias)*, eds W. J. Henney and S. Torres-Peimbert (Universidad Nacional Autónoma de México), 98–99.

Torrealba, J., Chavushyan, V., Cruz-González, I., Arshakian, T. G., Bertone, E., and Rosa-González, D. (2012). Optical Spectroscopic Atlas of the MOJAVE/2cm AGN Sample. *Rev. Mex. Astron. Astrofis.* 48, 9–40.

Torrealba, J., Chavushyan, V., Cruz-Gonzalez, I., Arshakian, T. G., Bertone, E., and Rosa-Gonzalez, D. (2014). VizieR online data catalog: mOJAVE/2cm AGN sample opt. spectros. atlas (Torrealba+, 2012). *VizieR Online Data Catalog* 80, 4802.

Tresse, L., Maddox, S., Loveday, J., and Singleton, C. (1999). Spectral analysis of the Stromlo-APM Survey - I. Spectral properties of galaxies. *Month. Notices R. Astronom. Soc.* 310, 262–280. doi: 10.1046/j.1365-8711.1999.02977.x

Urry, C. M., and Padovani, P. (1995). Unified Schemes for Radio-Loud Active Galactic Nuclei. *Publ. Astronom. Soc. Pacific* 107, 803. doi: 10.1086/133630

Vestergaard, M., and Wilkes, B. J. (2001). An empirical ultraviolet template for iron emission in quasars as derived from I Zwicky 1. *Astrophys. J. Suppl.* 134, 1–33. doi: 10.1086/320357

Warner, C., Hamann, F., and Dietrich, M. (2003). A Relation between Supermassive Black Hole Mass and Quasar Metallicity? *Astrophys. J.* 596, 72–84. doi: 10.1086/377710

Xu, Y., Bian, W.-H., Yuan, Q.-R., and Huang, K.-L. (2008). The origin and evolution of CIV Baldwin effect in QSOs from the Sloan Digital Sky Survey. *Month. Notices R. Astronom. Soc.* 389, 1703–1708. doi: 10.1111/j.1365-2966.2008.13545.x

York, D. G., Adelman, J., Anderson, J. E. Jr, Anderson, S. F., Annis, J., Bahcall, N. A., et al. (2000). The sloan digital sky survey: technical summary. *Astronom. J.* 120, 1579–1587. doi: 10.1086/301513

Zensus, J. A., Ros, E., Kellermann, K. I., Cohen, M. H., Vermeulen, R. C., and Kadler, M. (2002). Sub-milliarcsecond imaging of quasars and active galactic nuclei. II. Additional sources. *Astronom. J.* 124, 662. doi: 10.1086/341585

Zhang, K., Wang, T.-G., Gaskell, C. M., and Dong, X.-B. (2013). The baldwin effect in the narrow emission lines of active galactic nuclei. *Astrophys. J.* 762, 51. doi: 10.1088/0004-637X/762/1/51

Zheng, W., Kriss, G. A., and Davidsen, A. F. (1995). Relation between O VI Emission and X-Ray Intensity in Active Galactic Nuclei. *Astrophys. J.* 440, 606. doi: 10.1086/175299

Zheng, W., Kriss, G. A., Telfer, R. C., Grimes, J. P., and Davidsen, A. F. (1997). A Composite HST Spectrum of Quasars. *Astrophys. J.* 475, 469. doi: 10.1086/303560

Zheng, W., and Malkan, M. A. (1993). Does a luminosity-dependent continuum shape cause the baldwin effect? *Astrophys. J.* 415, 517. doi: 10.1086/173182

Conflict of Interest Statement: The authors declare that the research was conducted in the absence of any commercial or financial relationships that could be construed as a potential conflict of interest.

16

A Comparative Study of Divergence Cleaning Methods of Magnetic Field in the Solar Coronal Numerical Simulation

Man Zhang and Xueshang Feng *

SIGMA Weather Group, State Key Laboratory for Space Weather, National Space Science Center, Chinese Academy of Sciences, Beijing, China

This paper presents a comparative study of divergence cleaning methods of magnetic field in the solar coronal three-dimensional numerical simulation. For such purpose, the diffusive method, projection method, generalized Lagrange multiplier method and constrained-transport method are used. All these methods are combined with a finite-volume scheme in spherical coordinates. In order to see the performance between the four divergence cleaning methods, solar coronal numerical simulation for Carrington rotation 2056 has been studied. Numerical results show that the average relative divergence error is around $10^{-4.5}$ for the constrained-transport method, while about $10^{-3.1} - 10^{-3.6}$ for the other three methods. Although there exist some differences in the average relative divergence errors for the four employed methods, our tests show they can all produce basic structured solar wind.

Keywords: magnetic field divergence cleaning, three-dimensional MHD, solar corona, numerical simulation, solar wind

1. INTRODUCTION

Magnetohydrodynamics (MHD) equations are presently the only system available to self-consistently describe large-scale dynamics of space plasmas, and numerical MHD simulations has enabled us to capture the basic structures of the solar wind plasma flow and transient phenomena. The modern MHD codes can successfully solve both in time accurate and steady state problems involving all kinds of discontinuities. Different from the usual computational fluid mechanics, the MHD scheme has to be designed so as to guarantee the divergence free constraint of the magnetic field in two or three-dimensional MHD calculations. It is well-known that simply transferring conservation law methods for the Euler to the MHD equations can not be supposed to work at default in maintaining the divergence-free of magnetic field. The $\nabla \cdot \mathbf{B}$ error accumulated during the calculation may grow in an uncontrolled fashion, which can result in unphysical forces and numerical instability (Tóth, 2000; Jiang et al., 2012a).

Several methods have been proposed to satisfy the $\nabla \cdot \mathbf{B} = 0$ constraint in MHD calculations. The eight-wave formulation approach, suggested by Powell et al. (1993, 1999), is to solve the MHD equations with the additional source terms that are proportional to $\nabla \cdot \mathbf{B}$ without modifying the MHD solver. In this approach, divergence of the magnetic can be controlled to a truncation error and the robustness of a MHD code can be improved (Hayashi, 2005; Jiang et al., 2012a,b). The projection method was first proposed by Brackbill and Barnes (1980). In the projection method, the magnetic field $\mathbf{B}^*$ provided by the base scheme in the new time step $n + 1$ is projected onto the subspace of zero divergence solutions by a linear operator, and the magnetic field in the new time step $n + 1$ is completed by this projected magnetic field solution (Brackbill and Barnes, 1980; Tóth, 2000; Balsara and Kim, 2004; Hayashi, 2005; Feng et al., 2010). Some authors (e.g., Brandenburg et al., 2008; Manabu et al., 2009) modify the MHD equations with the help of vector potential $\mathbf{A}$ instead of the magnetic field $\mathbf{B} = \nabla \times \mathbf{A}$ to keep divergence cleaning

condition. In this case, $\nabla \cdot (\nabla \times \mathbf{A}) = 0$ is guaranteed mathematically, such that solving the time evolution of the vector potential $\mathbf{A}$ maintains the magnetic field divergence-free during the time evolution. The diffusive method is to add a source term $\eta \nabla (\nabla \cdot \mathbf{B})$ in the induction equation to reduce the numerical error of $\nabla \cdot \mathbf{B}$, so that the numerically generated divergence can be diffused away at the maximal rate limited by the CFL condition (van der Holst and Keppens, 2007; Feng et al., 2011; Shen et al., 2014). To guarantee the divergence

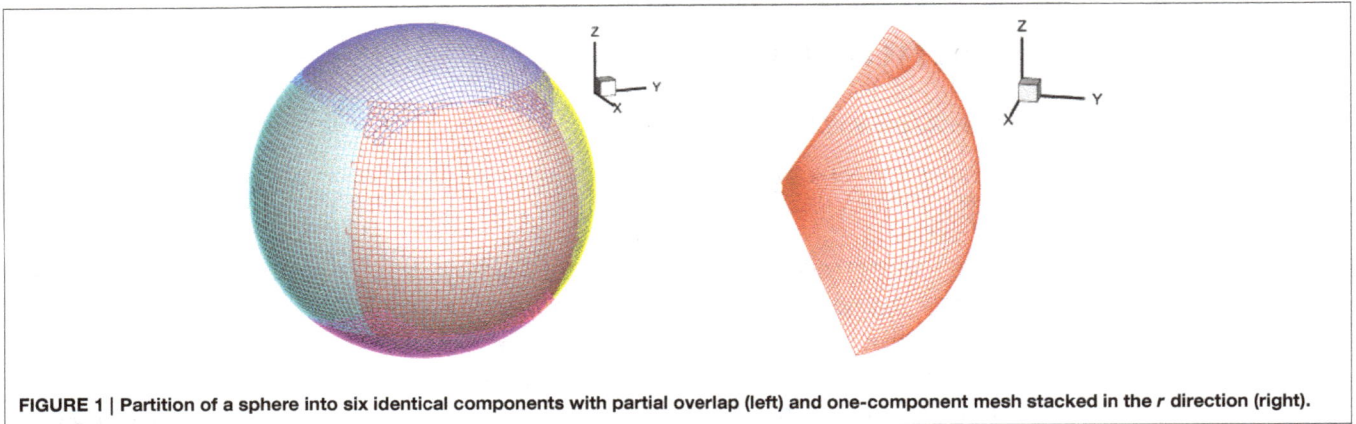

FIGURE 1 | Partition of a sphere into six identical components with partial overlap (left) and one-component mesh stacked in the r direction (right).

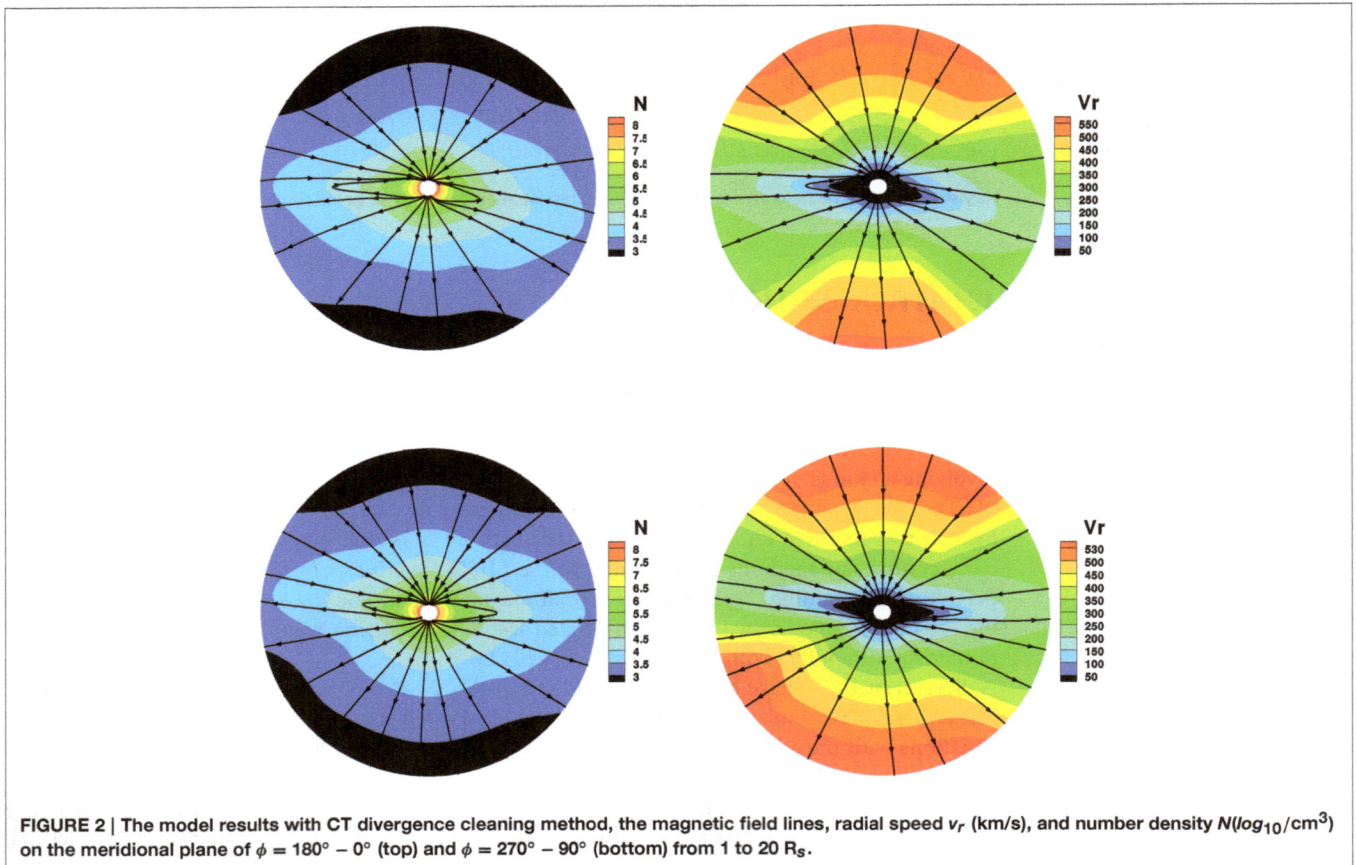

FIGURE 2 | The model results with CT divergence cleaning method, the magnetic field lines, radial speed v_r (km/s), and number density $N(log_{10}/cm^3)$ on the meridional plane of $\phi = 180° - 0°$ (top) and $\phi = 270° - 90°$ (bottom) from 1 to 20 R_s.

cleaning of the magnetic fields, Dedner et al. (2002) proposed the hyperbolic divergence cleaning approach by introducing a generalized Lagrange multiplier (GLM). In the GLM method, a newly transport variable ψ is introduced to the MHD system, which plays the role of convecting the local divergence error out of the computational domain (Dedner et al., 2002, 2003; Mignone and Tzeferacos, 2010; Mignone et al., 2010; Jiang et al., 2012a,b; Susanto et al., 2013). The constrained transport (CT) method is a different strategy to control $\nabla \cdot \mathbf{B}$ originally devised by Evans and Hawley (1988), in which the magnetic field is defined at face centers and the remaining fluid variables are provided at cell centers. In this approach, the electric field along the cell edges defining the boundary of the corresponding face is used to calculate the magnetic flux at cell faces. The CT method sustains a specified discretization of the magnetic field divergence around the machine round off error as long as the boundary and initial conditions are compatible with the constraints (Ziegler, 2011, 2012; Feng et al., 2014).

Since magnetic fields with a non-zero divergence can lead to severe artifacts in numerical simulations, keeping the magnetic field divergence-free is a curial problem in space plasma physics of solar and interplanetary phenomena. To say a few without exhausting, Linker et al. (1999) used the vector potential method to maintain the $\nabla \cdot \mathbf{B}$ constraint for global solar corona simulations. Hayashi (2005) simulate the solar corona and solar wind using the eight-wave method and the projection method to reduce the nonphysical effects of $\nabla \cdot \mathbf{B}$. Jiang et al. (2012a,b) simulated the coronal and chromospheric microflares by adopting the eight-wave method and the extended generalized Lagrange multiplier (EGLM) method to clean the divergence error. The GLM method was used in a nonlinear force-free field (NLFFF) study for the dynamics of solar active region (Inoue et al., 2015). The eight-wave method, the projection method, the CT method and the GLM method were implemented (Tóth, 2000; Tóth et al., 2005, 2012; Feng et al., 2010, 2011, 2014; Shen et al., 2014) for solar coronal and heliospheric studies, so as to maintain the solenoidal constraint.

In this paper, we give a comparative study of divergence cleaning methods of magnetic field in the solar coronal numerical simulation. The CT method, the diffusive method, the projection method and the GLM method are used to maintain divergence constraint respectively. The 3D solar wind model (Feng et al., 2014) is used for the experiments. The code employed a semi-discrete central scheme, designed by Ziegler (2011, 2012) within an finite volume (FV) framework without a Riemann solver or characteristic decomposition, and a composite grid system in spherical coordinates without polar singularities (Feng et al., 2010, 2011, 2014).

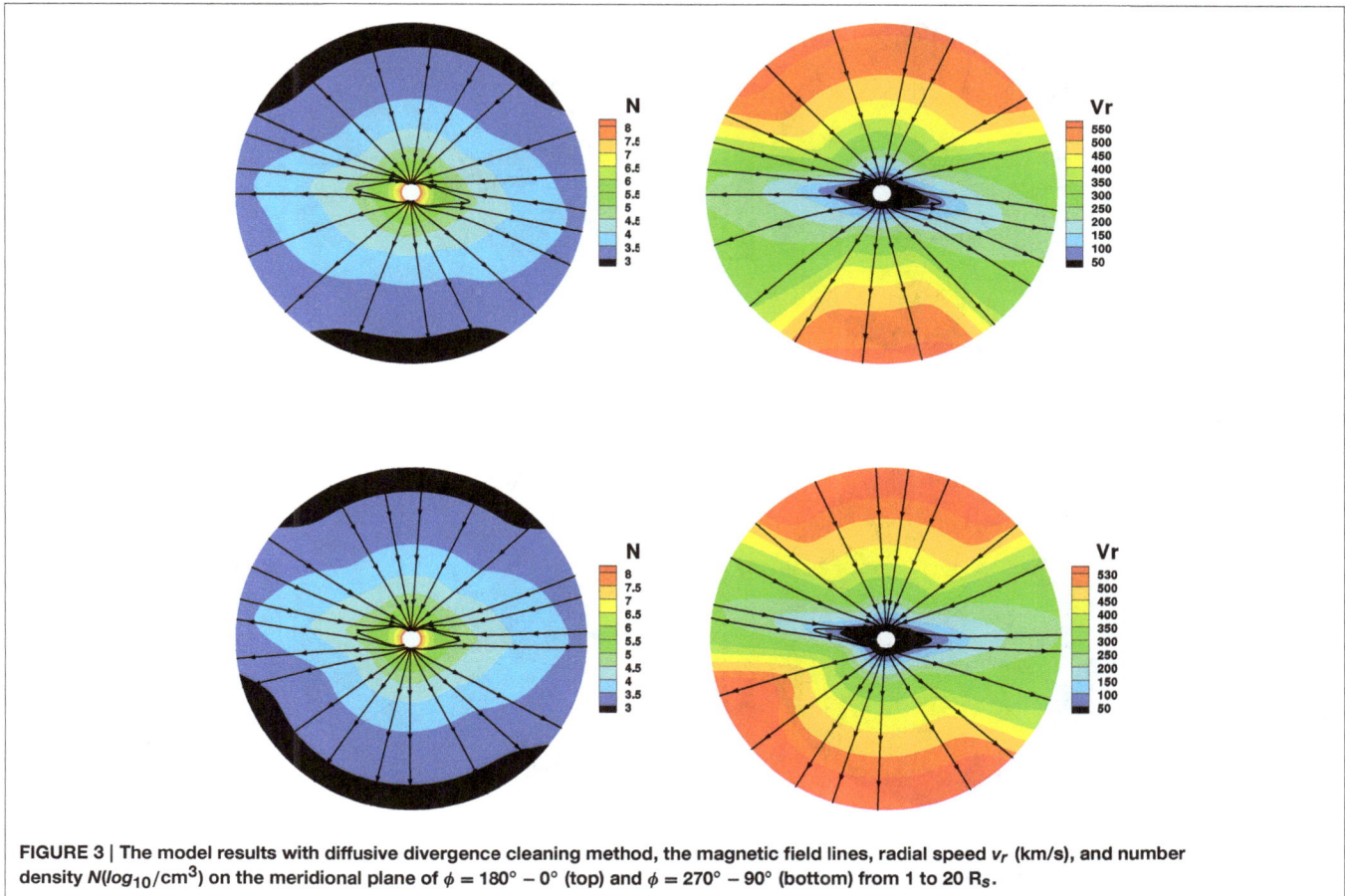

FIGURE 3 | The model results with diffusive divergence cleaning method, the magnetic field lines, radial speed v_r (km/s), and number density $N(log_{10}/cm^3)$ on the meridional plane of $\phi = 180° - 0°$ (top) and $\phi = 270° - 90°$ (bottom) from 1 to 20 R$_s$.

This paper proceeds as follows. In Section 2, model equations and grid system for solar wind plasma in spherical coordinates are described. Section 3 introduces the four methods to maintain the divergence cleaning constraint on the magnetic field. Section 4 gives the initial and boundary conditions in the code. Section 5 presents the numerical results for the steady-state solar wind structure of Carrington rotation (CR) 2056. Finally, we present some conclusions in Section 6.

2. GOVERNING EQUATIONS AND MESH GRID SYSTEM

2.1. Governing Equations

The magnetic field $\mathbf{B} = \mathbf{B}_1 + \mathbf{B}_0$ is splitted as a sum of a time-independent potential magnetic field $\mathbf{B}_0$ and a time-dependent deviation $\mathbf{B}_1$ (Feng et al., 2010, 2014). The MHD equations are splitted into the fluid and the magnetic parts. The governing equations have the same form as Feng et al. (2014). The fluid part of the vector $\mathbf{U} = \left(\rho, \rho v_r, \rho v_\theta, \rho v_\phi r \sin \theta, e \right)^T$ reads as follows:

$$\frac{\partial \mathbf{U}}{\partial t} + \frac{1}{r^2} \frac{\partial}{\partial r} r^2 \mathbf{F} + \frac{1}{r \sin \theta} \frac{\partial}{\partial \theta} \sin \theta \mathbf{G} + \frac{1}{r \sin \theta} \frac{\partial}{\partial \phi} \mathbf{H} = \mathbf{S} \quad (1)$$

The magnetic induction equation runs as follows:

$$\frac{\partial B_{1r}}{\partial t} + \frac{1}{r \sin \theta} \frac{\partial}{\partial \theta} (\sin \theta (v_\theta B_r - v_r B_\theta)) - \frac{1}{r \sin \theta} \frac{\partial}{\partial \phi}$$
$$(v_r B_\phi - v_\phi B_r) = 0 \quad (2)$$

$$\frac{\partial B_{1\theta}}{\partial t} - \frac{1}{r} \frac{\partial}{\partial r} (r(v_\theta B_r - v_r B_\theta)) + \frac{1}{r \sin \theta} \frac{\partial}{\partial \phi} (v_\phi B_\theta - v_\theta B_\phi) = 0 \quad (3)$$

$$\frac{\partial B_{1\phi}}{\partial t} + \frac{1}{r} \frac{\partial}{\partial r} (r(v_r B_\phi - v_\phi B_r)) - \frac{1}{r} \frac{\partial}{\partial \theta} (v_\phi B_\theta - v_\theta B_\phi) = 0 \quad (4)$$

Here, ρ is the mass density, $\mathbf{v} = (v_r, v_\theta, v_\phi)$ are the flow velocities in the frame rotating with the Sun, p is the thermal pressure. e stands for the modified total energy density consisting of the kinetic, thermal, and magnetic energy densities (written in terms of $\mathbf{B}_1$).

2.2. Mesh Grid System

Following Feng et al. (2010, 2012a,b,c), the computational domain is divided into a composite mesh consisting of six identical component meshes designed to envelop a spherical surface with partial overlap on their boundaries (**Figure 1**).

In the present work, the parallel implementation over the whole computational domain from 1 R_s to 26 R_s is realized

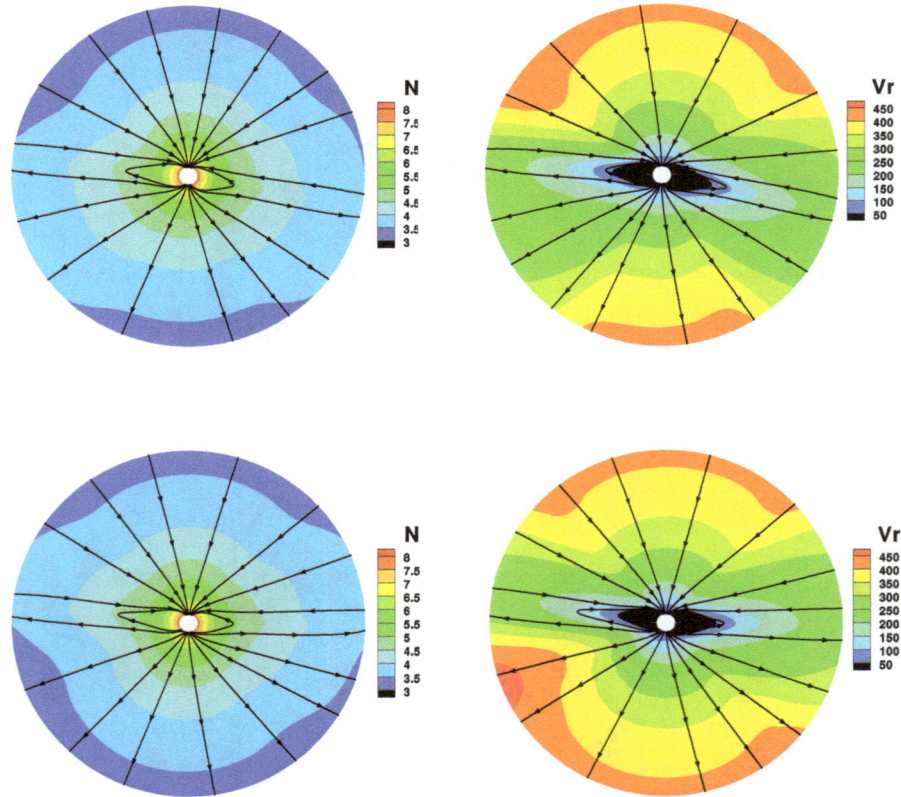

FIGURE 4 | The model results with projection divergence cleaning method, the magnetic field lines, radial speed v_r (km/s), and number density $N(log_{10}/cm^3)$ on the meridional plane of $\phi = 180° - 0°$ (top) and $\phi = 270° - 90°$ (bottom) from 1 to 20 R_s.

by domain decomposition of six-component grids based on the spherical surface and radial direction partition. The following grid partitions are employed: $N_\theta = N_\phi = 42. \Delta r(i) = 0.01$ R$_s$ if $r(i) < 1.1$ R$_s$; $\Delta r(i) = \min(A \times \log_{10}(r(i-1)), \Delta \theta \times r(i-1))$ with $A = 0.01/\log_{10}(1.09)$ if $r(i) < 3.5$ R$_s$; $\Delta r(i) = \Delta \theta \times r(i-1)$ if $r(i) > 3.5$ R$_s$.

3. NUMERICAL SCHEME FORMULATION

The following four subsections are devoted to the introduction of four methods to maintain the divergence cleaning constraint on the magnetic field.

3.1. CT Method

By the usage of a special discretization of the magnetic field Equations (2)–(4), CT technique imitates the analytical fact that $\frac{\partial \nabla \cdot \mathbf{B}}{\partial t} = \nabla \cdot \nabla \times (\mathbf{v} \times \mathbf{B}) = 0$. This discretization is routinely made on a particular stencil, therefore employs a staggered mesh, over which the solenoidal constraint up to the machine accuracy is satisfied on condition that initially $\nabla \cdot \mathbf{B} = 0$ is met in the whole computational domain. The hydrodynamic state variables are evaluated at the cell center, whereas magnetic field is evaluated at the cell faces and the electric field is at the cell edges. The origin of this technique is attributed to the staggered

divergence-free scheme formulated for electromagnetism by Yee (1966). For spatial discretization of our numerical scheme formulation, we strictly follow those of Feng et al. (2014) by using the FV discretization of Equation (1), and by averaging Equations (2)–(4) over facial areas to obtain the semi-integral forms of magnetic induction equations. Second-order accurate linear ansatz reconstruction are adopted.

3.2. Diffusive Method

The diffusive method in maintaining the divergence-free constraint runs as follows. As usual, regarding the coupling of fluids and magnetic fields as a whole system, then we have

$$\mathbf{U} = \left(\rho, \rho v_r, \rho v_\theta, \rho v_\phi r \sin \theta, e, B_{1r}, B_{1\theta}, B_{1\phi} \right)^T$$

$$\frac{\partial \mathbf{U}}{\partial t} + \frac{1}{r^2} \frac{\partial}{\partial r} r^2 \mathbf{F} + \frac{1}{r \sin \theta} \frac{\partial}{\partial \theta} \sin \theta \mathbf{G} + \frac{1}{r \sin \theta} \frac{\partial}{\partial \phi} \mathbf{H} = \mathbf{S} \quad (5)$$

with the symbols having their routine meanings, three variables added into Equation (1), and the first five variables keeping the same.

We use the diffusive method proposed to handle the $\nabla \cdot \mathbf{B}$ constraint. A source term $\eta \nabla (\nabla \cdot \mathbf{B})$ is introduced in the induction equation to reduce the numerical error of $\nabla \cdot \mathbf{B}$. The $\nabla \cdot \mathbf{B}$ error

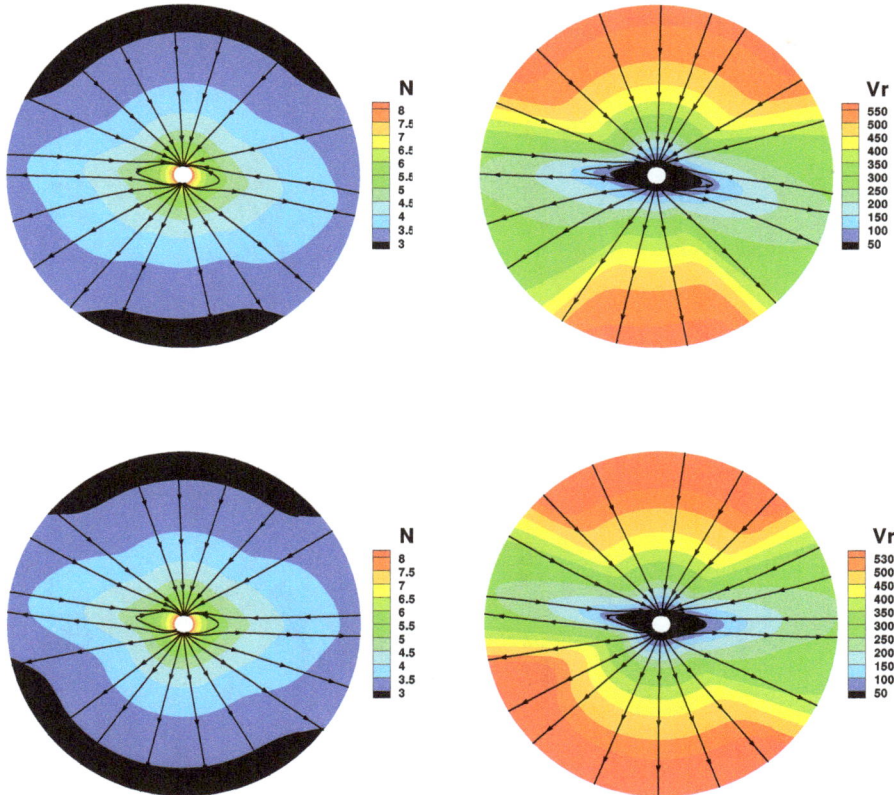

FIGURE 5 | The model results with GLM divergence cleaning method, the magnetic field lines, radial speed v_r (km/s), and number density $N(log_{10}/cm^3)$ on the meridional plane of $\phi = 180° - 0°$ (top) and $\phi = 270° - 90°$ (bottom) from 1 to 20 R$_s$.

produced by the diffusive method is controlled by iterating

$$\mathbf{B}^{n+1} = \mathbf{B}^{n+1} + \eta \Delta t \nabla (\nabla \cdot \mathbf{B}^{n+1})$$

$\eta \Delta t \leq C_d \left(\frac{1}{(\Delta r)^2} + \frac{1}{(r\Delta \theta)^2} + \frac{1}{(r \sin \theta \Delta \phi)^2} \right)^{-1}$, where Δr, $\Delta \theta$, $\Delta \phi$ are grid spacings in spherical coordinates. Here, we set $C_d = 1.3$ (van der Holst and Keppens, 2007; Rempel et al., 2009; Feng et al., 2011; Shen et al., 2014). This artificial diffusivity does not violate shock capturing property or second-order accuracy at least in smooth regions, but higher order accuracy may depend on the slope limiter used.

3.3. Projection Method

In the projection method formulation, the magnetic field $\mathbf{B}^*$ obtained by the base scheme using Equation (5) is projected onto the subspace of zero divergence solutions by a linear operator, and the magnetic field in the new time step $n+1$ is completed by this projected magnetic field solution $\mathbf{B}^{n+1}$. That is, the magnetic field can be decomposed by the sum of a curl and a gradient

$$\mathbf{B}^* = \nabla \times A + \nabla \phi$$

After taking the divergence of both sides one can achieve a Poisson equation

$$\nabla^2 \phi = \nabla \cdot \mathbf{B}^* \tag{6}$$

Then the magnetic field is corrected by

$$\mathbf{B}^{n+1} = \mathbf{B}^* - \nabla \phi \tag{7}$$

The numerical divergence of $\mathbf{B}^{n+1}$ can be exactly zero if the $\nabla^2 \phi$ in Equation (4) is evaluated as a divergence of the gradient with the same difference operators as used for calculating $\nabla \cdot \mathbf{B}^*$. In order to solve Equation (4), a pseudo-time derivative is introduced to the equation (Hayashi, 2005)

$$\frac{\partial \phi}{\partial \tau} = \nabla^2 \phi - \nabla \cdot \mathbf{B}^*$$

We adopt a first-order backward finite difference scheme for the pseudo-time derivative with the pseudo-time step $\Delta \tau$ ($< \Delta t$). If we want obtain an accurate transient solution, the pseudo-time (sub-iterations) must get converged at each physical time step. But this is too costly to make the sub-iteration procedure performed until convergence to machine precision. In this paper, besides setting up convergence criterion $\frac{\Delta \phi}{\Delta \tau} \leq 10^{-6}$ of the pseudo-time (sub-iterations), we also set up maximal sub-iterations 10 to avoid infinite iterations.

3.4. GLM Method

Using the GLM (Dedner et al., 2002), the divergence constraint is coupled with the conservation laws by introducing a newly

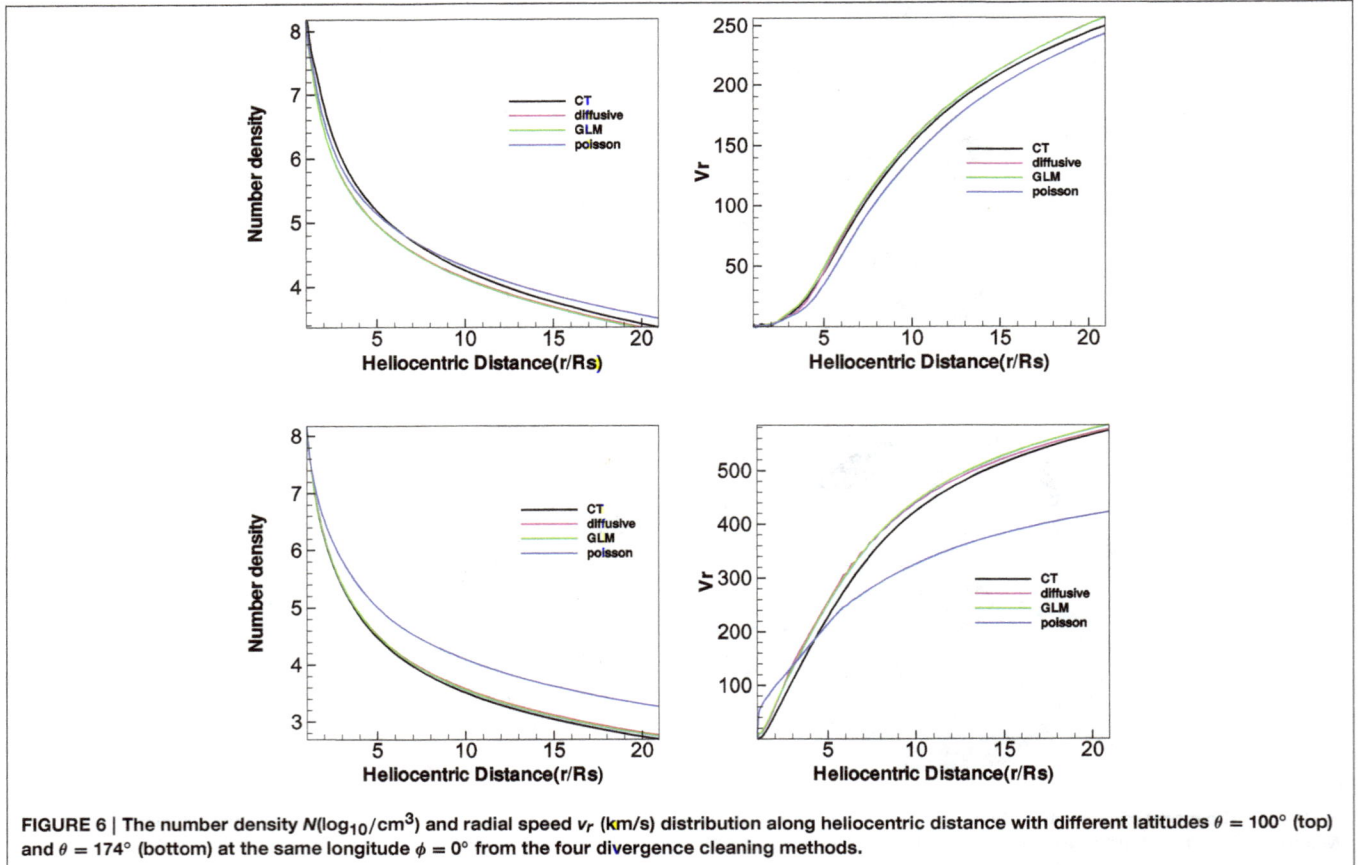

FIGURE 6 | The number density $N(\log_{10}/cm^3)$ and radial speed v_r (km/s) distribution along heliocentric distance with different latitudes $\theta = 100°$ (top) and $\theta = 174°$ (bottom) at the same longitude $\phi = 0°$ from the four divergence cleaning methods.

variable ψ. Now, the governing Equations (5) contain nine equations with the ninth equation written in the following form

$$\frac{\partial \psi}{\partial t} + c_h^2 \nabla \cdot \mathbf{B} = -\frac{c_h^2}{c_p^2} \psi$$

The fluxes for magnetic field have following forms $F_6 = c_h^2$, $G_7 = c_h^2$, $H_8 = c_h^2$, $S_6 = \frac{2\psi}{r}$, $S_7 = \frac{1}{r}(B_\theta v_r - B_r v_\theta) + \frac{\psi \cot \theta}{r}$. The c_h is often chosen to be the largest eigenvalue in the computational domain

$$c_h = \max_{i,j,k}\left(|v_r| + c_{fr}, |v_\theta| + c_{f\theta}, |v_\phi| + c_{f\phi}\right)$$

Here, c_{fr}, $c_{f\theta}$, and $c_{f\phi}$ are the fast magnetosonic speeds in the (r, θ, ϕ) directions, defined respectively by $c_{fr} = \frac{1}{\sqrt{2}}\sqrt{c_s^2 + c_A^2 + ((c_s^2 + c_A^2)^2 - 4c_s^2 \frac{B_r^2}{\mu\rho})^{\frac{1}{2}}}$, $c_{f\theta} = \frac{1}{\sqrt{2}}\sqrt{c_s^2 + c_A^2 + ((c_s^2 + c_A^2)^2 - 4c_s^2 \frac{B_\theta^2}{\mu\rho})^{\frac{1}{2}}}$, $c_{f\phi} = \frac{1}{\sqrt{2}}\sqrt{c_s^2 + c_A^2 + ((c_s^2 + c_A^2)^2 - 4c_s^2 \frac{B_\phi^2}{\mu\rho})^{\frac{1}{2}}}$, where $c_s = \sqrt{\frac{\gamma p}{\rho}}$ and $c_A = \sqrt{\frac{B_r^2 + B_\theta^2 + B_\phi^2}{\mu\rho}}$ are the sound and Alfvénic speeds. As for c_p, we follow Mignone and Tzeferacos (2010) and Mignone et al. (2010) by setting the parameter $\alpha = \Delta h c_h / c_p^2$,

$\Delta h = \min(\Delta r, r\Delta\theta, r\sin\theta\Delta\phi)$ and we choose $\alpha = 0.1$ in our code. Initially, ψ is set to 0. The ϕ at the inner and outer boundaries is fixed.

3.5. Time Integration

Time integration for the full system is implemented over time with a second-order Runge-Kutta scheme (Ziegler, 2004; Fuchs et al., 2009; Feng et al., 2014).

$$\overline{\mathbf{U}}^* = \overline{\mathbf{U}}^n + \Delta t \mathcal{R}_\mathbf{U}[\overline{\mathbf{U}}^n, \overline{\mathbf{B}}^n]$$
$$\overline{\mathbf{B}}^* = \overline{\mathbf{B}}^n + \Delta t \mathcal{R}_\mathbf{B}[\overline{\mathbf{U}}^n, \overline{\mathbf{B}}^n]$$
$$\overline{\mathbf{U}}^{n+1} = \frac{1}{2}\overline{\mathbf{U}}^n + \frac{1}{2}(\overline{\mathbf{U}}^* + \Delta t \mathcal{R}_\mathbf{U}[\overline{\mathbf{U}}^*, \overline{\mathbf{B}}^*]) \qquad (8)$$
$$\overline{\mathbf{B}}^{n+1} = \frac{1}{2}\overline{\mathbf{B}}^n + \frac{1}{2}(\overline{\mathbf{B}}^* + \Delta t \mathcal{R}_\mathbf{B}[\overline{\mathbf{U}}^*, \overline{\mathbf{B}}^*]) \qquad (9)$$

As usual, the time step length is limited by the Courant-Friedrichs-Lewy (CFL) stability condition:

$$\Delta t = \text{CFL}/\max\left(\sqrt{(\frac{|v_r| + c_{fr}}{\Delta r})^2 + (\frac{|v_\theta| + c_{f\theta}}{r\Delta\theta})^2 + (\frac{|v_\phi| + c_{f\phi}}{r\sin\theta\Delta\phi})^2}\right)$$

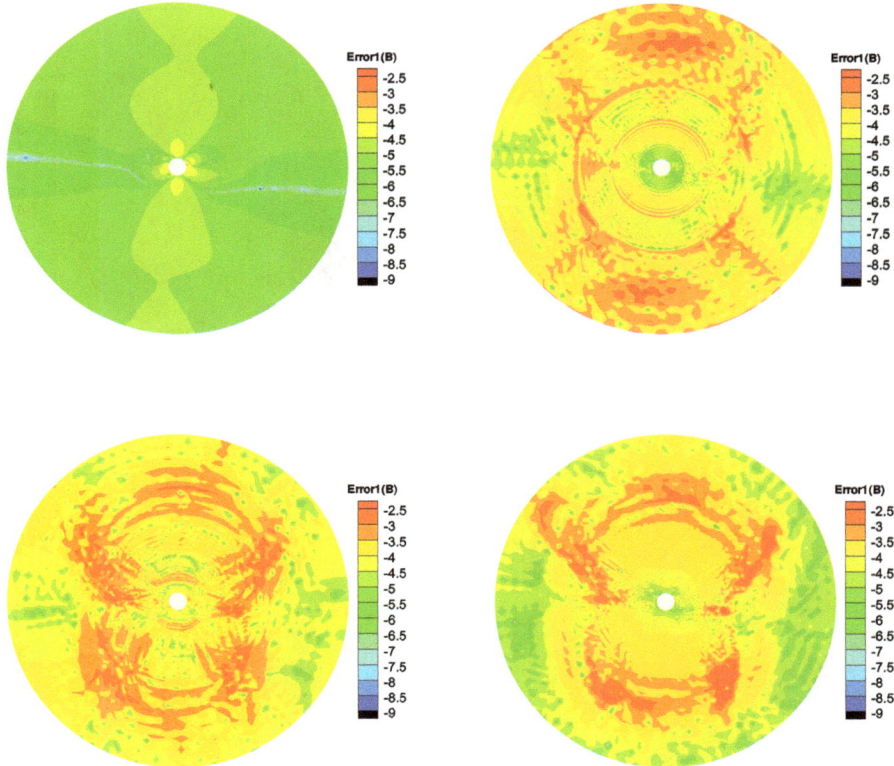

FIGURE 7 | The log_{10}Error1(B) in the calculation at $t = 5$ h on the meridional plane of $\phi = 180° - 0°$ from 1 to 20 R_s, the results from CT method (left) and diffusive method (right) are displayed in the top row, the bottom row is from the projection method (left) and GLM method (right).

Here, $\mathcal{R}_U[\overline{U}, \overline{B}]$ and $\mathcal{R}_B[\overline{U}, \overline{B}]$, denote the discretized fluxes moved to the right-hand sides of the governing Equations (1)–(4) and their corresponding source terms. In the following run we employ a simultaneous time integration with CFL = 0.5.

4. INITIAL-BOUNDARY VALUE CONDITIONS

Initially, the magnetic field is specified by using the potential field source surface (PFSS) model to produce a 3D global magnetic field in the computational domain with the line-of-sight photospheric magnetic data from the Wilcox Solar Observatory. **B** calculated by PFSS model inevitably can have a very small but non-zero $\nabla \cdot \mathbf{B}$ when evaluated in the discretized space. The initial profiles of flow parameters such as plasma density ρ, pressure p, and velocity **v** are given by Parker's solar wind flow solution (Parker, 1963).

In this paper, the inner boundary at 1 R_s is fixed for simplicity. The solar wind parameters at the outer boundary are imposed by linear extrapolation across the relevant boundary to the ghost node. The horizontal boundary values of each component grid in the (θ, ϕ) directions in the overlapping parts of the six-component system are determined by interpolation from the neighbor stencils lying in its neighboring component grid, which has been detailed (Feng et al., 2010, 2014).

5. NUMERICAL RESULTS

In this section, we present the numerical results from CR 2056 for the solar coronal numerical simulation with these four methods to maintain the divergence-free constraint.

To see the differences with the four divergence cleaning methods in solar corona simulation, **Figures 2–5** show the magnetic field lines, radial speed v_r, and number density N on two different meridional planes at $\phi = 180° - 0°$ and $\phi = 270° - 90°$ from 1 to 20 R_s, where the arrowheads on the black lines stand for the magnetic field directions. The four divergence cleaning methods can all produce structured solar wind. At high latitudes, the magnetic field lines extend into interplanetary space and the solar wind in this region has a faster speed and lower density. On the contrary, the slow solar wind and high density are located at lower latitudes around the Heliospheric current sheet (HCS). We can also see a helmet streamer stretched by the solar wind in this region. Above the streamer, a thin current sheet exists between different magnetic polarities.

Figure 6 presents the variation of number density N and radial speed v_r from 1 R_s to 20 R_s with the four divergence cleaning methods at different latitudes $\theta = 174°$ and $\theta = 100°$, where $\theta = 174°$ corresponds to the open field region while $\theta = 100°$ corresponds to the HCS region. Reasonably, the speed is larger in the open field region holding the fast solar wind, while the

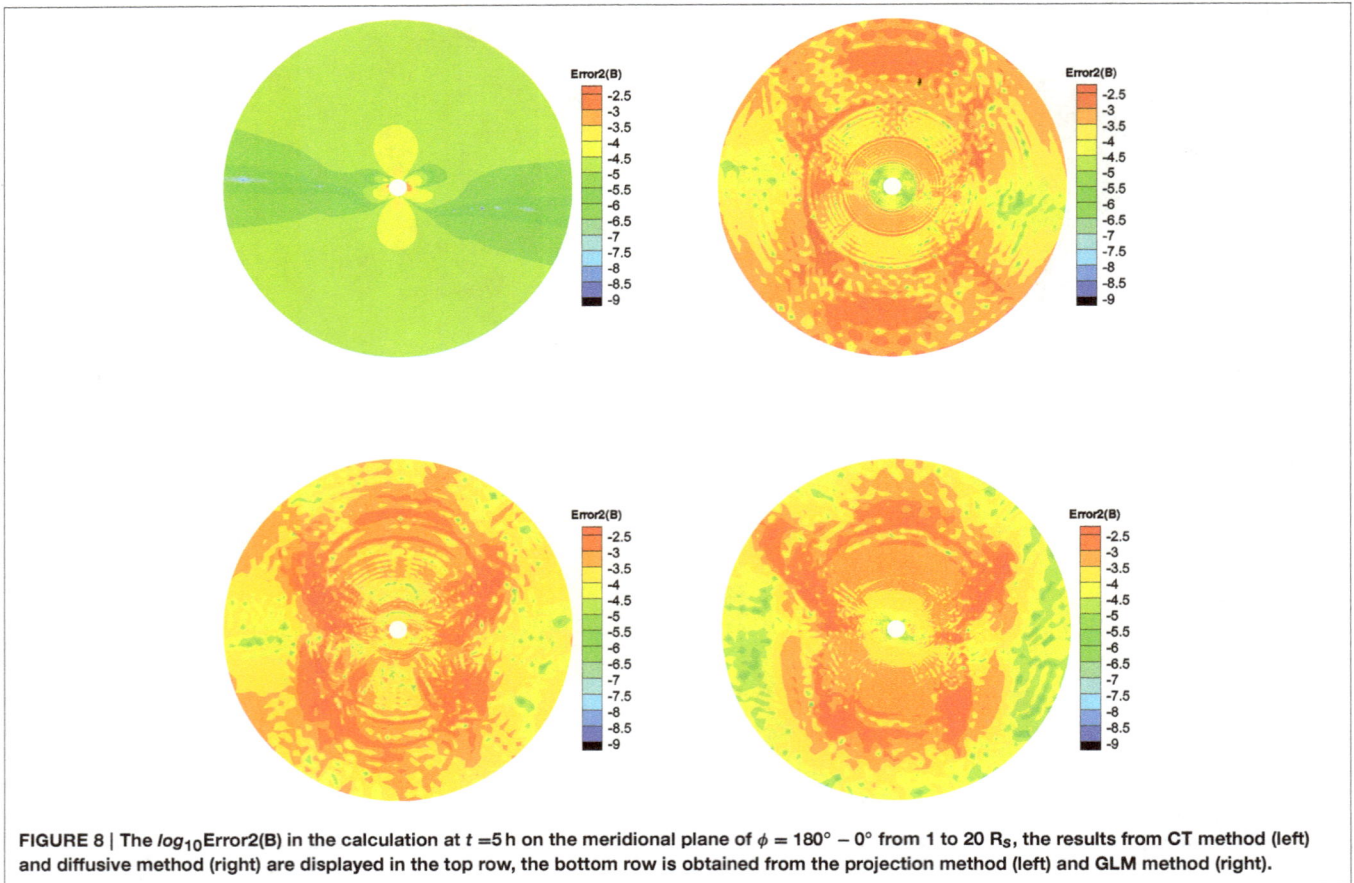

FIGURE 8 | The log_{10}Error2(B) in the calculation at t =5 h on the meridional plane of $\phi = 180° - 0°$ from 1 to 20 R_s, the results from CT method (left) and diffusive method (right) are displayed in the top row, the bottom row is obtained from the projection method (left) and GLM method (right).

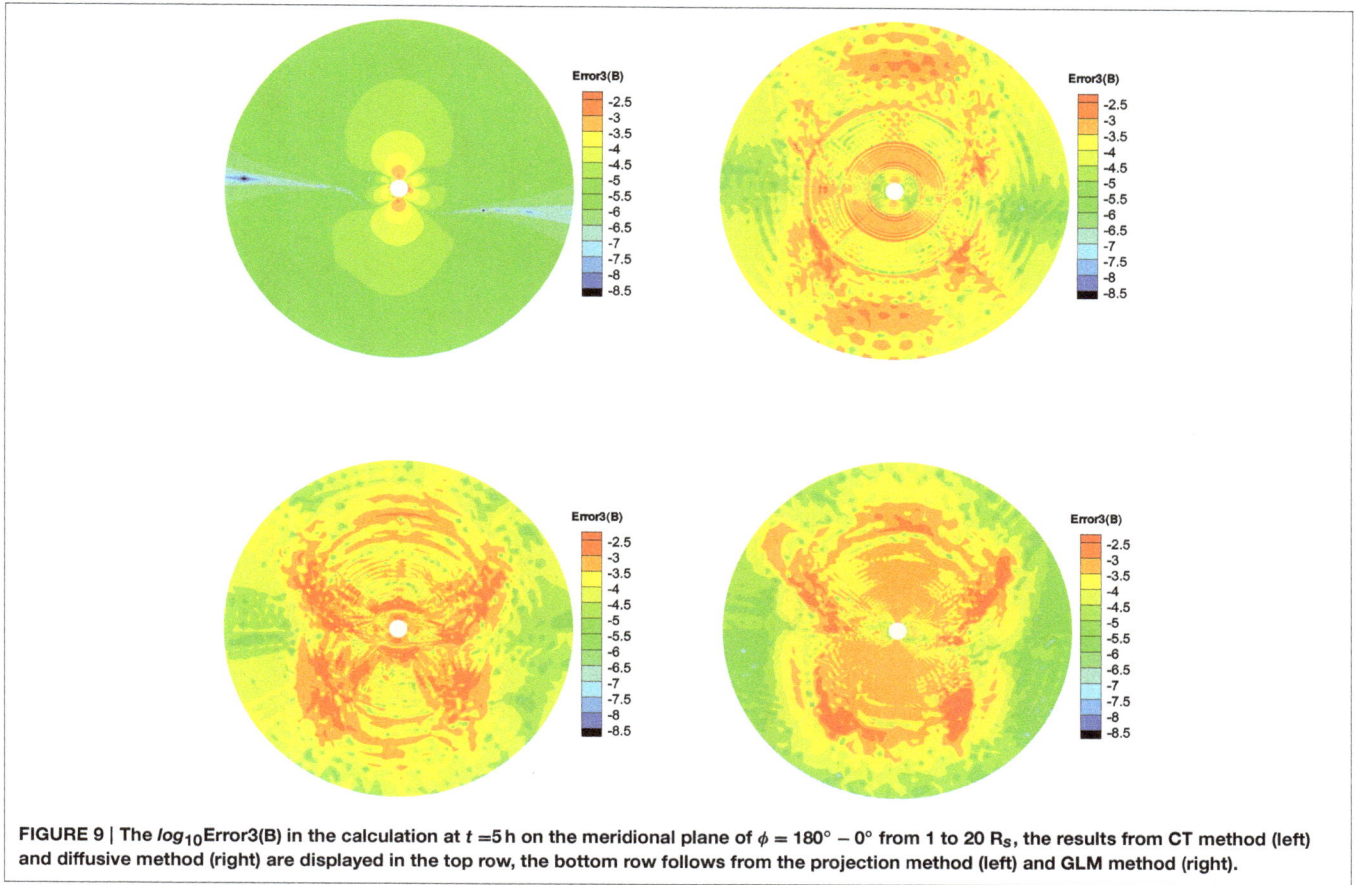

FIGURE 9 | The log_{10}Error3(B) in the calculation at $t = 5$ h on the meridional plane of $\phi = 180° - 0°$ from 1 to 20 R_s, the results from CT method (left) and diffusive method (right) are displayed in the top row, the bottom row follows from the projection method (left) and GLM method (right).

speed is smaller in the HCS region for the slow solar wind, and the number density changes contrarily to that of the speed. Overall, the four divergence cleaning methods can all produce large-scale solar wind.

To quantitatively see how $\nabla \cdot \mathbf{B}$ evolves, we define three relative divergence errors of the cell as Error1$(\mathbf{B}) = \frac{\left| \int_{V_k} \nabla \cdot \mathbf{B} dV \right|}{\int_{S_k} |\mathbf{B}| dS}$, Error2$(\mathbf{B}) = \frac{|\nabla \cdot \mathbf{B}| \cdot |R|}{|\mathbf{B}|}$, and Error3$(\mathbf{B}) = \frac{|\nabla \cdot \mathbf{B}| \cdot |R|}{\sqrt{2p}}$ (Powell et al., 1999; Pakmor and Springel, 2013; Mocz et al., 2014), where V_k is the kth sliding volume cell involved with the mesh grids, and S_k is the surface areas involved with V_k, and $R = \sqrt{\frac{3}{\frac{1}{(\Delta r)^2} + \frac{1}{(r\Delta\theta)^2} + \frac{1}{(r\sin\theta\Delta\phi)^2}}}$ is the characteristic size of the cell.

Figures 7–9 show the Error1$(\mathbf{B})$ and Error2$(\mathbf{B})$ and Error3$(\mathbf{B})$ of the four divergence cleaning methods at $t = 5$ h on the meridional plane of $\phi = 180° - 0°$. From these figures we can see that all the divergence cleaning methods can keep the $\nabla \cdot \mathbf{B}$ related errors under control, however, there are some differences in the relative magnitude of the resulting divergence errors. As is clearly visible in these figures, the divergence error is larger in the inner boundary for CT method compared to the other three methods, as to the outer region, that is on the contrary. That is because the local divergence error can be convected out of the domain using the diffusive method, the projection method or the GLM

method, and the CT method maintain the initial divergence error unchanged in computation.

Figures 10–12 show the Error1$(\mathbf{B})$, Error2$(\mathbf{B})$, and Error3$(\mathbf{B})$ of the four divergence cleaning methods at $t = 20$ h on the meridional plane of $\phi = 180° - 0°$. Compared to **Figures 7–9**, there have small difference of the four methods, which verify that these methods keeping the divergence error small and no obvious large error appears in computation. The divergence error for CT method stays almost the same after $t = 5$ h. As for GLM method, the divergence error is convected out of the domain. Overall, the spatial distribution of the errors for these four methods are very similar and the relative divergence errors are around $10^{-3} - 10^{-8}$.

Figure 13 gives the evolution of the average relative divergence errors as a function of time from the four methods in the calculation. The average relative divergence errors defined as Error1$(\mathbf{B})^{ave} = \sum_{k=1}^{M} \frac{\left| \int_{V_k} \nabla \cdot \mathbf{B} dV \right|}{\int_{S_k} |\mathbf{B}| dS} / M$, Error2$(\mathbf{B})^{ave} = \sum_{k=1}^{M} \frac{|\nabla \cdot \mathbf{B}| \cdot |R|}{|\mathbf{B}|} / M$, Error3$(\mathbf{B})^{ave} = \sum_{k=1}^{M} \frac{|\nabla \cdot \mathbf{B}| \cdot |R|}{\sqrt{2p}} / M$, where M is the total number of cells in the computational domain. From this figure we can see that Error1$(\mathbf{B})^{ave}$ for CT method is around $10^{-4.6}$, for diffusive method is around $10^{-3.7}$, for GLM method is around $10^{-3.6}$ and for projection method is around $10^{-3.2}$. The Error2$(\mathbf{B})^{ave}$ is larger than Error1$(\mathbf{B})^{ave}$ and Error3$(\mathbf{B})^{ave}$. The average relative divergence errors stay the

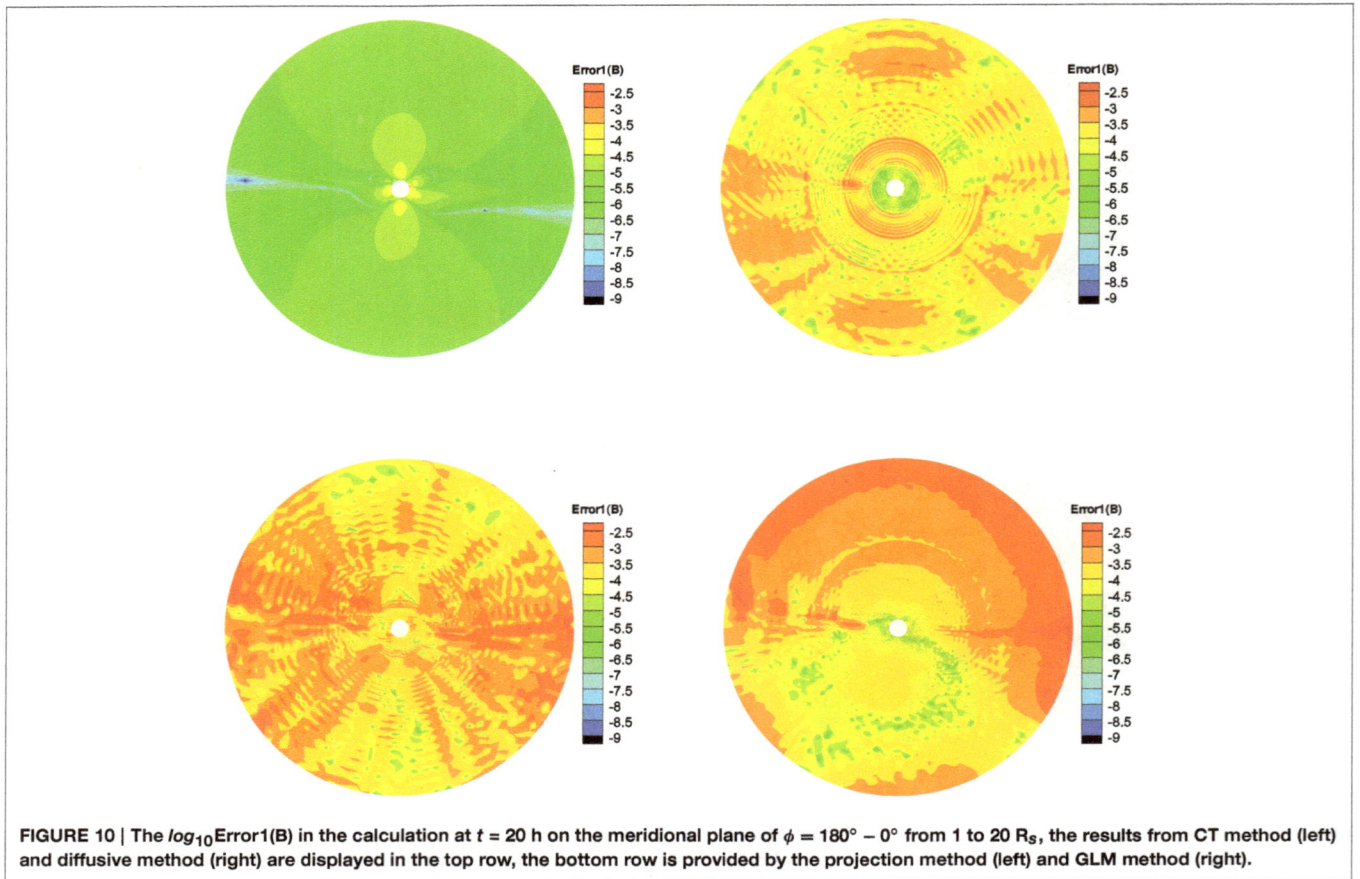

FIGURE 10 | The log_{10}Error1(B) in the calculation at t = 20 h on the meridional plane of ϕ = 180° − 0° from 1 to 20 R_s, the results from CT method (left) and diffusive method (right) are displayed in the top row, the bottom row is provided by the projection method (left) and GLM method (right).

same after 10 h and no obvious large error appears after a long run time. This verifies that the numerical error for the magnetic field divergence can continue to be acceptable during calculation. The CT method has the smallest average relative divergence errors compared to the other three methods and the errors stay the same as initial in calculation, the initial magnetic fields evaluated in the discretized space contributes significantly to the average relative divergence errors. The average relative divergence errors for diffusive method are smaller than GLM method or projection method. The relative divergence errors of diffusive method and projection method are affected by maximal sub-iterations, increasing maximal sub-iterations will make the relative divergence errors decrease but is time-consuming. **Figures 14, 15** shows the average relative divergence errors for diffusive method and projection method using maximal sub-iterations 30. The Error1(**B**)^{ave} for diffusive method is about $10^{-4.3}$, and for projection method is about $10^{-3.3}$, the errors become small compared to **Figure 13**. Since increasing the maximal sub-iterations will decrease the computation efficiency, and **Figure 13** also shows the results are acceptable without increasing the maximal sub-iterations. So we use sub-iteration 1 for diffusive method and maximal sub-iterations 10 for projection method in our code.

It is important to note that for all methods the average relative divergence errors are small, the spatial distribution of the errors

are very similar for them. Although the employed approach to limit divergence errors are significantly different and there have some differences in the average relative divergence errors as a function of time, there are excellent agreement for them in solar corona simulation, they can all produce structured solar wind.

6. CONCLUSIONS AND DISCUSSIONS

In this study, we employ four methods to maintain divergence cleaning constraint of magnetic field and compared the differences between them in solar corona simulation. All these algorithms are combined with a finite-volume scheme based on a six-component grid system in spherical coordinates (Ziegler, 2011, 2012; Feng et al., 2014), numerical results show that they can all produce large-scale solar wind though the relative divergence errors are different for them.

The CT method maintain the $\nabla \cdot \mathbf{B} = 0$ constraint by utilizing a special discretization on a staggered grid. This method evolves area-averaged magnetic field components at the cell faces rather than volume-averaged quantities as fluid part, and electric field components on cell edges are needed. The CT method is attractive from a physical point of view, however, requires the magnetic field variables to be treated differently from the fluid variables, which may be inconvenient

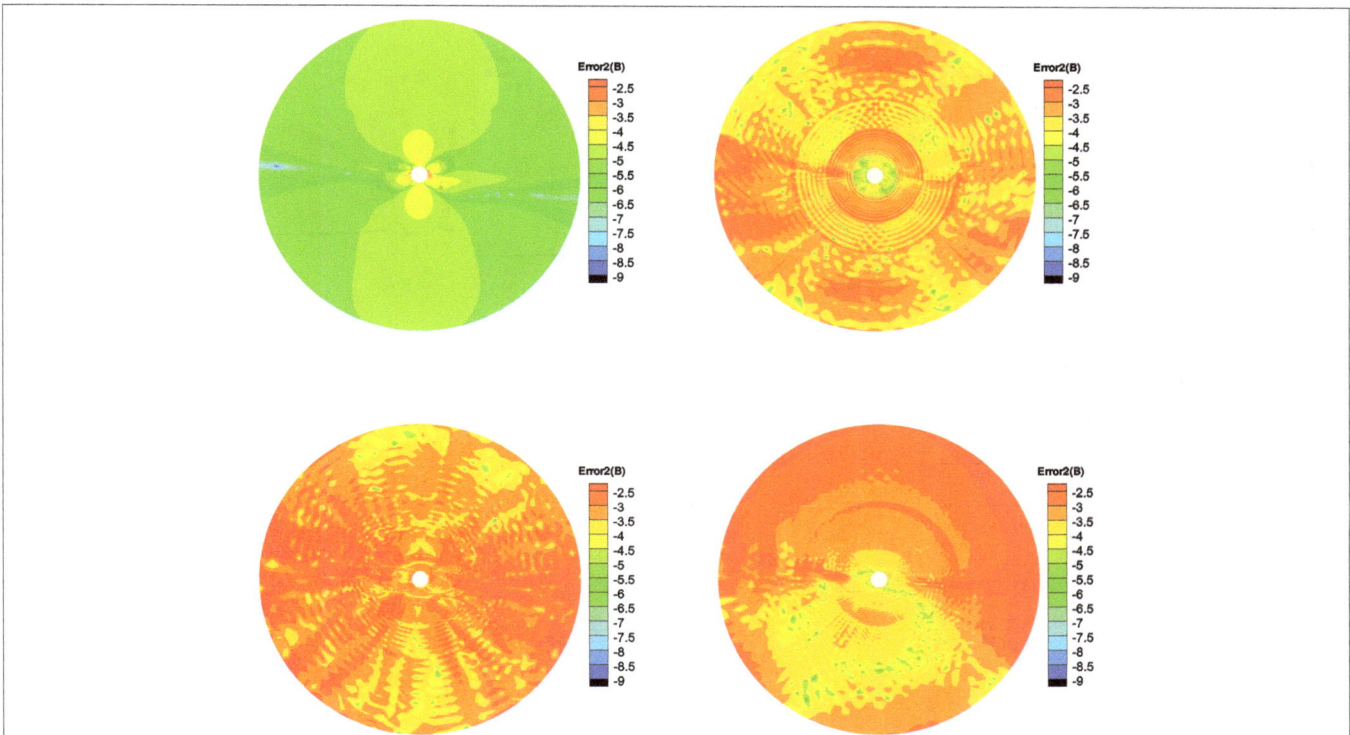

FIGURE 11 | The log_{10}Error2(B) in the calculation at $t = 20$ h on the meridional plane of $\phi = 180° - 0°$ from 1 to 20 R$_s$, the results from CT method (left) and diffusive method (right) are displayed in the top row, the bottom row is produced from the projection method (left) and GLM method (right).

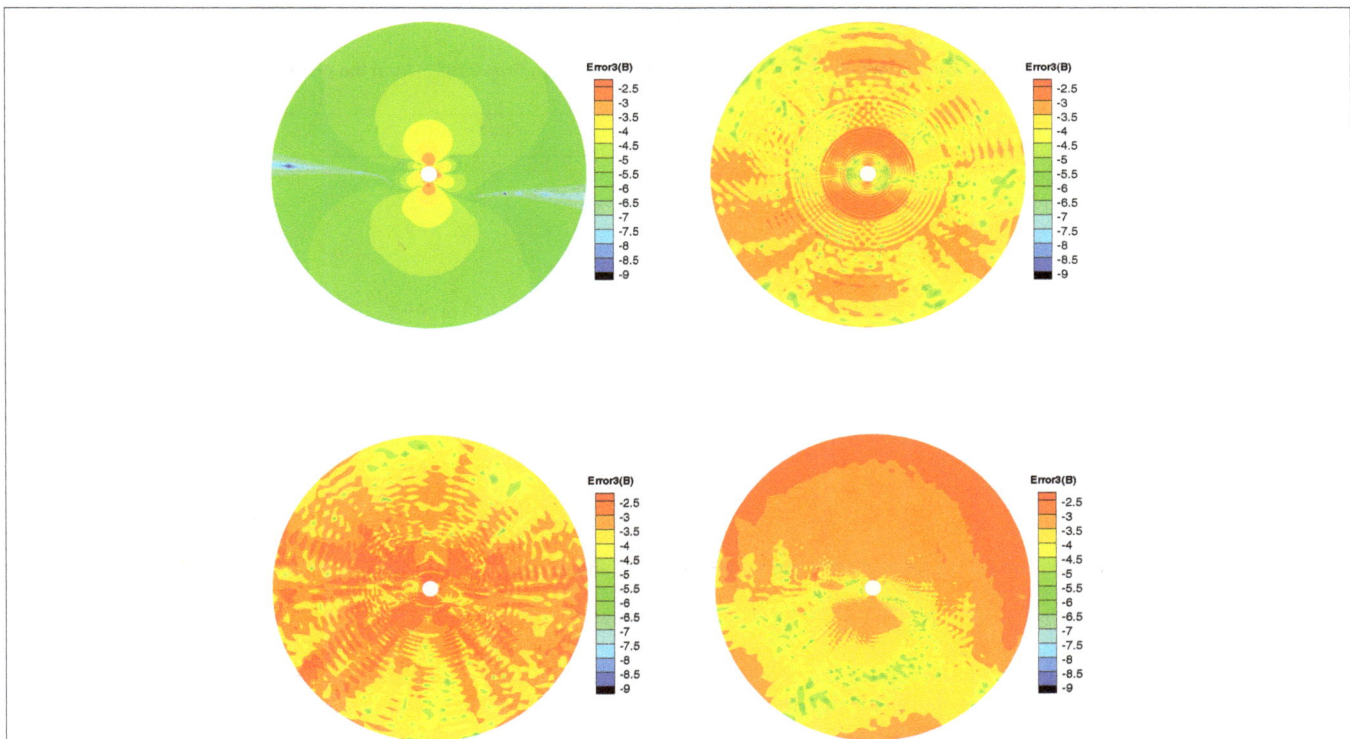

FIGURE 12 | The log_{10}Error3(B) in the calculation at $t = 20$ h on the meridional plane of $\phi = 180° - 0°$ from 1 to 20 R$_s$, the results from CT method (left) and diffusive method (right) are displayed in the top row, the bottom row is produced from the projection method (left) and GLM method (right).

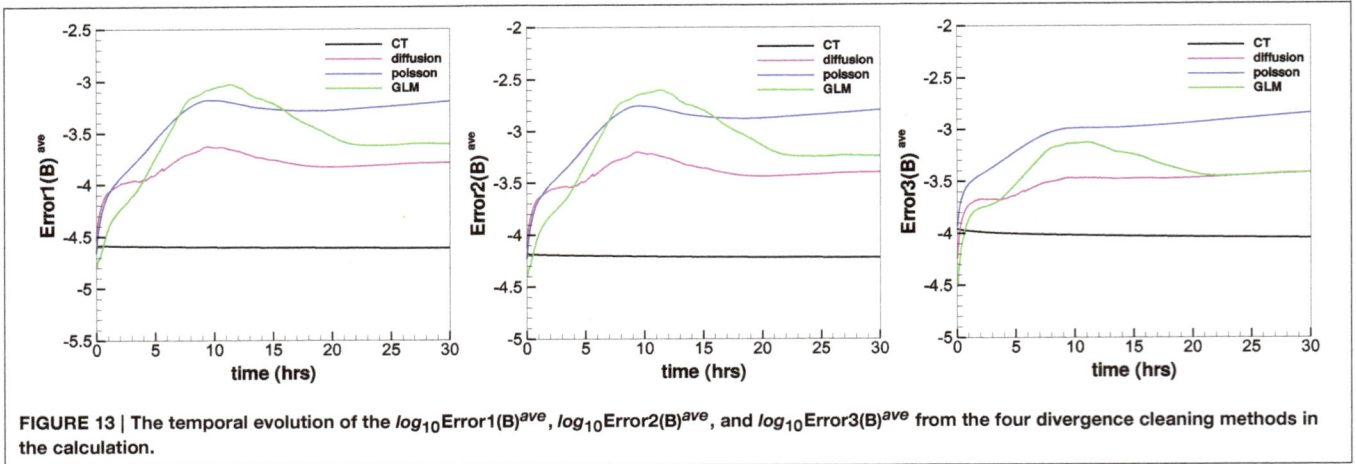

FIGURE 13 | The temporal evolution of the $log_{10}Error1(B)^{ave}$, $log_{10}Error2(B)^{ave}$, and $log_{10}Error3(B)^{ave}$ from the four divergence cleaning methods in the calculation.

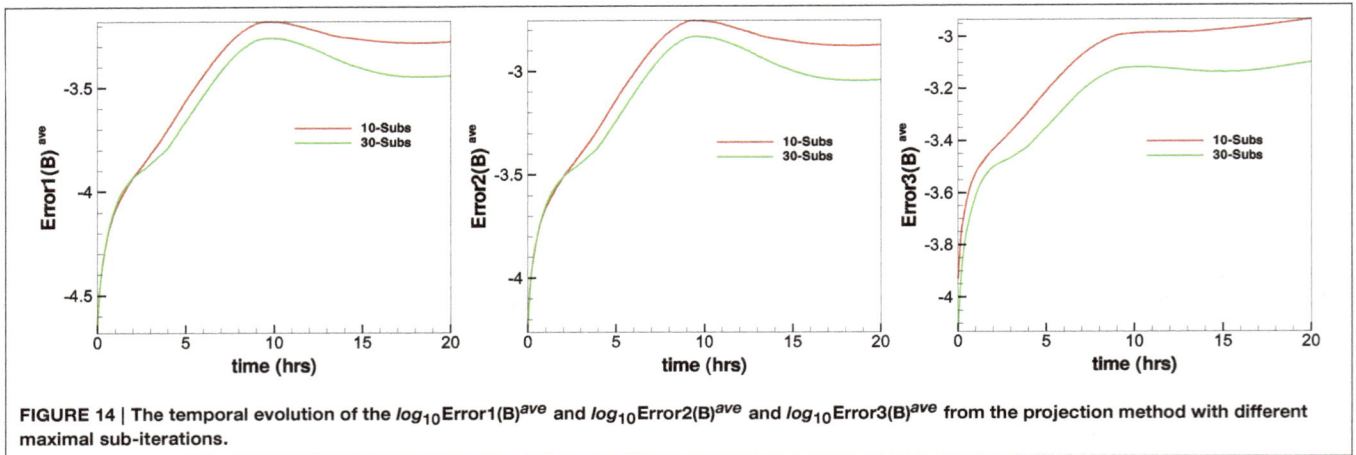

FIGURE 14 | The temporal evolution of the $log_{10}Error1(B)^{ave}$ and $log_{10}Error2(B)^{ave}$ and $log_{10}Error3(B)^{ave}$ from the projection method with different maximal sub-iterations.

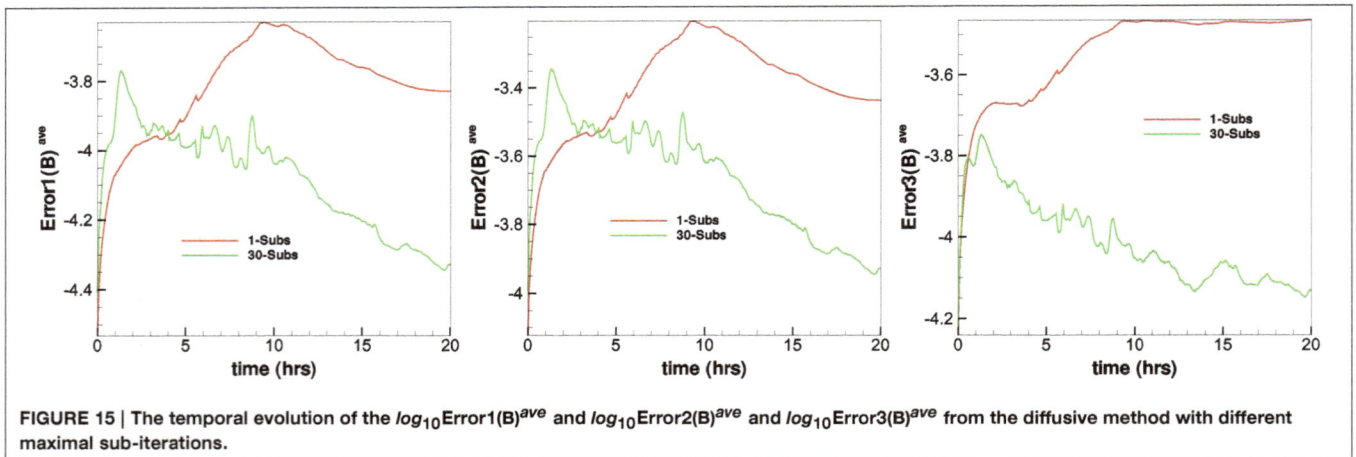

FIGURE 15 | The temporal evolution of the $log_{10}Error1(B)^{ave}$ and $log_{10}Error2(B)^{ave}$ and $log_{10}Error3(B)^{ave}$ from the diffusive method with different maximal sub-iterations.

for implementation. The diffusive method reduce the numerical error of $\nabla \cdot \mathbf{B}$ by adding a source term in the induction equation. The projection method involves the solution of a poisson equation after every time step to correct errors of $\nabla \cdot \mathbf{B}$, and thus can be coupled with any numerical scheme, but solving the additional Poisson equation can significantly increase the computational cost. The GLM method maintain the $\nabla \cdot \mathbf{B} = 0$ constraint by introducing a newly transport variable ψ is to the MHD system. The GLM method is fully conservative in mass, momentum, magnetic induction and energy, it is effective in controlling divergence error and can easily be applied on general grids. Our numerical results showed the CT method can maintain the average relative divergence error around $10^{-4.5}$. The diffusive method can maintain the average relative divergence error about $10^{-3.6}$, and we only use sub-iteration 1 in this paper, increasing maximal sub-iterations will decrease the relative divergence error. The average relative divergence error for GLM method is about $10^{-3.3}$ and for projection method is $10^{-3.1}$. So the CT method and the diffusive approach can maintain divergence cleaning constraint better, the diffusive method is a good choice by considering simplicity, and the CT method should be considered while we want to capture the discontinuity structure. The projection method in our paper is a preliminary try and we think the result can be better if we use multigrid in the future.

Although there have some differences in the average relative divergence errors for the four employed methods, the differences dose't effect the large-scale solar wind structure and they can all produce structured solar wind. They all produce many typical properties of the solar wind, such as an obvious slow speed area near the slightly tilted HCS plane and a fast speed area near the poles, and high density in the slow speed area and vice versa in both poles. Overall, our model can produce all the physical parameters everywhere within the computation domain.

AUTHOR CONTRIBUTIONS

MZ run the cases and plotted all the figures. Both MZ and XF are involved in the development of the three-dimensional MHD code, the analysis numerical results, the writing of the manuscript.

ACKNOWLEDGMENTS

The work is jointly supported by the National Basic Research Program of China (Grant No. 2012CB825601), the National Natural Science Foundation of China (Grant Nos. 41231068, 41504132, 41274192, and 41531073), the Knowledge Innovation Program of the Chinese Academy of Sciences (Grant No. KZZD-EW-01-4), and the Specialized Research Fund for State Key Laboratories. The numerical calculation has been completed on our SIGMA Cluster computing system. The Wilcox Solar Observatory is currently supported by NASA.

REFERENCES

Balsara, D. S., and Kim, J. (2004). A comparison between divergence-cleaning and staggered-mesh formulations for numerical magnetohydrodynamics. *Astrophys. J.* 602, 1079–1090. doi: 10.1086/381051

Brackbill, J. U., and Barnes, D. C. (1980). The effect of nonzero $\nabla \cdot \mathbf{B}$ on the numerical solution of the magnetohydrodynamic equations. *J. Comput. Phys.* 35, 426–430. doi: 10.1016/0021-9991(80)90079-0

Brandenburg, A., Rädler, K. H., Rheinhardt, M., and Käpyla, P. J. (2008). Magnetic diffusivity tensor and dynamo effects in rotating and shearing turbulence. *Astrophys. J.* 676, 740–751. doi: 10.1086/527373

Dedner, A., Kemm, F., Króner, D., Munz, C. D., Schnitzer, T., and Wesenberg, W. (2002). Hyperbolic divergence cleaning for the MHD equations. *J. Comput. Phys.* 175, 645–673. doi: 10.1006/jcph.2001.6961

Dedner, A., Rohde, C., and Wesenberg, M. (2003). "A new approach to divergence cleaning in magnetohydrodynamic simulations," in *Hyperbolic Problems: Theory, Numerics, Applications*, eds T. Y. Hou and E. Tadmor (Berlin; Heidelberg: Spring-Verlag), 509–518. doi: 10.1007/978-3-642-55711-8-47

Evans, C. R., and Hawley, J. F. (1988). Simulation of magnetohydrodynamic flows-a constrained transport method. *Astrophys. J.* 332, 659–677. doi: 10.1086/166684

Feng, X., Zhang, M., and Zhou, Y. (2014). A new three-dimensional solar wind model in spherical coordinates with a six-component grid. *Astrophys. J. Suppl. Ser.* 214, 6. doi: 10.1088/0067-0049/214/1/6

Feng, X. S., Jiang, C. W., Xiang, C. Q., Zhao, X. P., and Wu, S. T. (2012a). A data-driven model for the global coronal evolution. *Astrophys. J.* 758, 62. doi: 10.1088/0004-637X/758/1/62

Feng, X. S., Yang, L. P., Xiang, C. Q., Jiang, C. W., Ma, X. P., Wu, S. T., et al. (2012b). Validation of the 3D AMR SIP-CESE solar wind model for four Carrington rotations. *Solar Phys.* 279, 207–229. doi: 10.1007/s11207-012-9969-9

Feng, X. S., Yang, L. P., Xiang, C. Q., Liu, Y., Zhao, X. P., and Wu, S. T. (2012c). "Numerical study of the global corona for CR 2055 driven by daily updated synoptic magnetic field," in *Astronomical Society of the Pacific Conference Series*, Vol. 459 (San Francisco, CA), 202–208.

Feng, X. S., Yang, L. P., Xiang, C. Q., Wu, S. T., Zhou, Y. F., and Zhong, D. K. (2010). Three-dimensional solar wind modeling from the Sun to Earth by a SIP-CESE MHD model with a six-component grid. *Astrophys. J.* 723, 300–319. doi: 10.1088/0004-637X/723/1/300

Feng, X. S., Zhang, S. H., Xiang, C. Q., Yang, L. P., Jiang, C. W., and Wu, S. T. (2011). A hybrid solar wind model of the CESE+HLL method with a Yin-Yang overset grid and an AMR grid. *Astrophys. J.* 734, 50. doi: 10.1088/0004-637X/734/1/50

Fuchs, F. G., Mishra, S., and Risebro, N. H. (2009). Splitting based finite volume schemes for ideal MHD equations. *J. Comput. Phys.* 228, 641–660. doi: 10.1016/j.jcp.2008.09.027

Hayashi, K. (2005). Magnetohydrodynamic simulations of the solar corona and solar wind using a boundary treatment to limit solar wind mass flux. *Astrophys. J. Suppl. Ser.* 161, 480–494. doi: 10.1086/491791

Inoue, S., Hayashi, K., Magara, T., Choe, G. S., and Park, Y. D. (2015). Magnetohydrodynamic Simulation of the X2.2 Solar Flare on 2011 February 15. II. Dynamics Connecting the Solar Flare and the Coronal Mass Ejection. *Solar Stell. Astrophys.* 803, 73. doi: 10.1088/0004-637X/803/2/73

Jiang, R. L., Fang, C., and Chen, P. F. (2012a). A new MHD code with adaptive mesh refinement and parallelization for astrophysics. *Comput. Phys. Commun.* 183, 1617–1633. doi: 10.1016/j.cpc.2012.02.030

Jiang, R. L., Fang, C., and Chen, P. F. (2012b). Numerical simulation of solar microflares in a canopy-type magnetic configuration. *Astrophys. J.* 751, 152. doi: 10.1088/0004-637X/751/2/152

Linker, J. A., Mikić, Z., Biesecker, D. A., Forsyth, R. J., Gibson, S. E., Lazarus, A. J., et al. (1999). Magnetohydrodynamic modeling of the solar corona during whole Sun month. *J. Geophys. Res.* 104, 9809–9830. doi: 10.1029/1998JA900159

Manabu, Y., Kanako, S., and Yosuke, M. (2009). Development of a magnetohydrodynamic simulation code satisfying the solenoidal magnetic field condition. *Comput. Phys. Commun.* 180, 1550–1557. doi: 10.1016/j.cpc.2009.04.010

Mignone, A., and Tzeferacos, P. (2010). A second-order unsplit Godunov scheme for cell-centered MHD: the CTU-GLM scheme. *J. Comput. Phys.* 229, 2117–2138. doi: 10.1016/j.jcp.2009.11.026

Mignone, A., Tzeferacos, P., and Bodo, G. (2010). High-order conservative finite difference GLM-MHD schemes for cell-centered MHD. *J. Comput. Phys.* 229, 5896–5920. doi: 10.1016/j.jcp.2010.04.013

Mocz, P., Vogelsberger, M., and Hernquist, L. (2014). A constrained transport scheme for mhd on unstructured static and moving meshes. *Month. Notices R. Astron. Soc.* 442, 43–55. doi: 10.1093/mnras/stu865

Pakmor, R., and Springel, V. (2013). Simulations of magnetic fields in isolated disc galaxies. *Month. Notices R. Astron. Soc.* 432, 176–193. doi: 10.1093/mnras/stt428

Parker, E. N. (1963). *Interplanetary Dynamical Processes*. New York, NY: Interscience Publishers.

Powell, K. G., Roe, P. L., Linde, T. J., Gombosi, T. I., and De Zeeuw, D. L. (1999). A solution-adaptive upwind scheme for ideal magnetohydrodynamics. *J. Comput. Phys.* 154, 284–309. doi: 10.1006/jcph.1999.6299

Powell, K. G., Roe, P. L., and Quirk, J. (1993). "Adaptive-mesh algorithms for computational fluid dynamics," in *Algorithmic Trends in Computational Fluid Dynamics*, eds M. Y. Hussaini, A. Kumar and M. D. Salas (New York, NY: Springer), 303–337. doi: 10.1007/978-1-4612-2708-3-18

Rempel, M., Schüssler, M., and Knölker, M. (2009). Radiative magnetohydrodynamic simulation of sunspot structure. *Astrophys. J.* 691, 640–649. doi: 10.1088/0004-637X/691/1/640

Shen, F., Shen, C. L., Zhang, J., Hess, P., Wang, Y. M., Feng, X. S., et al. (2014). Evolution of the 12 July 2012 CME from the sun to the earth: data-constrained three-dimensional MHD simulations. *J. Geophys. Res. Space Phys.* 119, 7128–7141. doi: 10.1002/2014JA020365

Susanto, A., Ivan, L., Sterck, H. D., and Groth, C. P. T. (2013). High-order central ENO finite-volume scheme for ideal MHD. *J. Comput. Phys.* 250, 141–164. doi: 10.1016/j.jcp.2013.04.040

Tóth, G. (2000). The $\nabla \cdot \mathbf{B} = 0$ constraint in shock-capturing magnetohydrodynamics codes. *J. Comput. Phys.* 161, 605–652. doi: 10.1006/jcph.2000.6519

Tóth, G., Sokolov, I. V., Gombosi, T. I., Chesney, D. R., Clauer, C. R., Zeeuw, D. L. D., et al. (2005). Space weather modeling framework: a new tool for the space science community. *J. Geophys. Res.* 110, A12226. doi: 10.1029/2005JA011126

Tóth, G., van der Holst, B., Sokolov, I. V., De Zeeuw, D. L., Gombosi, T. I., Fang, F., et al. (2012). Adaptive numerical algorithms in space weather modeling. *J. Comput. Phys.* 231, 870–903. doi: 10.1016/j.jcp.2011.02.006

van der Holst, B., and Keppens, R. (2007). Hybrid block-AMR in Cartesian and curvilinear coordinates: MHD applications. *J. Comput. Phys.* 226, 925–946. doi: 10.1016/j.jcp.2007.05.007

Yee, K. (1966). Numerical solution of initial boundary value problems involving Maxwell's equations in isotropic media. *IEEE Trans. Anten. Propagat.* 14, 302–307. doi: 10.1109/TAP.1966.1138693

Ziegler, U. (2004). A central-constrained transport scheme for ideal magnetohydrodynamics. *J. Comput. Phys.* 196, 393–416. doi: 10.1016/j.jcp.2003.11.003

Ziegler, U. (2011). A semi-discrete central scheme for magnetohydrodynamics on orthogonal-curvilinear grids. *J. Comput. Phys.* 230, 1035–1063. doi: 10.1016/j.jcp.2010.10.022

Ziegler, U. (2012). Block-structured adaptive mesh refinement on curvilinear-orthogonal grids. *SIAM J. Sci. Comput.* 34, C102–C121. doi: 10.1137/110843940

Conflict of Interest Statement: The authors declare that the research was conducted in the absence of any commercial or financial relationships that could be construed as a potential conflict of interest.

Permissions

The contributors of this book come from diverse backgrounds, making this book a truly international effort. This book will bring forth new frontiers with its revolutionizing research information and detailed analysis of the nascent developments around the world.

We would like to thank all the contributing authors for lending their expertise to make the book truly unique. They have played a crucial role in the development of this book. Without their invaluable contributions this book wouldn't have been possible. They have made vital efforts to compile up to date information on the varied aspects of this subject to make this book a valuable addition to the collection of many professionals and students.

This book was conceptualized with the vision of imparting up-to-date information and advanced data in this field. To ensure the same, a matchless editorial board was set up. Every individual on the board went through rigorous rounds of assessment to prove their worth. After which they invested a large part of their time researching and compiling the most relevant data for our readers.

The editorial board has been involved in producing this book since its inception. They have spent rigorous hours researching and exploring the diverse topics which have resulted in the successful publishing of this book. They have passed on their knowledge of decades through this book. To expedite this challenging task, the publisher supported the team at every step. A small team of assistant editors was also appointed to further simplify the editing procedure and attain best results for the readers.

Apart from the editorial board, the designing team has also invested a significant amount of their time in understanding the subject and creating the most relevant covers. They scrutinized every image to scout for the most suitable representation of the subject and create an appropriate cover for the book.

The publishing team has been an ardent support to the editorial, designing and production team. Their endless efforts to recruit the best for this project, has resulted in the accomplishment of this book. They are a veteran in the field of academics and their pool of knowledge is as vast as their experience in printing. Their expertise and guidance has proved useful at every step. Their uncompromising quality standards have made this book an exceptional effort. Their encouragement from time to time has been an inspiration for everyone.

The publisher and the editorial board hope that this book will prove to be a valuable piece of knowledge for researchers, students, practitioners and scholars across the globe.

List of Contributors

Sandor M. Molnar
Institute of Astronomy and Astrophysics, Academia Sinica, Taipei, Taiwan

Nour E. Raouafi
The John Hopkins University Applied Physics Laboratory, Laurel, MD, USA

Pete Riley
Predictive Science Inc., San Diego, CA, USA

Sarah Gibson
High Altitude Observatory, National Center for Atmospheric Research, Boulder, CO, USA

Silvano Fineschi
The Astrophysical Observatory of Turin, National Institute for Astrophysics, Turin, Italy

Sami K. Solanki
Max-Planck-Institut für Sonnensystemforschung, Göttingen, Germany
School of Space Research, Kyung Hee University, Yongin, South Korea

Urszula Bąk-Stęślicka and Ewa Chmielewska
Astronomical Institute, University of Wrocław, Wrocław, Poland

Sarah E. Gibson
High Altitude Observatory, National Center for Atmospheric Research, Boulder, CO, USA

Patricia R. Jibben and Katharine K. Reeves
Harvard-Smithsonian Center for Astrophysics, Cambridge, MA, USA

Yingna Su
Key Laboratory for Dark Matter and Space Science, Purple Mountain Observatory, Chinese Academy of Sciences, Nanjing, China

Gabriel I. Dima and Jeffrey R. Kuhn
Institute for Astronomy, University of Hawaii, Pukalani, HI, USA

Svetlana V. Berdyugina
Institute for Astronomy, University of Hawaii, Pukalani, HI, USA
Kiepenheuer Institut fuer Sonnenphysik, Freiburg, Germany
Predictive Science Inc., San Diego, CA, USA

Ariel J. C. Dauzart, Joshua P. Vandenbrink and John Z. Kiss
Department of Biology, Graduate School, University of Mississippi, University, MS, USA

Kevin Dalmasse
CISL/HAO, National Center for Atmospheric Research, Boulder, CO, USA

Douglas W. Nychka and Natasha Flyer
CISL, National Center for Atmospheric Research, Boulder, CO, USA

Sarah E. Gibson and Yuhong Fan
HAO, National Center for Atmospheric Research, Boulder, CO, USA

Alessandra Celletti
Department of Mathematics, University of Rome Tor Vergata, Rome, Italy

Cătălin B. Galeş
Department of Mathematics, University of Iasi, Iasi, Romania

Zinovy Malkin
Department of Radio Astronomy Research, The Pulkovo Astronomical Observatory, St. Petersburg, Russia
Institute of Earth Sciences, St. Petersburg State University, St. Petersburg, Russia
Astronomy and Cosmic Geodesy Department, Kazan Federal University, Kazan, Russia

Sarah E. Gibson and Yuhong Fan
High Altitude Observatory, National Center for Atmospheric Research, Boulder, CO, USA

Therese A. Kucera
Goddard Space Flight Center, National Aeronautics and Space Administration (NASA), Greenbelt, MD, USA

Stephen M. White
Air Force Research Labs, Kirtland Air Force Base, Albuquerque, NM, USA

James B. Dove
Department of Physics, Metro State University Denver, Denver, CO, USA

Blake C. Forland
Department of Physics, Indiana University, Bloomington, IN, USA

Yuan-Kuen Ko, John M. Laming, Dennis G. Socker, Leonard Strachan, Samuel Tun Beltran and Clarence Korendyke
Space Science Division, Naval Research Laboratory, Washington, DC, USA

John D. Moses
Heliophysics Division, Science Mission Directorate, NASA, Washington, DC, USA

Steven Tomczyk, Sarah E. Gibson, Roberto Casini, Michael Knoelker, Scott W. McIntosh and Qian Wu
High Altitude Observatory, Boulder, CO, USA

Frédéric Auchère
Institut d'Astrophysique Spatiale, CNRS Université Paris-Sud, Orsay, France

Silvano Fineschi
INAF - National Institute for Astrophysics, Astrophysical Observatory of Torino, Pino Torinese, Italy

Marco Romoli
Department of Physics and Astronomy, University of Florence, Florence, Italy

Jan Rybak
Astronomical Institute, Slovak Academy of Sciences, Tatranska Lomnica, Slovakia

Angelos Vourlidas
Applied Physics Laboratory, Johns Hopkins University, Laurel, MD, USA

Jayashree Balakrishna
Department of Mathematics and Natural Sciences, College of Arts and Sciences, Harris-Stowe State University, St. Louis, MO, USA

Ruxandra Bondarescu
Department of Physics, University of Zurich, Zurich, Switzerland

Christine C. Moran
TAPIR, Department of Theoretical Astrophysics, California Institute of Technology, Pasadena, CA, USA

Haosheng Lin
Institute for Astronomy, University of Hawaii, Pukalani, HI, USA

Sergei M. Kopeikin
Department of Physics and Astronomy, University of Missouri, Columbia, MO, USA
Department of Physical Geodesy and Remote Sensing, Siberian State University of Geosystems and Technologies, Novosibirsk, Russia

Víctor M. Patiño-Álvarez, Janet Torrealba, Vahram Chavushyan and Jonathan León-Tavares
Instituto Nacional de Astrofísica, Óptica y Electrónica, Puebla, Mexico

Irene Cruz-González
Instituto de Astronomía, Universidad Nacional Autónoma de México, Mexico City, Mexico

Tigran Arshakian
Physikalisches Institut, Universität zu Köln, Köln, Germany
Byurakan Astrophysical Observatory, Byurakan, Armenia
Isaac Newton Institute of Chile in Estern Europe and Eurasia, Armenian Branch, Santiago, Chile

Luka Popović
Astronomical Observatory, Belgrade, Serbia

Katharine K. Reeves
Department of High Energy Astrophysics, Harvard-Smithsonian Center for Astrophysics, Cambridge, MA, USA

Laurel A. Rachmeler
Royal Observatory of Belgium, Brussels, Belgium
Marshall Space Flight Center, National Aeronautics and Space Administration (NASA), Huntsville, AL, USA

Cooper Downs
Predictive Science Inc., San Diego, CA, USA

Man Zhang and Xueshang Feng
SIGMA Weather Group, State Key Laboratory for Space Weather, National Space Science Center, Chinese Academy of Sciences, Beijing, China

Index

www.ingramcontent.com/pod-product-compliance
Lightning Source LLC
Chambersburg PA
CBHW080252230326
41458CB00097B/4381